Review of Gross Anatomy

Text and Illustrations by

BEN PANSKY, Ph.D., M.D.
Professor of Anatomy,
Medical College of Ohio at Toledo,
Toledo, Ohio

Review of
Gross Anatomy

FOURTH EDITION

Macmillan Publishing Co., Inc.
New York

Collier Macmillan Canada, Inc.
Toronto

Collier Macmillan Publishers
London

Macmillan Publishing Co., Inc.
866 Third Avenue, New York, New York 10022

Collier Macmillan Canada, Inc.

Collier Macmillan Publishers • London

Library of Congress Cataloging in Publication Data

Pansky, Ben.
  Review of gross anatomy.

  Bibliography: p.
  Includes index.
  1. Anatomy, Human—Outlines, syllabi, etc.
I. Title.
QM31.P28 1979   611   78-15783
ISBN 0-02-390630-8

Printing: 5 6 7 8   Year: 3 4 5 6 7 8

This book is dedicated to my beloved wife,
Julie,
without whose patience, inspiration, and constant encouragement
this fourth edition might never have become a reality!

Preface to the Fourth Edition

This book has been written for the medical student, dental student, physician, surgeon, physical therapist, nurse, and others who require a visual comprehension and review of the fundamentals of anatomy. The material is presented in a simplified, concise, and outline form but is, nevertheless, complete. The text is functionally oriented and clinically informative to emphasize the need for understanding the essentials of anatomy. In this fourth edition, additional special-feature anatomy and more clinical considerations have been included in each of the six units. Where repetition occurs, it does so to reemphasize particular points and to help the continuity of study. With only a few exceptions, illustrations appear on right-hand pages and corresponding text on opposite left-hand pages. The terminology used throughout is an anglicized version of the standard nomenclature (*Nomina Anatomica,* 4th ed.) approved at the Tenth International Congress of Anatomists, held at Tokyo in August, 1975.

Anatomy is a study of a three-dimensional body and, most teachers and students will agree, is best learned at the dissection table where the parts are visible and palpable. It is the visualization that is essential and plays a prime role in the understanding of the basic fundamentals. With this fact in mind, the book has been written around its pictures rather than having its pictures planned around the text. It contains over 1000 original line-cut illustrations, and in this fourth edition color has been added to many more of these line drawings. They have been conceived to give a simplified but three-dimensional aspect and truer visualization of the body than is often found in brief, or review-type, texts. In fact, most books tend to oversimplify the illustrations, which conveys a misconception of the complex structure of the human body, and the student fails to appreciate fully its beauty and function.

As the reader progresses from region to region—from the head and neck to the back, to the upper extremity, to the thorax, to the abdomen and pelvis, and, finally, to the lower extremity—he/she will note that a continuity exists between the regions. At the same time the illustrations demonstrate an overlapping of the structures, so as to enable the student to pass from one region to the next more easily. To facilitate the rapid location of each of the major units within the text, black tabs are printed in the margins of left-hand pages.

Since the physician initially considers the patient's external surface as a part of the physical diagnosis of disease, and because the student first approaches the superficial aspect of the body, a knowledge of the surface projections of underlying structures is basic. Therefore, numerous new photographs of topographic and surface anatomy have been added. Furthermore, inasmuch as radiographic anatomy is an essential tool that enables the physician to visualize areas within the body not readily accessible from the surface, new radiologic figures of normal anatomy have been included and old figures improved and updated.

Printed in the back matter of this fourth edition is an atlas of systemic anatomy, which further elaborates the beauty, complexity, and relationships of many structures found in the various regions of the body. It enables the student to visualize total systems and get a "bird's-eye view" of the whole, prior to approaching the specific regions of the body; it may also serve as an excellent, comprehensive review once the anatomy of the regions has been studied.

Whenever possible, the body has been discussed from its superficial layers to its deep structures, except for the osteology. Since the bones form the framework and lend themselves to the attachment of soft parts, one may find that if they are studied before further discussion of a particular region is undertaken, the relationship of the body soft parts is often more clearly understood. In this regard, a new series of photographs of actual bone specimens has been included at the beginning of each unit, along with the photographs of surface anatomy.

Through years of experience in teaching gross anatomy to medical students, interns, residents, and others, the author has become well aware of the fact that the subject, once learned and quickly memorized, is just as easily forgotten and, therefore, requires a constant and almost endless review; i.e., prolonged repetition is necessary to etch this mass of information in one's mind. Yet the student finds that "time" is his adversary; his other duties are so overwhelming that a repeated reading of the long and detailed textbooks of anatomy is just about impossible. Accordingly, the author is hopeful that this fourth edition of *Review of Gross Anatomy* will continue to fill the obvious gap in present-day medical literature.

The author would like to express his appreciation to the many colleagues, teachers, and, above all, students whose comments and suggestions have helped to maintain the accuracy and quality of this book. I am particularly grateful to John Doman, who photographed the models used in depicting areas of surface anatomy, and to Carol Perkins of the Audio-Visual Services of the Medical College of Ohio at Toledo, who prepared some of the photographic materials used in this fourth edition. In addition, I would like to thank Ms. Faye Keen, Library Media Technical Assistant in the Department of Radiology of the Medical College of Ohio, for many of the radiologic negatives used in this edition.

B.P.

Contents

Unit Three. UPPER EXTREMITY

Unit Four. THORAX

Unit Five. ABDOMEN AND PELVIS

Appendixes

UNIT ONE

Head and
Neck

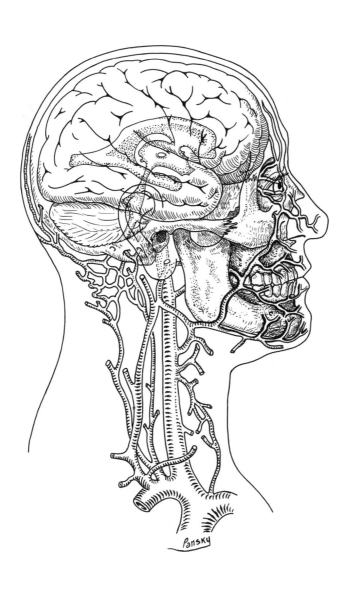

Pansky

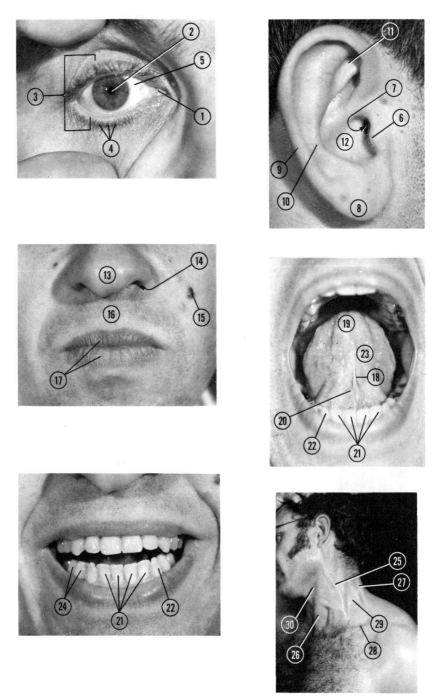

FIGURE 1. **Surface anatomy of head and neck.** *1,* Lacrimal caruncle; *2,* pupil; *3,* eyelids; *4,* eyelashes; *5,* conjunctiva; *6,* tragus; *7,* external auditory meatus; *8,* lobule; *9,* helix; *10,* antihelix; *11,* triangular fossa; *12,* concha; *13,* tip of nose; *14,* nares; *15,* nevus; *16,* philtrum; *17,* lips; *18,* frenulum; *19,* tip of tongue; *20,* sublingual caruncle; *21,* incisors; *22,* canine; *23,* fimbriated fold; *24,* premolars; *25,* external jugular vein; *26,* sternocleidomastoid m.; *27,* trapezius m.; *28,* clavicle; *29,* posterior triangle; *30,* carotid triangle.

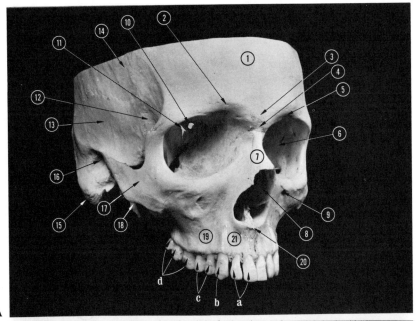

A

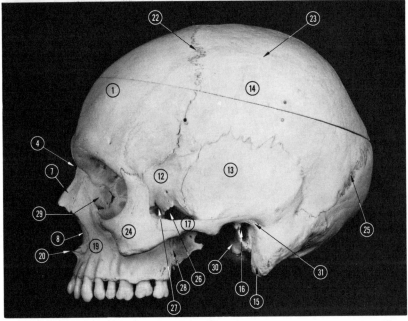

B

FIGURE 2. **A. Anterolateral aspect of the skull.** *1,* Frontal bone; *2,* superciliary arch; *3,* glabella; *4,* nasion; *5,* superior orbital notch; *6,* orbit; *7,* nasal bone; *8,* piriform (nasal) aperture; *9,* infraorbital foramen; *10,* optic foramen; *11,* superior orbital fissure; *12,* sphenoid bone (greater wing); *13,* temporal bone; *14,* parietal bone; *15,* mastoid process; *16,* external auditory meatus; *17,* zygomatic arch; *18,* styloid process; *19,* maxillary bone; *20,* anterior nasal spine; *21,* canine fossa. *a,* Incisors; *b,* canine; *c,* premolars; *d,* molars. **B. Lateral aspect of the skull.** *22,* Coronal suture; *23,* temporal lines; *24,* zygomatic bone; *25,* lambdoidal suture; *26,* infratemporal fossa; *27,* spine of sphenoid; *28,* lateral pterygoid plate; *29,* lacrimal bone; *30,* tympanic plate; *31,* suprameatal triangle.

– 3 –

1. ANTERIOR VIEW OF THE SKULL, MANDIBLE, AND HYOID

I. Bones of skull
A. FRONTAL: frontal eminence, superciliary arch with supraorbital notch or foramen, glabella, and zygomatic process, nasal part
B. NASAL
C. NASAL SEPTUM
D. MAXILLA: facial surface, incisive and canine fossae, infraorbital foramen, anterior nasal spine, nasal notch, alveolar process
E. ZYGOMATIC: frontal process, temporal process

II. Sutures of skull
A. FRONTONASAL: frontal and two nasal bones; nasion is the midpoint of the suture
B. ZYGOMATICOMAXILLARY: zygomatic and maxillary bones
C. FRONTOZYGOMATIC: zygomatic and frontal bones
D. INTERMAXILLARY: between maxillary bones

III. Hyoid: made up of 5 parts: body, 2 greater and 2 lesser cornua
A. BODY: quadrangular, convex ventrally
B. GREATER CORNUA: long, projecting backward from body; joined to body in a synostosis
C. LESSER CORNUA: short and conical processes projecting upward and backward from body; base attached at line of union of body and greater cornu by fibrous tissue (syndesmosis)

IV. Mandible: made up of body and 2 rami
A. BODY: convex anteriorly
   1. Externally: mental protuberance with mental tubercles on either side; mental foramen; oblique line; groove for facial artery
   2. Internally: mental spine; mylohyoid line; alveolar border; submandibular and sublingual fossae
B. RAMI, which form angles with the body
   1. Condyloid process, with rounded head for articulation
   2. Constricted neck
   3. Coronoid process: thin, separated from condyloid process by a wide mandibular notch and pointed at superior end
   4. On inner (medial) surface: mandibular foramen, the lingula and the mylohyoid groove

V. Clinical considerations
A. SKULL FRACTURE may be due to a blow on the head by a solid object, crushing of the head between two solid objects, or impaction of the moving head against a stationary solid object
   1. If the distortion of bone is sufficient, it will be fractured, the lines of force from the site of impact being transmitted through the vault and base of the skull
   2. Occasionally a fracture is contrecoup (opposite the point of impact), especially in an orbital roof in parietal or occipital fractures
B. FRACTURES take many forms and occur in many places
   1. Open type: bone fragments break the mucous membrane in the mouth
   2. Closed type: vertical breaks between ramus and body of mandible. The ramus fragment is pulled upward by the temporalis and masseter muscles and medially by the medial pterygoid muscle; the body fragment is pulled downward by gravity and the geniohyoid muscle and backward by the digastric and mylohyoid muscles

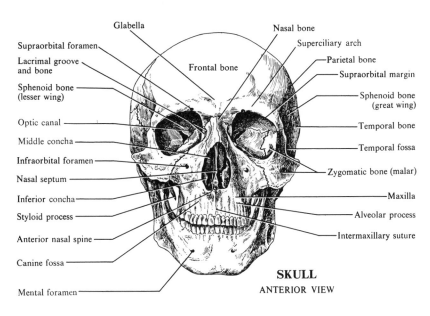

Glabella

Nasal bone

Superciliary arch

Supraorbital foramen

Lacrimal groove
and bone

Parietal bone

Supraorbital margin

Sphenoid bone
(lesser wing)

Frontal bone

Sphenoid bone
(great wing)

Optic canal

Temporal bone

Middle concha

Temporal fossa

Infraorbital foramen

Zygomatic bone (malar)

Nasal septum

Inferior concha

Maxilla

Styloid process

Alveolar process

Intermaxillary suture

Anterior nasal spine

Canine fossa

SKULL
ANTERIOR VIEW

Mental foramen

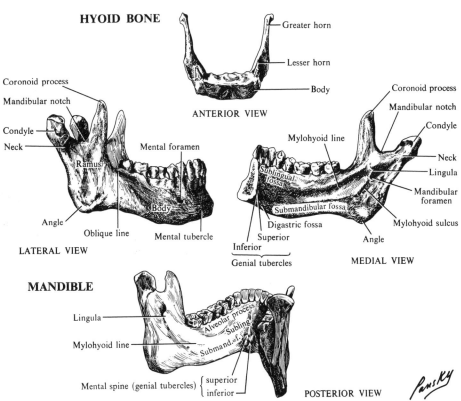

HYOID BONE

Greater horn

Lesser horn

Body

ANTERIOR VIEW

Coronoid process

Mandibular notch

Condyle

Neck

Ramus

Mental foramen

Coronoid process

Mandibular notch

Condyle

Mylohyoid line

Neck

Sublingual
fossa

Lingula

Mandibular
foramen

Body

Submandibular fossa

Mylohyoid sulcus

Angle

Oblique line

Mental tubercle

Digastric fossa

Superior

Angle

LATERAL VIEW

Inferior

Genial tubercles

MEDIAL VIEW

MANDIBLE

Lingula

Alveolar process

Subling

Mylohyoid line

Submand.f.

Mental spine (genial tubercles) { superior
{ inferior

POSTERIOR VIEW

Pansky

2. LATERAL VIEW OF THE SKULL

I. Bones
A. PARIETAL: part of roof and sides of cranium. Note superior and inferior temporal lines, and parietal foramen (inconstant)
B. FRONTAL (squamous part): anterior and anterolateral part of cranium. Note superciliary arch, the glabella (uniting the arches over the nose), zygomatic process, and temporal line
C. LACRIMAL, anteromedial wall of orbit: lacrimal sulcus and crest
D. GREAT WING of sphenoid and lateral pterygoid plate
E. TEMPORAL
    1. Squama, flat, forming part of side of cranium: note zygomatic process, articular tubercle, mandibular fossa, and suprameatal crest
    2. Mastoid, large portion behind ear: mastoid process and foramen
    3. Tympanic, surrounding external meatus
    4. Styloid process with its vaginal sheath
F. ZYGOMATIC, cheek bone: zygomaticofacial foramen
G. MAXILLA: anterior nasal spine and alveolar border
H. OCCIPITAL, squamous portion: external protuberance and superior nuchal line

II. Sutures of skull
A. CORONAL: frontal and parietal bones
B. SPHENOFRONTAL: great wing of sphenoid and frontal bones
C. SPHENOPARIETAL: great wing of sphenoid and parietal bones (pterion is the posterior end of the suture)
D. SQUAMOSAL: squama of temporal and parietal bones
E. PARIETOMASTOID: parietal and mastoid of temporal bone
F. LAMBDOIDAL: parietal and occipital bones
G. OCCIPITOMASTOID: occipital and mastoid of temporal bone: asterion is the point at which the lambdoidal and occipitomastoid sutures meet
H. SPHENOSQUAMOSAL: great wing of sphenoid and squamous temporal bones
I. TEMPOROZYGOMATIC: processes of zygomatic and temporal bones
J. FRONTOZYGOMATIC: zygomatic and frontal bones

III. Special features
A. TEMPORAL FOSSA. Bounded above and behind by temporal lines, in front by frontal and zygomatic bones, laterally by zygomatic arch, below by infratemporal crest
B. INFRATEMPORAL FOSSA. Bounded in front by maxilla, behind by articular tubercle of temporal and angular spine of sphenoid, above by great wing of sphenoid, below the infratemporal crest and alveolar border of maxilla, medially by lateral pterygoid plate
C. PTERYGOPALATINE FOSSA. Bounded above by body of sphenoid, in front by maxilla, behind by pterygoid process and great wing of sphenoid, and medially by palatine bone. Five foramina open into fossa: rotundum, pterygoid canal, pharyngeal canal, sphenopalatine, and pterygopalatine canal

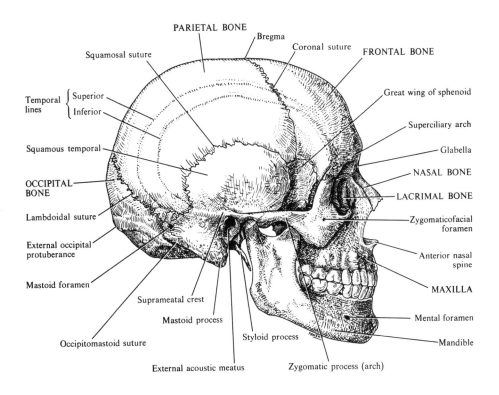

PARIETAL BONE

Bregma

Coronal suture

FRONTAL BONE

Squamosal suture

Great wing of sphenoid

Temporal lines
  Superior
  Inferior

Superciliary arch

Squamous temporal

Glabella

OCCIPITAL BONE

NASAL BONE

LACRIMAL BONE

Lambdoidal suture

Zygomaticofacial foramen

External occipital protuberance

Anterior nasal spine

Mastoid foramen

MAXILLA

Suprameatal crest

Mental foramen

Mastoid process

Mandible

Occipitomastoid suture

Styloid process

External acoustic meatus

Zygomatic process (arch)

**SKULL
LATERAL VIEW**

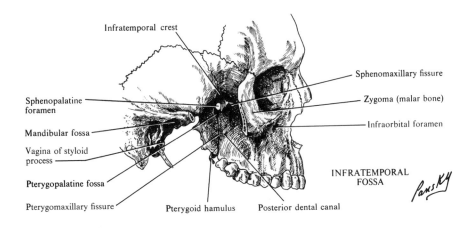

Infratemporal crest

Sphenomaxillary fissure

Zygoma (malar bone)

Sphenopalatine foramen

Infraorbital foramen

Mandibular fossa

Vagina of styloid process

Pterygopalatine fossa

**INFRATEMPORAL
FOSSA**

Pterygomaxillary fissure

Pterygoid hamulus

Posterior dental canal

3. SKULL FROM ABOVE AND BEHIND

I. Bones
A. FRONTAL: usually smooth and convex in adult. In infancy, 2 frontal bones are joined at frontal (metopic) suture. Lower end of this suture, above nose, sometimes seen in adults
B. PARIETAL: forms most of superior and lateral cranium; shows parietal foramen and tuberosity
C. OCCIPITAL: forming dorsal part of cranium, shows external occipital protuberance with external occipital crest running caudally and the supreme and superior nuchal lines arching laterally from it; the planum occipitale lies above highest line; the inferior nuchal lines run laterally from midpoint of the median crest; and planum nuchale is that part of bone below superior nuchal lines

II. Sutures of skull
A. SAGITTAL: between the parietal bones on either side
B. CORONAL: between frontal and the 2 parietal bones; bregma is that point of union of coronal and sagittal sutures
C. LAMBDOIDAL: between occipital and 2 parietal bones
D. FRONTAL (metopic) may not be visible in adults

III. Fontanelles: portions of the skull not ossified at birth, usually 6 in number
A. ANTERIOR (bregmatic) is largest and diamond-shaped, located at junctions of coronal, sagittal, and frontal sutures. Closes middle of second year
B. POSTERIOR: triangular in shape, is located at union of sagittal and lambdoidal sutures. Closes 2 months after birth
C. SPHENOIDAL: 1 on each side, is a small gap between parietal, great wing of sphenoid, and squamous temporal bones. Closes 1 to 2 months after birth
D. MASTOID: 1 on each side, is a small gap between parietal, temporal, and occipital bones. Closes 1 to 2 months after birth

IV. Clinical considerations
A. FRACTURE LINES follow the paths of least resistance: through the vault between the reinforcing pillars; through the base within the fossae, not at the junctions of the fossae; and along the suture lines. All lines of fracture directed toward the base will converge on the body of the sphenoid bone
B. IN BASAL SKULL FRACTURES, bleeding from the ear is common, and Battle's sign is discoloration of the skin along the course of the posterior auricular artery
C. A RING FRACTURE of the occipital bone about the foramen magnum may occur in injuries in which the base of the skull is jammed against the atlas
D. A DEPRESSED FRACTURE occurs when a piece of bone is displaced inward and presses on or lacerates the brain
E. A LINEAR FRACTURE consists of a single line of fracture without bone displacement
F. A COMPOSITE FRACTURE is one in which the fracture lines branch without bone displacement
G. SKULL FRACTURES may be either comminuted or compound
H. THE MOST COMMON CAUSE OF DEATH from a skull fracture is laceration of the brain; other causes include subdural hematoma, cerebral concussion, extradural hemorrhage, and meningitis
I. WHILE THE FONTANELLES ARE OPEN, the brain is very vulnerable. Care is needed to protect these regions, especially anteriorly, because this fontanelle does not close until the middle of the second year

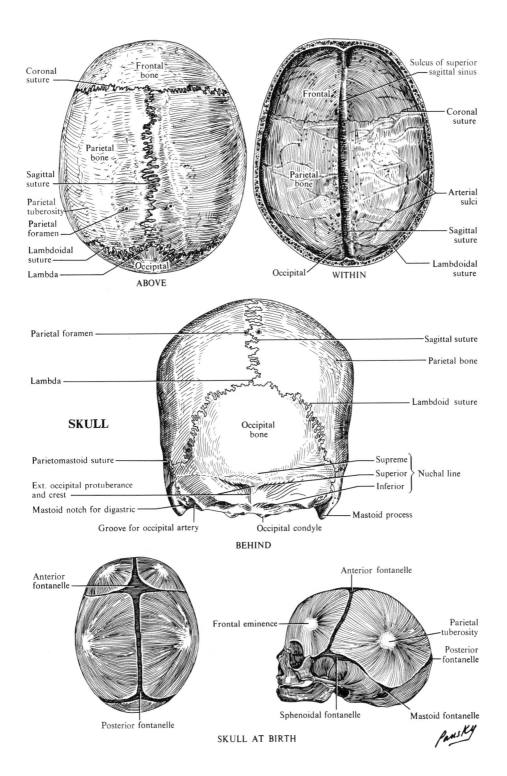

Coronal suture

Frontal bone

Parietal bone

Sagittal suture

Parietal tuberosity

Parietal foramen

Lambdoidal suture

Lambda

Occipital

ABOVE

Sulcus of superior sagittal sinus

Frontal

Coronal suture

Parietal bone

Arterial sulci

Sagittal suture

Lambdoidal suture

Occipital

WITHIN

Parietal foramen

Lambda

SKULL

Parietomastoid suture

Ext. occipital protuberance and crest

Mastoid notch for digastric

Groove for occipital artery

Sagittal suture

Parietal bone

Lambdoid suture

Occipital bone

Supreme
Superior } Nuchal line
Inferior

Mastoid process

Occipital condyle

BEHIND

Anterior fontanelle

Posterior fontanelle

Anterior fontanelle

Frontal eminence

Parietal tuberosity

Posterior fontanelle

Sphenoidal fontanelle

Mastoid fontanelle

SKULL AT BIRTH

pansky

-9-

4. BASAL VIEW OF THE SKULL

I. Bones

A. PALATINE PROCESSES OF MAXILLAE
1. Incisive foramen, behind incisor teeth
B. HORIZONTAL PLATE OF PALATINE: this, with A above, forms hard palate. Located here are the greater and lesser palatine foramina and posterior nasal spine
C. VOMER
D. MEDIAL PTERYGOID PLATE: scaphoid fossa located on lateral side of base; hamulus, at lower extremity
E. LATERAL PTERYGOID PLATE
F. GREATER WING OF SPHENOID: contains foramina ovale, rotundum, spinosum; angular spine; and sulcus for auditory tube
G. TEMPORAL BONE: mandibular and jugular fossae, styloid process, stylomastoid foramen, mastoid process with mastoid notch and occipital groove, tympanomastoid fissure, carotid canal, foramen lacerum, inferior tympanic canaliculus, and mastoid canaliculus
H. BASILAR PART OF OCCIPITAL: shows pharyngeal tubercle
I. LATERAL PART OF OCCIPITAL: condyles, condyloid fossa, and hypoglossal canal
J. SQUAMOUS PART OF OCCIPITAL: external occipital protuberance with the superior nuchal line, external occipital crest with the inferior nuchal line running laterally from its middle, and planum nuchale

II. Foramina (see p. 12). Only those not given elsewhere will be listed below

Name	Bone	Contents
Incisive canals	Palatine process of maxilla	Nasopal. n. Ant. brs. descending palat. a. & v.
Greater palatine	Palatine	Great. palat. n. Descend. palat. vessels
Lesser palatine	Palatine	Lesser palatine nn. & aa.
Stylomastoid	Temporal	Facial n. Stylomastoid a.
Carotid	Temporal	Int. carotid a. Carotid plexus
Condyloid	Occipital	Vein from transverse sinus
Tympanic canaliculus	Temporal	Tympanic br. of IX
Mastoid canaliculus	Temporal	Auricular br. of X
Tympanomastoid fissure	Temporal	Auricular br. of X

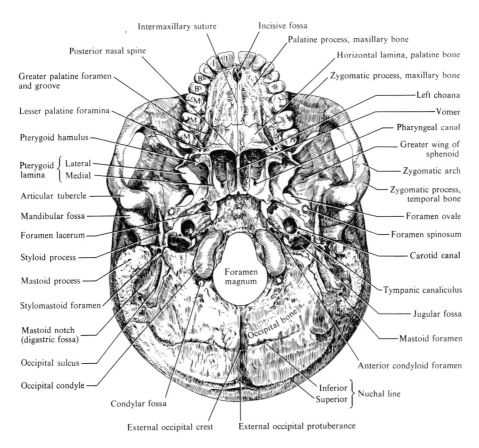

Intermaxillary suture
Incisive fossa
Palatine process, maxillary bone
Posterior nasal spine
Horizontal lamina, palatine bone
Greater palatine foramen and groove
Zygomatic process, maxillary bone
Left choana
Lesser palatine foramina
Vomer
Pharyngeal canal
Pterygoid hamulus
Greater wing of sphenoid
Pterygoid lamina { Lateral / Medial }
Zygomatic arch
Articular tubercle
Zygomatic process, temporal bone
Mandibular fossa
Foramen ovale
Foramen lacerum
Foramen spinosum
Styloid process
Carotid canal
Foramen magnum
Mastoid process
Tympanic canaliculus
Stylomastoid foramen
Jugular fossa
Occipital bone
Mastoid notch (digastric fossa)
Mastoid foramen
Occipital sulcus
Anterior condyloid foramen
Occipital condyle
Inferior } Nuchal line
Superior
Condylar fossa
External occipital crest
External occipital protuberance

BASE OF SKULL

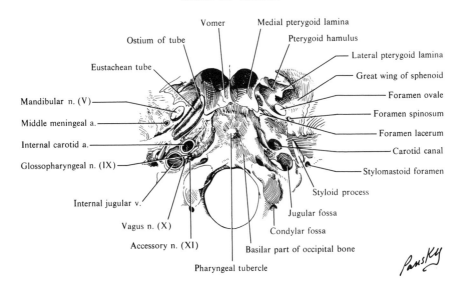

Vomer
Medial pterygoid lamina
Ostium of tube
Pterygoid hamulus
Lateral pterygoid lamina
Eustachean tube
Great wing of sphenoid
Foramen ovale
Mandibular n. (V)
Foramen spinosum
Middle meningeal a.
Foramen lacerum
Internal carotid a.
Carotid canal
Glossopharyngeal n. (IX)
Stylomastoid foramen
Styloid process
Internal jugular v.
Jugular fossa
Vagus n. (X)
Condylar fossa
Accessory n. (XI)
Basilar part of occipital bone
Pharyngeal tubercle

Pansky

5. SKULL INTERIOR: THE CRANIAL FOSSAE AND FORAMINA

I. Anterior cranial fossa

A. BOUNDARIES: anteriorly and laterally by squamous frontal bone, posteriorly by posterior margin of lesser wing of sphenoid and anterior margin of chiasmatic sulcus
B. FLOOR: orbital plate of frontal bone, cribriform plate of ethmoid bone, body and lesser wings of sphenoid bone
C. SPECIAL FEATURES: frontal crest, foramen cecum, crista galli, olfactory groove, anterior and posterior ethmoidal foramina, and grooves for meningeal vessels

II. Middle cranial fossa

A. BOUNDARIES: anterior, lesser wings of sphenoid, anterior clinoid processes, and anterior margin of chiasmatic groove; posterior, superior angles of petrous bone and dorsum sellae; lat., squamous temp., great wing of sphenoid and parietal bones
B. FLOOR: squamous temp., petrous bone, great wing of sphenoid, and sella turcica
C. SPECIAL FEATURES: sella turcica, posterior clinoid process, carotid sulcus, superior orbital fissure, foramen rotundum, foramen ovale, foramen spinosum, foramen lacerum, arcuate eminence, hiatus for greater petrosal n., grooves for greater and lesser petrosal nerves, and grooves for branches of middle meningeal artery

III. Posterior fossa

A. BOUNDARIES: anterior, dorsum sellae, basal occipital, and crest of petrous bone; lateral, parietal bone; posterior, squamous occipital bone
B. FLOOR: occipital and temporal bones
C. SPECIAL FEATURES: grooves for superior petrosal sinus, foramen magnum, hypoglossal canal, petro-occipital fissure, jugular foramen, internal acoustic meatus, vestibular aqueduct, internal occipital crest, grooves for the transverse sinuses, and mastoid foramen

IV. Foramina or other openings and their principal contents

Name	Fossa	Contents
Cribriform	Ant.	Olfactory nerve fibers
Ant. ethmoid	Ant.	Ant. ethmoid vessels and nerve
Post. ethmoid	Ant.	Post. ethmoid vessels and nerve
Optic	Mid.	Optic n. and ophthalmic a.
Sup. orbital fissure	Mid.	III, IV, VI, ophthalmic V nn.; sympathetic nn.; ophthalmic vv.
Rotundum	Mid.	Maxillary n.
Ovale	Mid.	Mandibular n., access. meningeal a.
Spinosum	Mid.	Middle meningeal a.
Hiatus for greater petrosal n.	Mid.	Greater petrosal n., petrosal br. middle meningeal a.
Magnum	Post.	Spinal cord, accessory n., vertebral aa., ant. and post. spinal aa.
Jugular	Post.	Inf. petrosal and trans. sinuses, meningeal brs. of occip. & ascend. pharyngeal aa.; IX, X, XI nn.
Hypoglossal	Post.	XII n., meningeal br. of ascend. pharyngeal a.
Int. aud. meatus	Post.	VII, VIII nn.; labyrinthine (int. auditory) a.

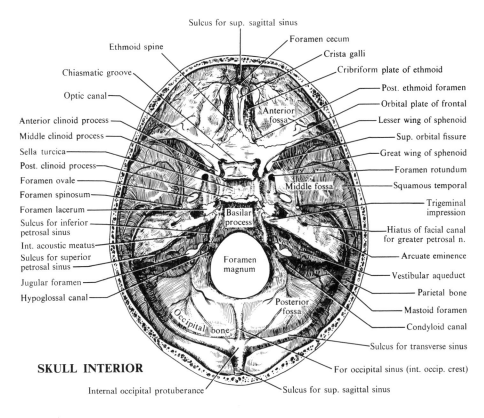

Sulcus for sup. sagittal sinus
Ethmoid spine
Chiasmatic groove
Optic canal
Anterior clinoid process
Middle clinoid process
Sella turcica
Post. clinoid process
Foramen ovale
Foramen spinosum
Foramen lacerum
Sulcus for inferior petrosal sinus
Int. acoustic meatus
Sulcus for superior petrosal sinus
Jugular foramen
Hypoglossal canal

Foramen cecum
Crista galli
Cribriform plate of ethmoid
Post. ethmoid foramen
Orbital plate of frontal
Lesser wing of sphenoid
Sup. orbital fissure
Great wing of sphenoid
Foramen rotundum
Squamous temporal
Trigeminal impression
Hiatus of facial canal for greater petrosal n.
Arcuate eminence
Vestibular aqueduct
Parietal bone
Mastoid foramen
Condyloid canal
Sulcus for transverse sinus
For occipital sinus (int. occip. crest)
Sulcus for sup. sagittal sinus

Anterior fossa
Middle fossa
Basilar process
Foramen magnum
Posterior fossa
Occipital bone

SKULL INTERIOR

Internal occipital protuberance

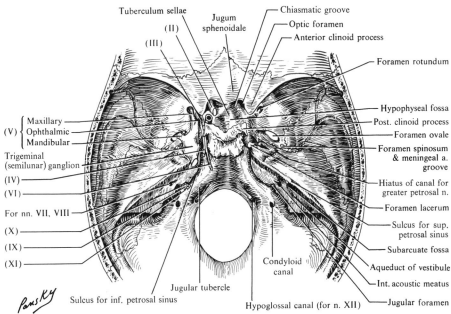

Tuberculum sellae
(II)
(III)
Jugum sphenoidale
Chiasmatic groove
Optic foramen
Anterior clinoid process
Foramen rotundum

Maxillary
(V) { Ophthalmic
Mandibular
Trigeminal (semilunar) ganglion
(IV)
(VI)
For nn. VII, VIII
(X)
(IX)
(XI)

Hypophyseal fossa
Post. clinoid process
Foramen ovale
Foramen spinosum & meningeal a. groove
Hiatus of canal for greater petrosal n.
Foramen lacerum
Sulcus for sup. petrosal sinus
Subarcuate fossa
Aqueduct of vestibule
Int. acoustic meatus
Jugular foramen

Pansky

Sulcus for inf. petrosal sinus
Jugular tubercle
Condyloid canal
Hypoglossal canal (for n. XII)

-13-

6. CUTANEOUS NERVES AND DERMATOMES OF THE HEAD

I. Cutaneous nerves of scalp
A. AREA IN FRONT OF EAR
1. Rostral and medial: supratrochlear and supraorbital branches of ophthalmic (V)
2. Lateral: zygomaticotemporal branch of maxillary (V), auriculotemporal branch of mandibular (V)
B. AREA BEHIND EAR
1. Lateral: great auricular and small occipital nerves from cervical plexus C2 and C3
2. Posterior and medial: great occipital nerve—the posterior division of C2; least (3rd) occipital nerve—posterior division of C3

II. Cutaneous nerves of face
A. NOSE: medially, supratrochlear, infratrochlear and nasal branches of ophthalmic (V); laterally, infraorbital branch of maxillary (V)
B. EYE
1. Upper lid: supratrochlear, infratrochlear, supraorbital, and lacrimal branches of ophthalmic (V)
2. Lower lid: infratrochlear of ophthalmic (V), infraorbital and zygomaticofacial branches of maxillary (V)
C. SKIN OVER UPPER JAW AND ZYGOMATIC AREA: infraorbital and zygomaticofacial branches of maxillary (V)
D. SKIN OVER BUCCINATOR: buccal branch of mandibular (V)
E. OVER LOWER JAW (front to back): mental, buccal and auriculotemporal branches of mandibular (V), great auricular of cervical plexus, C2 and C3

III. Cutaneous nerves of auricle (external ear)
A. LATERAL ASPECT: upper part, auriculotemporal branch of mandibular (V); lower, great auricular branch of cervical plexus C2 and C3
B. MEDIAL ASPECT: upper part, lesser occipital from cervical plexus, C2 and C3; lower part, great auricular branch of cervical plexus, C2 and C3

IV. Cutaneous nerves of neck
A. POSTERIOR (from cranial to caudal): third occipital and posterior divisions of C2–C6
B. LATERAL: great auricular and lesser occipital branches of cervical plexus, C2 and C3; intermediate supraclavicular branch of cervical plexus, C3 and C4
C. ANTERIOR: transverse cervical branch of cervical plexus, C2 and C3; medial supraclavicular branch of cervical plexus, C3 and C4

V. Dermatomes,
a term applied to those areas supplied by spinal nerves. Taking the back of the head and the neck as a unit, keeping in mind that the first cervical nerve has no cutaneous branch, one begins at the highest point with C2 and descends to the lowest point at the back of the root of the neck, with C6. On the back of the neck, the posterior divisions of these nerves have direct cutaneous branches. The lateral and anterior aspects, receiving fibers from C2–C4, are carried through branches of the cervical plexus

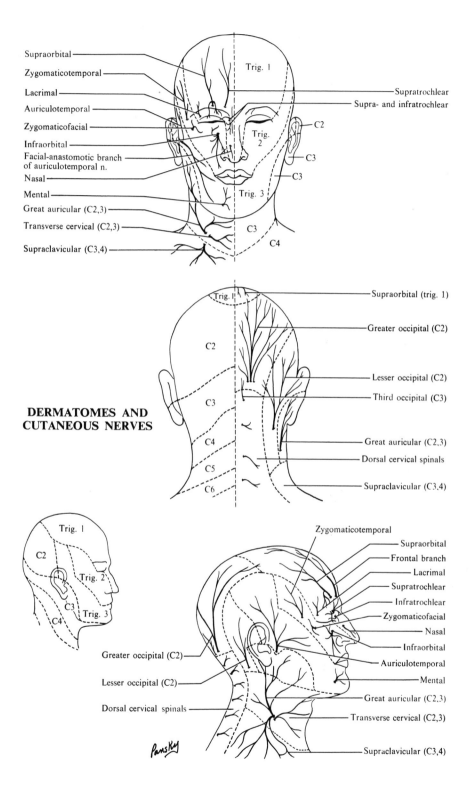

Supraorbital

Zygomaticotemporal

Lacrimal

Auriculotemporal

Zygomaticofacial

Infraorbital

Facial-anastomotic branch of auriculotemporal n.

Nasal

Mental

Great auricular (C2,3)

Transverse cervical (C2,3)

Supraclavicular (C3,4)

Trig. 1

Trig. 2

Trig. 3

Supratrochlear

Supra- and infratrochlear

C2

C3

C3

C3

C4

DERMATOMES AND CUTANEOUS NERVES

Trig. I

C2

C3

C4

C5

C6

Supraorbital (trig. 1)

Greater occipital (C2)

Lesser occipital (C2)

Third occipital (C3)

Great auricular (C2,3)

Dorsal cervical spinals

Supraclavicular (C3,4)

Trig. 1

C2

Trig. 2

C3

Trig. 3

C4

Zygomaticotemporal

Supraorbital

Frontal branch

Lacrimal

Supratrochlear

Infratrochlear

Zygomaticofacial

Nasal

Infraorbital

Auriculotemporal

Mental

Great auricular (C2,3)

Transverse cervical (C2,3)

Supraclavicular (C3,4)

Greater occipital (C2)

Lesser occipital (C2)

Dorsal cervical spinals

Pansky

7. THE SCALP

I. **Structure.** Composed of 5 layers
A. SKIN (s): thick, with many close-set hair follicles and their associated sebaceous and sweat glands. Firmly joined to next deeper layer
B. SUBCUTANEOUS TISSUE (C), SUPERFICIAL FASCIA: thick; strong with fiber bundles woven together, with fat interspaced
   1. Contains superficial vessels and nerves
   2. Hair follicles of skin project into this layer
C. MUSCULOAPONEUROTIC (A): represents the deep fascia
   1. In forehead and occipital regions the frontalis and occipitalis muscles are located here. In temporal region, auricular muscles are also in this layer
   2. Galea aponeurotica, a dense, thin, fibrous sheet that unites the frontal and occipital muscles of cranial vault
D. SUBAPONEUROTIC LAYER (L): very loose and scanty. Contains a few small vessels. The nature of this layer permits easy movement of layers A–C, which act as a unit, over the next layer
E. PERICRANIUM (P): the periosteum of the bones. Except at sutures, is poorly fixed to bone
If the above letters, *S C A L P*, are put together, they form the very word that each layer helps create.

II. **Arteries of the scalp**
A. OCCIPITAL: from external carotid artery to back of head
B. POSTERIOR AURICULAR: from external carotid artery to ear and scalp in posterior temporal region
C. SUPERFICIAL TEMPORAL: one of terminal branches of external carotid artery to lateral part of scalp in front of ear
D. SUPRAORBITAL: from ophthalmic artery to skin of forehead and rostral area of scalp
E. SUPRATROCHLEAR: from ophthalmic artery to most rostral and medial parts of scalp

III. **Cutaneous nerves**
A. ROSTRALLY: the supratrochlear and supraorbital branches of ophthalmic (V) supply regions corresponding to arteries of same name
B. LATERALLY: auriculotemporal branch of mandibular (V) follows superficial temporal artery
C. POSTERIORLY: the greater occipital (C2) accompany branches of occipital artery
D. POSTEROLATERALLY: the lesser occipital from cervical plexus (C2, 3); in general, covers same areas as posterior auricular artery

IV. **Clinical considerations**
A. DUE TO LOOSENESS of subaponeurotic layer, large amounts of blood can form enormous hematomas after blows on head
B. FOR SAME REASON, INFECTIONS entering this layer may also spread widely by way of:
   1. Emissary veins (see p. 76), which pass through bones directly to dura, can carry infections to meninges
   2. Diploic veins (see p. 76), located between bony tables of the skull
C. DUE TO THE THICKNESS AND EXTENT of the fibrous tissue of the second layer and extensive vascular anastomoses, wounds of the scalp tend to bleed profusely
D. THE PERICRANIUM possesses little osteogenic capacity as compared to most periosteum, and except over the sutures it is rather loosely attached to the bones of the calvaria
E. IT IS THE PULSATING DISTENTION of the arteries of the scalp that accounts for most of the pain in migraine headache

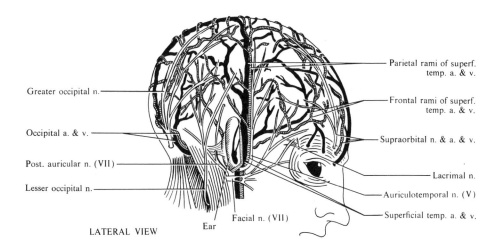

LATERAL VIEW

Greater occipital n.

Occipital a. & v.

Post. auricular n. (VII)

Lesser occipital n.

Ear

Facial n. (VII)

Parietal rami of superf. temp. a. & v.

Frontal rami of superf. temp. a. & v.

Supraorbital n. & a. & v.

Lacrimal n.

Auriculotemporal n. (V)

Superficial temp. a. & v.

SCALP VESSELS AND NERVES

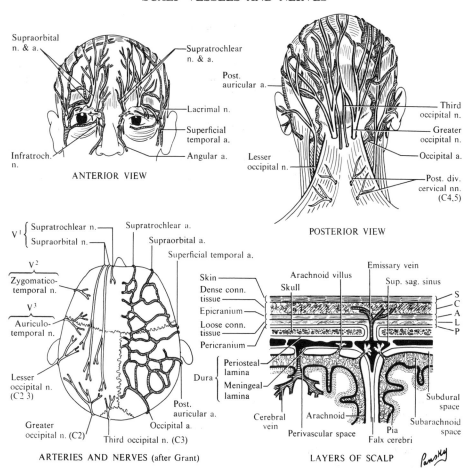

ANTERIOR VIEW

Supraorbital n. & a.

Supratrochlear n. & a.

Lacrimal n.

Superficial temporal a.

Angular a.

Infratroch. n.

POSTERIOR VIEW

Post. auricular a.

Third occipital n.

Greater occipital n.

Occipital a.

Post. div. cervical nn. (C4,5)

Lesser occipital n.

ARTERIES AND NERVES (after Grant)

V^1 { Supratrochlear n.
       Supraorbital n.

V^2

Zygomatico-temporal n.

V^3

Auriculo-temporal n.

Lesser occipital n. (C2,3)

Greater occipital n. (C2)

Third occipital n. (C3)

Supratrochlear a.

Supraorbital a.

Superficial temporal a.

Post. auricular a.

Occipital a.

LAYERS OF SCALP

Skin

Dense conn. tissue

Epicranium

Loose conn. tissue

Pericranium

Dura { Periosteal lamina
       Meningeal lamina

Cerebral vein

Perivascular space

Arachnoid

Pia

Falx cerebri

Arachnoid villus

Skull

Emissary vein

Sup. sag. sinus

S
C
A
L
P

Subdural space

Subarachnoid space

Pansky

8. SUPERFICIAL AND DEEP VEINS

    I. Supratrochlear vein. From middle scalp and forehead, near midline. Descends to medial angle of eye, where it joins supraorbital vein to form angular vein and communicates with branches of superficial temporal vein

    II. Supraorbital vein. From scalp and forehead. Sends branch through supraorbital notch and continues to join frontal vein forming angular vein

    III. Angular vein. Runs caudad at side of nose to level of lower orbit, where it becomes anterior facial vein. Receives tributaries from sides of nose and communicates, through nasofrontal vein at medial angle of orbit, with superior ophthalmic vein

    IV. Anterior facial vein. Continuation of angular vein caudally and runs beneath facial muscles but over masseter muscle, crosses mandible, runs caudally and posteriorly under platysma muscle to join ant. branch of retromandibular vein forming common facial vein. Has communication through deep facial vein with pterygoid plexus. Receives tributaries from eyelids, lips, cheeks, and masseter muscles while on face. In neck receives submental, submandibular, and palatine branches

    V. Superficial temporal vein. From side of head and scalp. Branches communicate with I and II, above. Is joined by middle temporal vein from temporalis muscle. Crosses zygomatic arch to enter parotid gland where it joins maxillary vein to form retromandibular vein. Tributaries from parotid gland, external ear, side of face

    VI. Pterygoid plexus. Network between temporalis and pterygoid muscles. Receives middle meningeal, sphenopalatine, deep temporal, alveolar, palatine, and muscular veins. Communicates with the anterior facial vein through inferior orbital fissure via the inferior orbital vein and with cavernous sinus through veins in foramina ovale and lacerum

    VII. Maxillary vein. Short vein draining pterygoid plexus and joining superficial temporal to form posterior facial vein

    VIII. Retromandibular vein. Formed by union of superficial temporal and maxillary veins. Descends in parotid, lateral to external carotid artery but deep to facial nerve. Divides into 2 branches
    A. ANTERIOR, running rostrally to join ant. facial vein to form common facial vein
    B. POSTERIOR, which is joined by post. auricular vein to form external jugular vein

    IX. Posterior auricular vein. From side of head, behind ear, descends behind auricle and joins posterior branch of posterior facial to form external jugular vein

    X. Occipital vein. From back of head to deep cervical and vertebral veins

    XI. External jugular vein. Formed by VIIIB and IX above, in parotid gland at level of angle of mandible. Descends vertically to enter subclavian triangle, where it pierces fascia and terminates in subclavian vein. Communicates with internal jugular vein; receives trans. cervical, trans. scapular, and anterior jugular veins

    XII. Posterior external jugular vein. From occipital region and upper posterior side of neck to external jugular vein

    XIII. Anterior jugular vein. From submandibular region, descends near midline to enter external jugular vein. Receives some drainage from larynx and thyroid. Above sternum, veins of 2 sides are joined through venous jugular arch

    XIV. Clinical considerations. Infections on the face (above mouth) can be carried into the cavernous dural sinus via the facial, angular, and sup. ophthalmic vv

SUPERFICIAL AND DEEP VEINS

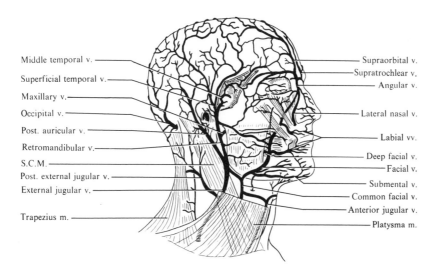

Middle temporal v.
Superficial temporal v.
Maxillary v.
Occipital v.
Post. auricular v.
Retromandibular v.
S.C.M.
Post. external jugular v.
External jugular v.
Trapezius m.

Supraorbital v.
Supratrochlear v.
Angular v.
Lateral nasal v.
Labial vv.
Deep facial v.
Facial v.
Submental v.
Common facial v.
Anterior jugular v.
Platysma m.

SUPERFICIAL VEINS

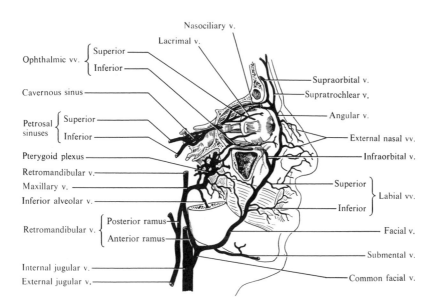

Nasociliary v.
Lacrimal v.
Ophthalmic vv. { Superior
                 Inferior
Cavernous sinus
Petrosal sinuses { Superior
                   Inferior
Pterygoid plexus
Retromandibular v.
Maxillary v.
Inferior alveolar v.
Retromandibular v. { Posterior ramus
                     Anterior ramus
Internal jugular v.
External jugular v.

Supraorbital v.
Supratrochlear v.
Angular v.
External nasal vv.
Infraorbital v.
Superior } Labial vv.
Inferior }
Facial v.
Submental v.
Common facial v.

FACIAL AND OPHTHALMIC VEINS

9. SUPERFICIAL LYMPHATICS OF HEAD AND NECK

I. **Lymphatic vessels**
   A. LYMPHATICS OF SCALP. Three drainage areas: frontal, ending in anterior auricular and parotid nodes; temporal and parietal, ending in parotid and retroauricular nodes; occipital, ending in occipital and deep cervical nodes
   B. LYMPHATICS OF EXTERNAL EAR. Drain into pre- and retroauricular nodes and superficial and deep cervical nodes
   C. LYMPHATICS OF FACE
      1. Eyelids and conjunctiva: to submandibular and parotid nodes
      2. Cheek: to parotid and submandibular nodes
      3. Side of nose, upper lip, and lateral lower lip: to submandibular nodes
      4. Medial lower lip: into submental nodes
      5. Temporal and infratemporal fossae: into deep facial and deep cervical nodes

II. **Lymph nodes of head**
   A. OCCIPITAL: back of head close to edge of trapezius. Afferents from scalp at back of head; efferents to superior deep cervical
   B. RETROAURICULAR: at insertion of sternocleidomastoid on mastoid. Afferents from posterior temporal and parietal regions; efferents to superior deep cervicals
   C. PREAURICULAR: in front of tragus of ear. Afferents from pinna and temporal region; efferents to superior deep cervical
   D. PAROTID: two sets, either embedded in gland or just below it. Afferents from root of nose, eyelids, anterior temporal region and external auditory meatus; efferents to superior deep cervical
   E. FACIAL: three sets: infraorbital, buccal, and mandibular. Afferents from eyelids, conjunctiva, skin of nose, nasal mucosa, and cheek; efferents to submandibular nodes

III. **Lymph nodes of neck**
   A. SUBMANDIBULAR: under body of mandible. Afferents from cheek, nose, upper lip, lower lip, facial and submental nodes; efferents to superior deep cervical
   B. SUBMENTAL: between anterior bellies of digastric. Afferents from central lower lip, floor of mouth; efferents to submandibular and deep cervical
   C. SUPERFICIAL CERVICAL: along external jugular vein. Afferents from ear and parotid region; efferents to superior deep cervical
   D. DEEP CERVICAL: along carotid sheath
      1. Superior, under sternocleidomastoid muscle, along accessory nerve and internal jugular vein. Afferents from back of head and neck, tongue, larynx, thyroid, palate, nose, esophagus, and all nodes except inferior deep cervical; efferents to inferior deep cervical nodes and jugular trunk
      2. Inferior, extending below border of sternocleidomastoid, close to subclavian vein. Afferents from dorsum of scalp and neck, superficial pectoral region, part of arm, and superior deep cervical; efferents join efferents of superior deep nodes to form jugular trunk

IV. **Jugular trunk terminates:** on right, at junction of internal jugular and subclavian veins; on left, in thoracic duct

V. **Clinical considerations**
   A. Since lymph nodes act as filters for lymph, and both infectious material and metastatic cells are carried in lymph, one must know the position as well as the drainage areas of node groups. In case of cancer, the swelling of regional nodes may be the first sign of a malignancy in some deep-lying structure

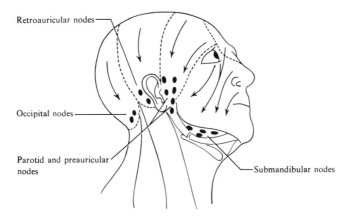

Retroauricular nodes

Occipital nodes

Parotid and preauricular
nodes

Submandibular nodes

LYMPH DRAINAGE OF HEAD AND NECK

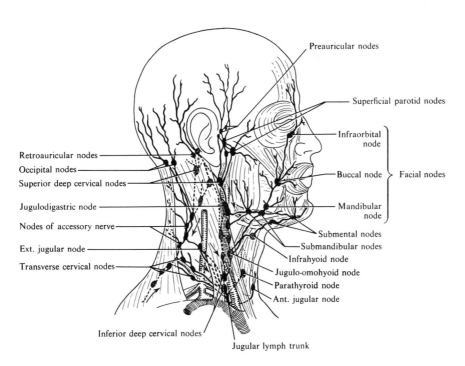

Preauricular nodes

Superficial parotid nodes

Retroauricular nodes
Occipital nodes
Superior deep cervical nodes

Jugulodigastric node

Nodes of accessory nerve

Ext. jugular node

Transverse cervical nodes

Infraorbital
node

Buccal node   Facial nodes

Mandibular
node

Submental nodes
Submandibular nodes
Infrahyoid node
Jugulo-omohyoid node
Parathyroid node
Ant. jugular node

Inferior deep cervical nodes

Jugular lymph trunk

PRINCIPAL NODE GROUPS AND DRAINAGE

Pansky

-21-

10. MUSCLES OF FACIAL EXPRESSION

I. General considerations. All are skin (cutaneous) muscles. They lie in superficial fascia and may arise from either fascia or bone. They insert into skin. Many individual muscles are indistinctly separated from those closely adjoining. All are innervated by the facial nerve (VII). They are grouped according to location or area of principal action.

II. Groupings, with principal actions only
A. SCALP
    1. Frontalis: raises brows and wrinkles forehead as in surprise
    2. Occipitalis: with above in raising brows
B. EAR
    1. Anterior auricular: draws ear up and forward
    2. Superior auricular: draws ear up
    3. Posterior auricular: draws ear back
C. EYE
    1. Orbicularis oculi: sphincter of eye. Closes lids; compresses lacrimal sac
    2. Corrugator: draws eyebrows downward and medially as in frowning or suffering
D. NOSE
    1. Procerus: draws medial angle of eyebrow downward
    2. Anterior and posterior dilator nares: enlarge nares in hard breathing and anger
    3. Depressor septi: constricts nares
    4. Nasalis: draws wings toward septum and depresses cartilage
E. MOUTH
    1. Levator labii superioris: raises upper lip
    2. Levator labii superioris alaeque nasi: raises lip and dilates nares
    3. Zygomaticus minor: with 1 and 2 above, forms nasolabial furrow; deepened when in sorrow
    4. Levator anguli oris: with 1–3 above, expresses contempt or disdain
    5. Zygomaticus major: draws angle of mouth up and back, as in laughing
    6. Risorius: retracts angle of mouth
    7. Depressor labii inferioris: draws lip down and back, as in irony
    8. Depressor anguli oris: depresses angle of mouth
    9. Mentalis: raises and protrudes lower lip
    10. Platysma: retracts and depresses angle of mouth (see p. 40)
    11. Orbicularis oris: a complex muscle with layers, some parts intrinsic to the lips, the others derived from the following facial muscles: buccinator, levator anguli oris, depressor anguli oris, zygomaticus major and minor. Closes lips, protrudes lips, and presses lips to teeth
    12. Buccinator: compresses cheek, holds food under teeth in mastication, important in blowing when cheeks are distended with air

III. Clinical considerations
A. DAMAGE TO THE FACIAL NERVE OR ITS BRANCHES produces variable amounts of weakness or paralysis of facial muscles, sometimes called "Bell's palsy." Weakness is particularly noticeable about the mobile mouth, where it may be evidenced even in repose by a sagging corner and becomes very obvious as a result of the asymmetry attending an attempt to show the teeth or smile. Upper facial weakness can similarly be brought out by having a patient attempt to frown, raise his eyebrows, or close his eyes tightly
B. PARALYSIS of an entire side of the face is an indication that the facial nerve as a whole has been damaged along its course
C. WEAKNESS rather than complete paralysis of a group of muscles typically results from injury to facial nerve branches, because of the overlap of their distribution

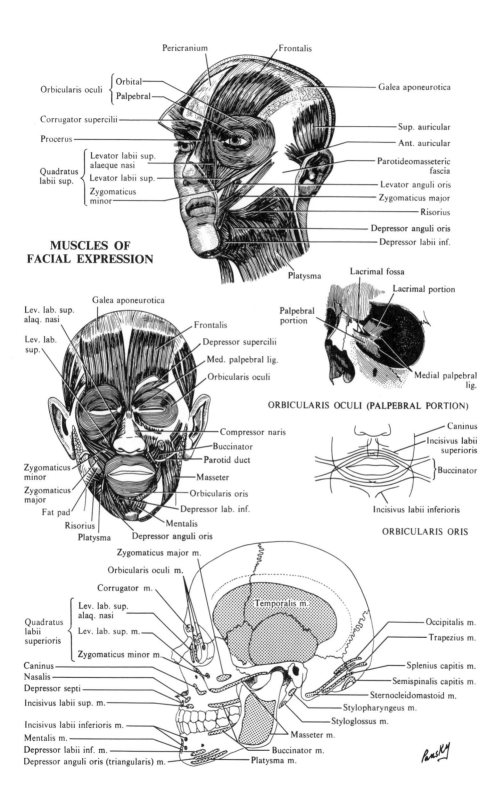

Pericranium · Frontalis

Orbicularis oculi { Orbital / Palpebral

Galea aponeurotica

Corrugator supercilii

Procerus

Sup. auricular

Ant. auricular

Quadratus labii sup. { Levator labii sup. alaeque nasi / Levator labii sup. / Zygomaticus minor

Parotideomasseteric fascia

Levator anguli oris

Zygomaticus major

Risorius

Depressor anguli oris

Depressor labii inf.

MUSCLES OF FACIAL EXPRESSION

Platysma

Galea aponeurotica

Lev. lab. sup. alaq. nasi

Lev. lab. sup.

Frontalis

Depressor supercilii

Med. palpebral lig.

Orbicularis oculi

Lacrimal fossa

Lacrimal portion

Palpebral portion

Medial palpebral lig.

ORBICULARIS OCULI (PALPEBRAL PORTION)

Compressor naris

Buccinator

Parotid duct

Masseter

Orbicularis oris

Depressor lab. inf.

Mentalis

Depressor anguli oris

Zygomaticus minor

Zygomaticus major

Fat pad

Risorius

Platysma

Caninus

Incisivus labii superioris

Buccinator

Incisivus labii inferioris

ORBICULARIS ORIS

Zygomaticus major m.

Orbicularis oculi m.

Corrugator m.

Lev. lab. sup. alaq. nasi

Quadratus labii superioris { Lev. lab. sup. m.

Zygomaticus minor m.

Caninus

Nasalis

Depressor septi

Incisivus labii sup. m.

Incisivus labii inferioris m.

Mentalis m.

Depressor labii inf. m.

Depressor anguli oris (triangularis) m.

Temporalis m.

Occipitalis m.

Trapezius m.

Splenius capitis m.

Semispinalis capitis m.

Sternocleidomastoid m.

Stylopharyngeus m.

Styloglossus m.

Masseter m.

Buccinator m.

Platysma m.

Pansky

11. MUSCLES OF MASTICATION

I. General considerations. All these muscles are inserted upon the mandible and are concerned in the process of biting and chewing. All are innervated by the mandibular division of the trigeminal (V) nerve

II. Arrangement
A. TEMPORALIS (see p. 32)
  1. Origin: from temporal fascia and entire temporal fossa from temporal lines to infratemporal crest
  2. Insertion: coronoid process and anterior border of ramus of mandible
  3. Action: closes jaw, posterior part retracts jaw
B. MASSETER (see p. 32)
  1. Origin
    a. Superficial part: lower border of zygomatic arch
    b. Deep part: posterior and medial side of zygomatic arch
  2. Insertion
    a. Superficial part: angle and lower lateral side of ramus of mandible
    b. Deep part: upper lateral ramus and condyloid process of mandible
  3. Action: closes jaw
C. MEDIAL PTERYGOID (INTERNAL PTERYGOID)
  1. Origin: medial side of lateral pterygoid plate of sphenoid, pyramidal process of palatine and tuberosity of maxilla
  2. Insertion: lower and posterior part of medial surface of ramus of mandible
  3. Action: closes jaw
D. LATERAL PTERYGOID (EXTERNAL PTERYGOID)
  1. Origin
    a. Upper part: from lower lateral great wing of sphenoid and infratemporal crest
    b. Lower part: from lateral surface of lateral pterygoid plate
  2. Insertion: neck of mandibular condyle and articular disk of temporomandibular joint
  3. Action: depresses, protrudes, and moves mandible from side to side

III. Clinical considerations
A. LESIONS OF THE MANDIBULAR DIVISION OF THE TRIGEMINAL NERVE will cause unilateral paralysis of muscles of mastication followed by atrophy. This results in a sunken-in appearance along the ramus of the mandible and above the zygomatic arch
B. THE MASSETER, TEMPORALIS, AND MEDIAL PTERYGOID MUSCLES are powerful closers of the jaw, accounting for the strength of the bite
C. THE MASSETER AND TEMPORALIS MUSCLES both abduct (deviate) the jaw to the same side, but the medial pterygoid abducts to the opposite side. Thus, with proper synchronization, these muscles can produce the grinding movement of chewing (the tongue and the buccinator muscle also aid in mastication, but in different ways; the tongue positions the food on the teeth, while the buccinator helps to maintain it there during chewing)
D. THE POSTERIOR FIBERS OF THE TEMPORALIS MUSCLE are the chief retractor of the mandible and are also responsible for maintaining the resting position of closure of the mouth
E. ALTERNATING ACTION OF THE LATERAL PTERYGOIDS can move the jaw from side to side, but the important action of these muscles is to act bilaterally and help to open the mouth

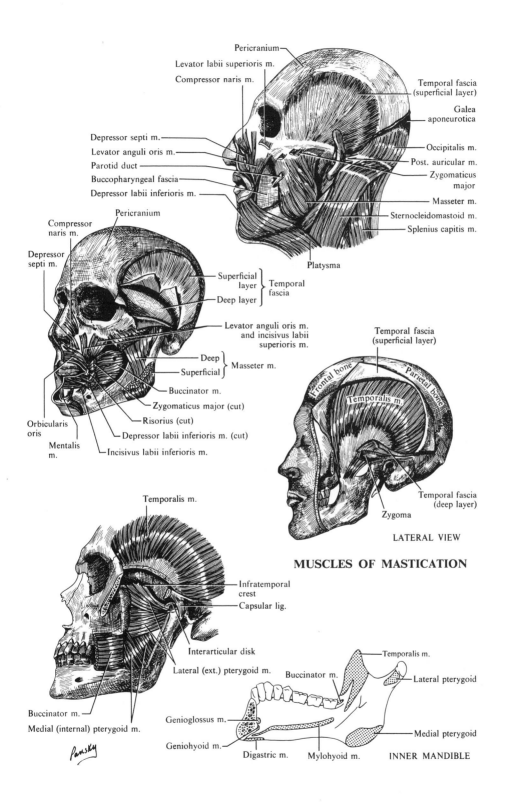

Pericranium
Levator labii superioris m.
Compressor naris m.
Temporal fascia (superficial layer)
Galea aponeurotica
Depressor septi m.
Levator anguli oris m.
Parotid duct
Buccopharyngeal fascia
Depressor labii inferioris m.
Occipitalis m.
Post. auricular m.
Zygomaticus major
Masseter m.
Sternocleidomastoid m.
Splenius capitis m.
Platysma

Pericranium
Compressor naris m.
Depressor septi m.
Superficial layer
Deep layer
} Temporal fascia
Levator anguli oris m. and incisivus labii superioris m.
Deep
Superficial
} Masseter m.
Buccinator m.
Zygomaticus major (cut)
Risorius (cut)
Depressor labii inferioris m. (cut)
Incisivus labii inferioris m.
Orbicularis oris
Mentalis m.

Temporal fascia (superficial layer)
Frontal bone
Parietal bone
Temporalis m.
Temporal fascia (deep layer)
Zygoma

LATERAL VIEW

MUSCLES OF MASTICATION

Temporalis m.
Infratemporal crest
Capsular lig.
Interarticular disk
Lateral (ext.) pterygoid m.
Buccinator m.
Medial (internal) pterygoid m.
Pansky

Temporalis m.
Buccinator m.
Lateral pterygoid
Genioglossus m.
Geniohyoid m.
Medial pterygoid
Digastric m.
Mylohyoid m.

INNER MANDIBLE

-25-

12. THE FACIAL (VII) NERVE

I. **Roots:** 2 in number; a large voluntary motor and a smaller mixed sensory and parasympathetic (nervus intermedius)

II. **Course**
A. Leaves cranial fossa with acoustic nerve (VIII) via internal acoustic meatus
B. Leaves nerve (VIII) to enter facial canal, which passes laterally and then posteriorly along the medial wall of the tympanic cavity above oval window, then turns caudally to exit from skull through stylomastoid foramen
C. Ganglion: geniculate, at acute bend of facial canal
D. Branches
1. Great petrosal nerve: mixed, with parasympathetic and sensory fibers. Leaves region of geniculate ganglion through hiatus for greater petrosal nerve to run in sulcus on petrous bone. Unites with deep petrosal nerve to form nerve of pterygoid canal. The parasympathetic fibers synapse in pterygopalatine ganglion whose postganglionic fibers reach lacrimal gland and glands of nose and palate. Sensory fibers from these same areas merely pass through pterygopalatine ganglion to reach geniculate ganglion
2. Nerve to stapedius muscle: very small, leaves facial nerve in its descending part to supply muscle
3. Chorda tympani nerve: leaves facial nerve in descending part of facial canal, arches upward and rostrally, enters tympanic cavity by crossing its lateral wall, enters the petrous bone, and finally emerges from the skull on the medial surface of angular spine of sphenoid. Descends to join the posterior border of the lingual nerve. Its sensory fibers are those of taste from the anterior two thirds of tongue, and these follow the general sensory branches of the lingual nerve. The parasympathetic fibers continue along the lingual nerve and leave the latter by parasympathetic roots, which terminate by synapse in the submandibular ganglion. From the latter, postganglionic glandular rami go to the submandibular and sublingual glands
4. Muscular branches supply posterior belly of digastric, stylohyoid muscle and all muscles of facial expression. The latter branches are designated as follows: temporal, zygomatic, buccal, marginal mandibular, and cervical, depending on region of face supplied

III. **Clinical considerations**
A. After leaving stylomastoid foramen, the nerve runs rostrally in substance of parotid gland. In surgical operations on parotid this must be noted
B. Paralysis: if lesions occur in either the facial nucleus or its peripheral fibers, there is total unifacial paralysis; if lesions occur in either the facial area of cerebral cortex or the descending corticobulbar fibers, only muscles below the eye will be paralyzed since those above the eye receive cortical fibers from both sides of the brain. Bell's facial paralysis is sometimes caused by nerve irritations and may therefore be only temporary
C. Loss of taste on anterior two thirds of tongue indicates chorda tympani involvement
D. The exact distribution and function of many of the sensory fibers of the facial nerve are not known, but some are thought to be concerned with deep pain from the face; some of them apparently are distributed to a small part of the soft palate; a few may reach the middle ear cavity; and the few cutaneous fibers that the nerve contains are distributed to skin on the posterior surface of the ear, along with similar fibers from the ninth and tenth nerves. The best-known sensory fibers in this nerve are those for taste on the anterior two-thirds of the tongue

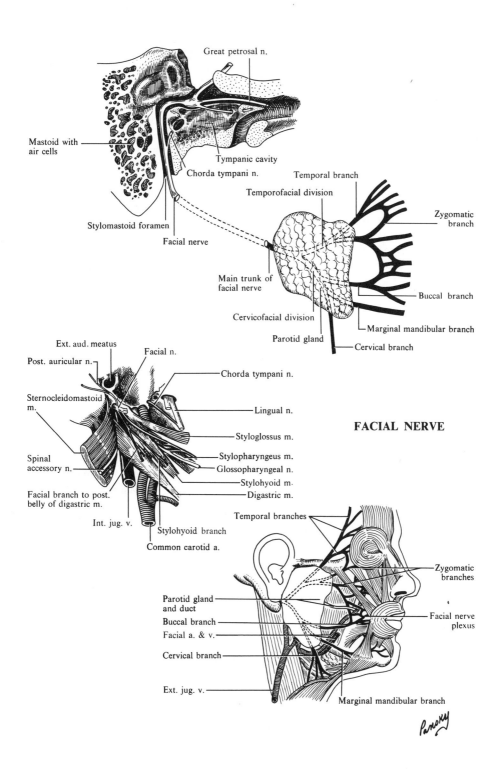

Great petrosal n.

Mastoid with
air cells

Tympanic cavity

Chorda tympani n.

Temporal branch

Temporofacial division

Zygomatic
branch

Stylomastoid foramen

Facial nerve

Main trunk of
facial nerve

Buccal branch

Cervicofacial division

Marginal mandibular branch

Parotid gland

Cervical branch

Ext. aud. meatus

Facial n.

Post. auricular n.

Chorda tympani n.

Sternocleidomastoid
m.

Lingual n.

FACIAL NERVE

Styloglossus m.

Stylopharyngeus m.

Spinal
accessory n.

Glossopharyngeal n.

Stylohyoid m.

Facial branch to post.
belly of digastric m.

Digastric m.

Int. jug. v.

Stylohyoid branch

Temporal branches

Common carotid a.

Zygomatic
branches

Parotid gland
and duct

Buccal branch

Facial nerve
plexus

Facial a. & v.

Cervical branch

Ext. jug. v.

Marginal mandibular branch

-27-

13. THE PAROTID GLAND

    I. Size: largest of major salivary glands; weight, 14–28 g

    II. Type: compound, branched tubuloalveolar gland; serous

  III. Location and relations
    A. ANTERIOR SURFACE: lies against posterior border of ramus of mandible, with rostral projections on both medial and lateral surface of the bone
      1. Medial projection: extension between pterygoid mm. medial to mandible
      2. Lateral projection: lies on external surface of masseter muscle, with small piece, often detached, just below zygomatic arch (accessory part)
    B. POSTERIOR SURFACE: on external auditory meatus and sternocleidomastoid muscle
    C. SUPERFICIAL SURFACE: lobulated, covered by skin, fascia, lymph nodes, and facial branches of great auricular nerve
    D. DEEP SURFACE: has 2 extensions; most posterior part lies on styloid process and its muscles as well as under mastoid and sternocleidomastoid muscle; more cephalic part enters mandibular fossa
      1. Relations: internal and external carotid arteries, internal jugular vein, vagus and glossopharyngeal nerves, and pharyngeal wall

  IV. Capsule: continuous with deep cervical fascia, the superficial layer being dense and closely united with gland (parotidiomasseteric fascia). Below, this fascia forms, between styloid process and angle of mandible, the stylomandibular ligament, which separates parotid from submandibular gland

    V. Duct (Stensen's): from rostral border of gland, crosses masseter muscle, turns inward to pierce the fat pad of cheek and then the buccinator muscle, to open into mouth opposite second upper molar tooth

  VI. Innervaton
    A. PARASYMPATHETIC: preganglionics arising in the inferior salivatory nucleus pass by way of the tympanic branch of the glossopharyngeal nerve, through the tympanic plexus and into the lesser petrosal nerve. This nerve ends by synapsing in the otic ganglion. Postganglionic fibers enter the mandibular (V) nerve, pass through its auriculotemporal branch to reach the gland
    B. SYMPATHETICS: preganglionics from the intermediolateral cell column of the upper thoracic cord, through ventral roots and white rami to synapse in the superior cervical ganglion. Postganglionic fibers run in a plexus on the external carotid artery to reach gland

  VII. Structures embedded in gland
    A. EXTERNAL CAROTID ENTERS GLAND, gives posterior auricular branch and its terminal branches, superficial temporal and maxillary; the transverse facial artery also arises in gland
    B. VEINS ARE MORE SUPERFICIAL BUT ARE ALSO IN GLAND: maxillary and superficial temporal join to form retromandibular vein in gland; in lower part, latter divides into branches to the common facial and external jugular veins
    C. MOST SUPERFICIAL ARE BRANCHES of facial and auriculotemporal nerves

 VIII. Clinical considerations
    A. VIRAL INFLAMMATION OF THE PAROTID GLAND (mumps) causes it to swell, resulting in pain on movement of jaw
    B. ABSCESSES OR CYSTS OF THE GLAND may result in pressure on facial nerve

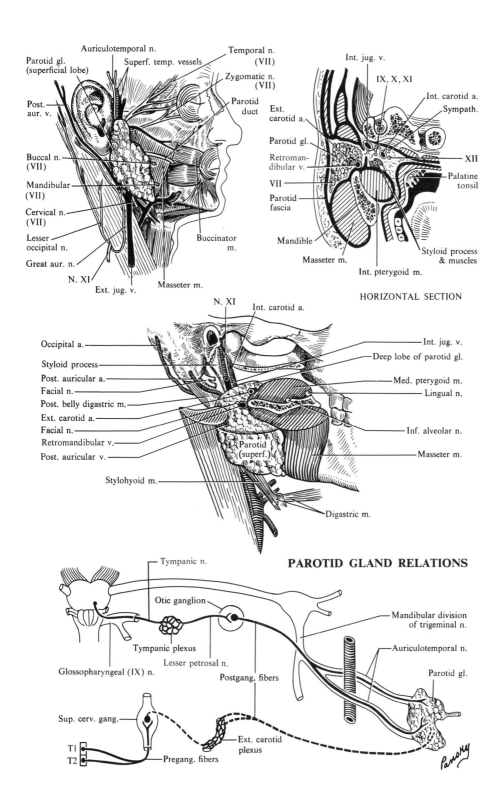

Auriculotemporal n.
Parotid gl. (superficial lobe)
Superf. temp. vessels
Temporal n. (VII)
Zygomatic n. (VII)
Post. aur. v.
Parotid duct
Ext. carotid a.
Buccal n. (VII)
Mandibular (VII)
Cervical n. (VII)
Lesser occipital n.
Great aur. n.
N. XI
Ext. jug. v.
Masseter m.
Buccinator m.

Int. jug. v.
IX, X, XI
Int. carotid a.
Sympath.
Parotid gl.
Retromandibular v.
XII
VII
Palatine tonsil
Parotid fascia
Mandible
Masseter m.
Int. pterygoid m.
Styloid process & muscles

HORIZONTAL SECTION

N. XI
Int. carotid a.
Occipital a.
Int. jug. v.
Deep lobe of parotid gl.
Styloid process
Post. auricular a.
Med. pterygoid m.
Facial n.
Lingual n.
Post. belly digastric m.
Ext. carotid a.
Facial n.
Inf. alveolar n.
Retromandibular v.
Masseter m.
Post. auricular v.
Parotid (superf.)
Stylohyoid m.
Digastric m.

PAROTID GLAND RELATIONS

Tympanic n.
Otic ganglion
Mandibular division of trigeminal n.
Tympanic plexus
Auriculotemporal n.
Glossopharyngeal (IX) n.
Lesser petrosal n.
Postgang. fibers
Parotid gl.
Sup. cerv. gang.
T1
T2
Pregang. fibers
Ext. carotid plexus

Parsley

-29-

14. THE TRIGEMINAL (V) NERVE

I. **Trigeminal nerve,** largest of cranial nerves

A. ROOTS: sensory (portio major), larger; motor (portio minor)

B. GANGLION: semilunar (gasserian), lies in trigeminal impression lateral to the cavernous sinus and internal carotid artery (see p. 90)

C. DIVISIONS

1. Ophthalmic: all sensory; runs through the lateral border of cavernous sinus and enters orbit through the superior orbital fissure. Distribution: (see p. 112)

2. Maxillary: all sensory; runs through lower lateral wall of cavernous sinus, under dura to foramen rotundum, crosses pterygopalatine fossa, enters orbit via inferior orbital fissure, and runs in groove on floor of orbit as the *infraorbital nerve.* Distribution: (for branches to skin, see p. 14)

   a. Middle meningeal nerve to dura

   b. Pterygopalatine nerves, greater and lesser palatine nerves (see p. 66)

   c. Posterior superior nasal branch to mucous membrane of nose

   d. Pharyngeal nerve to nasopharynx and auditory tube

   e. Alveolar branches: posterior superior to gums and 3 molar teeth; middle superior to gums and 2 premolar teeth; anterior superior to nasal cavity, canine and incisor teeth. The middle and anterior superior alveolar nerves are branches of the infraorbital nerve

3. Mandibular: largest; is a mixed nerve (motor and sensory); leaves skull through foramen ovale. This trunk divides into anterior and posterior divisions

   a. Main trunk branches: meningeal to dura; medial pterygoid to medial pterygoid, tensor veli palatini, and tensor tympani muscles

   b. Anterior division (for branches to skin, see p. 14): masseter nerve to masseter muscle; deep temporal nerves (2) to temporalis muscle; lateral pterygoid nerve to lateral pterygoid muscle; and buccal nerve to mucous membrane of mouth

   c. Posterior division: auriculotemporal nerve (see p. 14); lingual nerve (see p. 60); inferior alveolar nerve to mylohyoid and anterior belly of digastric muscles; to all lower teeth

   d. Autonomic connections

      i. To ophthalmic: sympathetic fibers from the cavernous plexus, to vessels and lacrimal gland of the orbit; parasympathetics (VII) from the pterygopalatine ganglion through the zygomatic branch of the maxillary nerve pass to the lacrimal branch of the ophthalmic nerve to the lacrimal gland

      ii. To maxillary: postganglionic parasympathetics from the pterygopalatine ganglion via the pterygopalatine nerves for vessels and glands of the nasal cavity, palate (palatine and nasal branches of the pterygopalatine nerves)

      iii. To mandibular: postganglionic parasympathetics (IX) from otic ganglion through the auriculotemporal nerve to the parotid gland (see p. 28); to lingual nerve; preganglionic parasympathetic fibers via the chorda tympani nerve leave lingual nerve to synapse in the submandibular ganglion; postganglionics go to the submandibular and sublingual glands

II. **Clinical considerations**

A. TIC DOULOUREUX: condition of unknown cause associated with excruciating pain, especially along the maxillary and mandibular roots of the trigeminal nerve. It may be relieved by alcohol injections either into the trigeminal ganglion via the foramen ovale or along the root involved as it leaves the skull. It is sometimes necessary to transect the entire sensory part of the nerve between the ganglion and pons

B. WHILE ALL 4 OF THE CRANIAL PARASYMPATHETIC GANGLIA are located close to or on some branch of the 5th nerve, the preganglionic fibers to these ganglia actually come from other nerves: the 3rd, 7th, and 9th. The postganglionic fibers join and are distributed with peripheral branches of the trigeminal nerve

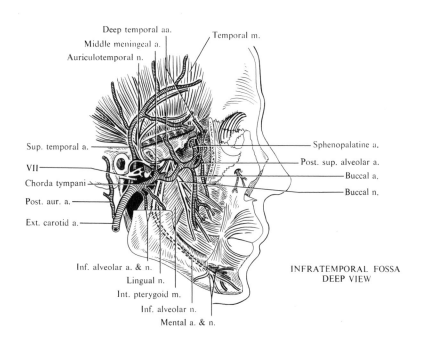

Deep temporal aa.
Middle meningeal a.
Auriculotemporal n.
Temporal m.

Sup. temporal a.
VII
Chorda tympani
Post. aur. a.
Ext. carotid a.

Sphenopalatine a.
Post. sup. alveolar a.
Buccal a.
Buccal n.

Inf. alveolar a. & n.
Lingual n.
Int. pterygoid m.
Inf. alveolar n.
Mental a. & n.

INFRATEMPORAL FOSSA
DEEP VIEW

TRIGEMINAL NERVE

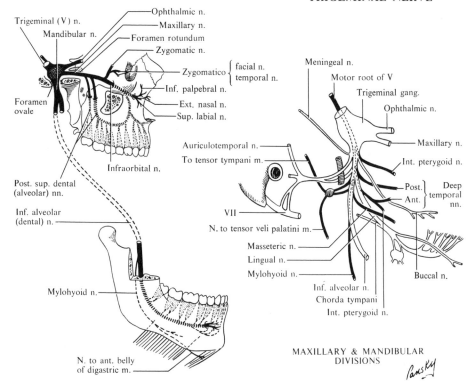

Trigeminal (V) n.
Mandibular n.

Ophthalmic n.
Maxillary n.
Foramen rotundum
Zygomatic n.
Zygomatico { facial n.
              temporal n.
Inf. palpebral n.
Ext. nasal n.
Sup. labial n.

Foramen
ovale

Infraorbital n.

Post. sup. dental
(alveolar) nn.

Inf. alveolar
(dental) n.

Mylohyoid n.

N. to ant. belly
of digastric m.

Meningeal n.
Motor root of V
Trigeminal gang.
Ophthalmic n.

Maxillary n.
Int. pterygoid n.
Post. } Deep
Ant. } temporal
        nn.

Auriculotemporal n.
To tensor tympani m.

VII
N. to tensor veli palatini m.
Masseteric n.
Lingual n.
Mylohyoid n.

Inf. alveolar n.
Chorda tympani
Int. pterygoid n.

Buccal n.

MAXILLARY & MANDIBULAR
DIVISIONS

Pansky

-31-

15. THE TEMPOROMANDIBULAR JOINT AND RELATED STRUCTURES

I. **Type:** combined ginglymus and arthrodial

II. **Movements:** opening and closing jaws, protrusion, retraction, lateral displacement

III. **Bones:** condyle of mandible, the mandibular fossa and articular tubercle of temporal bone

IV. **Ligaments**
   A. CAPSULE: from rim of mandibular fossa and articular tubercle to neck of condyloid process of mandible
   B. LATERAL (TEMPOROMANDIBULAR): from zygomatic arch to neck of mandible
   C. SPHENOMANDIBULAR: from angular spine of sphenoid to lingula of mandible
   D. ARTICULAR DISK: divides joint cavity into upper and lower parts; undersurface concave to conform to shape of mandibular condyle; upper surface, convex to fit into mandibular fossa. Is attached to capsule and to tendon of lateral pterygoid muscle
   E. STYLOMANDIBULAR: apex of styloid to angle and posterior border of mandible

V. **Muscles acting on joint**

Open	Close	Protrude	Retract	Lateral Displacement
Lateral pterygoid	Masseter	Lateral pterygoids (together)	Posterior fibers of temporalis	Lateral pterygoids (individually)
Digastric	Medial pterygoid			
Mylohyoid Geniohyoid	Temporalis			

VI. **Important relations**
   A. MAXILLARY VESSELS pass between sphenomandibular ligament and neck of condyle. At lower level, part of parotid gland and inferior alveolar vessels and nerve lie between this ligament and ramus of mandible
   B. STYLOMANDIBULAR LIGAMENT separates submandibular gland from the parotid
   C. EXTERNAL CAROTID ARTERY DIVIDES INTO TERMINAL BRANCHES: superficial temporal and maxillary arteries behind neck of mandible
   D. INFERIOR ALVEOLAR NERVE RUNS DOWNWARD with its artery and enters mandibular foramen. Gives off the mylohyoid branch just before it enters the bone
      1. Mylohyoid nerve descends in groove on ramus of mandible to reach mylohyoid muscle and runs on its outer surface. Supplies this and anterior belly of digastric muscle
   E. LINGUAL NERVE descends parallel to inferior alveolar nerve but is medial and rostral to it, between the medial pterygoid muscle and ramus of mandible

VII. **Clinical considerations**
   A. DISLOCATIONS at this joint can occur. Care should be taken in returning jaw to normal position since the muscles of mastication exert great force
   B. BLOWS AGAINST THE JAW may drive the condyle upward and backward resulting in injury to the external auditory meatus and may cause bleeding
   C. THE PRESSURE produced by the closers of the jaw is normally borne almost entirely by the molar teeth and the synovial membrane of the joint, which extends in part over the articular surfaces. Thus, malocclusion, or any factor that leads to spastic contraction of the muscles (trismus), may cause pain

TEMPEROMANDIBULAR JOINT

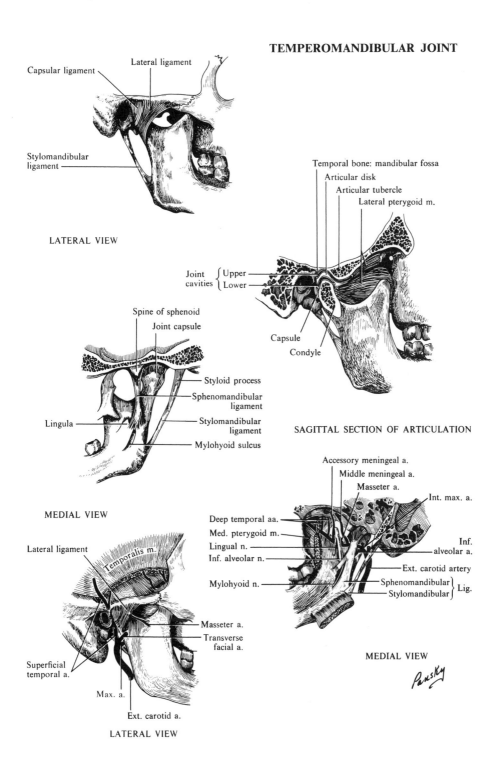

Lateral ligament

Capsular ligament

Stylomandibular ligament

LATERAL VIEW

Temporal bone: mandibular fossa
Articular disk
Articular tubercle
Lateral pterygoid m.

Joint cavities { Upper / Lower }

Capsule
Condyle

Spine of sphenoid
Joint capsule

Styloid process
Sphenomandibular ligament
Stylomandibular ligament
Mylohyoid sulcus

Lingula

SAGITTAL SECTION OF ARTICULATION

MEDIAL VIEW

Accessory meningeal a.
Middle meningeal a.
Masseter a.
Int. max. a.

Deep temporal aa.
Med. pterygoid m.
Lingual n.
Inf. alveolar n.
Inf. alveolar a.
Ext. carotid artery
Sphenomandibular } Lig.
Stylomandibular }
Mylohyoid n.

Lateral ligament
Temporalis m.

Masseter a.
Transverse facial a.

Superficial temporal a.

Max. a.

Ext. carotid a.

LATERAL VIEW

MEDIAL VIEW

Pansky

-33-

16. SUPERFICIAL AND DEEP ARTERIES OF THE FACE

I. Facial
A. ORIGIN: from external carotid in carotid triangle
B. COURSE: beneath digastric and stylohyoid muscles, crosses posterior surface of sub-mandibular gland, curves over body of mandible. Proceeds upward and forward to angle of mouth, ascending along side of nose to terminate at medial angle of eye as *angular artery*
C. BRANCHES
    1. Cervical: *ascending palatine, tonsillar, glandular* (to submandibular gland), *submental,* and *muscular*
    2. Facial: *inferior labial, superior labial, lateral nasal, muscular,* and *angular* (terminal branch)

II. Superficial temporal artery, smaller of terminal branches of the external carotid artery
A. COURSE: ascends in substance of parotid gland, crosses zygomatic process, 5 cm (2 in.) above which it terminates in frontal and parietal branches
B. BRANCHES: *transverse facial, medial temporal, anterior auricular, frontal,* and *parietal*

III. Maxillary artery, the larger of the terminal branches of the external carotid artery
A. COURSE: runs rostrally in parotid gland, then between mandible and sphenomandibular ligament, crosses external pterygoid muscle either medially or laterally, and enters pterygopalatine fossa
B. PARTS AND THEIR BRANCHES
    1. First part lies between mandible and sphenomandibular ligament. Branches: *anterior tympanic, deep auricular, middle meningeal, accessory meningeal,* and *inferior alveolar*
    2. Second part runs medial to the ramus of mandible and tendon of the temporalis muscle, lying on the external pterygoid muscle, and passes between heads of the latter. Branches: *2 deep temporals, pterygoid, masseteric, buccal*
    3. Third part lies in the pterygopalatine fossa lateral to the ganglion. Branches: *posterior superior alveolar, infraorbital, descending palatine, artery of pterygoid canal, pharyngeal and sphenopalatine*
C. SPECIAL FEATURE: it should be noted that there are 5 branches from each part of the maxillary artery. The first 5 reach their destinations by entering foramina, the middle 5 supply soft tissues and use no foramina, and the last 5 also use foramina (the infraorbital can be considered as the termination of the artery)

IV. Clinical considerations
A. WOUNDS of either the face or scalp bleed profusely because of multiple anastomoses, both from side to side and front to back
B. THERE ARE ANASTOMOSES in the head, between branches of the internal and external carotid arteries, especially through the dura mater

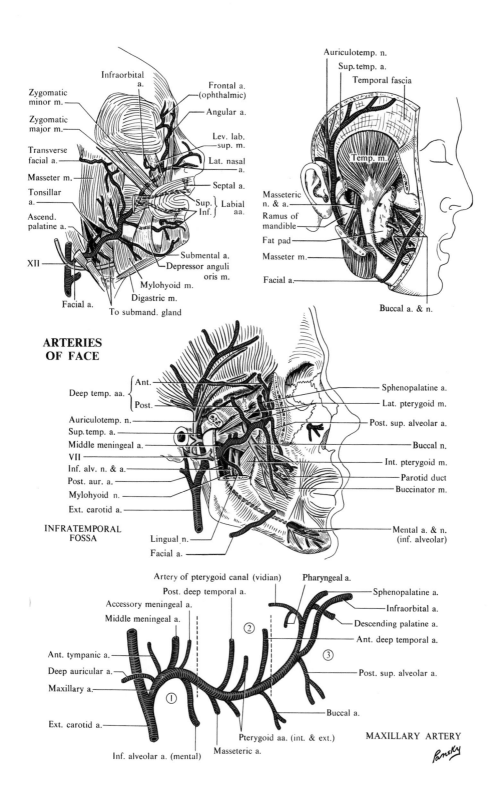

Top left figure labels:

Infraorbital a.

Zygomatic minor m.

Zygomatic major m.

Transverse facial a.

Masseter m.

Tonsillar a.

Ascend. palatine a.

XII

Facial a.

Frontal a. (ophthalmic)

Angular a.

Lev. lab. sup. m.

Lat. nasal a.

Septal a.

Sup. } Labial
Inf. } aa.

Submental a.

Depressor anguli oris m.

Mylohyoid m.

Digastric m.

To submand. gland

Top right figure labels:

Auriculotemp. n.

Sup. temp. a.

Temporal fascia

Temp. m.

Masseteric n. & a.

Ramus of mandible

Fat pad

Masseter m.

Facial a.

Buccal a. & n.

ARTERIES OF FACE

Middle figure labels:

Deep temp. aa. { Ant. / Post.

Auriculotemp. n.

Sup. temp. a.

Middle meningeal a.

VII

Inf. alv. n. & a.

Post. aur. a.

Mylohyoid n.

Ext. carotid a.

INFRATEMPORAL FOSSA

Lingual n.

Facial a.

Sphenopalatine a.

Lat. pterygoid m.

Post. sup. alveolar a.

Buccal n.

Int. pterygoid m.

Parotid duct

Buccinator m.

Mental a. & n. (inf. alveolar)

Bottom figure labels:

Artery of pterygoid canal (vidian)

Pharyngeal a.

Post. deep temporal a.

Accessory meningeal a.

Middle meningeal a.

Sphenopalatine a.

Infraorbital a.

Descending palatine a.

Ant. deep temporal a.

Ant. tympanic a.

Deep auricular a.

Maxillary a.

Post. sup. alveolar a.

Ext. carotid a.

Buccal a.

Inf. alveolar a. (mental)

Masseteric a.

Pterygoid aa. (int. & ext.)

① ② ③

MAXILLARY ARTERY

Pansky

17. THE CERVICAL TRIANGLES AND FASCIA

I. Cervical triangles

A. ANTERIOR: midline ventrally; sternocleidomastoid laterally; body of mandible cephalically. This is further subdivided into:
   1. Submandibular: cephalically, the body of the mandible; anteriorly and below, anterior belly of digastric muscle; posteriorly and below, posterior belly of digastric and stylohyoid muscles
   2. Carotid: cephalically, posterior belly of digastric and stylohyoid muscles; caudally, superior belly of omohyoid muscle; posteriorly, sternocleidomastoid muscle
   3. Suprahyoid (submental): laterally, anterior belly of digastric muscle; caudally, body of the hyoid bone; medially, midline of the neck
   4. Muscular: posteriorly and below, sternocleidomastoid muscle; posteriorly and above, superior belly of omohyoid muscle; medially, midline, from hyoid to sternum
B. POSTERIOR: anteriorly, sternocleidomastoid muscle; posteriorly, trapezius muscle; caudally, clavicle. Further subdivided:
   1. Occipital: posteriorly and anteriorly, as above; caudally, inferior belly of omohyoid
   2. Omoclavicular (subclavian): anteriorly, sternocleidomastoid muscle; cephalically, inferior belly of omohyoid muscle; caudally, clavicle

II. Cervical fascia

A. SUPERFICIAL: covers the ant. and post. triangles but splits to enclose sternocleidomastoid and trapezius. Attachments: posteriorly, ext. occipital protuberance, ligamentum nuchae, spine of C7; cephalically, sup. nuchal line, mastoid process and mandible; inferiorly, clavicle, manubrium of sternum, acromion and spine of scapula
B. FASCIA OF THE INFRAHYOID MUSCLES. Consists of 2 layers: superficial, which encloses the sternohyoid and omohyoid muscles, and a deep, which invests the sternothyroid and thyrohyoid muscles
C. VISCERAL: encloses pharynx, larynx, trachea, esophagus, and thyroid. Has 2 parts:
   1. Pretracheal: covers larynx and trachea; splits to enclose thyroid. Cephalically is attached to hyoid bone and thyroid cartilage; laterally is continuous with the next layer; caudally enters the thorax to join fascia of aorta and the pericardium
   2. Buccopharyngeal: covers buccinator muscle and posterior side of esophagus. Cephalically is attached to pharyngeal tubercle and medial pterygoid plate, covers the superior pharyngeal constrictor, and joins the pterygomandibular raphé
D. PREVERTEBRAL: encloses vertebral column and its muscles. Covers prevertebral muscles and forms floor of the post. triangle. Attachments: to the transverse processes of the cervical vertebrae laterally; the occipital bone near jugular foramen, the sup. nuchal line, and mastoid process cephalically; continues into mediastinum caudally
   1. Suprapleural membrane (Sibson's fascia): fascia deep to the scalene muscles. Covers cervical pleura

III. Carotid sheath: an investment of the internal and common carotid arteries, the internal jugular vein, and the vagus nerve. It is adherent to the thyroid sheath and the fascia under the sternocleidomastoid. Cephalically it is attached to the bone around the jugular foramen and carotid canal; caudally it continues into the thorax

IV. Fascial spaces

A. RETROPHARYNGEAL. Between buccopharyngeal and prevertebral fascia. Above ends at skull; laterally is closed by carotid sheath; below continues into mediastinum
   1. Alar fascia: thin layer of fascia which subdivides this space. It is attached in the midline to the buccopharyngeal fascia; laterally it joins the carotid sheath
B. SUPRASTERNAL. This is an interval between the layers of the superficial fascia which has split to attach to the posterior and anterior sides of the manubrium

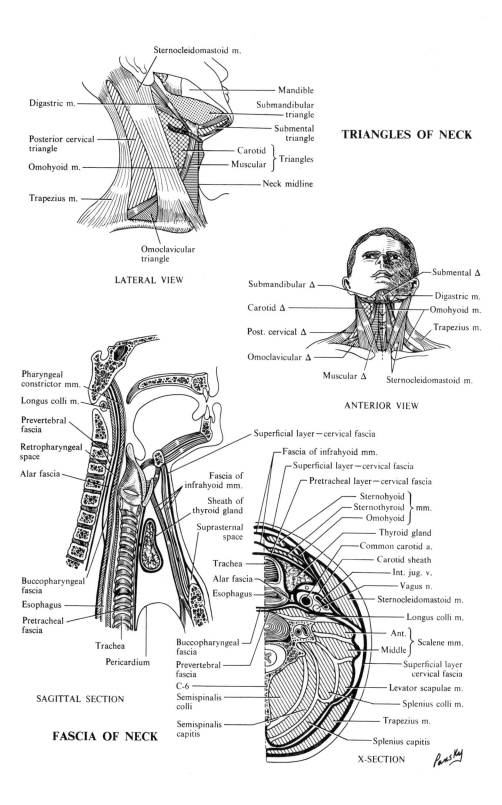

Sternocleidomastoid m.

Digastric m.

Posterior cervical triangle

Omohyoid m.

Trapezius m.

Mandible

Submandibular triangle

Submental triangle

Carotid

Muscular

} Triangles

Neck midline

Omoclavicular triangle

LATERAL VIEW

TRIANGLES OF NECK

Submandibular Δ

Carotid Δ

Post. cervical Δ

Omoclavicular Δ

Muscular Δ

Submental Δ

Digastric m.

Omohyoid m.

Trapezius m.

Sternocleidomastoid m.

ANTERIOR VIEW

Pharyngeal constrictor mm.

Longus colli m.

Prevertebral fascia

Retropharyngeal space

Alar fascia

Buccopharyngeal fascia

Esophagus

Pretracheal fascia

Trachea

Pericardium

Fascia of infrahyoid mm.

Sheath of thyroid gland

Suprasternal space

Trachea

Alar fascia

Esophagus

Buccopharyngeal fascia

Prevertebral fascia

C-6

Semispinalis colli

Semispinalis capitis

SAGITTAL SECTION

FASCIA OF NECK

Superficial layer — cervical fascia

Fascia of infrahyoid mm.

Superficial layer — cervical fascia

Pretracheal layer — cervical fascia

Sternohyoid

Sternothyroid

Omohyoid

} mm.

Thyroid gland

Common carotid a.

Carotid sheath

Int. jug. v.

Vagus n.

Sternocleidomastoid m.

Longus colli m.

Ant.

Middle

} Scalene mm.

Superficial layer cervical fascia

Levator scapulae m.

Splenius colli m.

Trapezius m.

Splenius capitis

X-SECTION

Pansky

-37-

18. VEINS OF THE NECK

I. Internal jugular
A. ORIGIN: in jugular fossa, as a continuation of transverse sinus
B. COURSE: in carotid sheath, first with internal and then common carotid artery
C. TERMINATION: joins subclavian vein just lateral to sternoclavicular joint to form brachiocephalic vein
   1. One inch above termination has pair of valves below which is a dilatation, the *inferior bulb*
D. TRIBUTARIES
   1. At origin: inferior petrosal sinus and a meningeal vein
   2. Veins from pharyngeal plexus, near angle of jaw
   3. Common facial (formed from anterior branch of retromandibular and facial, see p. 18) enters at level of hyoid bones
   4. Lingual, from tongue, may enter with or just below common facial, drains tongue and sublingual area
   5. Superior thyroid, from upper thyroid gland, enters with or just below common facial
   6. Middle thyroid vein, from the lateral lobe of gland

II. External jugular, formed by posterior branch of retromandibular vein and posterior auricular vein, terminates in subclavian vein behind middle of clavicle (see p. 18)
A. TRIBUTARIES
   1. Posterior auricular, occipital, retromandibular, and anterior jugular (see p. 18)
   2. Posterior external jugular from cephalic part of back of neck
   3. Transverse scapular from scapular region
   4. Transverse cervical (inconstant) from root of the neck. Most often this ends in subclavian vein but may join the external jugular vein

III. Subclavian
A. ORIGIN: continuation of axillary vein, beginning at lateral border of first rib
B. COURSE: medial, anterior, and slightly inferior to subclavian artery. Its second part is separated from the artery by the anterior scalene muscle
C. TERMINATION: joins internal jugular to form brachiocephalic vein just lateral to sternoclavicular joint
   1. Contains pair of valves just before ending
D. TRIBUTARIES
   1. External jugular (see II, above)
   2. Transverse cervical from root of neck (see II, A4 above)
   3. Thoracoacromial from shoulder and upper pectoral region

IV. Clinical considerations
A. THE LOSS OF A MAJOR VEIN such as the internal jugular is well tolerated because of interconnections between the veins of the two sides, both within and outside of the skull

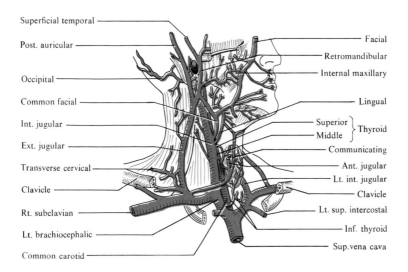

Superficial temporal
Post. auricular
Occipital
Common facial
Int. jugular
Ext. jugular
Transverse cervical
Clavicle
Rt. subclavian
Lt. brachiocephalic
Common carotid

Facial
Retromandibular
Internal maxillary
Lingual
Superior ⎱ Thyroid
Middle ⎰
Communicating
Ant. jugular
Lt. int. jugular
Clavicle
Lt. sup. intercostal
Inf. thyroid
Sup. vena cava

DEEP VEINS OF NECK AND HEAD

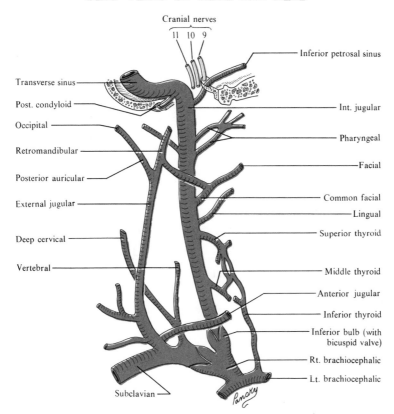

Cranial nerves
11 10 9

Inferior petrosal sinus

Transverse sinus
Post. condyloid
Occipital
Retromandibular
Posterior auricular
External jugular
Deep cervical
Vertebral

Int. jugular
Pharyngeal
Facial
Common facial
Lingual
Superior thyroid
Middle thyroid
Anterior jugular
Inferior thyroid
Inferior bulb (with bicuspid valve)
Rt. brachiocephalic
Lt. brachiocephalic

Subclavian

Pansky

19. THE ANTEROLATERAL MUSCLES OF THE NECK

Name	Origin	Insertion	Action	Nerve
Platysma	Deltoid and pectoral fascia	Mandible & skin	Depresses angle of mouth, opens jaw	Facial (VII)
Sternocleido-mastoid	Sternum, clavicle	Mastoid process	Bends head to same side, rotates head, raises chin to opposite side, together bend head forward & elevate chin	Accessory (XI), C2 & C3
Digastric				
Ant. belly	Lower border mandible	Body & great cornu of hyoid	Opens jaw, draws hyoid forward	Trigeminal (V)
Post. belly	Mastoid notch of temporal bone	Body & great cornu of hyoid	Draws hyoid back, together raise hyoid	Facial (VII)
Stylohyoid	Styloid process	Body of hyoid	Draws hyoid up & back	Facial (VII)
Mylohyoid (see p. 57)	Mylohyoid line of mandible	Body of hyoid, median raphé	Raises hyoid	Trigeminal (V)
Geniohyoid (see p. 57)	Inf. mental spine behind symphysis	Body of hyoid	Draws hyoid and tongue forward	C1, via ansa cervicalis
Sternohyoid (see p. 57)	Medial end of clavicle, post. manubrium	Body of hyoid	Depresses hyoid	C1–3, via ansa cervicalis
Sternothyroid (see p. 57)	Post. manubrium	Oblique line on lamina of thyroid cartilage	Depresses thyroid cartilage	C1–3, via ansa cervicalis
Thyrohyoid (see p. 48)				
Omohyoid (see p. 57)				
Inf. belly	Scapula, sup. transverse lig.	By tendon to clavicle	Depresses hyoid	C1–3, via ansa cervicalis
Sup. belly	Tendon suspended from clavicle	Body of hyoid bone	Depresses hyoid	C1–3, via ansa cervicalis

Trapezius, levator scapulae, and scalene mm. (see p. 57)

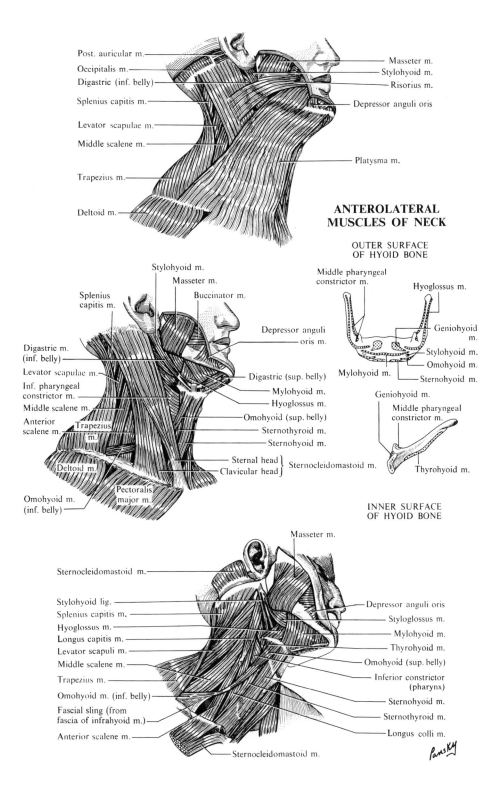

Post. auricular m.
Occipitalis m.
Digastric (inf. belly)
Splenius capitis m.
Levator scapulae m.
Middle scalene m.
Trapezius m.
Deltoid m.

Masseter m.
Stylohyoid m.
Risorius m.
Depressor anguli oris
Platysma m.

**ANTEROLATERAL
MUSCLES OF NECK**

OUTER SURFACE
OF HYOID BONE

Stylohyoid m.
Masseter m.
Buccinator m.
Splenius capitis m.
Digastric m. (inf. belly)
Levator scapulae m.
Inf. pharyngeal constrictor m.
Middle scalene m.
Anterior scalene m.
Trapezius m.
Deltoid m.
Omohyoid m. (inf. belly)
Pectoralis major m.

Depressor anguli oris m.
Digastric (sup. belly)
Mylohyoid m.
Hyoglossus m.
Omohyoid (sup. belly)
Sternothyroid m.
Sternohyoid m.
Sternal head
Clavicular head
Sternocleidomastoid m.

Middle pharyngeal constrictor m.
Hyoglossus m.
Geniohyoid m.
Stylohyoid m.
Omohyoid m.
Sternohyoid m.
Mylohyoid m.

Geniohyoid m.
Middle pharyngeal constrictor m.
Thyrohyoid m.

INNER SURFACE
OF HYOID BONE

Masseter m.
Sternocleidomastoid m.
Stylohyoid lig.
Splenius capitis m.
Hyoglossus m.
Longus capitis m.
Levator scapuli m.
Middle scalene m.
Trapezius m.
Omohyoid m. (inf. belly)
Fascial sling (from fascia of infrahyoid m.)
Anterior scalene m.
Sternocleidomastoid m.

Depressor anguli oris
Styloglossus m.
Mylohyoid m.
Thyrohyoid m.
Omohyoid (sup. belly)
Inferior constrictor (pharynx)
Sternohyoid m.
Sternothyroid m.
Longus colli m.

Pansky

-41-

20. THE ANTERIOR TRIANGLE AND ROOT OF NECK

I. Superior carotid (for boundaries, see p. 36)
  A. ROOF: skin, superficial fascia, platysma muscle, deep fascia
  B. FLOOR: thyrohyoid, hyoglossus, middle and inferior pharyngeal constrictor muscles
  C. CONTENTS: bifurcation of *common carotid artery; superior thyroid, lingual, facial, occipital,* and *ascending pharyngeal arteries; internal jugular vein; ansa cervicalis; cervical sympathetic trunk; hypoglossal nerve; superior laryngeal nerve*

II. Inferior carotid (muscular) (for boundaries, see p. 36)
  A. ROOF: skin, superficial fascia, platysma, deep fascia
  B. CONTENTS: sternohyoid and thyroid mm., isthmus of thyroid, larynx, and trachea

III. Submandibular (digastric) (for boundaries, see p. 36)
  A. ROOF: skin, superficial fascia, platysma muscle, and deep fascia
  B. FLOOR: mylohyoid and hyoglossus muscles
  C. DIVISIONS: stylomandibular ligament divides it into rostral and posterior parts
  D. CONTENTS
    1. Rostral part: *submandibular gland, facial artery, facial vein, submental artery, mylohyoid artery and nerve*
    2. Posterior part: *parotid gland* containing *external carotid artery,* crossed laterally by *facial nerve, internal jugular vein,* and *vagus nerve*

IV. Suprahyoid (submental) (for boundaries, see p. 36)
  A. ROOF: skin, superficial fascia, deep fascia
  B. FLOOR: Mylohyoid muscle
  C. CONTENTS: One or two lymph nodes, small tributaries of anterior jugular vein

V. Structures beneath sternocleidomastoid muscle
  A. ANSA CERVICALIS on surface of carotid sheath
  B. CAROTID SHEATH, containing vagus n., int. jugular v., and common carotid a.
  C. CERVICAL SYMPATHETIC TRUNK is embedded in dorsomedial wall of sheath

VI. Phrenic nerve
  A. ORIGIN: fourth, but may receive fibers from third and fifth cervical nerves
  B. COURSE, IN NECK: on ventral surface of anterior scalene muscle, and is crossed by inferior belly of omohyoid muscle, transverse (superficial) cervical and suprascapular arteries, and passes between subclavian artery and vein to reach thorax

VII. Termination of thoracic duct. Arches above left clavicle. Lies ventral to subclavian artery, vertebral artery and vein, thyrocervical trunk and prevertebral fascia, which separates duct from phrenic nerve and anterior scalene muscle. Lies behind left common carotid artery, internal jugular vein, and vagus nerve. Terminates at junction of left internal jugular and subclavian veins

VIII. Branches of external carotid artery

Anterior	Posterior	Ascending	Terminal
Sup. thyroid	Occipital	Ascend. pharyngeal	Superfic. temp.
Lingual	Posterior auricular		Maxillary
Facial			

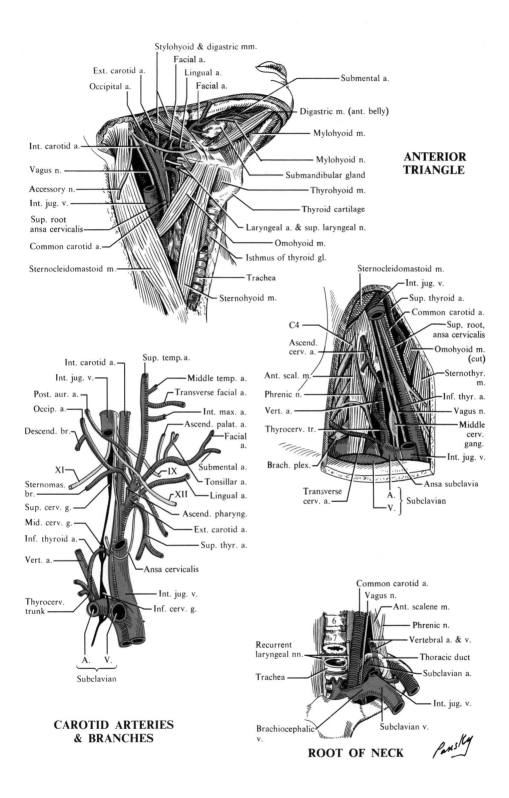

Stylohyoid & digastric mm.
Facial a.
Ext. carotid a.
Lingual a.
Occipital a.
Facial a.
Submental a.
Digastric m. (ant. belly)
Mylohyoid m.
Int. carotid a.
Mylohyoid n.
Vagus n.
Submandibular gland
Accessory n.
Thyrohyoid m.
Int. jug. v.
Thyroid cartilage
Sup. root ansa cervicalis
Laryngeal a. & sup. laryngeal n.
Common carotid a.
Omohyoid m.
Isthmus of thyroid gl.
Sternocleidomastoid m.
Trachea
Sternohyoid m.

ANTERIOR TRIANGLE

Sternocleidomastoid m.
Int. jug. v.
Sup. thyroid a.
Common carotid a.
C4
Sup. root, ansa cervicalis
Ascend. cerv. a.
Omohyoid m. (cut)
Ant. scal. m.
Sternothyr. m.
Phrenic n.
Inf. thyr. a.
Vert. a.
Vagus n.
Thyrocerv. tr.
Middle cerv. gang.
Int. jug. v.
Brach. plex.
Ansa subclavia
Transverse cerv. a.
A.
V. } Subclavian

Int. carotid a.
Sup. temp. a.
Int. jug. v.
Middle temp. a.
Post. aur. a.
Transverse facial a.
Occip. a.
Int. max. a.
Descend. br.
Ascend. palat. a.
Facial a.
XI
IX
Submental a.
Sternomas. br.
Tonsillar a.
Sup. cerv. g.
XII
Lingual a.
Mid. cerv. g.
Ascend. pharyng.
Inf. thyroid a.
Ext. carotid a.
Vert. a.
Sup. thyr. a.
Ansa cervicalis
Thyrocerv. trunk
Int. jug. v.
Inf. cerv. g.
A.   V.
Subclavian

CAROTID ARTERIES & BRANCHES

Common carotid a.
Vagus n.
Ant. scalene m.
6
Phrenic n.
7
Vertebral a. & v.
Recurrent laryngeal nn.
Thoracic duct
Subclavian a.
Trachea
Int. jug. v.
Brachiocephalic v.
Subclavian v.

ROOT OF NECK

Pansky

-43-

21. THE POSTERIOR TRIANGLE AND
SUBCLAVIAN ARTERY

I. **Posterior triangle** (for boundaries, see p. 36)
   A. OCCIPITAL DIVISION
      1. Floor; splenius capitis, levator scapulae, posterior and middle scalene muscles
         a. Structures crossing floor: accessory nerve and brachial plexus
      2. Structures piercing roof (see p. 14): lesser occipital, great auricular, cervical cutaneous, and supraclavicular nerves. All emerge just behind the posterior border of the sternocleidomastoid muscle
   B. SUBCLAVIAN DIVISION
      1. Floor: first rib and first part of serratus anterior muscle
         a. Structures crossing floor: subclavian artery and vein, part of brachial plexus
      2. Structures crossing through: transverse scapular and superficial cervical vessels
      3. Structures piercing roof: external jugular vein

II. **Subclavian artery**
   A. ORIGIN: on right, from brachiocephalic trunk; on left, from arch of aorta
   B. PARTS: first, from origin to medial border of anterior scalene muscle; second, lies behind anterior scalene muscle; third, extends from lateral border of that muscle to lateral border of first rib
   C. RELATIONS (OTHER THAN MUSCLE OR FASCIA) AND BRANCHES OF PARTS:
      1. Part I: right side
         a. Relations: crossed by internal jugular and vertebral veins, vagus nerve, cardiac branches of the vagus nerve, and the subclavian loop of the sympathetic chain. Behind and below: pleura, sympathetic trunk, and recurrent nerve
         b. Branches: vertebral, thyrocervical trunk and internal thoracic arteries
      2. Part I: left side
         a. Relations: crossed by vagus, cardiac, and phrenic nerves; common carotid artery; int. jugular and vertebral veins; and origin of brachiocephalic vein. Behind: esophagus, thoracic duct, recurrent nerve, and inferior cervical ganglion. Medial: esophagus, trachea, thoracic duct, and recurrent nerve. Lateral: pleura and lung
         b. Branches: same as right side plus the costocervical trunk
      3. Part II
         a. Relations of the two sides are the same. Behind and below, related to pleura; above is the brachial plexus
         b. Branches: sometimes, descending (transverse) scapular artery; costocervical trunk on the right side
      4. Part III
         a. Relations of the two sides are the same. In front: external jugular vein, venous plexus, transverse scapular vessels and subclavian vein. Behind: lowest trunk of brachial plexus. Above and lateral: upper and middle trunks of plexus. Below: upper surface of first rib
         b. Branches: descending (transverse) scapular artery

III. **Arteries crossing triangle of neck**
   A. SUPRASCAPULAR from thyrocervical trunk, which goes over transverse scapular ligament to supraspinous fossa
   B. TRANSVERSE CERVICAL from thyrocervical trunk. It divides into a *superficial* and a *deep* branch (which descends along vertebral border of scapula). It may arise from 2 arteries: a *superficial cervical* from the thyrocervical trunk and a *descending (dorsal) scapular,* from the third part of the subclavian

POSTERIOR TRIANGLE

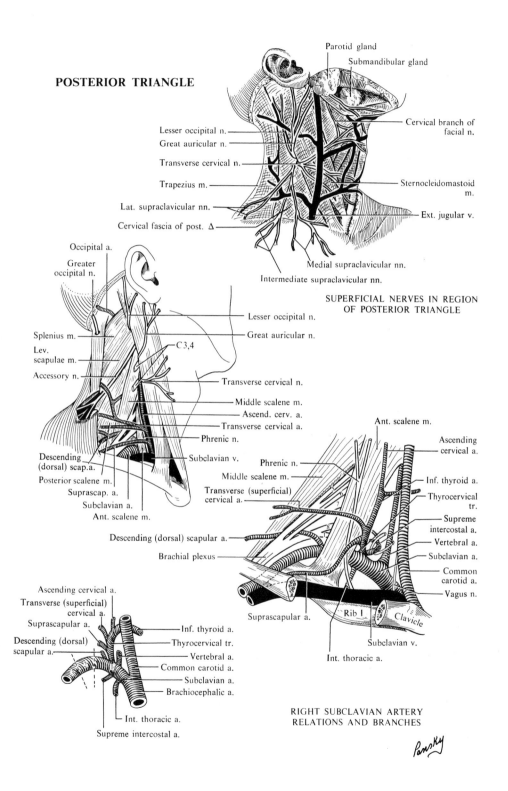

Parotid gland
Submandibular gland
Cervical branch of facial n.
Lesser occipital n.
Great auricular n.
Transverse cervical n.
Trapezius m.
Sternocleidomastoid m.
Lat. supraclavicular nn.
Ext. jugular v.
Cervical fascia of post. Δ
Medial supraclavicular nn.
Intermediate supraclavicular nn.

SUPERFICIAL NERVES IN REGION
OF POSTERIOR TRIANGLE

Occipital a.
Greater occipital n.
Lesser occipital n.
Great auricular n.
Splenius m.
Lev. scapulae m.
C 3,4
Accessory n.
Transverse cervical n.
Middle scalene m.
Ascend. cerv. a.
Transverse cervical a.
Phrenic n.
Subclavian v.
Ant. scalene m.
Descending (dorsal) scap.a.
Posterior scalene m.
Suprascap. a.
Subclavian a.
Ant. scalene m.
Phrenic n.
Middle scalene m.
Transverse (superficial) cervical a.
Descending (dorsal) scapular a.
Brachial plexus
Ascending cervical a.
Inf. thyroid a.
Thyrocervical tr.
Supreme intercostal a.
Vertebral a.
Subclavian a.
Common carotid a.
Vagus n.
Suprascapular a.
Rib 1
Clavicle
Subclavian v.
Int. thoracic a.

Ascending cervical a.
Transverse (superficial) cervical a.
Suprascapular a.
Descending (dorsal) scapular a.
Inf. thyroid a.
Thyrocervical tr.
Vertebral a.
Common carotid a.
Subclavian a.
Brachiocephalic a.
Int. thoracic a.
Supreme intercostal a.

RIGHT SUBCLAVIAN ARTERY
RELATIONS AND BRANCHES

Pansky

-45-

22. THYROID AND PARATHYROID GLANDS

I. Thyroid gland
A. PARTS: 2 lateral lobes interconnected by an *isthmus*
B. POSITION, RELATIONS, AND SIZE OF LOBES
    1. Apex of lateral lobes reaches junction of middle and lower thirds of thyroid cartilage. Base at level of fifth (or sixth) tracheal ring
    2. Lateral surface: convex; covered from superficial to deep by skin, superficial and deep fascia, sternocleidomastoid muscle, superior belly of omohyoid muscle, sternothyroid and sternohyoid muscles, and visceral (pretracheal) fasia, which forms a capsule
    3. Medial surface: trachea, inferior pharyngeal constrictor muscle, posterior part of cricothyroid muscle, esophagus, superior and inferior thyroid ateries, and the recurrent laryngeal nerve
    4. Posterior border: common carotid artery and parathyroid glands
C. POSITION AND RELATIONS OF ISTHMUS
    1. Connects lower thirds of the lateral lobes at the level of tracheal rings 2 and 3
    2. Covered by skin, fascia, and sternohyoid muscles
    3. A communicating artery between the two superior thyroid arteries runs on its cephalic border and the inferior thyroid veins are located at its lower border
    4. Pyramidal lobe, if present, is attached to the left side of the superior border
D. ARTERIAL BLOOD SUPPLY
    1. Superior thyroid artery from external carotid artery
    2. Inferior thyroid artery from thyrocervical trunk
    3. Thyroidea ima (sometimes present) from brachiocephalic trunk
E. VENOUS DRAINAGE: starts from a plexus on surface of gland and trachea
    1. Superior and middle arise in plexus and end in internal jugular vein
    2. Inferior terminates in brachiocephalic veins of either side
F. LYMPHATIC DRAINAGE: from a subcapsular plexus to superior deep cervical, pretracheal, paratracheal, retropharyngeal, and inferior deep cervical nodes

II. Parathyroids
A. PARTS: 4 separate glands, arranged into superior and inferior pairs
B. POSITION AND RELATIONS: They lie between the posterior border of thyroid and its capsule; the superior at the level of lower cricoid cartilage; the inferior near lower end of lateral lobes
C. ARTERIAL BLOOD: supplied from the superior thyroid artery to the superior parathyroids and from the inferior thyroid artery to inferior pair
D. VEINS AND LYMPHATICS: same as thyroid

III. Special features
A. THE THYROID GLAND secretes thyroxin, which is important in metabolism
B. THE SUPERIOR THYROID ARTERY accompanies the superior thyroid vein, but the inferior thyroid artery is accompanied by the middle thyroid vein. The thyroidea ima, when present, is accompanied by the inferior thyroid vein
C. THE PARATHYROIDS secrete parathormone, important in calcium metabolism

IV. Clinical considerations
A. A PATHOLOGIC ENLARGEMENT of the thyroid gland is termed *goiter*
B. A THYROGLOSSAL DUCT can persist between foramen cecum and pyramidal lobe
C. IN THYROIDECTOMY, great care must be taken to avoid the recurrent laryngeal nerves, and to preserve some parathyroid tissue

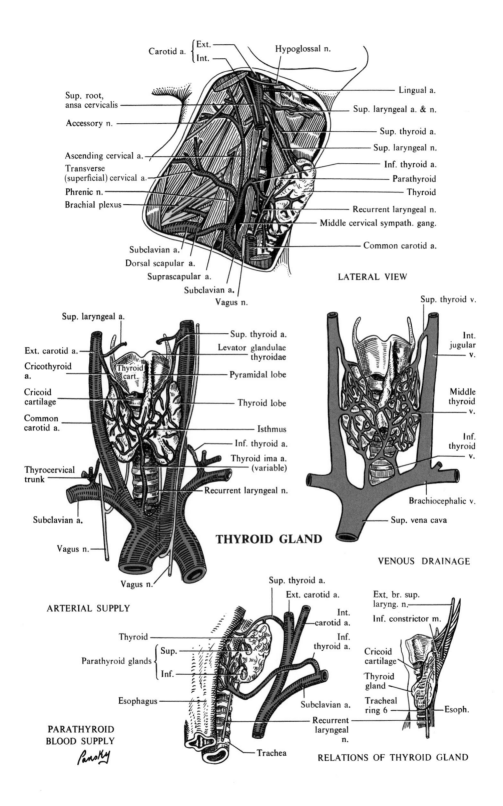

Carotid a. { Ext. / Int. }
Hypoglossal n.
Sup. root, ansa cervicalis
Accessory n.
Ascending cervical a.
Transverse (superficial) cervical a.
Phrenic n.
Brachial plexus
Subclavian a.
Dorsal scapular a.
Suprascapular a.
Subclavian a.
Vagus n.

Lingual a.
Sup. laryngeal a. & n.
Sup. thyroid a.
Sup. laryngeal n.
Inf. thyroid a.
Parathyroid
Thyroid
Recurrent laryngeal n.
Middle cervical sympath. gang.
Common carotid a.

LATERAL VIEW

Sup. laryngeal a.
Ext. carotid a.
Cricothyroid a.
Cricoid cartilage
Common carotid a.
Thyrocervical trunk
Subclavian a.
Vagus n.
Vagus n.

Thyroid cart.

Sup. thyroid a.
Levator glandulae thyroidae
Pyramidal lobe
Thyroid lobe
Isthmus
Inf. thyroid a.
Thyroid ima a. (variable)
Recurrent laryngeal n.

ARTERIAL SUPPLY

THYROID GLAND

Sup. thyroid v.
Int. jugular v.
Middle thyroid v.
Inf. thyroid v.
Brachiocephalic v.
Sup. vena cava

VENOUS DRAINAGE

Sup. thyroid a.
Ext. carotid a.
Int. carotid a.
Inf. thyroid a.

Thyroid
Parathyroid glands { Sup. / Inf. }
Esophagus

PARATHYROID
BLOOD SUPPLY
Panaky

Subclavian a.
Recurrent laryngeal n.
Trachea

Ext. br. sup. laryng. n.
Inf. constrictor m.
Cricoid cartilage
Thyroid gland
Tracheal ring 6
Esoph.

RELATIONS OF THYROID GLAND

-47-

23. MUSCLES OF NECK, PHARYNX, AND TONGUE

I.

Name	Origin	Insertion	Action	Nerve
Buccinator (see p. 22)	Maxilla, mandible, pterygomandibular raphé	Muscles in lips	Compresses cheek, holds food between teeth	Facial (VII)
Geniohyoid	Inf. mental spine	Hyoid bone	Raises hyoid, depresses mand.	Ansa cervicalis (C1)
Genioglossus	Sup. mental spine	Hyoid bone, underside of tongue	Post. protrudes tongue, ant. retracts and depresses tongue	Hypoglossal (XII)
Hyoglossus	Body and great cornu of hyoid	Under lateral surface of tongue	Depresses tongue, draws sides down	Hypoglossal (XII)
Styloglossus	Styloid proc. stylomandib. lig.	Dorsal lat. tongue	Raises and retracts tongue	Hypoglossal (XII)
Thyrohyoid	Oblique line on thyroid cart.	Lat. side of great cornu of hyoid	Depresses hyoid, raises thyroid cart.	Sup. root, ansa cervicalis (C1–2)
Cricothyroid	Cricoid cart.	Inf. cornu & lower border thyroid cart.	Tilts lamina of cricoid cart. dorsally, tenses vocal cords	Ext. br. of sup. laryngeal n. (X)
Stylopharyngeus	Styloid process	Pharyngeal mm., thyroid cart.	Raises pharynx, expands sides of pharynx	Glossopharyngeal (IX)
Salpingopharyngeus	Auditory tube	Post. lat. pharynx	Raises upper and lat. pharynx	Pharyngeal plexus
Pharyngeal constrictors				
Superior	Med. pterygoid plate, hamulus, pterygomand. raphé, mandible	Pharyngeal tubercle of occipital bone, median raphé	Constricts upper pharynx	Pharyngeal plexus (IX?, X)
Middle	Great cornu of hyoid, lesser cornu	Median raphé	Constricts middle pharynx	Pharyngeal plexus (IX?, X)
Inferior	Sides of cricoid and thyroid cartilages	Median raphé	Constricts lower pharynx	Pharyngeal plexus (IX?, X)

II. Special features

A. PHARYNGOBASILAR FASCIA serves as the submucosa of the pharynx, located superiorly above the border of the superior constrictor, underlies all the pharyngeal muscles

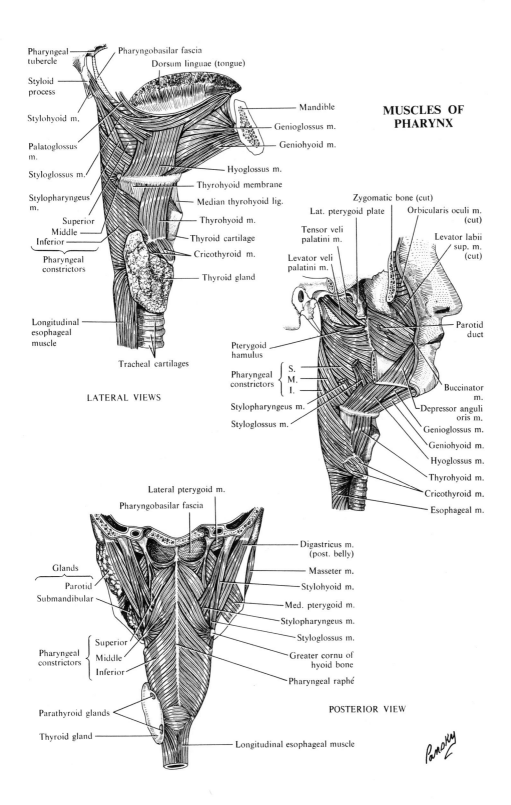

Pharyngeal tubercle
Pharyngobasilar fascia
Dorsum linguae (tongue)
Styloid process
Stylohyoid m.
Mandible
Genioglossus m.
Geniohyoid m.
Palatoglossus m.
Styloglossus m.
Stylopharyngeus m.
Hyoglossus m.
Thyrohyoid membrane
Median thyrohyoid lig.
Superior
Middle
Inferior
Thyrohyoid m.
Thyroid cartilage
Cricothyroid m.
Pharyngeal constrictors
Thyroid gland
Longitudinal esophageal muscle
Tracheal cartilages

LATERAL VIEWS

MUSCLES OF PHARYNX

Zygomatic bone (cut)
Lat. pterygoid plate
Orbicularis oculi m. (cut)
Tensor veli palatini m.
Levator labii sup. m. (cut)
Levator veli palatini m.
Parotid duct
Pterygoid hamulus
Pharyngeal constrictors { S. M. I. }
Stylopharyngeus m.
Styloglossus m.
Buccinator m.
Depressor anguli oris m.
Genioglossus m.
Geniohyoid m.
Hyoglossus m.
Thyrohyoid m.
Cricothyroid m.
Esophageal m.

Lateral pterygoid m.
Pharyngobasilar fascia
Glands { Parotid Submandibular
Pharyngeal constrictors { Superior Middle Inferior }
Parathyroid glands
Thyroid gland
Digastricus m. (post. belly)
Masseter m.
Stylohyoid m.
Med. pterygoid m.
Stylopharyngeus m.
Styloglossus m.
Greater cornu of hyoid bone
Pharyngeal raphé
Longitudinal esophageal muscle

POSTERIOR VIEW

24. LARYNX: MUSCLES, VESSELS, AND NERVES

I. Muscles

Name	Origin	Insertion	Action
Cricothyroid			
Oblique	Arch. of cricoid	Inf. cornu and lamina of	Draws up arch
Straight	Same	thyroid cartilage	Tenses vocal cords
Post. crico-arytenoid	Cricoid lamina	Muscular process of arytenoid	Separates vocal cords
Lat. crico-arytenoid	Arch of cricoid cartilage	Muscular process of arytenoid	Closes ant. part of rima glottidis
Arytenoid			
Oblique	Dorsal and lat. borders of arytenoid	Apex of opposite arytenoid	Closes rima glottidis, especially dorsal part
Transverse	Same	Dorsal surface of opposite arytenoid	Same
Thyroarytenoid (vocalis)	Angle of thyroid cartilage	Vent. base of arytenoid, vocal process	Shortens and relaxes vocal cords

II. Nerves
A. MOTOR
    1. Ext. laryngeal branch of the sup. laryngeal n. of vagus to cricothyroid m.
    2. Recurrent laryngeal nerve to other intrinsic muscles of larynx
B. SENSORY
    1. Internal laryngeal branch of superior laryngeal nerve from the vagus nerve

III. Arteries
A. SUPERIOR LARYNGEAL ARTERY from the superior thyroid artery. Accompanies internal laryngeal branch of superior laryngeal nerve
B. INFERIOR LARYNGEAL ARTERY, a branch of the inf. thyroid a., ascends on back of larynx, under inf. constrictor of pharynx. Accompanied by recurrent nerve

IV. Veins follow same pattern as arteries

V. Lymphatic drainage: to infrahyoid, superior deep cervical, prelaryngeal, pretracheal, and paratracheal nodes

VI. General functions of larynx
A. IN RESPIRATION, framework is rigid to maintain passage. Even when vocal cords are closely approximated for extreme changes in pitch of voice, the glottis respiratoria, between the vocal processes of the arytenoids, remains open
    1. To prevent substances from entering the trachea, both parts of the rima glottidis can be closed by the lateral cricoarytenoid and arytenoid muscles
    2. In forced respiration, the rima glottidis can be widely opened by action of the posterior cricoarytenoid muscles
B. IN VOCALIZATION, pitch is changed through changes in tension on the vocal cords
    1. Increased tension: through action of the cricothyroid muscles
    2. Decreased tension: through action of the thyroarytenoid muscles

LARYNX

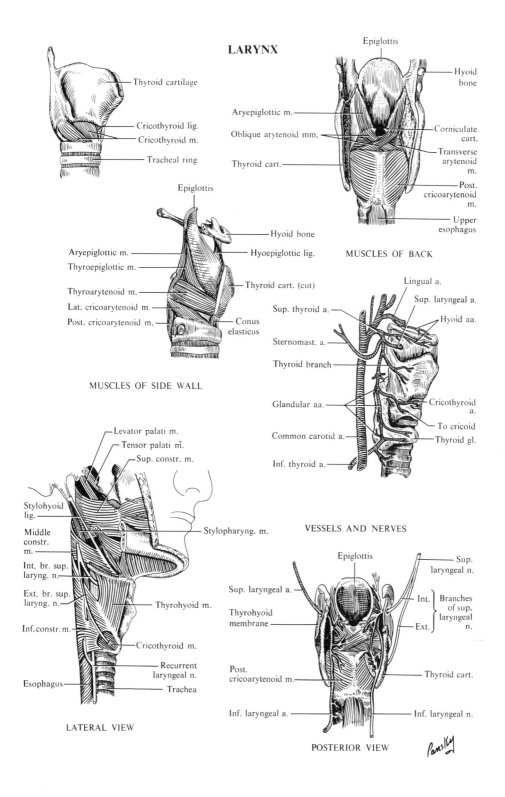

Thyroid cartilage
Cricothyroid lig.
Cricothyroid m.
Tracheal ring

Epiglottis
Hyoid bone

Aryepiglottic m.
Oblique arytenoid mm.
Thyroid cart.

Corniculate cart.
Transverse arytenoid m.
Post. cricoarytenoid m.
Upper esophagus

MUSCLES OF BACK

Epiglottis
Aryepiglottic m.
Thyroepiglottic m.
Thyroarytenoid m.
Lat. cricoarytenoid m.
Post. cricoarytenoid m.

Hyoid bone
Hyoepiglottic lig.
Thyroid cart. (cut)
Conus elasticus

MUSCLES OF SIDE WALL

Lingual a.
Sup. laryngeal a.
Sup. thyroid a.
Hyoid aa.
Sternomast. a.
Thyroid branch
Glandular aa.
Cricothyroid a.
To cricoid
Common carotid a.
Thyroid gl.
Inf. thyroid a.

VESSELS AND NERVES

Levator palati m.
Tensor palati m.
Sup. constr. m.
Stylohyoid lig.
Middle constr. m.
Int. br. sup. laryng. n.
Ext. br. sup. laryng. n.
Inf. constr. m.
Stylopharyng. m.
Thyrohyoid m.
Cricothyroid m.
Recurrent laryngeal n.
Esophagus
Trachea

LATERAL VIEW

Epiglottis
Sup. laryngeal n.
Sup. laryngeal a.
Thyrohyoid membrane
Int. }
Ext. } Branches of sup. laryngeal n.
Post. cricoarytenoid m.
Thyroid cart.
Inf. laryngeal a.
Inf. laryngeal n.

POSTERIOR VIEW

Pansky

25. LARYNX: FRAMEWORK AND STRUCTURE

    I. Location: opposite fourth, fifth, and sixth cervical vertebrae

    II. Size: 44 mm long, 43 mm wide, and 36 mm deep

    III. Cartilages: 9 in number
      A. THYROID (largest and single), composed of *laminae,* one on either side, which meet at an acute angle in ventral midline of neck; *superior horn,* directed cephalically from posterior border of each lamina; *inferior horn,* directed caudally from posterior border of each lamina; *superior tubercle,* at root of superior cornu; *inferior tubercle,* on lower border; and *oblique line,* on lateral side of the lamina
      B. CRICOID (single), composed of *lamina,* posteriorly, 2 to 3 cm long; *arch,* a narrow, convex ring of cartilage. *Articular facets: lateral,* at junction of lamina and arch; and *superior,* at border of lateral corners of lamina
      C. EPIGLOTTIS: yellow elastic cartilage; single, leaf-shaped piece
      D. ARYTENOIDS (paired): each consists of a pyramidal piece with *3 surfaces,* a *base,* and an *apex.* At caudal end of base is the *articular surface.* The *muscular process* projects backward and the *vocal process* anteriorly from the base
      E. CORNICULATE AND CUNEIFORM: small paired pieces of yellow elastic cartilage rostral to the arytenoids

    IV. Ligaments
      A. EXTRINSIC. Connect laryngeal cartilages and hyoid bone or trachea: *thyrohyoid membrane,* between thyroid cartilage and hyoid bone; *middle thyroid ligament,* thickened middle part of membrane; the *lateral thyrohyoid ligaments,* thickened, cordlike posterior edge of the membrane; *hyoepiglottic ligament,* from epiglottis to hyoid bone; and *cricotracheal ligament,* from cricoid cartilage to first tracheal ring
      B. INTRINSIC. Connect laryngeal cartilages together: *elastic membrane,* cephalic part between arytenoid and epiglottic cartilages; caudal part, *conus elasticus,* interconnecting thyroid, cricoid, and arytenoid cartilages. The middle *cricothyroid ligament,* is the anterior part of the conus. The lateral part of the conus, the *thyroepiglottic ligament,* is continuous with the *vocal ligaments*

    V. Articulations: cricothyroid, between inferior cornu of thyroid and lateral facet of cricoid; cricoarytenoid, between cricoid and base of arytenoid cartilage

    VI. Interior of larynx: extends from entrance to lower border of cricoid cartilage
      A. AROUND THE ENTRANCE OF THE LARYNX
        1. Three glossoepiglottic folds, 1 median and 2 lateral. Between these folds are the *valleculae*
        2. Between the aryepiglottic folds and thyroid cartilage are the *piriform sinuses*
      B. ENTRANCE: bounded anteriorly by epiglottis, posteriorly by apices of arytenoid cartilages and corniculate cartilages and laterally by the aryepiglottic folds
      C. LARYNGEAL CAVITY: divided by the *true vocal folds* into the *vestibule* above the folds, which contains the *vestibular or false vocal cords,* and the *ventricle.* Below the true vocal folds, the larynx is continuous with the trachea

    VII. True vocal cords (folds): mucous membrane covering the vocal ligaments which extend from the vocal processes of the arytenoids to the thyroid cartilage

    VIII. Rima glottidis: the fissure between the true vocal folds, including the vocal processes. Has two parts: intramembranous, ventral three fifths between folds (glottis vocalis), and intercartilaginous, posterior two fifths between arytenoid cartilages (glottis respiratoria)

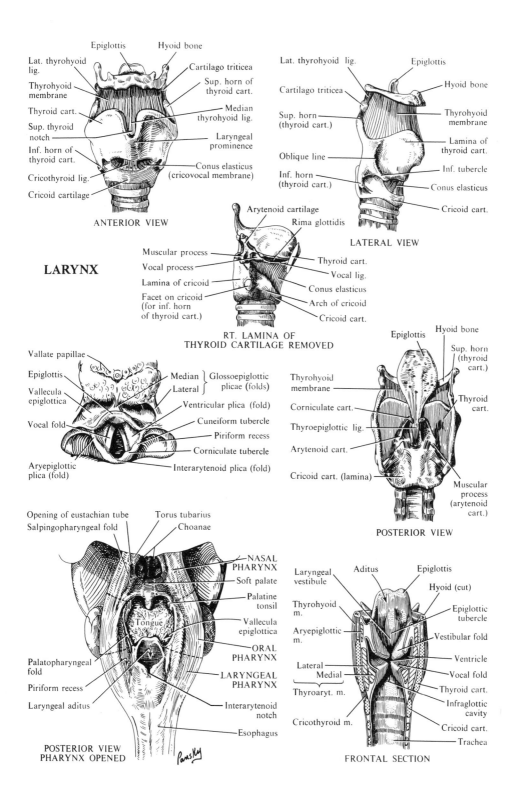

LARYNX

ANTERIOR VIEW

Epiglottis
Hyoid bone
Lat. thyrohyoid lig.
Cartilago triticea
Thyrohyoid membrane
Sup. horn of thyroid cart.
Thyroid cart.
Median thyrohyoid lig.
Sup. thyroid notch
Laryngeal prominence
Inf. horn of thyroid cart.
Cricothyroid lig.
Conus elasticus (cricovocal membrane)
Cricoid cartilage

LATERAL VIEW

Lat. thyrohyoid lig.
Epiglottis
Cartilago triticea
Hyoid bone
Sup. horn (thyroid cart.)
Thyrohyoid membrane
Oblique line
Lamina of thyroid cart.
Inf. horn (thyroid cart.)
Inf. tubercle
Conus elasticus
Cricoid cart.

RT. LAMINA OF THYROID CARTILAGE REMOVED

Arytenoid cartilage
Rima glottidis
Muscular process
Vocal process
Thyroid cart.
Lamina of cricoid
Vocal lig.
Facet on cricoid (for inf. horn of thyroid cart.)
Conus elasticus
Arch of cricoid
Cricoid cart.

Vallate papillae
Epiglottis
Vallecula epiglottica
Vocal fold
Aryepiglottic plica (fold)
Median / Lateral Glossoepiglottic plicae (folds)
Ventricular plica (fold)
Cuneiform tubercle
Piriform recess
Corniculate tubercle
Interarytenoid plica (fold)

POSTERIOR VIEW

Epiglottis
Hyoid bone
Sup. horn (thyroid cart.)
Thyrohyoid membrane
Corniculate cart.
Thyroid cart.
Thyroepiglottic lig.
Arytenoid cart.
Cricoid cart. (lamina)
Muscular process (arytenoid cart.)

POSTERIOR VIEW PHARYNX OPENED

Opening of eustachian tube
Torus tubarius
Salpingopharyngeal fold
Choanae
NASAL PHARYNX
Soft palate
Palatine tonsil
Tongue
Vallecula epiglottica
ORAL PHARYNX
Palatopharyngeal fold
LARYNGEAL PHARYNX
Piriform recess
Laryngeal aditus
Interarytenoid notch
Esophagus
Pansky

FRONTAL SECTION

Laryngeal vestibule
Aditus
Epiglottis
Hyoid (cut)
Thyrohyoid m.
Epiglottic tubercle
Aryepiglottic m.
Vestibular fold
Lateral
Medial
Ventricle
Thyroaryt. m.
Vocal fold
Thyroid cart.
Infraglottic cavity
Cricothyroid m.
Cricoid cart.
Trachea

26. CERVICAL NERVES, PLEXUS, AND CERVICAL SYMPATHETIC TRUNK

I. **Cervical plexus:** ventral primary divisions of cervical nerves 1–4, forming 3 loops
   A. LOCATION: opposite cervical vertebrae 1–3, ventral and lateral to levator scapulae and middle scalene muscles
   B. DISTRIBUTION
      1. To cranial nerves X and XII, mainly from C1 and C2; to XI, mainly from C2–4
      2. To skin (see p. 14)
      3. To muscles, directly: rectus capitis anterior and lateralis; longus capitis and cervicis; levator scapulae and middle scalene
      4. To muscles, through communication or special branches: nerves to sternocleido-mastoid from C2 and C3, and to trapezius from C3 and C4, may go with the spinal accessory nerve (XI)
         a. Communications with the hypoglossal from C1 and C2 run with the hypoglossal for a short distance, then leave this trunk as the *superior root of the ansa cervicalis.* This forms a loop with the *inferior root* from C2 and C3. This loop is the *ansa cervicalis,* which supplies the geniohyoid, thyrohyoid, omohyoid, sternothyroid, and sternohyoid muscles

II. **Cervical sympathetic system** consists of 3 ganglia with intervening cords
   A. LOCATION: ventral to transverse processes of vertebrae, embedded in dorsal part of carotid sheath
   B. COMPOSITION: pre- and postganglionic autonomic fibers
   C. ORIGIN OF PREGANGLIONIC FIBERS: from the intermediolateral gray column of spinal cord segments T1 through T5. White rami communicantes leave the ventral roots of corresponding thoracic nerves and ascend in the trunk

III. **Ganglia**
   A. SUPERIOR. Largest; lies on transverse process of C2 (and C3). Branches: internal carotid nerve to cranial nerves IX, X, and XII; gray rami to C1 through C4; pharyngeal to pharyngeal plexus; to internal carotid plexus; intercarotid to carotid body and sinus; superior cardiac nerve; to larynx and thyroid
   B. MIDDLE. Smallest (sometimes absent); lies ventral to transverse process of C6 or C7. Branches: gray rami to C5 and C6; middle cardiac nerve; thyroid nerves to form plexus on inferior thyroid artery
   C. INFERIOR. Lies between base of transverse process of C7 and neck of first rib, may be fused with first thoracic ganglion. Branches: gray rami to C7, C8, and T1; inferior cardiac nerve; and vertebral nerves along vertebral arteries

IV. **Special features**
   A. TRUNK between middle and inferior cervical ganglia splits to enclose the subclavian artery. The ventral part is longer and forms a loop, the *ansa subclavia*
   B. WHEN THE INFERIOR CERVICAL and first thoracic ganglia are fused, this mass is the cervicothoracic (stellate) ganglion

V. **Clinical considerations**
   A. INTERRUPTION OF CERVICAL SYMPATHETICS: when accomplished above C8 or T1, all sympathetic autonomic control for head, neck, and upper extremity is lost. This will result in Horner's syndrome, which shows narrowed palpebral fissure, constriction of pupil, flushing of skin and lack of sweating. If bilateral, there is no acceleration of heart rate. Thus, when performed for the relief of Raynaud's disease, which is characterized by abnormal vasoconstriction of the upper limb, it is preferable to avulse the anterior roots of T2 and T3 or, still better, the white rami at those levels

-54-

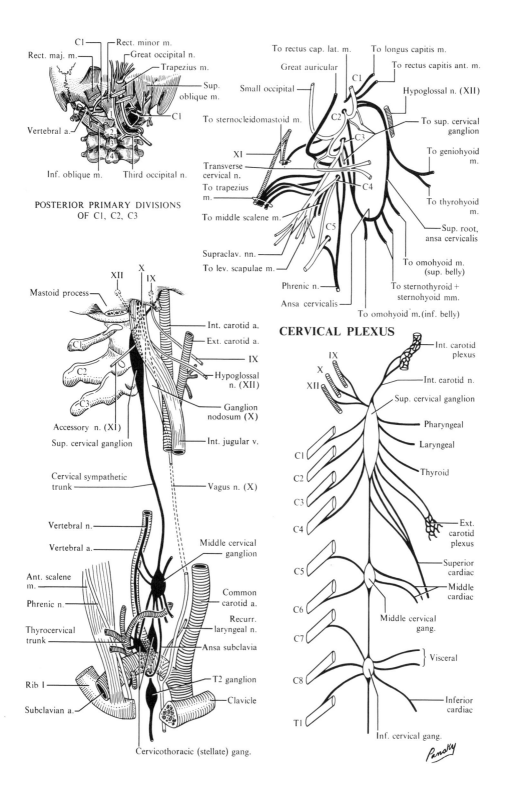

POSTERIOR PRIMARY DIVISIONS
OF C1, C2, C3

CERVICAL PLEXUS

-55-

27. THE FLOOR OF THE MOUTH, THE SUPRAHYOID, INFRAHYOID, AND ANTERIOR VERTEBRAL MUSCLES

I. The floor of the mouth. The mylohyoid and geniohyoid muscles, which form the floor of the mouth, are described elsewhere (see p. 40)

II. The strap muscles, attached to hyoid bone and thyroid cartilage (see p. 40)

III. The anterior vertebral muscles

Name	Origin	Insertion	Action	Nerve
Rectus capitis				
Anterior	Root of trans. proc. of atlas	Basilar part of occipital bone	Flexes head	C1, 2
Lateralis	Upper trans. process of atlas	Lower jugular process of temporal bone	Bends head to side	C1, 2
Longus capitis	Ant. tubercle, trans. proc. vertebrae C3–6	Basilar part of occipital bone	Flexes head	C1, 2, 3
Longus colli				
Sup. oblique	Ant. tubercle, trans. proc. vertebrae C3–5	Tubercle on ant. arch of atlas	Flexes neck, slight rotation of cervical part	C2–7
Inf. oblique	Bodies of vertebrae T1–3	Ant. tubercle, trans. proc. C5, C6	Slight rotation of cervical part	C2–7
Vertical	Bodies of vertebrae C5–7, T1–3	Bodies of C2–4	Flexes neck	C2–7
Anterior scalene	Ant. tubercle, trans. proc. vertebrae C3–6	Scalene tubercle & ridge on upper first rib	Raises first rib, bends neck	C5–8
Middle scalene	Post. tubercle, trans. proc. vertebrae C2–7	Upper first rib, behind subclav. groove	Raises first rib, bends neck	C5–8
Posterior scalene	Post. tubercle, trans. proc. vertebrae C5–7	Outer second rib	Raises second rib, bends neck	C6–8

IV. Special features
- A. RELATIONS OF ATTACHMENT of anterior scalene muscle to first rib and subclavian vessels
    1. Subclavian vein crosses first rib in front of insertion of anterior scalene muscle
    2. Subclavian artery crosses first rib in subclavian groove behind insertion of anterior scalene muscle
- B. The fascia on the deep surface of the scalene muscles forms a conical fibrous dome called suprapleural membrane (Sibson's fascia) which arches over the cupula of the pleura

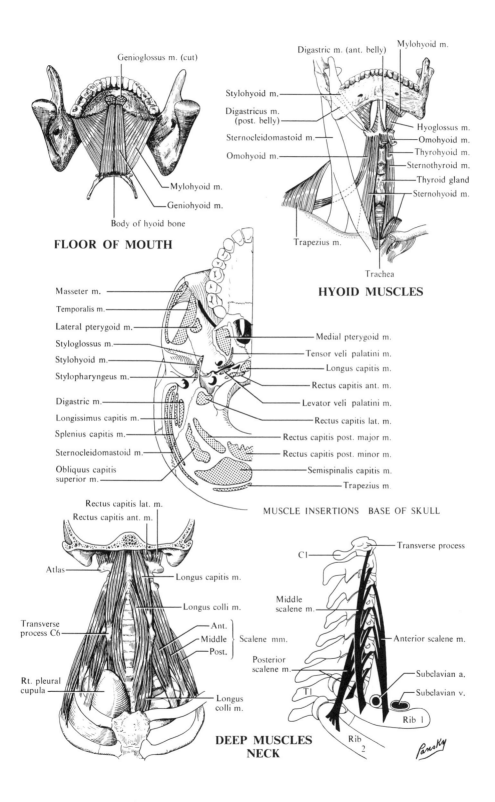

Genioglossus m. (cut)

Mylohyoid m.
Geniohyoid m.
Body of hyoid bone

FLOOR OF MOUTH

Digastric m. (ant. belly)
Mylohyoid m.
Stylohyoid m.
Digastricus m. (post. belly)
Sternocleidomastoid m.
Omohyoid m.

Hyoglossus m.
Omohyoid m.
Thyrohyoid m.
Sternothyroid m.
Thyroid gland
Sternohyoid m.

Trapezius m.

Trachea

HYOID MUSCLES

Masseter m.
Temporalis m.
Lateral pterygoid m.
Styloglossus m.
Stylohyoid m.
Stylopharyngeus m.
Digastric m.
Longissimus capitis m.
Splenius capitis m.
Sternocleidomastoid m.
Obliquus capitis superior m.

Medial pterygoid m.
Tensor veli palatini m.
Longus capitis m.
Rectus capitis ant. m.
Levator veli palatini m.
Rectus capitis lat. m.
Rectus capitis post. major m.
Rectus capitis post. minor m.
Semispinalis capitis m.
Trapezius m.

MUSCLE INSERTIONS   BASE OF SKULL

Rectus capitis lat. m.
Rectus capitis ant. m.
Atlas
Transverse process C6
Rt. pleural cupula

Longus capitis m.
Longus colli m.
Ant.
Middle } Scalene mm.
Post.
Longus colli m.

DEEP MUSCLES NECK

C1
Transverse process
Middle scalene m.
Posterior scalene m.
Anterior scalene m.
Subclavian a.
Subclavian v.
T1
Rib 1
Rib 2

Pansky

28. SUBMANDIBULAR AND SUBLINGUAL GLANDS

I. Submandibular
A. SIZE: second largest of major salivary glands, size of walnut
B. TYPE: compound, tubuloalveolar; mixed serous and mucous
C. LOCATION AND RELATIONS. Mainly in submandibular triangle (see p. 36)
   1. Superficial surface: cephalic part touches submandibular depression of mandible and internal pterygoid muscle; caudal part is covered by skin, superficial fascia, platysma and deep cervical fascia
      a. Facial vein and small branches of facial nerve cross it; submandibular lymph nodes lie against it
   2. Deep surface: lies on mylohyoid, hyoglossus, styloglossus, stylohyoid, and posterior belly of digastric muscles
      a. Mylohyoid nerve and vessels are in contact with it
   3. Posterior border: facial artery
   4. Deep process: a projection from its deep surface is bounded below and laterally by mylohyoid muscle; medially, by hyoglossus and styloglossus muscles; above, by lingual nerve and submandibular ganglion; below by hypoglossal nerve
D. DUCT (WHARTON'S): formed on deep surface, runs rostrally between mylohyoid muscle laterally and hyoglossus and genioglossus muscles medially. Opens at caruncle at side of frenulum of tongue. In part of course is between lingual and hypoglossal nerves; finally is crossed by lingual nerve

II. Sublingual
A. SIZE: smallest of 3 major salivary glands
B. TYPE: compound, tubuloalveolar; mixed serous and mucous
C. LOCATION AND RELATIONS: beneath mucous membrane of floor of mouth
   1. Above: floor of mouth
   2. Below: mylohyoid muscle
   3. Behind: deep part, submandibular gland
   4. Laterally: sublingual depression of mandible
   5. Medially: genioglossus muscle
      a. Lingual nerve and submandibular duct lie between gland and muscle
D. DUCTS
   1. Small (of Rivinus): some join submandibular duct, others open separately in floor of mouth
   2. Large (Bartholin): 1 or 2 duct branches join submandibular duct

III. Innervation of both glands
A. PARASYMPATHETIC: preganglionic fibers, from superior salivatory nucleus through nervus intermedius of facial nerve: from facial via chorda tympani to lingual nerve, through parasympathetic root to submandibular ganglion, where they terminate in synapses. Postganglionic branches from ganglion go to both glands

IV. Vessels
A. ARTERIES: branches of the facial artery to submandibular gland, sublingual branch of lingual artery supplies sublingual
B. VEINS: follow the arteries
C. LYMPHATIC DRAINAGE: to submandibular nodes

SUBLINGUAL AND SUBMANDIBULAR GLANDS

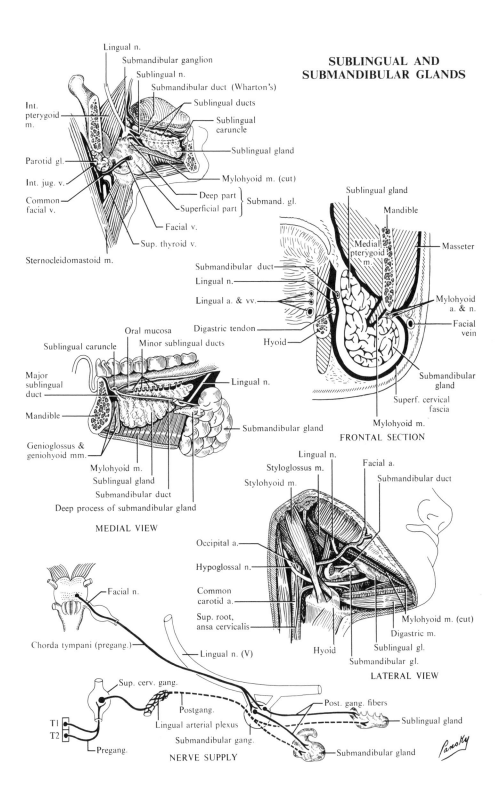

Lingual n.
Submandibular ganglion
Sublingual n.
Submandibular duct (Wharton's)
Sublingual ducts
Sublingual caruncle
Int. pterygoid m.
Sublingual gland
Parotid gl.
Mylohyoid m. (cut)
Int. jug. v.
Common facial v.
Deep part
Superficial part } Submand. gl.
Facial v.
Sup. thyroid v.
Sternocleidomastoid m.

Sublingual gland
Mandible
Medial pterygoid m.
Masseter
Submandibular duct
Lingual n.
Lingual a. & vv.
Mylohyoid a. & n.
Facial vein
Digastric tendon
Hyoid
Submandibular gland
Superf. cervical fascia
Mylohyoid m.

FRONTAL SECTION

Oral mucosa
Minor sublingual ducts
Sublingual caruncle
Major sublingual duct
Lingual n.
Mandible
Genioglossus & geniohyoid mm.
Submandibular gland
Mylohyoid m.
Sublingual gland
Submandibular duct
Deep process of submandibular gland

MEDIAL VIEW

Lingual n.
Styloglossus m.
Facial a.
Stylohyoid m.
Submandibular duct
Occipital a.
Hypoglossal n.
Common carotid a.
Sup. root, ansa cervicalis
Mylohyoid m. (cut)
Digastric m.
Hyoid
Sublingual gl.
Submandibular gl.

LATERAL VIEW

Facial n.
Chorda tympani (pregang.)
Lingual n. (V)
Sup. cerv. gang.
Postgang.
Post. gang. fibers
Lingual arterial plexus
Sublingual gland
T1
T2
Submandibular gang.
Pregang.
Submandibular gland

NERVE SUPPLY

-59-

29. THE TONGUE

   I. Functions: chief organ for taste; important in speech, mastication, and deglutition

  II. Parts: root, apex, and body

 III. Surfaces
   A. INFERIOR: mucous membrane continuous with that of gums and floor of mouth. *Frenulum linguae,* in midline
   B. DORSAL: convex. Median sulcus divides tongue into lateral halves and ends posteriorly in a depression, the *foramen cecum.* The latter is the apex for 2 diverging grooves, the *terminal sulci.* Two thirds of the tongue lies rostral to the sulcus and is rough and covered with papillae, while the caudal third is more smooth and contains glands and crypts of the *lingual tonsil*

  IV. Papillae are of 3 basic types: *vallate,* largest, 8 to 10 in number, lying in a row beginning just rostral to the foramen cecum and along the terminal sulci; *foliate* at sides and *fungiform* scattered over dorsum; and the *filiform,* smallest and most numerous, covering all of dorsum

   V. Taste buds: sensory organs of taste, scattered over mucous membrane of mouth and tongue and especially numerous on and around vallate papillae

  VI. Nerves of tongue
   A. SENSORY
     1. General sensation: lingual branch of mandibular (V) for anterior two thirds; lingual branch of glossopharyngeal (IX) for caudal one third; and lingual branches from superior laryngeal of vagus (X) for root near epiglottis
     2. Special sensation (taste): taste buds on anterior two thirds, through lingual, chorda tympani, and nervus intermedius by way of facial (VII); taste buds on caudal one third, through lingual branch of glossopharyngeal (IX); and taste buds in epiglottic region by sup. laryngeal br. of vagus (X)
   B. MOTOR
     1. Hypoglossal nerve. From nucleus in medulla, nerve emerges through hypoglossal canal, becomes closely united with vagus nerve, passes between internal carotid artery and internal jugular vein, crosses superficial to both to hook around occipital artery, crosses lateral to external carotid and lingual arteries, curves upward above hyoid bone, deep to digastric and stylohyoid muscles, and then between mylohyoid and hyoglossus muscles to apex of tongue. Supplies both intrinsic and extrinsic muscles of tongue

 VII. Arteries of tongue
   A. MAJOR: *lingual artery* from external carotid artery. Runs deep, along inferior tongue surface to end as *deep lingual artery.* The *dorsal lingual arteries,* 2 or 3 branches, ascend to dorsum of tongue
   B. MINOR: lingual branch of inferior alveolar artery, tonsillar branch of facial artery, and branches from ascending pharyngeal artery

 VIII. Veins
   A. LINGUAL: from dorsum, sides, and undersurface to internal jugular vein
   B. VENA COMITANS OF HYPOGLOSSAL N. (RANINE): begin near apex and run with hypoglossal nerve to terminate in lingual or common facial veins

  IX. Lymphatic drainage (see p. 64)

   X. Modalities of taste: sweet at tip, bitter toward root, salty at sides and tip, and sour at sides

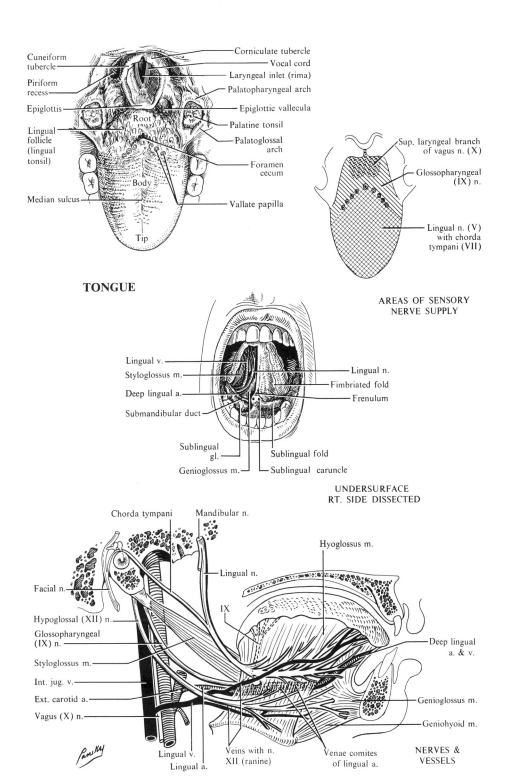

Cuneiform tubercle

Piriform recess

Epiglottis

Lingual follicle (lingual tonsil)

Median sulcus

Root

Body

Tip

Corniculate tubercle

Vocal cord

Laryngeal inlet (rima)

Palatopharyngeal arch

Epiglottic vallecula

Palatine tonsil

Palatoglossal arch

Foramen cecum

Vallate papilla

TONGUE

Sup. laryngeal branch of vagus n. (X)

Glossopharyngeal (IX) n.

Lingual n. (V) with chorda tympani (VII)

AREAS OF SENSORY NERVE SUPPLY

Lingual v.

Styloglossus m.

Deep lingual a.

Submandibular duct

Sublingual gl.

Genioglossus m.

Lingual n.

Fimbriated fold

Frenulum

Sublingual fold

Sublingual caruncle

UNDERSURFACE
RT. SIDE DISSECTED

Chorda tympani

Mandibular n.

Hyoglossus m.

Facial n.

Lingual n.

IX

Hypoglossal (XII) n.

Glossopharyngeal (IX) n.

Styloglossus m.

Int. jug. v.

Ext. carotid a.

Vagus (X) n.

Deep lingual a. & v.

Genioglossus m.

Geniohyoid m.

Lingual v.

Lingual a.

Veins with n. XII (ranine)

Venae comites of lingual a.

NERVES & VESSELS

30. MUSCLES OF THE TONGUE AND SWALLOWING

I. **Extrinsic** (see p. 48)
A. GENIOGLOSSUS, HYOGLOSSUS, STYLOGLOSSUS, AND PALATOGLOSSUS MUSCLES

II. **Intrinsic:** all of which change the shape of the tongue
A. SUPERIOR LONGITUDINAL: on the dorsum of the tongue. Shortens tongue and raises tip and sides
B. INFERIOR LONGITUDINAL: on underside of tongue. Shortens tongue and depresses tip
C. TRANSVERSUS: narrows and lengthens tongue
D. VERTICALIS: makes tongue flat and broadens it

III. **Swallowing:** a complex of muscular action involving several activities, all in a proper sequence. This requires the propulsion of the bolus of food, enlargement of the alimentary canal in front of the bolus, closure of the tube behind the bolus, and closure of the other systems such as the nasopharynx and larynx

IV. **Actions involved in swallowing**
A. PROPULSION
   1. With the jaws and lips closed, the tongue is voluntarily raised against the hard palate, using the genioglossus, mylohyoid, and geniohyoid muscles, while the palatoglossus muscle relaxes. This moves the bolus to the level of the soft palate
   2. The soft palate is raised to enlarge the space by the levator and tensor veli palatini muscles
   3. At the same time, the base of the tongue is raised and drawn back, forcing the bolus farther down the tube. The styloglossus muscle is important in this function
   4. The pharynx, attached to the larynx, is raised and expanded to receive the bolus. With the jaw fixed, many muscles accomplish this action: the mylohyoid, geniohyoid, stylohyoid, posterior belly of the digastric, stylopharyngeus, salpingopharyngeus, and palatopharyngeus muscles. The stylopharyngeus also widens and enlarges the pharynx
   5. The bolus now lies in the region of the pharyngeal constrictors. When the superior contracts, the middle and inferiors relax. As the bolus passes to the next field, the middle constrictor then contracts, etc. After the inferior contracts, the bolus has reached the upper esophagus, where a peristaltic-like wave carries it down
B. WHEN THE BOLUS passes through the isthmus of the pharynx, the palatoglossus muscles contract. This closes off the mouth from the pharynx
C. CLOSURE OF THE OTHER SYSTEMS
   1. The nasopharynx is closed: by raising the soft palate through the levator muscle and by contraction of the 2 palatopharyngeus muscles causing them to approach each other medially
   2. The larynx is closed:
      a. Superiorly, by the epiglottis being folded back under the retracted tongue as the larynx is raised; the epiglottis is thus folded over the laryngeal orifice
      b. Inferiorly, by the closure of the rima glottidis through the action of the arytenoid, lateral cricoarytenoid, and thyroarytenoid muscles

V. **Clinical considerations**
A. LESION OF THE HYPOGLOSSAL NERVE causes deviation of the protruded tongue toward the side of the lesion. This is due to the lack of function of the genioglossus muscle on the diseased side

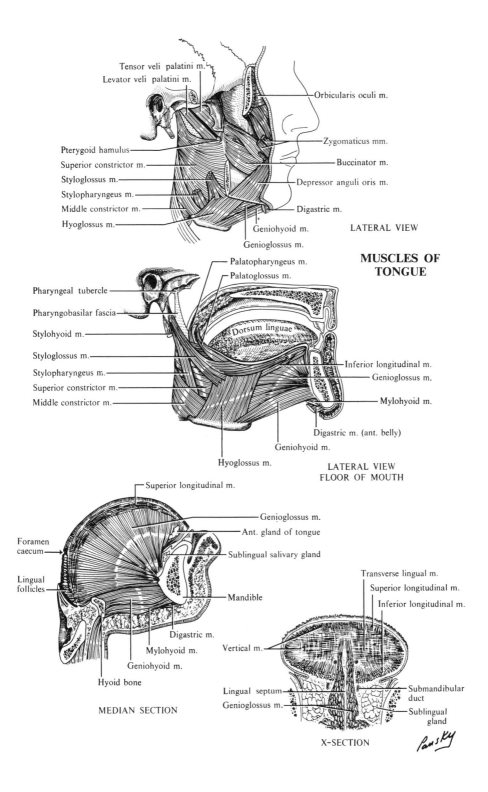

Tensor veli palatini m.
Levator veli palatini m.
Orbicularis oculi m.
Pterygoid hamulus
Superior constrictor m.
Styloglossus m.
Stylopharyngeus m.
Middle constrictor m.
Hyoglossus m.
Zygomaticus mm.
Buccinator m.
Depressor anguli oris m.
Digastric m.
Geniohyoid m.
Genioglossus m.

LATERAL VIEW

MUSCLES OF TONGUE

Palatopharyngeus m.
Palatoglossus m.
Pharyngeal tubercle
Pharyngobasilar fascia
Stylohyoid m.
Styloglossus m.
Stylopharyngeus m.
Superior constrictor m.
Middle constrictor m.
Dorsum linguae
Inferior longitudinal m.
Genioglossus m.
Mylohyoid m.
Digastric m. (ant. belly)
Geniohyoid m.
Hyoglossus m.

LATERAL VIEW
FLOOR OF MOUTH

Superior longitudinal m.
Genioglossus m.
Ant. gland of tongue
Sublingual salivary gland
Foramen caecum
Lingual follicles
Mandible
Digastric m.
Mylohyoid m.
Geniohyoid m.
Hyoid bone

MEDIAN SECTION

Transverse lingual m.
Superior longitudinal m.
Inferior longitudinal m.
Vertical m.
Lingual septum
Genioglossus m.
Submandibular duct
Sublingual gland

X-SECTION

Pansky

31. LYMPHATICS OF THE NOSE AND TONGUE

I. Lymphatic drainage of the nasal cavities and nasopharynx
A. NASAL CAVITY
  1. Anterior part communicates with the vessels of the skin of the nose. These usually drain through the mandibular nodes to the submandibular and superior deep cervical nodes
  2. Posterior part drains into the retropharyngeal nodes located in the fascia posterior to the pharynx. The efferents from these go to the deep cervical nodes. Part of this drainage may go directly to the superior deep cervical nodes. A few channels from this same area may also go to the subparotid nodes first and then to the superior deep cervical nodes
  3. From floor: vessels may enter the parotid set and then follow the courses given above
  4. From the nasopharynx: there is also drainage to both the retropharyngeal and parotid nodes
  5. The paranasal sinuses drain partly into the retropharyngeal and partly by direct paths to the superior deep cervical nodes

II. Palatine tonsil
A. FROM THE TONSIL, the lymphatic channels drain directly into the most cephalic of the superior deep cervical nodes through 3 to 5 vessels

III. Tongue. It is usually divided into 4 main drainage areas
A. APEX drains into the suprahyoid and principal node* of the tongue
B. LATERAL (margins) drain to the submandibular and superior deep cervical nodes
C. BASAL (near vallate papillae) drain to the superior deep cervical nodes
D. MEDIAN drain to the submandibular and superior deep cervical nodes
E. CROSS DRAINAGE: channels from the opposite sides of the tongue cross the midline. Some of these enter the superior deep cervical nodes, and others enter the inferior deep cervical nodes

IV. Clinical considerations
A. JUGULODIGASTRIC NODE: because of the inaccessibility of the posterior third of the tongue and the tonsil, sometimes the first positive sign of carcinoma in these regions will be a swelling of this node, sometimes called the "main gland of the tonsil"
B. IN CONSEQUENCE OF THE LYMPHATIC DRAINAGE OF THE TONGUE, metastatic carcinoma may be widely disseminated through the submental and submandibular regions and along the internal jugular vein. The operation designed to remove such metastatic lesions is called "radical neck dissection" or "block dissection of the neck"
C. CARCINOMA OF THE TONGUE affects males chiefly, and the edges of the tongue are most often affected. The tumor is an epidermoid carcinoma and may be of any degree of malignancy. Metastasis is usually slow, involving the lymph nodes of the neck; rarely there may be rapid, widespread metastases

*The "principal" node is a single constant node of the deep cervical group, lying at the bifurcation of the common carotid artery, which receives a large number of vessels from the tongue.

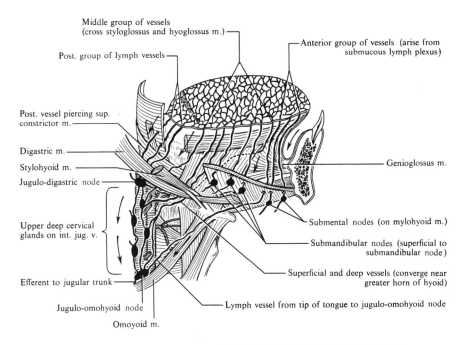

Middle group of vessels
(cross styloglossus and hyoglossus m.)

Anterior group of vessels (arise from
submucous lymph plexus)

Post. group of lymph vessels

Post. vessel piercing sup.
constrictor m.

Digastric m.

Stylohyoid m.

Genioglossus m.

Jugulo-digastric node

Upper deep cervical
glands on int. jug. v.

Submental nodes (on mylohyoid m.)

Submandibular nodes (superficial to
submandibular node)

Superficial and deep vessels (converge near
greater horn of hyoid)

Efferent to jugular trunk

Jugulo-omohyoid node

Lymph vessel from tip of tongue to jugulo-omohyoid node

Omoyoid m.

LYMPH VESSELS OF TONGUE
AND NASAL CAVITY

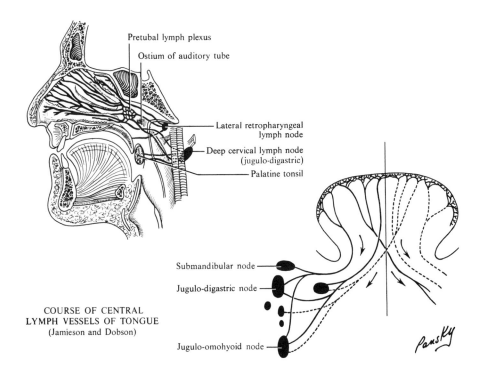

Pretubal lymph plexus

Ostium of auditory tube

Lateral retropharyngeal
lymph node

Deep cervical lymph node
(jugulo-digastric)

Palatine tonsil

Submandibular node

Jugulo-digastric node

COURSE OF CENTRAL
LYMPH VESSELS OF TONGUE
(Jamieson and Dobson)

Jugulo-omohyoid node

Pansky

-65-

32. TEETH AND PALATE

I. Teeth

A. VARIETIES: incisors, canines, premolars, and molars
B. SETS
    1. Deciduous, 20: 2 incisors, 1 canine, 2 molars*
    2. Permanent, 32: 2 incisors, 1 canine, 2 premolars, 3 molars*
C. GENERAL PARTS for all teeth: *crown,* projecting above gum; *root,* embedded in alveolus of bone; and *neck,* constriction between crown and root
D. SURFACES: *labial* (or *buccal*), next to lip or cheek; *lingual,* toward tongue; and *contact edge* touching adjoining teeth
E. SPECIAL FEATURES
    1. Premolars, or bicuspids: free surface of crown shows 2 eminences or cusps (a labial and lingual) separated by a groove. Root is usually single
    2. Molars: have several cusps, separated by grooves
        a. Upper, have 3 roots—2 buccal, 1 lingual; first molar has 4 cusps; second has 3 or 4 cusps; and third has 3 cusps
        b. Lower, have 2 roots—anterior and posterior; first molar has 5 cusps; second and third have 4 or 5 cusps
F. STRUCTURE, from within, outward: *pulp cavity,* containing loose connective tissue, vessels, and nerves; *dentin* (ivory), modified bone that forms bulk of tooth; *enamel,* covers exposed part; and *cement,* thin layer of bone, superficial to dentin, in root
G. AGE at time of eruption of permanent teeth

1st molar	6 years	2nd premolar	10 years
Central incisors	7 years	Canine	11–12 years
Lat. incisors	8 years	2nd molars	12–13 years
1st premolar	9 years	3rd molars	17–25 years

H. VESSELS: posterior, middle, and superior alveolar branches of maxillary artery for upper teeth; inferior alveolar for lower teeth. Veins run with arteries
I. NERVES (see p. 30)

II. Palate

A. DIVISIONS
    1. Hard: roof of mouth, separating nasal and oral cavities (see p. 10)
    2. Soft: suspended from dorsal edge of hard palate. Consists of mucous membrane containing muscles (see p. 70), glands, etc. *Uvula* hangs from its caudal border
B. INNERVATION
    1. Sensory: *greater (ant.) palatine nerve* via the pterygopalatine branches of the maxillary (V) nerve, mainly on hard palate and gums; *nasopalatine nerve* from posterior superior nasal branch of the pterygopalatine nerve to rostral hard palate; and *lesser (post.) palatine nerve,* from same source, to soft palate
    2. Motor: to tensor veli palatini muscle via trigeminal (V) nerve and levator veli palatini via X and XI through the pharyngeal plexus
    3. Autonomic. Parasympathetic: preganglionic from the sup. salivatory nucleus through the nervus intermedius, great petrosal, and nerve of pterygoid canal, to pterygopalatine ganglion. Postganglionics run with sensory fibers. Sympathetic: preganglionics from upper thoracic cord synapse in superior cervical ganglion. Postganglionics, through deep petrosal nerve, then nerve of pterygoid canal through pterygopalatine ganglion to be distributed as above

---

*Dental formula refers to number and variety of teeth in one half of each jaw, thus, in adult man: 2-1-2-3.

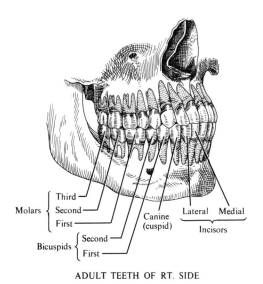

ADULT TEETH OF RT. SIDE

Molars { Third / Second / First
Bicuspids { Second / First
Canine (cuspid)
Lateral Medial
Incisors

TEETH

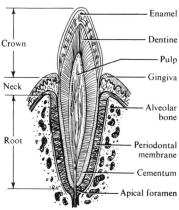

VERTICAL SECTION
OF A TOOTH

Enamel
Dentine
Pulp
Gingiva
Alveolar bone
Periodontal membrane
Cementum
Apical foramen

Crown
Neck
Root

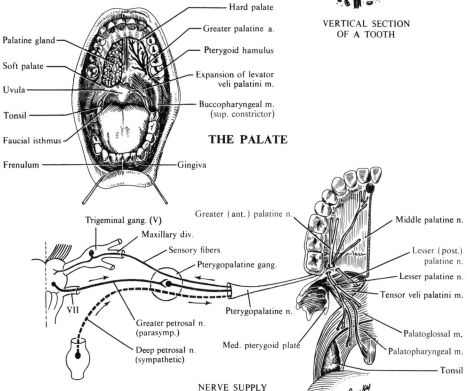

Palatine gland
Soft palate
Uvula
Tonsil
Faucial isthmus
Frenulum

Hard palate
Greater palatine a.
Pterygoid hamulus
Expansion of levator veli palatini m.
Buccopharyngeal m. (sup. constrictor)

Gingiva

THE PALATE

Trigeminal gang. (V)
Maxillary div.
Sensory fibers
Pterygopalatine gang.
VII
Greater petrosal n. (parasymp.)
Deep petrosal n. (sympathetic)
Pterygopalatine n.
Med. pterygoid plate

Greater (ant.) palatine n.
Middle palatine n.
Lesser (post.) palatine n.
Lesser palatine n.
Tensor veli palatini m.
Palatoglossal m.
Palatopharyngeal m.
Tonsil

NERVE SUPPLY

33. PALATINE TONSIL

I. **Location.** In triangular depression (tonsillar fossa), bounded rostrally by glossopalatine arch, caudally by dorsum of tongue, posteriorly by palatopharyngeal arch
   A. SUPRATONSILLAR FOSSA: small area above cephalic pole of tonsil

II. **Structure**
   A. SHAPE: oval
   B. SIZE: variable, relatively and often actually larger in children
   C. MEDIAL OR FREE SURFACE
      1. Covered by stratified squamous epithelium
      2. Surface epithelium extends inward at from 12 to 15 points. These orifices open into branching crypts
   D. LYMPH NODULES lie beneath epithelium of surface and along crypts
   E. LATERAL OR DEEP SURFACE is covered by a fibrous connective tissue capsule

III. **Relations**
   A. MEDIAL SURFACE: free, faces pharynx
   B. LATERAL SURFACE: separated from constrictor muscle by loose connective tissue. This muscle separates tonsil from facial artery and its tonsillar and ascending palatine branches. Internal carotid artery is about 1 in. behind and lateral
   C. ROSTRAL BORDER: glossopalatine arch containing glossopalatine muscle
   D. POSTERIOR BORDER: palatopharyngeal arch containing palatopharyngeal muscle

IV. **Arterial supply.** Tonsillar branches of:
   A. DORSAL LINGUAL ARTERY from lingual artery
   B. ASCENDING PALATINE AND TONSILLAR ARTERIES from facial artery
   C. ASCENDING PHARYNGEAL from external carotid artery
   D. DESCENDING PALATINE from maxillary artery
   E. ACCESSORY MENINGEAL from maxillary artery

V. **Veins:** from venous plexus lateral to tonsil, which enters pterygoid plexus of veins (see p. 18)

VI. **Tonsillar ring (Waldeyer's) of lymphatic tissue,** said to "guard" openings into respiratory and digestive systems. Ring composed of, caudally, the lingual tonsil on root of tongue, laterally the palatine and tubal tonsils (at orifice of auditory tube), and dorsally the pharyngeal tonsil (adenoid)

VII. **Clinical considerations**
   A. THE TONSILS AND ADENOIDS tend to hypertrophy most commonly in childhood due to a general lymphatic hyperplasia
   B. THE PALATINE TONSILS are subject to infection as a result of the numerous crypts on their surface
   C. ADENOID HYPERTROPHY can block the nasopharynx, creating problems in breathing
   D. BECAUSE OF THE TREMENDOUS VASCULAR SUPPLY of the palatine tonsils, extreme care must be used in tonsillectomy
   E. BECAUSE OF THE THIN LATERAL WALL behind which lie many major vessels, excision must be done with great care
   F. THE PHARYNGEAL TONSILS posteriorly and above, the palatine tonsils laterally, and the lingual tonsil anteriorly and below form an oblique ring of lymphoid tissue around the pharynx

PALATINE TONSIL

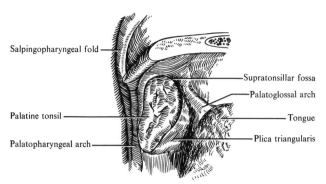

Salpingopharyngeal fold

Supratonsillar fossa

Palatoglossal arch

Palatine tonsil

Tongue

Plica triangularis

Palatopharyngeal arch

MEDIAL VIEW AND RELATED ARCHES

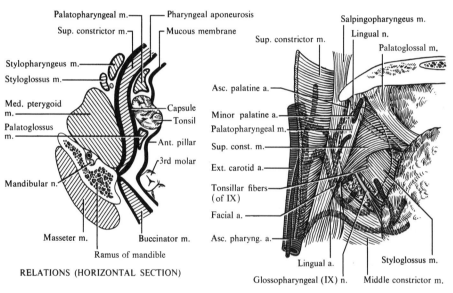

Palatopharyngeal m.

Sup. constrictor m.

Pharyngeal aponeurosis

Mucous membrane

Stylopharyngeus m.

Styloglossus m.

Med. pterygoid m.

Palatoglossus m.

Capsule

Tonsil

Ant. pillar

3rd molar

Mandibular n.

Masseter m.

Buccinator m.

Ramus of mandible

RELATIONS (HORIZONTAL SECTION)

Sup. constrictor m.

Salpingopharyngeus m.

Lingual n.

Palatoglossal m.

Asc. palatine a.

Minor palatine a.

Palatopharyngeal m.

Sup. const. m.

Ext. carotid a.

Tonsillar fibers (of IX)

Facial a.

Asc. pharyng. a.

Lingual a.

Glossopharyngeal (IX) n.

Styloglossus m.

Middle constrictor m.

TONSILLAR BED

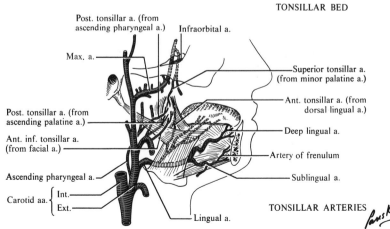

Post. tonsillar a. (from ascending pharyngeal a.)

Infraorbital a.

Max. a.

Superior tonsillar a. (from minor palatine a.)

Ant. tonsillar a. (from dorsal lingual a.)

Post. tonsillar a. (from ascending palatine a.)

Deep lingual a.

Ant. inf. tonsillar a. (from facial a.)

Artery of frenulum

Ascending pharyngeal a.

Sublingual a.

Carotid aa. { Int. Ext.

Lingual a.

TONSILLAR ARTERIES

Pansky

34. MUSCLES OF THE PALATE

I.

Name	Origin	Insertion	Action	Nerve
Levator veli palatini	Petrous bone Medial auditory tube	Velum of palate	Raises soft palate	Pharyngeal plexus (X & XI)
Tensor veli palatini	Scaphoid fossa Angular spine Lateral auditory tube	Aponeurosis of palate	Tenses soft palate	Trigeminal (V)
Musculus uvulae	Post. nasal spine Palatine apo- neurosis	Uvula	Raises uvula	Pharyngeal plexus (X & XI)
Palato- pharyngeus	Soft palate	Lat. post. pharnyx Post. thyroid cart.	Raises pharynx, helps close nasopharynx	Pharyngeal plexus (X & XI)
Palatoglossus	Anterior soft palate	Dorsolateral tongue	Raises tongue	Pharyngeal plexus (X & XI)
Salpingo- pharyngeus (see p. 48)				
Pharyngeal constrictors (see p. 48)				

II. Special features
A. PALATINE APONEUROSIS: a fibrous sheet extending posteriorly from caudal edge of palatine bone into the soft palate to support and give attachment to muscles of that structure
B. MOUTH AND PHARYNX unite at palatoglossal arch.* The opening between the 2 arches is called the *isthmus of fauces*. The arch consists of mucous membrane over the palatoglossus muscle
C. PALATOPHARYNGEAL ARCH,† larger than above, produced by the membrane overlying the palatopharyngeus muscle, is separated from the palatoglossal arch by a triangular space containing the *palatine tonsil*
D. VERTICAL DESCENT of the fleshy part of the tensor veli palatini muscle is along the medial pterygoid plate and ends in a tendon that bends at right angles around hamulus to join the palatine aponeurosis
E. THERE IS A SEPARATION of the fibers of the palatopharyngeus muscle into anterior and posterior layers by the levator veli palatini muscle

*Also called anterior pillar of fauces.
†Also called posterior pillar of fauces.

MUSCLES OF PALATE

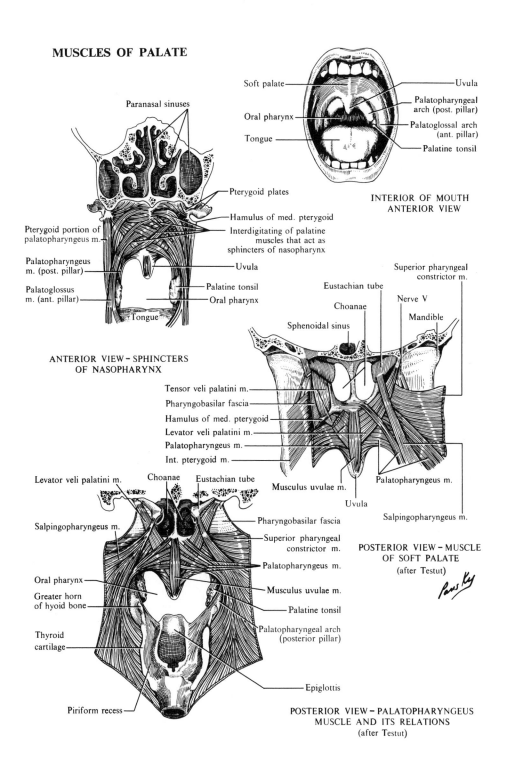

Paranasal sinuses

Soft palate

Oral pharynx

Tongue

Uvula

Palatopharyngeal arch (post. pillar)

Palatoglossal arch (ant. pillar)

Palatine tonsil

INTERIOR OF MOUTH ANTERIOR VIEW

Pterygoid plates

Hamulus of med. pterygoid

Interdigitating of palatine muscles that act as sphincters of nasopharynx

Pterygoid portion of palatopharyngeus m.

Palatopharyngeus m. (post. pillar)

Palatoglossus m. (ant. pillar)

Uvula

Palatine tonsil

Oral pharynx

Tongue

ANTERIOR VIEW – SPHINCTERS OF NASOPHARYNX

Superior pharyngeal constrictor m.

Eustachian tube

Choanae

Nerve V

Sphenoidal sinus

Mandible

Tensor veli palatini m.

Pharyngobasilar fascia

Hamulus of med. pterygoid

Levator veli palatini m.

Palatopharyngeus m.

Int. pterygoid m.

Musculus uvulae m.

Uvula

Palatopharyngeus m.

Salpingopharyngeus m.

POSTERIOR VIEW – MUSCLE OF SOFT PALATE
(after Testut)

Levator veli palatini m.

Choanae

Eustachian tube

Salpingopharyngeus m.

Pharyngobasilar fascia

Superior pharyngeal constrictor m.

Palatopharyngeus m.

Musculus uvulae m.

Palatine tonsil

Palatopharyngeal arch (posterior pillar)

Oral pharynx

Greater horn of hyoid bone

Thyroid cartilage

Piriform recess

Epiglottis

POSTERIOR VIEW – PALATOPHARYNGEUS MUSCLE AND ITS RELATIONS
(after Testut)

35. AUDITORY TUBE, NASOPHARYNX, AND ORAL PHARYNX

I. Auditory tube (see p. 117)
A. EXTENT: 36 mm
B. COURSE: medially, rostrally, and caudally from middle ear to nasopharynx
C. PARTS
    1. Osseous: proximal one third
    2. Cartilaginous: distal two thirds
D. SPECIAL FEATURES
    1. Torus tubarius (cushion): base of cartilage of the tube, forming an elevation dorsal to its orifice in the nasopharynx
    2. Tubal tonsil: an accumulation of lymphoid tissue around the orifice of the auditory tube

II. Nasopharynx (entirely respiratory)
A. BOUNDARIES: rostrally, the choanae; posteriorly, the pharyngeal wall against the occipital bone, the anterior arch of vertebra C1 and body of C2; caudally, opens into the oral pharynx at border of soft palate
B. SPECIAL FEATURES
    1. Orifice of auditory tube with its torus and tubal tonsil
    2. Salpingopharyngeal fold: a fold of mucous membrane due to the muscle of the same name. The fold extends from the torus to the lateral pharyngeal wall
    3. Salpingopalatine fold: a fold of mucous membrane overlying the salpingopalatine muscle from the torus to the soft palate
    4. Pharyngeal recess: a recess behind the salpingopharyngeal fold
    5. Pharyngeal tonsil (adenoid): in upper end of the posterior wall
    6. Pharyngeal bursa: a small diverticulum lying just rostral to the pharyngeal tonsil, in the midline

III. Oral pharynx (respiratory and digestive)
A. BOUNDARIES: anteriorly, the dorsum of the tongue and the palatoglossal fold, which marks the limit of the oral cavity; caudally, the hyoid bone
B. SPECIAL FEATURES
    1. Palatine tonsils: in the lateral wall (see p. 68)
    2. Palatopharyngeal fold: due to the palatopharyngeal muscle

IV. Clinical considerations
A. ONE OF THE FUNCTIONS OF THE AUDITORY TUBE is to equalize air pressure on both sides of the tympanic membrane. Swallowing helps to open the tube (when changes in external pressure cause a disagreeable sensation) due to the action of the salpingopharyngeus muscle, which originates on the tube, and the dilator tubae, which arises on the cartilage of the tube and blends with the tensor veli palatini, both of which contract in swallowing

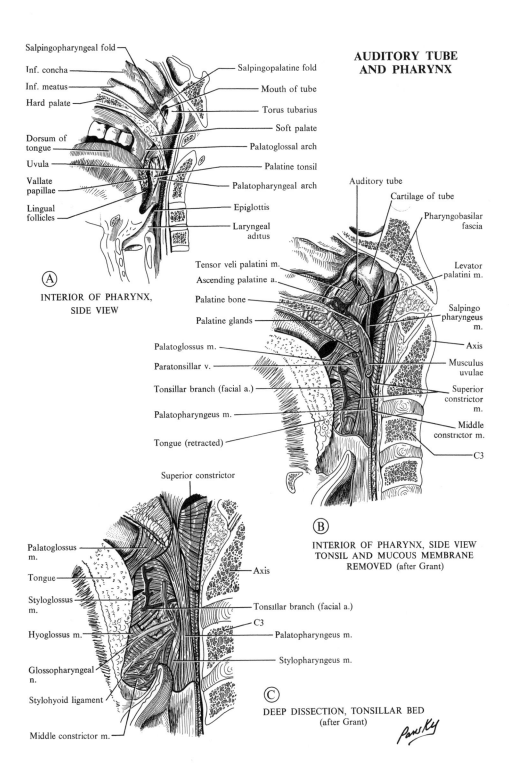

AUDITORY TUBE AND PHARYNX

A Salpingopharyngeal fold — Inf. concha — Inf. meatus — Hard palate — Dorsum of tongue — Uvula — Vallate papillae — Lingual follicles

Salpingopalatine fold — Mouth of tube — Torus tubarius — Soft palate — Palatoglossal arch — Palatine tonsil — Palatopharyngeal arch — Epiglottis — Laryngeal aditus

A

INTERIOR OF PHARYNX, SIDE VIEW

B Auditory tube — Cartilage of tube — Pharyngobasilar fascia — Levator palatini m. — Salpingo pharyngeus m. — Axis — Musculus uvulae — Superior constrictor m. — Middle constrictor m. — C3

Tensor veli palatini m. — Ascending palatine a. — Palatine bone — Palatine glands — Palatoglossus m. — Paratonsillar v. — Tonsillar branch (facial a.) — Palatopharyngeus m. — Tongue (retracted)

B

INTERIOR OF PHARYNX, SIDE VIEW
TONSIL AND MUCOUS MEMBRANE
REMOVED (after Grant)

C Superior constrictor

Palatoglossus m. — Tongue — Styloglossus m. — Hyoglossus m. — Glossopharyngeal n. — Stylohyoid ligament — Middle constrictor m.

Axis — Tonsillar branch (facial a.) — C3 — Palatopharyngeus m. — Stylopharyngeus m.

C

DEEP DISSECTION, TONSILLAR BED
(after Grant)

Pansky

36. MENINGES OF THE BRAIN AND THE DURAL REFLECTIONS

I. Meninges, the fibrous coverings of the brain
A. PIA: innermost, thin, vascular, and intimately related to brain
B. ARACHNOID: middle, thin; forms outer wall of subarachnoid space
C. DURA: outermost layer, is very tough; is separated from arachnoid by slight, subdural space. Has 2 layers: the outer, endostial layer is lightly adherent to bones, for which it serves as periosteum; inner, meningeal layer, is smooth with a mesothelial layer on its inner surface

II. Reflections or reduplication of the meningeal layer of dura
A. FALX CEREBRI extends caudally between cerebral hemispheres
   1. Attachments: to skull from crista galli to internal occipital protuberance, where it joins tentorium cerebelli. Its inferior border is free
B. TENTORIUM CEREBELLI extends horizontally between cerebellum and cerebrum
   1. Attachments: posterior and lateral to bone along transverse sinus; rostral and lateral to ridge of petrous bone and posterior clinoid process. Its free margin surrounds cerebral peduncles—this opening is the *incisura tentorii*. The falx cerebri is attached to its cephalic surface in the midline
C. FALX CEREBELLI projects between the cerebellar hemispheres
   1. Attachments: to internal occipital crest from protuberance to foramen magnum
D. DIAPHRAGMA SELLAE forms roof of sella turcica
   1. Attachments: to clinoid processes. Has opening for infundibulum

III. Dural venous sinuses (see p. 76)

IV. Arteries of meninges

Name	Source	Area	Entry
Meningeal br.	Occipital	Post. fossa	Jugular foramen
Post. meningeal	Ascend. pharyngeal	Post. fossa	Jugular foramen
Meningeal br.	Ascend. pharyngeal	Post. fossa	Hypoglossal foramen
Meningeal br.	Vertebral	Post. fossa	Foramen magnum
Meningeal br.	Ascend. pharyngeal	Middle fossa	Foramen lacerum
Middle meningeal	Maxillary	Middle fossa	Foramen spinosum
Access. meningeal	Maxillary	Middle fossa	Foramen ovale
Recurrent br.	Lacrimal	Ant. fossa	Sup. orbital fissure
Meningeal br.	Ant. ethmoid	Ant. fossa	Ant. ethmoid canal
Meningeal br.	Post. ethmoid	Ant. fossa	Post. ethmoid canal

V. Nerves of meninges

Name	Nerve	Distribution
Meningeal br.	Semilunar ganglion (V)	Tentorium and middle fossa
Recurrent br.	Ophthalmic (V)	Middle fossa
Tentorial br.	Ophthalmic (V)	Tentorium
Middle meningeal	Maxillary (V)	Middle fossa and ant. fossa
Meningeal ramus	Mandibular (V)	Middle and post. fossa
Meningeal brs.	Hypoglossal (C1 & C2)	Post. fossa

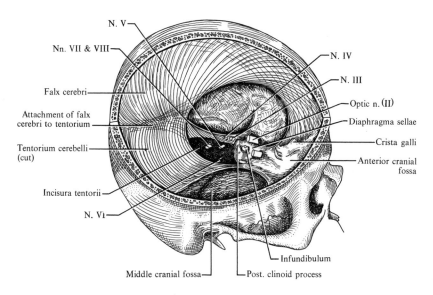

N. V

Nn. VII & VIII

Falx cerebri

Attachment of falx
cerebri to tentorium

Tentorium cerebelli
(cut)

Incisura tentorii

N. VI

N. IV

N. III

Optic n. (II)

Diaphragma sellae

Crista galli

Anterior cranial
fossa

Middle cranial fossa

Post. clinoid process

Infundibulum

CRANIAL DURAL PROCESSES

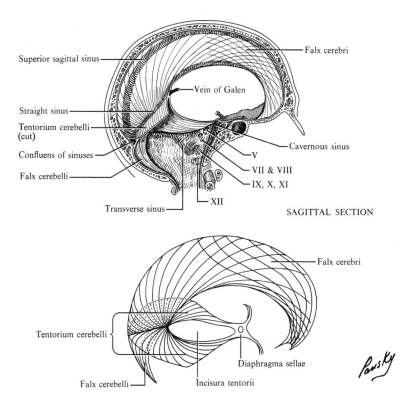

Superior sagittal sinus

Straight sinus

Tentorium cerebelli
(cut)

Confluens of sinuses

Falx cerebelli

Transverse sinus

Falx cerebri

Vein of Galen

Cavernous sinus

V

VII & VIII

IX, X, XI

XII

SAGITTAL SECTION

Falx cerebri

Tentorium cerebelli

Falx cerebelli

Incisura tentorii

Diaphragma sellae

DIAGRAMMATIC ILLUSTRATION OF CRANIAL DURA

37. VENOUS SINUSES AND DRAINAGE OF HEAD AND FACE

I. **Diploic veins:** a network between tables of the flat bones of skull. Communicate with meningeal veins and dural sinuses internally, veins of the scalp externally. Found in frontal, anterior and posterior temporal and occipital regions

II. **Emissary veins:** direct connections between dural sinuses and veins on outside of bone. Some major ones pass through foramen ovale, mastoid and parietal foramina, hypoglossal and condyloid canals

III. **Dural sinuses**
   A. SUPERIOR SAGITTAL: in convexity of falx cerebri from foramen cecum to internal occipital protuberance, where it deviates to right transverse sinus
      1. Receives superior cerebral veins, venous lacunae, arachnoid granulations, diploic and dural veins, and parietal emissary veins
   B. INFERIOR SAGITTAL: in free edge of falx cerebri. Ends in straight sinus
      1. Receives veins from falx cerebri and medial side of hemispheres
   C. STRAIGHT: along line of attachment of falx cerebri to tentorium cerebelli. Terminates in left transverse sinus
      1. Receives inferior sagittal sinus, great cerebral vein (of Galen), left transverse sinus and superior cerebellar veins
   D. TRANSVERSE: one on each side; begin at internal occipital protuberance; pass laterally in attached margin of tentorium to petrous bone, then bend caudally and medially as the sigmoid sinus to enter jugular foramina
      1. Receive superior sagittal, straight, and superior petrosal sinuses, mastoid and condyloid emissary veins, inferior cerebral and cerebellar veins, and diploic veins
   E. OCCIPITAL: in attached margin of falx cerebelli; begins at foramen magnum; ends at confluence of sinuses
   F. CONFLUENCE OF SINUSES: at internal occipital protuberance where superior sagittal, straight, occipital, and transverse sinuses meet
   G. CAVERNOUS: lie on each side of body of sphenoid bone
      1. Receive superior ophthalmic and cerebral veins
      2. Communicate with transverse sinus via superior petrosal sinus, and internal jugular vein via inferior petrosal sinus. With the internal carotid venous plexus and pterygoid venous plexus through foramina of Vesalii, ovale, and lacerum. And with the angular vein, through superior ophthalmic vein
   H. INTERCAVERNOUS: anterior and posterior, connecting cavernous sinuses, rostral and dorsal, to pituitary
   I. SUPERIOR PETROSAL: along superior border of petrous bone in attached margin of tentorium; join cavernous and transverse sinuses
      1. Receive cavernous sinus; cerebellar, inferior cerebral, and tympanic veins
   J. INFERIOR PETROSAL: along suture between basilar portion of occipital and petrous temporal bones; join cavernous sinus and internal jugular vein
      1. Receive cavernous sinus, internal auditory vein, inferior cerebellar veins, and veins from medulla
   K. BASILAR PLEXUS: network on basilar portion of occipital bone; interconnects inferior petrosal sinuses
      1. Communicates with anterior vertebral plexus

IV. **Clinical considerations**
   A. IN LIFE, blood flows freely through the sinuses. They are a surgical hazard at operation; e.g., the superior petrosal sinus is located immediately superior to the sensory root of the 5th cranial nerve. Surgical procedures on that nerve in this region must take this sinus into account

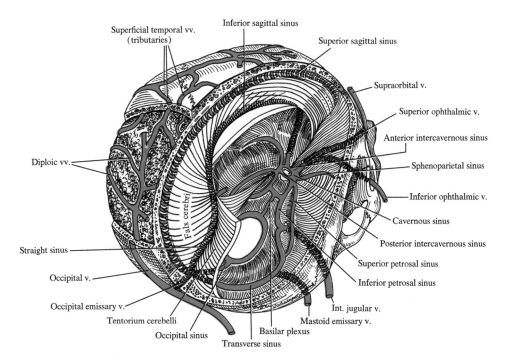

Superficial temporal vv. (tributaries)

Inferior sagittal sinus

Superior sagittal sinus

Supraorbital v.

Superior ophthalmic v.

Anterior intercavernous sinus

Sphenoparietal sinus

Inferior ophthalmic v.

Cavernous sinus

Posterior intercavernous sinus

Superior petrosal sinus

Inferior petrosal sinus

Diploic vv.

Falx cerebri

Straight sinus

Occipital v.

Occipital emissary v.

Tentorium cerebelli

Occipital sinus

Basilar plexus

Transverse sinus

Mastoid emissary v.

Int. jugular v.

VENOUS DRAINAGE OF HEAD

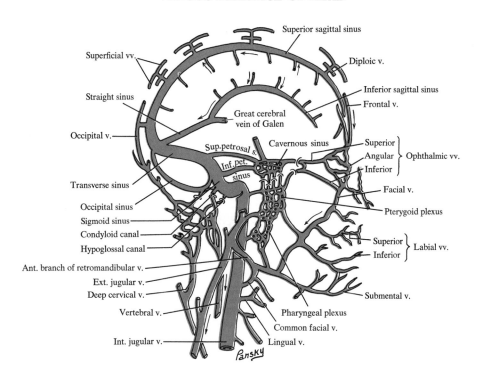

Superior sagittal sinus

Superficial vv.

Diploic v.

Straight sinus

Inferior sagittal sinus

Frontal v.

Great cerebral vein of Galen

Occipital v.

Sup. petrosal s.

Cavernous sinus

Superior

Angular

Ophthalmic vv.

Inferior

Inf. pet. sinus

Transverse sinus

Facial v.

Occipital sinus

Pterygoid plexus

Sigmoid sinus

Condyloid canal

Hypoglossal canal

Superior

Labial vv.

Inferior

Ant. branch of retromandibular v.

Ext. jugular v.

Deep cervical v.

Vertebral v.

Submental v.

Pharyngeal plexus

Common facial v.

Int. jugular v.

Lingual v.

Pansky

38. VENOUS DRAINAGE OF BRAIN AND SUBARACHNOID SPACE

I. **Cerebral veins:** thin-walled with no valves

A. EXTERNAL

1. Superior cerebral: 8–12 mainly in sulci; drain superior, medial, and lateral surfaces. End in superior sagittal sinus
2. Middle cerebral: begins on lateral side, runs in lateral fissure, and ends in cavernous sinus
3. Inferior cerebral: drains basal areas of hemispheres. From frontal lobe, passes to superior veins and superior sagittal sinus; from temporal lobe, joins middle cerebral and basilar veins
4. Basilar: formed by confluence of anterior cerebral vein, deep middle cerebral vein from insular area, and inferior striate veins from corpus striatum. Ends in internal cerebral veins

B. INTERNAL CEREBRAL VEINS: 2 in number. Formed by junction of terminal and choroid veins; run in roof of third ventricle to splenium of corpus callosum; join together to form the great cerebral vein (of Galen), which ends in straight sinus.

1. Terminal vein begins in groove between corpus striatum and thalamus
2. Choroid vein runs along choroid plexus and drains corpus callosum, fornix, and hippocampus

II. **Cerebellar veins:** in 2 groups

A. SUPERIOR

1. Superior cerebellar veins run rostrally and medially to straight sinus or internal cerebral veins; laterally, to transverse and superior petrosal sinuses

B. INFERIOR

1. Inferior cerebellar veins to transverse, occipital, and superior petrosal sinuses

III. **Subarachnoid space** (see p. 94): the space between arachnoid and pia, which receives cerebrospinal fluid from the fourth ventricle via the foramina of Magendie and Luschka

A. CISTERNS: areas of the subarachnoid space where the brain contour is greatly changed such that the pia and arachnoid diverge from each other

1. Cisterna cerebellomedullaris (magna): between undersurface of cerebellum and dorsum of medulla
2. Cisterna pontis: at lower border of pons
3. Interpeduncular cistern: between cerebral peduncles beneath midbrain
4. Superior cistern: between anterior cerebellum, tentorium, and roof of midbrain

IV. **Clinical considerations**

A. THE CISTERNA MAGNA is used as a site for puncture to remove cerebrospinal fluid samples

B. IN THE AREAS OF THESE CISTERNS, the increase in amount of fluid offers protection for brain parts

C. WHEN THE SUBARACHNOID SPACE IS OPENED, cerebrospinal fluid leaks out because the fluid is normally under pressure. After a surgical procedure, it is not possible to close the subarachnoid space by suturing the arachnoid because it is flimsy and transparent. Leakage of cerebrospinal fluid is prevented by carefully suturing the dura mater

D. THE CISTERNS can be studied by injecting air or radiopaque material into them (encephalography)

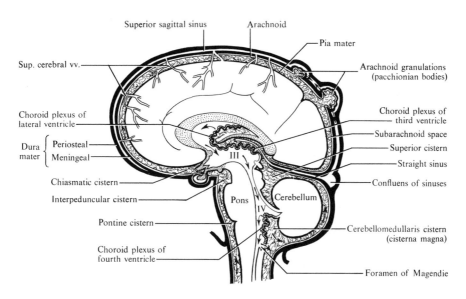

Superior sagittal sinus

Arachnoid

Pia mater

Sup. cerebral vv.

Arachnoid granulations (pacchionian bodies)

Choroid plexus of lateral ventricle

Choroid plexus of third ventricle

Dura mater { Periosteal / Meningeal }

Subarachnoid space

Superior cistern

Straight sinus

III

Chiasmatic cistern

Confluens of sinuses

Interpeduncular cistern

Pons

Cerebellum

Pontine cistern

IV

Cerebellomedullaris cistern (cisterna magna)

Choroid plexus of fourth ventricle

Foramen of Magendie

RELATIONS OF MENINGES TO BRAIN, CORD, AND CEREBROSPINAL FLUIDS
(from Rasmussen,
PRINCIPAL NERVOUS PATHWAYS)

Superior sagittal sinus

Ascending vv.

Ascending vv.

VENOUS DRAINAGE OF BRAIN

Middle cerebral v.

Transverse sinus

Descending vv.

Inf. anastomosing v. (Labbé)

VENOUS DRAINAGE – EXTERNAL BRAIN SURFACE

Ant. communicating v.

Ant. communicating a.

Ant. cerebral v.

Ant. cerebral a.

Int. carotid a.

Insular v.

Middle cerebral a.

Post. communicating v.

Post. communicating a.

Post. cerebral a.

Basilar v.

Basilar a.

Internal cerebral v.

Great cerebral v. (vein of Galen)

CIRCLE OF WILLIS AND ASSOCIATED VEINS

Pansky

-79-

39. BASAL VIEW OF THE BRAIN

I. Telencephalon hemispheres:
A. FRONTAL LOBE
   1. Frontal pole
   2. Orbital gyri
B. TEMPORAL LOBE
   1. Temporal pole
   2. Temporal gyri
C. RHINENCEPHALON (OLFACTORY BRAIN)
   1. Olfactory bulb
   2. Olfactory tract
   3. Anterior perforated substance

II. Diencephalon
A. TUBER CINEREUM
B. MAMMILLARY BODIES
C. HYPOPHYSIS
D. OPTIC CHIASMA
E. OPTIC TRACT

III. Mesencephalon
A. CEREBRAL PEDUNCLES
   1. Interpeduncular fossa

IV. Metencephalon
A. PONS
   1. Basilar sulcus
   2. Middle cerebellar peduncle (brachium pontis)
B. CEREBELLAR HEMISPHERES AND FLOCCULUS

V. Myelencephalon (medulla)
A. SULCI: anteromedian and anterolateral
B. PYRAMIDS
   1. Decussation of pyramids
   2. Inferior olive

VI. Nerves
A. OLFACTORY (I) attached to olfactory bulb
B. OPTIC (II)
C. OCULOMOTOR (III) emerging from the interpeduncular fossa
D. TROCHLEAR (IV) descends lateral to the cerebral peduncles
E. TRIGEMINAL (V) emerging from lateral side of pons
F. ABDUCENS (VI) emerging from the groove between pons and medulla
G. FACIAL (VII) AND VESTIBULOCOCHLEAR (VIII) join brain stem at the cerebellopontine angle
H. GLOSSOPHARYNGEAL (IX) AND VAGUS (X) emerge from lateral sulcus of medulla caudal to the facial (VII) nerve
I. SPINAL AND BULBAR PARTS OF ACCESSORY (XI) NERVE
   1. The bulbar part is associated with the vagus (X)
   2. The spinal part extends upward along the side of medulla
J. HYPOGLOSSAL (XII) emerges from sulcus between the inferior olive and the pyramid

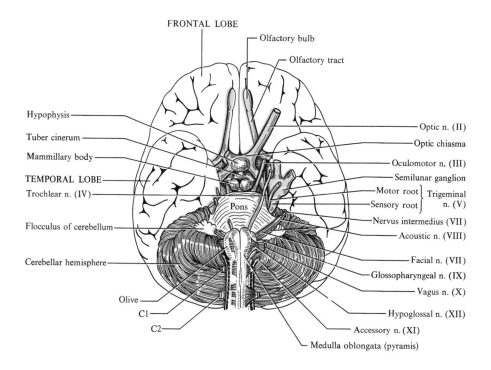

FRONTAL LOBE

Olfactory bulb

Olfactory tract

Hypophysis

Tuber cinerum

Mammillary body

TEMPORAL LOBE

Trochlear n. (IV)

Flocculus of cerebellum

Cerebellar hemisphere

Pons

Olive

C1

C2

Optic n. (II)

Optic chiasma

Oculomotor n. (III)

Semilunar ganglion

Motor root ⎱ Trigeminal
Sensory root ⎰ n. (V)

Nervus intermedius (VII)

Acoustic n. (VIII)

Facial n. (VII)

Glossopharyngeal n. (IX)

Vagus n. (X)

Hypoglossal n. (XII)

Accessory n. (XI)

Medulla oblongata (pyramis)

BRAIN - BASAL VIEW

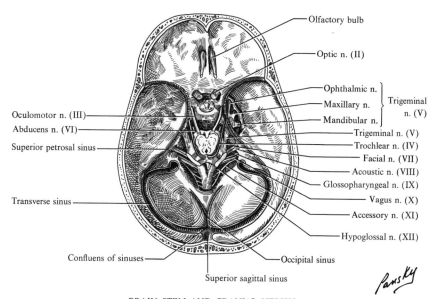

Olfactory bulb

Optic n. (II)

Ophthalmic n.

Maxillary n. ⎱ Trigeminal
 n. (V)

Mandibular n. ⎰

Oculomotor n. (III)

Abducens n. (VI)

Superior petrosal sinus

Transverse sinus

Trigeminal n. (V)

Trochlear n. (IV)

Facial n. (VII)

Acoustic n. (VIII)

Glossopharyngeal n. (IX)

Vagus n. (X)

Accessory n. (XI)

Hypoglossal n. (XII)

Confluens of sinuses

Occipital sinus

Superior sagittal sinus

Pansky

BRAIN STEM AND CRANIAL NERVES
IN RELATION TO SKULL

40. LATERAL VIEW OF THE BRAIN

I. **Poles:** frontal, temporal, and occipital

II. **Principal fissures or sulci**
A. LATERAL FISSURE (Sylvian): between temporal, frontal, and parietal lobes
1. Anterior part, the stem
2. Long posterior part, the posterior ramus
a. Near junction of 1 and 2, two short branches, anterior horizontal and anterior ascending, extend into frontal lobe
B. CENTRAL SULCUS (of Rolando) runs obliquely downward, from superior margin of brain, almost to lateral fissure

III. **Lobes**
A. FRONTAL LOBE, from frontal pole to central sulcus
1. Precentral sulcus, rostral and parallel to central. Superior and inferior frontal sulci run rostrally toward frontal pole
2. Precentral gyrus, between central and precentral sulci. Superior frontal gyrus lies above superior frontal sulcus; middle frontal gyrus, between superior and inferior sulci; inferior frontal gyrus, below inferior frontal sulcus. The anterior rami of lateral fissure usually divide this into triangular, opercular, and orbital portions
B. PARIETAL LOBE extends from the central sulcus to parieto-occipital fissure, and a line extending from this to the preoccipital notch
1. Postcentral sulcus lies parallel and dorsal to central sulcus. Intraparietal sulcus extends dorsally from middle of this to the transverse occipital sulcus
2. Postcentral gyrus lies between central and posterior central sulci. Superior and inferior parietal lobules lie above and below intraparietal sulcus
a. In the superior lobule, the parieto-occipital gyrus arches over the parieto-occipital sulcus
b. In the inferior lobule, the *supramarginal gyrus* caps upper end of lateral fissure; *angular gyrus* arches over superior temporal sulcus
C. OCCIPITAL LOBE, from occipital pole to parieto-occipital fissure and its extension to preoccipital notch
D. TEMPORAL LOBE, from temporal pole to line extending from the parieto-occipital fissure to preoccipital notch
1. Superior and middle temporal sulci divide lobe into superior, middle, and inferior temporal gyri
2. On the upper surface of superior temporal gyrus are the transverse temporal gyri
E. INSULA, at bottom of lateral fissure, covered by opercula, from frontal, parietal, and temporal lobes
1. Encircled by circular sulcus
2. Crossed by long and short gyri

IV. **Special features**
A. THE PRIMARY VOLUNTARY MOTOR CORTEX lies along the posterior part of the precentral gyrus adjoining the central sulcus
B. THE PRIMARY SENSORY CORTEX for the body (pain, temperature, touch) lies in the postcentral gyrus, some actually in the central sulcus
C. THE PRIMARY AUDITORY AREA lies on the cephalic border of the superior temporal gyrus in the depths of the lateral fissure
D. THE PRIMARY VISUAL CORTEX, as seen from the lateral side, lies at the occipital pole

BRAIN
LATERAL VIEW

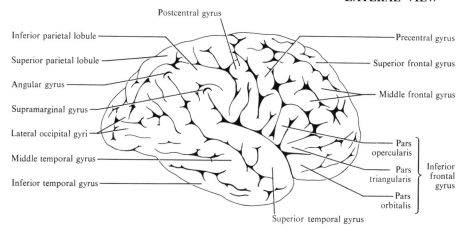

Postcentral gyrus

Inferior parietal lobule

Superior parietal lobule

Angular gyrus

Supramarginal gyrus

Lateral occipital gyri

Middle temporal gyrus

Inferior temporal gyrus

Precentral gyrus

Superior frontal gyrus

Middle frontal gyrus

Pars opercularis

Pars triangularis

Pars orbitalis

Inferior frontal gyrus

Superior temporal gyrus

PRINCIPAL GYRI

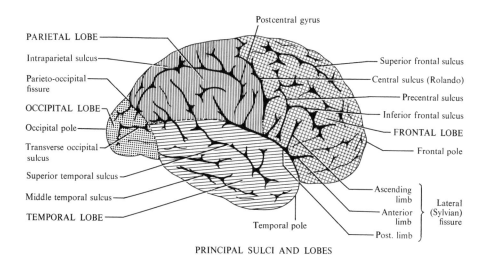

Postcentral gyrus

PARIETAL LOBE

Intraparietal sulcus

Parieto-occipital fissure

OCCIPITAL LOBE

Occipital pole

Transverse occipital sulcus

Superior temporal sulcus

Middle temporal sulcus

TEMPORAL LOBE

Superior frontal sulcus

Central sulcus (Rolando)

Precentral sulcus

Inferior frontal sulcus

FRONTAL LOBE

Frontal pole

Ascending limb

Anterior limb

Post. limb

Lateral (Sylvian) fissure

Temporal pole

PRINCIPAL SULCI AND LOBES

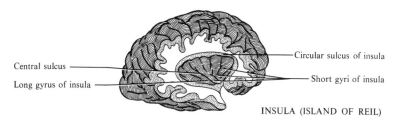

Central sulcus

Long gyrus of insula

Circular sulcus of insula

Short gyri of insula

INSULA (ISLAND OF REIL)

Pansky

41. MEDIAL VIEW OF THE BRAIN

I. Telencephalon (hemisphere)
A. OUTSTANDING LANDMARKS
  1. Corpus callosum with its genu, body, and splenium
  2. The fornix
  3. The septum pellucidum, stretched between 1 and 2
  4. Anterior commissure
B. PRINCIPAL FISSURES AND SULCI
  1. Calcarine fissure starts below splenium of corpus callosum and curves upward and posteriorly toward the occipital pole
  2. Cingulate sulcus curves parallel to the body of the corpus callosum to a point dorsal to the central sulcus and then curves cephalically to its superior margin
  3. Parieto-occipital begins near the middle of the calcarine fissure and runs upward to its superior margin
  4. Inferior temporal sulcus lies between the fusiform and inferior temporal gyri
  5. Collateral fissure runs from near occipital pole, parallel to brain margin almost to the temporal pole
C. LOBES
  1. Frontal bounded posteriorly by a line drawn obliquely downward and rostrally from the upper end of the central sulcus to the middle of the corpus callosum. Includes part of the cingulate gyrus and paracentral lobule
  2. Parietal bounded rostrally by the posterior line of the frontal lobe and posteriorly by the parieto-occipital fissure
  3. Occipital bounded rostrally by the parieto-occipital fissure and a line drawn caudally to the preoccipital notch
     a. The posterior part of the calcarine fissure divides its medial surface into the *cuneus* and *lingual* gyri
  4. Temporal bounded dorsally by the rostral boundary of the occipital lobe
     a. Fusiform gyrus lies between the inferior temporal sulcus and the collateral sulcus
     b. Medial to the collateral fissure are the lingual gyrus and the hippocampus
     c. Parahippocampal gyrus lies medial to the collateral fissure and is continuous with the cingular gyrus through the isthmus, which lies below the splenium of the corpus callosum. Rostrally, it is curved around as the uncus
        i. The hippocampal fissure extends from the splenium of the corpus callosum to the medial side of the uncus

II. Diencephalon: the following structures should be noted, namely, the choroid plexus of the third ventricle, pineal body, posterior commissure, interthalamic adhesion, hypothalamic sulcus, optic recess, optic chiasma, infundibulum, thalamus, hypothalamus, and third ventricle

III. Mesencephalon: the following structures should be noted, namely, the corpora quadrigemina (superior and inferior colliculi), cerebral aqueduct, and cerebral peduncles

IV. Metencephalon: the following structures should be noted, namely, the superior medullary velum, cerebellum, fourth ventricle, and pons

V. Myelencephalon: the following structures should be noted, namely, the medulla and the inferior medullary velum

VI. Special feature: primary visual cortex lies on both sides of the entire length of calcarine fissure

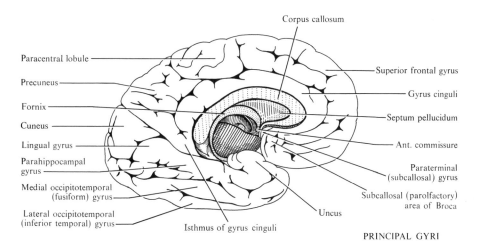

Corpus callosum

Paracentral lobule

Precuneus

Fornix

Cuneus

Lingual gyrus

Parahippocampal gyrus

Medial occipitotemporal (fusiform) gyrus

Lateral occipitotemporal (inferior temporal) gyrus

Isthmus of gyrus cinguli

Uncus

Superior frontal gyrus

Gyrus cinguli

Septum pellucidum

Ant. commissure

Paraterminal (subcallosal) gyrus

Subcallosal (parolfactory) area of Broca

PRINCIPAL GYRI

BRAIN — MEDIAL VIEW

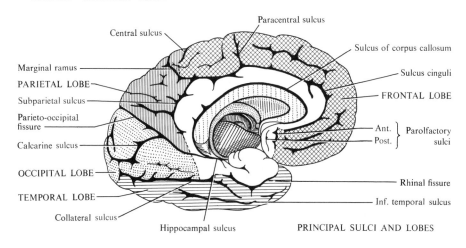

Paracentral sulcus

Central sulcus

Sulcus of corpus callosum

Marginal ramus

PARIETAL LOBE

Subparietal sulcus

Parieto-occipital fissure

Calcarine sulcus

OCCIPITAL LOBE

TEMPORAL LOBE

Collateral sulcus

Hippocampal sulcus

Sulcus cinguli

FRONTAL LOBE

Ant. } Parolfactory
Post. } sulci

Rhinal fissure

Inf. temporal sulcus

PRINCIPAL SULCI AND LOBES

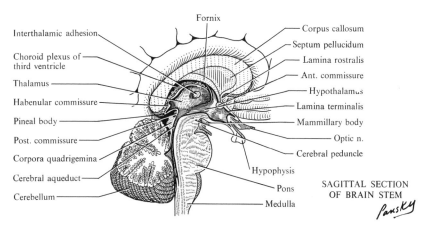

Fornix

Interthalamic adhesion

Choroid plexus of third ventricle

Thalamus

Habenular commissure

Pineal body

Post. commissure

Corpora quadrigemina

Cerebral aqueduct

Cerebellum

Corpus callosum

Septum pellucidum

Lamina rostralis

Ant. commissure

Hypothalamus

Lamina terminalis

Mammillary body

Optic n.

Cerebral peduncle

Hypophysis

Pons

Medulla

SAGITTAL SECTION
OF BRAIN STEM

Pansky

42. PITUITARY GLAND

I. Position, size, and relations

A. LOCATION: in the sella turcica (hypophysial fossa) of sphenoid bone, attached to the hypothalamus by a stalk (infundibulum). A shelf formed by the meningeal layer of dura stretches between the clinoid processes to cover the sella (diaphragma sellae) except for a small hole for the stalk. Periosteal layer of dura lines the sella

B. SIZE: larger in females, increases in pregnancy, and is largest in multiparous women
   1. Length (rostrocaudal): 1.0 cm
   2. Width (transverse): 1.2–1.5 cm
   3. Thickness: 0.5 cm
   4. Weight: 0.5–0.6 g

C. RELATIONS
   1. Laterally: internal carotid arteries
   2. Above: intercavernous (circular) sinuses in the diaphragma sellae and the base of the diencephalon
   3. In front and above: optic chiasma and optic tracts
   4. In front, below: sphenoid air sinus

II. Blood supply

A. ARTERIES
   1. Superior hypophysial arteries from internal carotid and posterior communicating arteries to stalk and adjacent anterior lobe
   2. Inferior hypophysial arteries from the internal carotid artery, passing through cavernous sinus, to posterior lobe
   3. Hypophysial portal veins carrying blood from stalk, lower hypothalamus, and pars tuberalis to most of anterior lobe

B. VEINS (LATERAL HYPOPHYSIAL) drain into cavernous and intercavernous sinuses

III. Parts

A. ANTERIOR LOBE: largest (75%), formed from buccal ectoderm
   1. Developmentally this includes pars distalis (anterior), pars tuberalis, and pars intermedia. All three parts comprise the adenohypophysis

B. POSTERIOR LOBE: formed from neural ectoderm of the floor of the diencephalon
   1. So-called neurohypophysis includes infundibular process (posterior lobe), neural stalk (tuber cinereum and infundibular stem), and the median eminence of the hypothalamus

IV. Principal hormones and their effects

A. ANTERIOR LOBE: somatotropic—promotes body growth; thyrotropic—stimulates thyroid secretion; adrenocorticotropic (ACTH)—stimulates secretion of most of adrenal steroid hormones; two gonadotropic hormones, FSH and LH, which stimulate development of ovarian follicles and secretion of corpora lutea, respectively; prolactin—promotes secretion of mammary gland

B. POSTERIOR LOBE: oxytocin—stimulates contraction of smooth muscle, especially of uterus; vasopressin—raises blood pressure and inhibits diuresis

V. Clinical considerations

A. GROSS ENLARGEMENT (TUMOR): can put pressure directly on internal carotid arteries, giving symptoms of occlusion; direct pressure on optic chiasma, causing bitemporal hemianopsia. Oversecretion of anterior lobe can cause gigantism, with hyperglycemia; overactivity of adrenals, leading to Cushing's syndrome; overactivity of thyroid, which may be a factor in Graves' disease

B. INSUFFICIENCY: may lead to dwarfism; lower metabolism due to low thyroid activity; symptoms of adrenal insufficiency; and underdevelopment of the genital system

PITUITARY GLAND

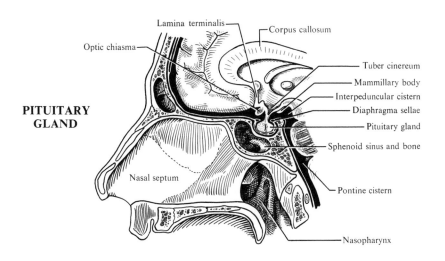

Lamina terminalis — Corpus callosum
Optic chiasma —
Tuber cinereum
Mammillary body
Interpeduncular cistern
Diaphragma sellae
Pituitary gland
Sphenoid sinus and bone
Nasal septum
Pontine cistern
Nasopharynx

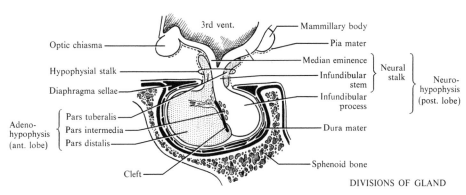

3rd vent.
Optic chiasma ——— Mammillary body
Pia mater
Median eminence
Hypophysial stalk ——— Infundibular stem — Neural stalk — Neuro-hypophysis (post. lobe)
Diaphragma sellae ——— Infundibular process
Adeno-hypophysis (ant. lobe) — Pars tuberalis
Pars intermedia
Pars distalis — Dura mater
Cleft — Sphenoid bone

DIVISIONS OF GLAND

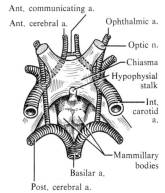

Ant. communicating a.
Ant. cerebral a. — Ophthalmic a.
Optic n.
Chiasma
Hypophysial stalk
Int. carotid a.
Mammillary bodies
Basilar a.
Post. cerebral a.

RELATIONSHIP TO
MAJOR VESSELS

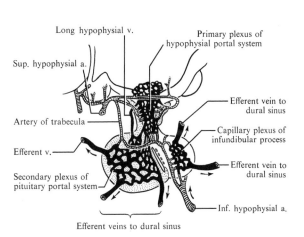

Long hypophysial v. — Primary plexus of hypophysial portal system
Sup. hypophysial a. — Efferent vein to dural sinus
Artery of trabecula — Capillary plexus of infundibular process
Efferent v. — Efferent vein to dural sinus
Secondary plexus of pituitary portal system — Inf. hypophysial a.
Efferent veins to dural sinus

VESSELS OF PITUITARY

-87-

43. BRAIN STEM (BASAL GANGLIA TO MEDULLA)

I. Diencephalon
A. OBSERVE caudally directed pineal body, pulvinar (large bilateral projections of dorsal thalamus), and medial geniculate bodies beneath the pulvinar

II. Mesencephalon
A. OBSERVE the tectum (roof), composed of 2 pairs of swellings—the corpora quadrigemina (the cephalic pair are the superior colliculi, the caudal pair are the inferior colliculi); the brachium of the inferior colliculus; and the brachium conjunctivum

III. Metencephalon (with cerebellum cut away)
A. OBSERVE the three brachia of cerebellum, namely, the superior (conjunctivum), middle (pontis), and inferior (restiform body)
B. STRUCTURES IN THE FLOOR OF THE FOURTH VENTRICLE: median fissure, sulcus limitans, medial eminence, facial colliculus, locus ceruleus, medullary stria and vestibular area

IV. Myelencephalon
A. STRUCTURES IN THE FLOOR OF THE FOURTH VENTRICLE: median fissure, sulcus limitans, vestibular area, hypoglossal trigone, and vagal trigone (ala cinerea)
B. OBSERVE the *taenia,* attachments of the ventricular roof to the brain stem with their horizontal part and caudally directed apex (the *calamus scriptorius*)
C. IN THE REGION CAUDAL TO THE TAENIA, OBSERVE:
1. Sulci: posterior median, intermediate, and lateral
2. Fasciculi: gracilis, between median and intermediate sulci, ending rostrally in a swelling, the tubercle of nucleus gracilis; and the cuneatus, between the intermediate and lateral sulci, ending rostrally in an enlargement, the *cuneate tubercle* (tubercle of cuneate nucleus)

V. Nerves
A. TROCHLEAR (IV), emerging from roof, caudal to inferior colliculus
B. GLOSSOPHARYNGEAL (IX)
C. VAGUS (X)
D. ACCESSORY (XI)

VI. Cerebellum
A. LIES BELOW the posterior portions of the cerebral hemispheres and is separated from them by the tentorium cerebelli
B. CONSISTS OF paired lateral parts, *the cerebellar hemispheres* (consisting of an anterior and posterior lobe); and a smaller midline portion, *the vermis*
C. LIES IN the posterior cranial fossa
D. HAS A CORTEX, with folia and fissures, that covers a much larger center of white matter. It also has certain masses of gray matter, *the cerebellar nuclei,* embedded in it
E. NO CRANIAL NERVE is directly attached to the cerebellum

VII. Internal capsule: this is a large, fan-shaped band of fibers passing to and from the hemispheres. The thalamus lies to its medial side and the lenticular nucleus on its lateral side

VIII. Clinical considerations
A. TRAUMA OR DISEASE IN THE MYELENCEPHALON is often fatal because vital centers, such as those controlling circulation and respiration, are located here
B. A SINGLE LESION IN THE INTERNAL CAPSULE can result in complete unilateral motor and sensory loss
C. DAMAGE TO THE CEREBELLUM may result in disturbances of voluntary movement

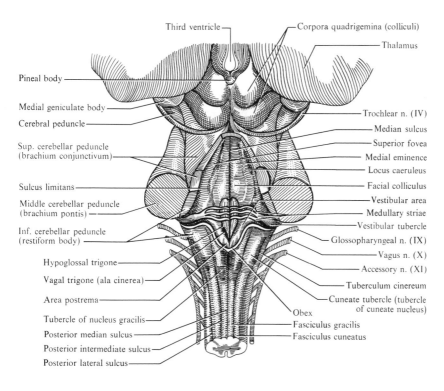

Third ventricle — — Corpora quadrigemina (colliculi)

— Thalamus

Pineal body —

Medial geniculate body —

Cerebral peduncle —

Sup. cerebellar peduncle
(brachium conjunctivum) —

Sulcus limitans —

Middle cerebellar peduncle
(brachium pontis) —

Inf. cerebellar peduncle
(restiform body) —

Hypoglossal trigone —

Vagal trigone (ala cinerea) —

Area postrema —

Tubercle of nucleus gracilis —

Posterior median sulcus —

Posterior intermediate sulcus —

Posterior lateral sulcus —

— Trochlear n. (IV)

— Median sulcus

— Superior fovea

— Medial eminence

— Locus caeruleus

— Facial colliculus

— Vestibular area

— Medullary striae

— Vestibular tubercle

— Glossopharyngeal n. (IX)

— Vagus n. (X)

— Accessory n. (XI)

— Tuberculum cinereum

— Cuneate tubercle (tubercle
of cuneate nucleus)

Obex

— Fasciculus gracilis

— Fasciculus cuneatus

DORSAL VIEW (CEREBELLUM REMOVED
TO SHOW RHOMBOID FOSSA)

BRAIN STEM

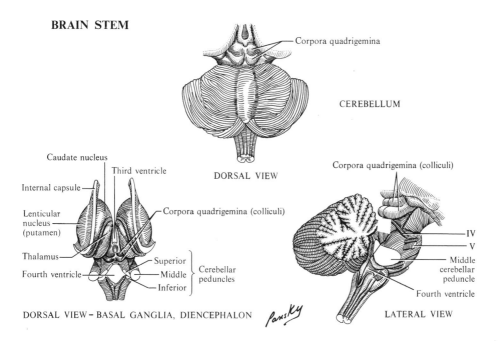

— Corpora quadrigemina

CEREBELLUM

DORSAL VIEW

Caudate nucleus

Internal capsule —

Lenticular
nucleus —
(putamen)

Thalamus —

Fourth ventricle —

Third ventricle

— Corpora quadrigemina (colliculi)

Superior
Middle — Cerebellar
Inferior — peduncles

DORSAL VIEW — BASAL GANGLIA, DIENCEPHALON

Pansky

Corpora quadrigemina (colliculi)

— IV
— V
— Middle
cerebellar
peduncle

Fourth ventricle

LATERAL VIEW

-89-

44. ARTERIES OF THE BRAIN

I. Arteries
A. INTERNAL CAROTIDS enter middle cranial fossa to lie in grooves on lateral sides of sphenoid. Branches:
  1. Anterior cerebral runs rostrally over optic tract to longitudinal fissure
    a. Anterior communicating join in fissure
    b. Branches to medial side of frontal and parietal lobes; anterior perforated substance; body and rostrum of corpus callosum and septum pellucidum
    c. Medial striate (inconstant) to caudate nucleus, putamen, and anterior limb of internal capsule
  2. Middle cerebral (largest) runs in lateral cerebral fissure over insula and is distributed to lateral surface of hemisphere
    a. Lateral striate to basal ganglia and internal capsule
  3. Posterior communicating runs dorsally to join posterior cerebral branch of basilar
    a. Branches pass medially to internal capsule, thalamus, and walls of third ventricle
  4. Anterior choroidal pass along optic tract, around cerebral peduncles to level of lateral geniculate bodies, then enter choroid plexus of lateral ventricle
    a. Branches to optic tract, hippocampus, caudate nucleus, internal capsule, and cerebral peduncles
B. VERTEBRAL ARTERIES enter posterior fossa through foramen magnum. Join to form basilar artery just below pons. Branches:
  1. Anterior spinal descend on cord ventrally
  2. Posterior spinal descend on cord dorsally
  3. Posterior inferior cerebellar: to cerebellum and choroid plexus of fourth ventricle
C. BASILAR ARTERY formed by union of vertebrals and runs in basilar groove of pons. Branches:
  1. Pontine to pons
  2. Labyrinthine (internal auditory) to inner ear
  3. Anterior inferior cerebellar to cerebellum
  4. Superior cerebellar to upper cerebellum, corpora quadrigemina, pineal body, and choroid plexus of third ventricle
  5. Posterior cerebral: terminal branch of basilar, receives posterior communicating from internal carotid; supplies inferior and medial surfaces of occipital and temporal lobes
    a. Posterior choroid branches to choroid plexuses of third and lateral ventricles

II. Circle of Willis. A circular anastomosis formed by the posterior communicating arteries, the posterior cerebral arteries, the anterior cerebral arteries, and the anterior communicating arteries

III. Clinical considerations
A. NERVOUS TISSUE extremely sensitive to lack of oxygen
B. ON SURFACE OF BRAIN, arterial anastomoses are numerous; in substance of central nervous system, anastomoses are rare, and those present are small. Thus, occlusion or rupture of a vessel can lead to widespread and permanent destruction of nervous tissue. The damage is permanent, since nerve cells do not reproduce
C. THE STRIATE ARTERIES are frequently involved in cerebrovascular accidents
D. "CONGENITAL" ANEURYSM (berry or miliary aneurysm) occurs usually on the circle of Willis, especially the anterior communicating artery, the middle or anterior cerebral arteries, or the basilar artery, usually at or near a point of bifurcation. The media of the vessel is normally defective. In 20% of cases, they are multiple. Rupture is found in about 75% of cases of cerebral arterial aneurysm

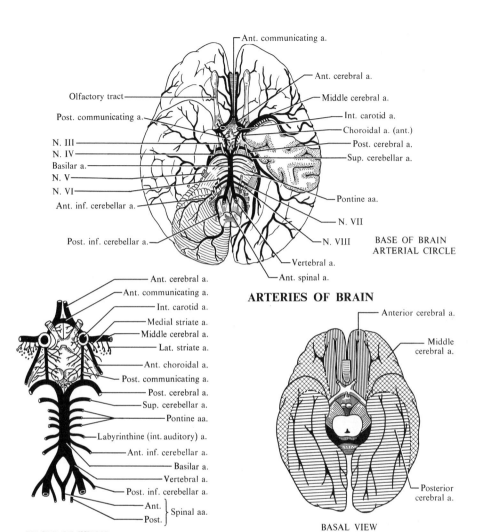

Ant. communicating a.

Olfactory tract

Post. communicating a.

N. III
N. IV
Basilar a.
N. V
N. VI
Ant. inf. cerebellar a.

Post. inf. cerebellar a.

Ant. cerebral a.
Middle cerebral a.
Int. carotid a.
Choroidal a. (ant.)
Post. cerebral a.
Sup. cerebellar a.

Pontine aa.

N. VII

N. VIII

Vertebral a.

Ant. spinal a.

BASE OF BRAIN
ARTERIAL CIRCLE

ARTERIES OF BRAIN

Ant. cerebral a.
Ant. communicating a.
Int. carotid a.
Medial striate a.
Middle cerebral a.
Lat. striate a.
Ant. choroidal a.
Post. communicating a.
Post. cerebral a.
Sup. cerebellar a.
Pontine aa.
Labyrinthine (int. auditory) a.
Ant. inf. cerebellar a.
Basilar a.
Vertebral a.
Post. inf. cerebellar a.
Ant.
Post. } Spinal aa.

CIRCLE OF WILLIS

Anterior cerebral a.

Middle
cerebral a.

Posterior
cerebral a.

BASAL VIEW

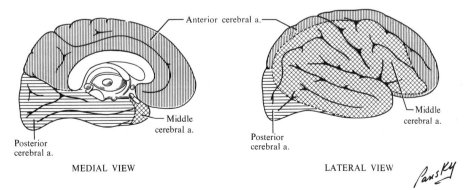

Anterior cerebral a.

Middle
cerebral a.

Posterior
cerebral a.

Middle
cerebral a.

Posterior
cerebral a.

MEDIAL VIEW

LATERAL VIEW

Pansky

45. OCCLUSION OF MAJOR ARTERIES TO BRAIN

Artery and Findings	Area Involved in Lesion
I. Anterior cerebral artery	
A. CONTRALATERAL MONOPLEGIA (LEG)	Paracentral lobule
B. CONTRALATERAL SENSORY LOSS	Thalamocortical radiations
C. IF ON DOMINANT SIDE	
1. Mental confusion	Frontal lobe
2. Apraxia, aphasia	Corpus callosum, cortical speech area
II. Anterior choroidal artery	
A. HOMONYMOUS HEMIANOPSIA	Geniculocalcarine tract (optic radiation)
B. CONTRALATERAL	
1. Hemiplegia	Corticospinal fibers of internal capsule
2. Hemianesthesia	Posterior limb of internal capsule
III. Middle cerebral artery	
A. HOMONYMOUS HEMIANOPSIA	Optic tract
B. CONTRALATERAL	
1. Hemiplegia and hemianesthesia	Ant. and post. limbs of internal capsule
C. IF ON DOMINANT SIDE	
1. Global aphasia	Motor and sensory speech areas
IV. Posterior cerebral artery	
A. HOMONYMOUS HEMIANOPSIA	Optic radiations
B. CONTRALATERAL	
1. Hemiplegia, ataxia	Internal capsule, spinocerebellar tract
2. Impaired sensation	Posterolateral ventral nucleus of thalamus
3. Burning pain	Dorsal nucleus of thalamus
4. Choreoathetoid movements	Red nucleus
V. Superior cerebellar artery	
A. HOMOLATERAL	
1. Cerebellar ataxia	Spinocerebellar tract
2. Choreiform movements	Red nucleus
3. Horner's syndrome	Reticular formation
B. CONTRALATERAL	
1. Loss of pain and temperature, face and body	Spinal nucleus of V. and lemniscus system
2. Central facial weakness	Corticobulbar fibers to nucleus of VII
3. Partial deafness	Lateral lemniscus
VI. Anterior inferior cerebellar artery	
A. HOMOLATERAL	
1. Cerebellar ataxia, deafness	Spinocerebellar tract, cochlear nuclei
2. Loss of sensation, face	Spinal tract and nucleus of V
B. CONTRALATERAL	
1. Loss of pain and temp., body	Lemniscus system (spinothalamics)
VII. Posterior inferior cerebellar artery	
A. HOMOLATERAL	
1. Cerebellar ataxia, nystagmus	Spinocerebellar tract
2. Horner's syndrome	Reticular formation
3. Loss of sensation, face	Spinal tract and nucleus of V
4. Dysphagia and dysphonia	Nucleus ambiguus, to IX and to X
B. CONTRALATERAL	
1. Loss of pain and temp., body	Lateral spinothalamic tract

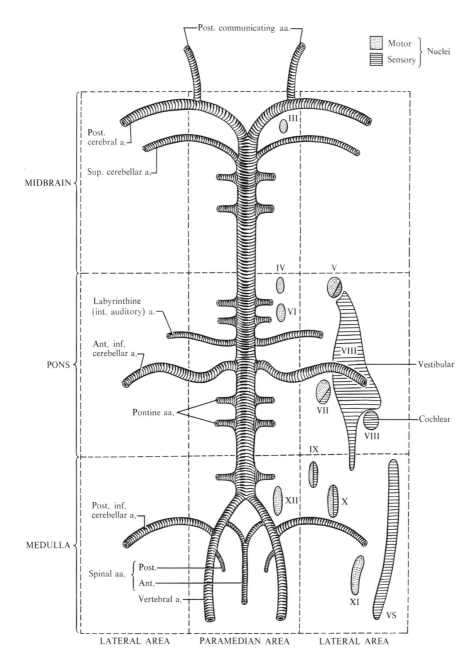

SCHEME OF VASCULAR SUPPLY
AT BASE OF BRAIN WITH CRANIAL NUCLEI
(AFTER HOLTZMAN, PANIN, AND EBEL)

46. VENTRICULAR SYSTEM AND CEREBROSPINAL FLUID

I. The ventricular system
A. LATERAL (one in each cerebral hemisphere)
1. Anterior horn: rostral to interventricular foramen, between septum pellucidum medially and caudate nucleus laterally; roofed by corpus callosum
2. Body (pars centralis): posterior from interventricular foramen to level of splenium of corpus callosum; medial wall and roof, as above. Laterally are caudate nucleus, stria terminalis, dorsal thalamus, choroid plexus, and fornix
3. Collateral trigone: area at end of body from which the two horns arise
4. Post. horn: extends posteriorly into occipital lobe. Roof and lat. wall formed by tapetum of corpus callosum. Medial wall shows bulb of post. horn and calcar avis
5. Inferior horn curves around and extends rostrally into temporal lobe. Roof is white matter of lobe, the stria terminalis, and tail of caudate nucleus. The amygdala is a swelling in rostral end of horn. Medial wall formed by collateral eminence, hippocampus, choroid plexus, and fimbria
6. Openings: interventricular foramina from lateral ventricles to 3rd ventricle
B. THIRD VENTRICLE
1. Single, narrow cavity, bridged by interthalamic adhesion of thalamus. Lateral wall is formed by thalamus and hypothalamus. Roof is formed by choroid tela, with choroid plexus. Floor contains optic chiasma, infundibulum, tuber cinereum, mammillary bodies, and subthalamus
2. Openings: 2 interventricular foramina laterally; caudally the cerebral aqueduct
C. CEREBRAL AQUEDUCT (of Sylvius)
1. A triangular canal in the midbrain. Begins at 3rd and ends in 4th ventricle
D. FOURTH VENTRICLE
1. Between cerebellum posteriorly and medulla and pons anteriorly. Its floor is the rhomboid fossa, bounded by sup. and inf. cerebellar peduncles, cuneate tubercles, and clavae. For structures in floor, see p. 89. The roof is formed by sup. medullary velum, cerebellum, inf. medullary velum, and choroid tela
2. Openings: rostrally the cerebral aqueduct; caudally closed part of medulla; foramina of Luschka (lateral apertures of 4th ventricle) from lateral angles of ventricles to subarachnoid space; and foramen of Magendie (median aperture of 4th ventricle) a median opening to subarachnoid space

II. Choroid tela—fusion of neural ependymal lining of the ventricle with pia mater

III. Choroid plexus: network of enlarged blood vessels in tela, which invaginate into the ventricles
A. LOCATIONS: into each of lateral ventricles, through choroid fissure and through roof, into 3rd and 4th ventricles

IV. Cerebrospinal fluid
A. FORMATION: through hydrostatic pressure in highly convoluted choroid plexus, passes caudally through ventricles and out via foramina of Luschka and Magendie into subarachnoid space
B. DRAINAGE: into arachnoid villi to discharge into sup. sagittal sinus
C. FUNCTION: mainly protection, especially supporting the basal aspects of brain
D. CLINICAL IMPORTANCE
1. Substances can be added to it via puncture for anesthesia
2. Fluid can be withdrawn for diagnostic purposes
3. Hydrocephalus: overproduction or failure of the cerebrospinal fluid to drain results in distention of the ventricles and subsequent enlargement of the head

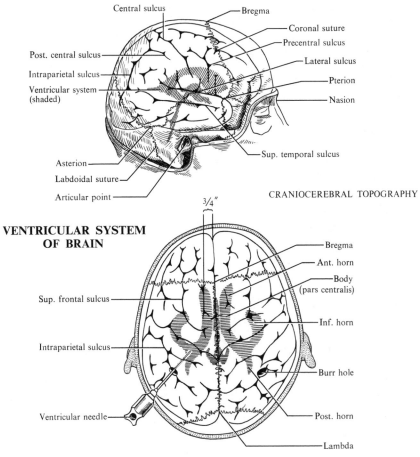

VENTRICULAR SYSTEM OF BRAIN

Central sulcus
Bregma
Coronal suture
Precentral sulcus
Lateral sulcus
Pterion
Nasion
Post. central sulcus
Intraparietal sulcus
Ventricular system (shaded)
Asterion
Labdoidal suture
Articular point
Sup. temporal sulcus

CRANIOCEREBRAL TOPOGRAPHY

3/4″
Bregma
Ant. horn
Body (pars centralis)
Sup. frontal sulcus
Inf. horn
Intraparietal sulcus
Burr hole
Ventricular needle
Post. horn
Lambda

DORSAL VIEW – SURFACE PROJECTION
OF VENTRICULAR SYSTEM

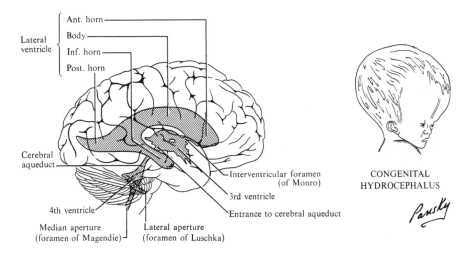

Ant. horn
Body
Lateral ventricle
Inf. horn
Post. horn
Cerebral aqueduct
Interventricular foramen (of Monro)
3rd ventricle
Entrance to cerebral aqueduct
4th ventricle
Median aperture (foramen of Magendie)
Lateral aperture (foramen of Luschka)

CONGENITAL
HYDROCEPHALUS

Pansky

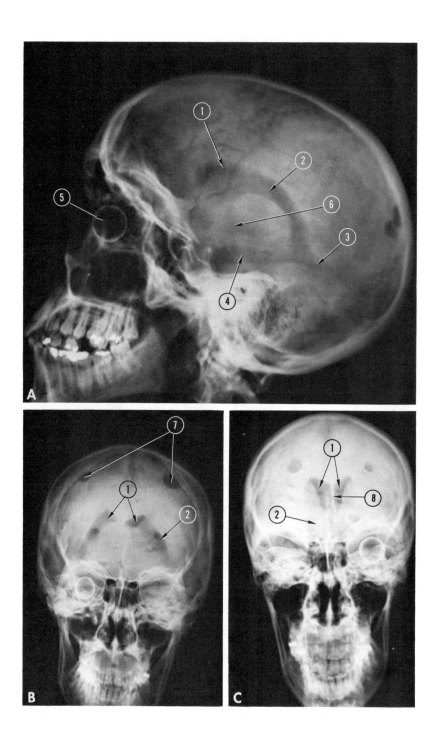

FIGURE 3. **Pneumoencephalogram. A, Lateral view; B, anterior view; C, posterior view.**
1, Anterior horn of lateral ventricle; *2,* body of lateral ventricle; *3,* posterior horn of lateral ventricle; *4,* inferior horn of lateral ventricle; *5,* glass eye; *6,* third ventricle; *7,* burr holes; *8,* septum pellucidum.

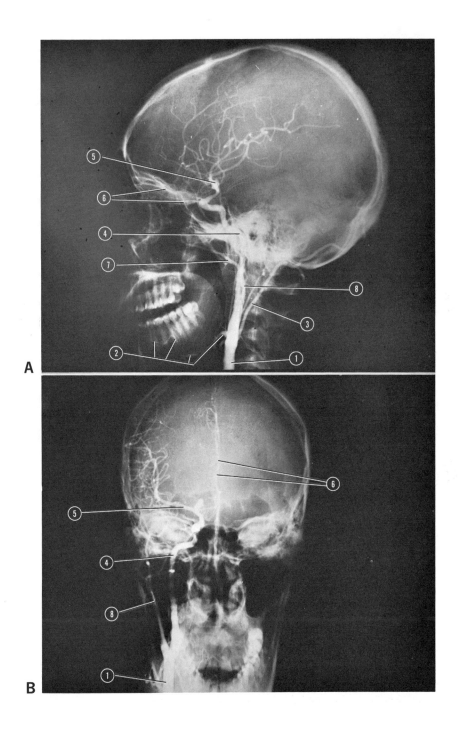

FIGURE 4. **Carotid arteriogram. A, Lateral view; B, anterior view.** *1*, Common carotid a.; *2*, facial a.; *3*, occipital a.; *4*, internal carotid a.; *5*, middle cerebral a.; *6*, anterior cerebral a.; *7*, maxillary a.; *8*, external carotid a.

47. SURFACE ANATOMY OF HEAD AND NECK

I. Vessels

A. ARTERIES
1. Common carotid: on a line from upper border of sternal end of clavicle to a point midway between apex of mastoid and angle of mandible
2. Subclavian: indicated by an arch, the medial end at the sternoclavicular articulation and the lateral end at middle of clavicle
3. Facial (on face): on a line from the facial groove on the inferior border of the mandible, 1 in. from the angle, to the medial corner of the eye, the line passing ½ in. lateral to the angle of the mouth
4. Superficial temporal: runs upward just in front of the ear, crossing posterior root of the zygomatic process
5. Middle meningeal and its anterior branch: by a line beginning at midpoint of the zygomatic arch, curving slightly forward, then back through the pterion,* then upward and backward toward the vertex

B. VEINS
1. Internal jugular: follows same line as internal carotid artery
2. Superior sagittal sinus: from nasion to external occipital protuberance, running in the midsagittal plane
3. Transverse sinus: horizontal part—on a line from the external occipital protuberance laterally, approximating the superior nuchal line, to a point just posterosuperior to the external auditory meatus: sigmoid part—begins with the line above and is continued vertically downward to the level of lower border of external auditory meatus

II. Nerves

A. VAGUS: same line as internal carotid artery
B. ACCESSORY: passes under anterior border of sternocleidomastoid, 3.75 cm (1.5 in.) below tip of mastoid; emerges from posterior border of that muscle at junction of upper and middle thirds; passes obliquely downward and backward across posterior triangle to pass under anterior border of trapezius, 5 cm (2 in.) above clavicle
C. PHRENIC: begins at level of middle of lamina of thyroid cartilage, and its caudal course is indicated by a line down the middle of the sternocleidomastoid, parallel to the direction of the muscle

III. Viscera and Sinuses

A. BRAIN
1. Lateral fissure begins at pterion* with the posterior ramus extending upward and backward to the parietal eminence
2. Central sulcus: a line extending from a point halfway between nasion and external occipital protuberance downward and forward to a point 1 in. behind the pterion*
3. Base of cerebrum: this lies one fingerbreadth above Reid's line, a line drawn between lower border of the orbit and the auricular point (center of external auditory meatus)
B. THYROID GLAND: upper pole at junction of middle and caudal thirds of lamina of thyroid cartilage, lateral and caudal to prominence; caudal pole at level of 5th or 6th tracheal ring; isthmus covering tracheal rings 2–4 across the midline.
C. AIR SINUSES OR PARANASAL SINUSES: vary greatly in size, shape, and position. The *frontal sinus* occupies the area in the bone deep to the medial part of the superciliary ridge. The *maxillary sinus* occupies the body of the maxilla, the area between the orbit, nasal cavity, and upper teeth

---

*Pterion—located 35 mm behind and 12 mm above the frontozygomatic suture.

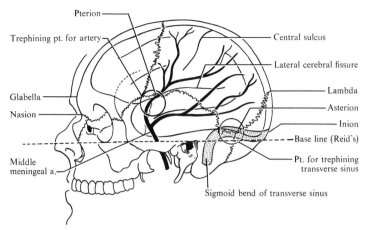

Pterion
Trephining pt. for artery
Central sulcus
Lateral cerebral fissure
Glabella
Nasion
Lambda
Asterion
Inion
Base line (Reid's)
Middle meningeal a.
Pt. for trephining transverse sinus
Sigmoid bend of transverse sinus

SURFACE RELATIONS

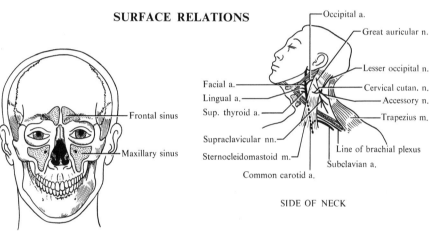

Frontal sinus
Maxillary sinus

Occipital a.
Great auricular n.
Lesser occipital n.
Facial a.
Lingual a.
Sup. thyroid a.
Cervical cutan. n.
Accessory n.
Trapezius m.
Supraclavicular nn.
Sternocleidomastoid m.
Line of brachial plexus
Subclavian a.
Common carotid a.

SIDE OF NECK

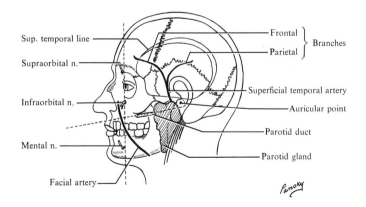

Sup. temporal line
Supraorbital n.
Infraorbital n.
Mental n.
Facial artery
Frontal
Parietal
} Branches
Superficial temporal artery
Auricular point
Parotid duct
Parotid gland

Pansky

48. EYELID AND LACRIMAL GLAND

I. Eye, from in front
A. PARTS OF THE EYEBALL: pupil and iris are seen through the transparent cornea, which is surrounded by the white sclera
B. EYELIDS with *cilia, palpebral fissure* (rima) between lids, *medial and lateral commissures* (*canthi*) at respective corners; medial is prolonged toward nose, leaving triangular space between lids, the *lacus lacrimalis* (lacrimal lake). On each lid, at margin of lacus is an elevation, the *lacrimal papilla,* surmounted with small orfice, the *lacrimal punctum*

II. Structure of lid
A. GROSS: eyelashes (cilia), short hairs, set in double or triple rows; near attachment of lash, openings of ciliary glands (Zeis or Moll) also in rows; along border of lid, behind cilia and glands, single row of openings of tarsal (meibomian) glands. *Lacrimal caruncle* in lacus, with *plica semilunaris* of conjunctiva lateral to it
B. MICROSCOPIC, from without inward
  1. Skin very thin, continuous with conjunctiva at margins
  2. Loose connective tissue, without fat
  3. Orbicularis oculi, skeletal (voluntary) muscle running longitudinally
  4. Tarsus with orbital septum
     a. Tarsus (tarsal plates): thin plates of dense connective tissue; upper is larger, attached to palpebral ligaments laterally and medially
     b. Orbital septum: fibrous membrane attached to margin of bony orbit where it is continuous with periorbita (periosteum). In upper lid, joins tendon of levator palpebrae muscle and attaches to superior tarsus. In lower lid, also attaches to tarsus. Pierced by vessels and nerves from orbit to face
  5. Tarsal glands (meibomian): between tarsi and conjunctiva; modified sebaceous glands
  6. Conjunctiva, mucous membrane

III. Lacrimal gland (p. 103) is located in lacrimal fossa of roof of orbit. Has a superior part attached to roof and lies on superior and lateral recti muscles; and an inferior part in lateral part of upper lid. In structure, like serous salivary glands. It has 6 to 12 excretory ductules which open into superior fornix of conjunctiva

IV. Clinical considerations
A. STY: an inflammation at the edge of lid due to infection of the ciliary glands
B. CHALAZION: due to an obstruction of the ducts of the meibomian glands causing a cyst or enlargement of the glands
C. CONJUNCTIVITIS (pink-eye): inflammation of conjunctiva
D. PTOSIS: drooping of upper lid usually due to an involvement of the oculomotor (III) nerve
E. LESIONS OF THE FACIAL NERVE: eliminate the blink reflex of the eye
F. A CHARACTERISTIC FEATURE of mongoloid facies is the absence of the upper lid fold, known to some as the orbitopalpebral sulcus—the skin of the eyelid does not slip beneath the shelter of the superior orbital margin but continues flat to the forehead
G. THE EYELID was known to the Greeks as the *blepharon*—hence, the term *blepharitis,* meaning inflammation of the eyelids
H. MOST OF THE LYMPHATIC DRAINAGE OF THE LIDS is downward and backward toward the parotid nodes, but some of it goes downward along the angular and facial vessels to the submandibular lymph nodes

EYELID AND
LACRIMAL GLAND

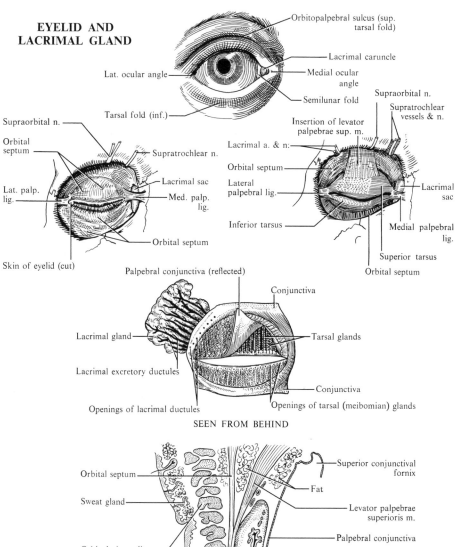

Orbitopalpebral sulcus (sup. tarsal fold)

Lacrimal caruncle

Lat. ocular angle

Medial ocular angle

Semilunar fold

Tarsal fold (inf.)

Supraorbital n.

Supratrochlear vessels & n.

Supraorbital n.

Orbital septum

Supratrochlear n.

Lacrimal sac

Lat. palp. lig.

Med. palp. lig.

Insertion of levator palpebrae sup. m.

Lacrimal a. & n.

Orbital septum

Lateral palpebral lig.

Lacrimal sac

Inferior tarsus

Medial palpebral lig.

Superior tarsus

Orbital septum

Orbital septum

Skin of eyelid (cut)

Palpebral conjunctiva (reflected)

Conjunctiva

Lacrimal gland

Tarsal glands

Lacrimal excretory ductules

Conjunctiva

Openings of lacrimal ductules

Openings of tarsal (meibomian) glands

SEEN FROM BEHIND

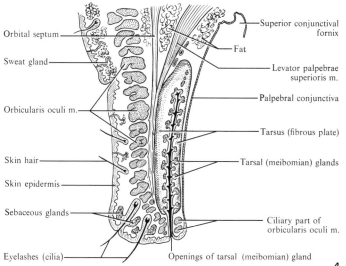

Orbital septum

Superior conjunctival fornix

Fat

Sweat gland

Levator palpebrae superioris m.

Palpebral conjunctiva

Orbicularis oculi m.

Tarsus (fibrous plate)

Skin hair

Tarsal (meibomian) glands

Skin epidermis

Sebaceous glands

Ciliary part of orbicularis oculi m.

Eyelashes (cilia)

Openings of tarsal (meibomian) gland

Pansky

VERTICAL SECTION (after Netter)

49. THE LACRIMAL APPARATUS

I. Conjunctiva, the mucous membrane of eye
A. PALPEBRAL PORTION: thick, very vascular, lines inside of eyelid
   1. At margin of lid, continuous with skin; at medial angle forms a crescentic fold, plica semilunaris
   2. Above and below leaves lid to reach eyeball—these are the superior and inferior conjunctival fornices
B. BULBAR PORTION: thin, transparent, slightly vascular over sclera; on the cornea only the epithelial part is present
C. CONJUNCTIVAL SAC: when lids are closed, a space lined with conjunctiva lies in front of eye

II. Lacrimal gland
A. LOCATION: see page 100
B. EXCRETORY DUCTULES: number from 6 to 12, open into upper lateral part of superior fornix of conjunctival sac

III. Lacrimal canals (ducts)
A. LOCATION: 1 at the medial end of each lid
B. ORIGIN: at puncta lacrimalia, which open on summit of lacrimal papilla, at lateral end of lacus lacrimalis
C. COURSE
   1. Superior: first upward, then medially and downward
   2. Inferior: first descends, then directly medially
D. TERMINATION: both in lacrimal sac

IV. Lacrimal sac, cephalic, dilated end of nasolacrimal duct
A. LOCATION: lacrimal groove formed by lacrimal bone and frontal process of maxillary bone
   1. Covered anteriorly by expansion of medial palpebral ligament and posteriorly by fibers of orbicularis oculi muscle
B. TERMINATION: caudally into nasolacrimal duct

V. Nasolacrimal duct
A. LOCATION AND COURSE: lies in bony canal formed by maxilla, lacrimal bone, and inferior nasal concha
B. TERMINATION: inferior meatus of nose

VI. Function
A. PURPOSE: (1) to prevent drying of eyeball, (2) to wash out foreign bodies that might damage eyeball
B. BLINKING OF EYE by alternations of contraction of levator palpebrae and orbicularis oculi muscles helps (1) to spread secretion throughout conjunctival sac, (2) action of orbicularis oculi, because of relations to lacrimal ducts and sac, "pumps" tears through duct system

VII. Clinical considerations
A. DESTRUCTION OF THE SENSORY ROOT OF THE TRIGEMINAL NERVE or its ophthalmic division may lead to ulcerations of the cornea of eye because of the loss of the afferent limb of the tearing reflex. The conjunctiva becomes dry and is constantly irritated by foreign substances which abrade the surface as the eye and lid move

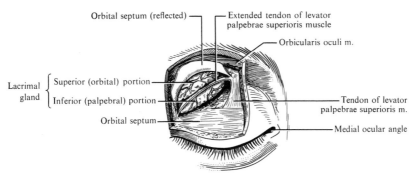

Orbital septum (reflected)
Extended tendon of levator palpebrae superioris muscle
Orbicularis oculi m.
Lacrimal gland { Superior (orbital) portion
Inferior (palpebral) portion }
Orbital septum
Tendon of levator palpebrae superioris m.
Medial ocular angle

LACRIMAL APPARATUS

LACRIMAL GLAND

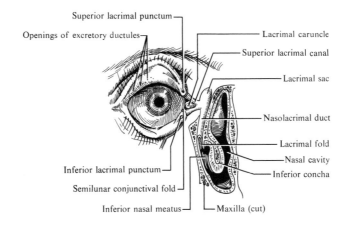

Superior lacrimal punctum
Openings of excretory ductules
Lacrimal caruncle
Superior lacrimal canal
Lacrimal sac
Nasolacrimal duct
Lacrimal fold
Nasal cavity
Inferior concha
Inferior lacrimal punctum
Semilunar conjunctival fold
Inferior nasal meatus
Maxilla (cut)

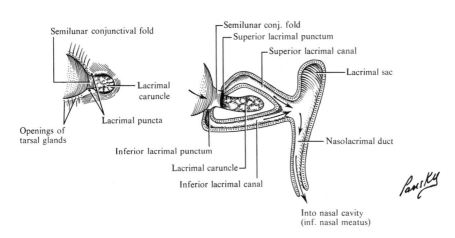

Semilunar conjunctival fold
Semilunar conj. fold
Superior lacrimal punctum
Superior lacrimal canal
Lacrimal sac
Lacrimal caruncle
Lacrimal puncta
Openings of tarsal glands
Inferior lacrimal punctum
Nasolacrimal duct
Lacrimal caruncle
Inferior lacrimal canal

Into nasal cavity (inf. nasal meatus)

Pansky

50. THE ORBIT AND FASCIA BULBI

I. **Bony orbit:** a pyramidal cavity, with base, apex, roof, floor, medial and lateral walls. Apex points dorsomedially; base directed rostrolaterally
   A. APEX: optic foramen for optic nerve
   B. FLOOR: orbital part of maxilla, orbital process of zygomatic bone, orbital process of palatine bone. Medially, opening of nasolacrimal duct; laterally, depression for origin of inferior oblique muscle; in floor, groove for infraorbital nerve
   C. ROOF: orbital plate of frontal bone and small wing of sphenoid. Medially, trochlea for superior oblique muscle; laterally, lacrimal fossa
   D. MEDIAL: frontal process of maxilla, lacrimal bone, orbital lamina (lamina papyracea) of ethmoid, body of sphenoid bone, lacrimal groove with posterior lacrimal crest near rostral end of wall, and near roof, anterior and posterior ethmoidal foramina
   E. LATERAL: orbital process of zygomatic bone and orbital part of great wing of sphenoid bone. Orifices for branches of zygomatic nerve, superior orbital fissure, inferior orbital fissure
   F. BASE: above, supraorbital arch of frontal bone with its supraorbital notch (foramen); laterally, zygomatic and zygomatic process of frontal bone; below, zygomatic and maxillary bones; medially, frontal bone and frontal process of maxillary bone

II. **Orbital septum:** continuous with the periorbita (the periosteum of the bony orbit) at margins of base of orbit; in upper lid joins tendon of levator palpebrae muscle; in both lids attaches to tarsal plates

III. **Vagina bulbi (fascia bulbi, Tenon's capsule)**
   A. THIN MEMBRANE to envelop eyeball from optic nerve to level of ciliary muscle. Surrounded by periorbital fat
   B. SEPARATED FROM SCLERA by episcleral space—continuous with subdural and subarachnoid spaces
   C. FUSES WITH SHEATH OF OPTIC NERVE and with sclera at entrance of this nerve
   D. FUSES WITH BULBAR CONJUNCTIVA
   E. AT TENDONS OF EXTRINSIC MUSCLES is reflected back, along muscles as sheath
      1. Extension of this, from superior rectus muscle, unites with tendon of levator palpebrae muscle
      2. Extension from inferior rectus muscle joins inferior tarsus
      3. From medial and lateral rectus muscles, fascia extends to lacrimal and zygomatic bones (these are strong and known as *medial and lateral check ligaments* [*lacertus of medial and lateral recti mm.*], for they are thought to check action of respective muscles)

IV. **Special features**
   A. THE ORBIT IS LINED by periosteum that is given the special name *"periorbita."* It is continuous over the rim of the orbit and through the inferior orbital fissure with the periosteum of the outer surface of the skull, and through the superior orbital fissure and the optic canal with the periosteal or outer layer of the dura mater

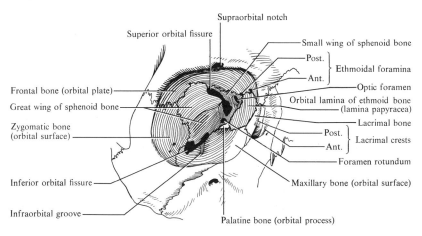

Supraorbital notch

Superior orbital fissure

Small wing of sphenoid bone

Post.

} Ethmoidal foramina

Ant.

Optic foramen

Frontal bone (orbital plate)

Orbital lamina of ethmoid bone (lamina papyracea)

Great wing of sphenoid bone

Lacrimal bone

Zygomatic bone (orbital surface)

Post.

} Lacrimal crests

Ant.

Foramen rotundum

Inferior orbital fissure

Maxillary bone (orbital surface)

Infraorbital groove

Palatine bone (orbital process)

BONY ORBIT

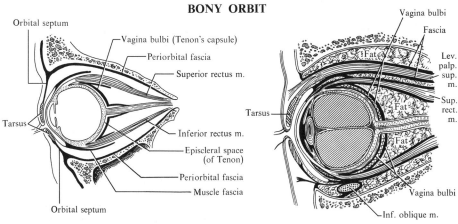

Orbital septum

Vagina bulbi (Tenon's capsule)

Periorbital fascia

Superior rectus m.

Tarsus

Inferior rectus m.

Episcleral space (of Tenon)

Periorbital fascia

Muscle fascia

Orbital septum

Vagina bulbi

Fascia

Fat

Lev. palp. sup. m.

Tarsus

Fat

Sup. rect. m.

Fat

Vagina bulbi

Inf. oblique m.

FASCIA BULBI - TENON'S FASCIA

SAGITTAL
SECTION

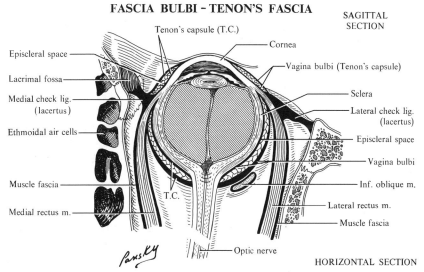

Tenon's capsule (T.C.)

Cornea

Episcleral space

Lacrimal fossa

Vagina bulbi (Tenon's capsule)

Medial check lig. (lacertus)

Sclera

Lateral check lig. (lacertus)

Ethmoidal air cells

Episcleral space

Vagina bulbi

Muscle fascia

Inf. oblique m.

T.C.

Lateral rectus m.

Medial rectus m.

Muscle fascia

Pansky

Optic nerve

HORIZONTAL SECTION

51. THE EXTRINSIC MUSCLES OF THE EYE

I. **Extrinsic muscles:** 7 in orbit; 6 move eyeball, 1 moves lid. The first 6 listed below have their origins from bone or a ring of fibrous tissue (common tendinous ring) around the optic nerve. The seventh, inferior oblique muscle, has a different origin

A. SUPERIOR RECTUS inserts into sclera between equator and corneal margin. Turns eye up and slightly medially; slight medial rotation. Nerve: oculomotor (III)

B. INFERIOR RECTUS inserts in sclera between equator and corneal margin. Turns eye down and medially; slight lateral rotation. Nerve: oculomotor (III)

C. MEDIAL RECTUS inserts in sclera, between equator and corneal margin. Turns eye medially. Nerve: oculomotor (III)

D. LATERAL RECTUS has 2 heads of origin, from upper and from lower parts of common ring. Inserts in sclera between equator and corneal margin. Turns eye laterally. Nerve: abducens (VI)

E. SUPERIOR OBLIQUE after its tendon bends through the trochlea (pulley) at angle of 50° inserts in sclera behind equator. Turns eye down and laterally, with medial rotation. Nerve: trochlear (IV)

F. LEVATOR PALPEBRAE arises as A–E, above; inserts in upper eyelid. Raises lid. Nerve: oculomotor (III)

G. INFERIOR OBLIQUE arises from medial edge of orbital floor and inserts in sclera behind equator posterolaterally. Turns eye up and laterally with lateral rotation. Nerve: oculomotor (III)

II. **Nerves**

A. MOTOR: to extrinsic muscles
1. Oculomotor divides into 2 branches, superior and inferior, before entering orbit through superior orbital fissure; passes between heads of lateral rectus
   a. Superior branch, above optic nerve to superior rectus and levator palpebrae muscles
   b. Inferior branch, below optic nerve, to inferior and medial recti and inferior oblique muscles
2. Trochlear (IV) enters orbit through superior orbital fissure above lateral rectus muscle, rises to roof of orbit and then passes medially to superior oblique muscle
3. Abducens (VI) enters orbit through superior orbital fissure between heads of the lateral rectus muscle, but below inferior branch of the oculomotor nerve, and runs along the inner surface of the lateral rectus muscle, which it supplies

B. SENSORY (see p. 112)

III. **Clinical considerations**

A. DESTRUCTION OF A NERVE or any nerve involvement results in abnormal deviations of the eye and faulty eye movements. For example, if the oculomotor nerve (III) is destroyed while the trochlear (IV) and abducens (VI), which supply the superior oblique and lateral rectus muscles, respectively, are functional, the eye will look lateralward and downward because these muscles are now unopposed by those normally supplied by the oculomotor nerve

B. IT SHOULD BE POINTED OUT that under normal conditions both eyes work together (*conjugate movement*)

C. JUST BEFORE THE FASCIAL SHEATHS of the 4 rectus muscles blend with the bulbar sheath, they expand laterally to fuse with each other, thus forming what is called *the intermuscular membrane.* Tumors or other masses lying internal to this membrane and the rectus muscles may not be visible unless the space among the muscles is explored, and therefore the orbit is frequently described as being subdivided into 2 spaces, 1 within and 1 outside the muscle cone

EXTRINSIC EYE MUSCLES

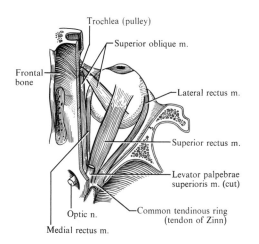

Trochlea (pulley)

Superior oblique m.

Frontal bone

Lateral rectus m.

Superior rectus m.

Levator palpebrae superioris m. (cut)

Optic n.

Common tendinous ring (tendon of Zinn)

Medial rectus m.

SUPERIOR VIEW

Superior oblique m.

Levator palpebrae superioris m. (cut)

Trochlea (pulley)

Superior rectus m.

Optic n.

Medial   Lateral

Recti mm.

Inferior oblique m.

Inferior rectus m.

LATERAL VIEW

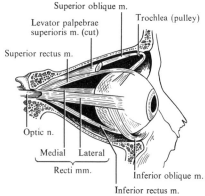

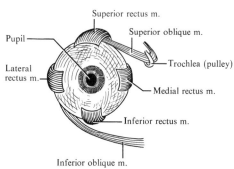

Superior rectus m.

Pupil

Superior oblique m.

Lateral rectus m.

Trochlea (pulley)

Medial rectus m.

Inferior rectus m.

Inferior oblique m.

ANTERIOR VIEW

Supraorbital n.

Trochlear (IV) n.

Superior rectus m.

Levator palpebrae superioris m.

Lacrimal n.

Oculomotor (III) n.

Superior oblique m.

Lateral rectus m.

Nasociliary n.

Abducens (VI) n.

Medial rectus m.

Oculomotor (III) n.

Ophthalmic a.

Optic (II) n.

Ciliary ganglion

Inferior rectus m.

Inferior oblique m.

Infraorbital n.

Pansky

ORIGINS OF OCULAR MUSCLES;
NERVES ENTERING ORBIT

52. STRUCTURE OF THE EYE

I. The wall of the eyeball
A. SCLERA, THE OUTER FIBROUS COAT: made up of densely packed collagenous fibers
   1. Cornea, the anterior part of sclera, consists of transparent connective tissue
   2. Cribriform plate weakest point. Circular area perforated by optic nerve
B. VASCULAR TUNIC, THE MIDDLE COAT
   1. Choroid, the posterior two thirds of coat, composed of blood vessels for nutrition of retina and pigment
   2. Ciliary body with the ciliary muscle of involuntary (smooth) type arranged in 2 planes and the ciliary processes
   3. Iris: covered anteriorly by endothelium; filled with connective tissue, pigment cells, and the sphincter and dilator pupillae muscles. *Pupil* is in its center
C. RETINA, THE INNER COAT consists of two primary layers, the innermost called the retina proper; the outer, the pigment layer. In addition, the retina (including both layers) is divisible into pars optica and pars caeca
   1. Pars optica: light sensitive; extends from posterior pole. The fovea centralis is a thin area where the receptors are easily exposed to light
   2. Pars caeca: the nonsensitive portion lying rostral to the ora serrata
      a. Pars ciliaris: has thin inner epithelial layer and an outer pigment layer
      b. Pars iridica: where most of pigment layer has formed smooth muscle, while the inner layer now contains pigment

II. Divisions of the eyeball
A. OCULAR CHAMBER lies in front of the lens and suspensory ligament. This is further subdivided into *anterior* and *posterior chambers* by the iris
B. VITREOUS BODY lies behind lens and its ligament

III. The refractive media
A. CORNEA: the transparent, strongly curved, rostral part of sclera
B. AQUEOUS HUMOR: the fluid of the anterior and posterior chambers
C. CRYSTALLINE LENS: composed of elastic capsule containing lens fibers. The lens is suspended from the ciliary processes by a *suspensory ligament* (*zonular fibers*)
D. VITREOUS BODY: a transparent jelly contained in thin, hyaloid membrane

IV. Formation and circulation of aqueous humor
A. FLUID LEAVES CAPILLARY NET in ciliary processes of post. chamber, flows medially to edge of pupil, where it enters ant. chamber. Here it flows laterally to the iridocorneal angle, to enter a meshwork of spaces (Fontana) of the angle. From these, fluids enter the scleral venous sinus (canal of Schlemm)

V. Accommodation (for near vision)
A. WHEN OBJECT is close to eye, ciliary muscle contracts, easing tension on suspensory lig. of lens. This reduces tension on capsule, which permits lens to expand

VI. Clinical considerations
A. OPAQUE CORNEA: corneal fibers lose transparency
B. IRREGULARITIES OF CORNEA result in astigmatism or ametropia
C. CATARACT: opacity of lens necessitating its removal
D. GLAUCOMA: accumulation of aqueous humor resulting in increased ocular pressure
E. HYPEROPIA (farsightedness): axis of eyeball too short, focus behind retina
F. MYOPIA (nearsightedness): axis of eyeball too long, focus in front of retina
G. RETINAL SEPARATION: separation of sensory from pigment layer of retina

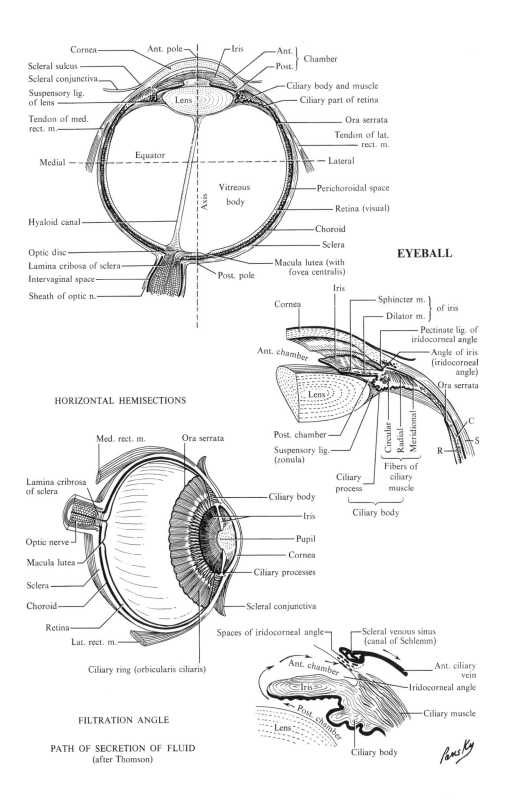

Cornea — Ant. pole — Iris — Ant. ⎱ Chamber
Scleral sulcus — Post. ⎰ Chamber
Scleral conjunctiva —
Suspensory lig. of lens —
Lens
Tendon of med. rect. m. —
Medial —
Equator
Hyaloid canal —
Optic disc —
Lamina cribosa of sclera —
Intervaginal space —
Sheath of optic n. —

Ciliary body and muscle
Ciliary part of retina
Ora serrata
Tendon of lat. rect. m.
Lateral
Perichoroidal space
Retina (visual)
Choroid
Sclera

Vitreous body
Axis
Post. pole
Macula lutea (with fovea centralis)

EYEBALL

HORIZONTAL HEMISECTIONS

Cornea
Ant. chamber
Lens
Post. chamber —
Suspensory lig. (zonula)
Ciliary process
Fibers of ciliary muscle
Ciliary body

Iris
Sphincter m. ⎱ of iris
Dilator m. ⎰
Pectinate lig. of iridocorneal angle
Angle of iris (iridocorneal angle)
Ora serrata
C
S
R
Circular
Radial
Meridional

Med. rect. m.
Ora serrata
Lamina cribrosa of sclera
Optic nerve —
Macula lutea —
Sclera —
Choroid —
Retina —
Lat. rect. m. —

Ciliary body
Iris
Pupil
Cornea
Ciliary processes
Scleral conjunctiva

Ciliary ring (orbicularis ciliaris)

FILTRATION ANGLE

PATH OF SECRETION OF FLUID
(after Thomson)

Spaces of iridocorneal angle
Ant. chamber
Iris
Post. chamber
Lens
Ciliary body

Scleral venous sinus (canal of Schlemm)
Ant. ciliary vein
Iridocorneal angle
Ciliary muscle

Pansky

- 109 -

53. BLOOD VESSELS OF THE ORBIT AND EYE

I. Ophthalmic artery

A. ORIGIN: from internal carotid artery at end of cavernous sinus

B. COURSE: through optic foramen, below and lateral to optic nerve; in orbit, crosses above optic nerve to medial wall of orbit, then passes rostrally to divide into supratrochlear and dorsal nasal arteries

C. BRANCHES

  1. In orbit, for surrounding parts

    a. Lacrimal near optic foramen, along border of lateral rectus muscle to lacrimal gland, eyelids, and conjunctiva

    b. Supraorbital passes rostrally, along medial border of superior rectus muscle, through supraorbital notch to forehead

    c. Anterior and posterior ethmoidal to sinuses and nasal cavity

    d. Medial palpebral arteries: 1 to each lid

    e. Supratrochlear: leaves orbit with supratrochlear nerve to forehead

    f. Dorsal nasal to outer surface of nose

  2. In orbit, for eyeball

    a. Central artery of retina leaves main channel as it crosses optic nerve, pierces the latter, and runs in its center to spread over retina

    b. Ciliary arteries arranged in 3 groups

      i. Short posterior ciliary: 6–12, pierce sclera around entrance of optic nerve; to choroid and ciliary processes

      ii. Long posterior ciliary: 2 enter sclera on either side of the optic nerve; run between choroid and sclera to ciliary body, where their branches form anterior major arterial circle

      (a) Send branches inward, which form anterior minor circle

      iii. Anterior ciliary artery: from muscular branches running with tendons of recti muscles to form vascular zone under conjunctiva; pierce sclera to join major circle

II. Distribution of central artery to retina

A. IN EYEBALL, immediately bifurcates giving upper and lower branches

B. BOTH UPPER AND LOWER BRANCHES divide into medial and lateral branches

C. AT FIRST, ARTERIES are between hyaloid membrane and nervous layer of retina but soon pierce latter, forming fine capillary net. Do not go deeper than inner nuclear layer

D. MACULA gets 2 small twigs from temporal branch and a few directly from central artery

III. Venous drainage of eyeball

A. RETINA DRAINED BY VEINS that accompany branches and trunk of central artery

B. OUTER COATS DRAINED BY VORTICOSE VEINS in outer layer of choroid. These converge into 4 or 5 trunks, pierce sclera between optic nerve and corneoscleral junction to drain into superior ophthalmic vein

IV. Clinical considerations

A. EYEGROUNDS: a great deal can be learned from the appearance of the retinal arteries as observed by ophthalmoscopic examination

B. BLIND SPOT: the area of retina where the optic nerve fibers leave eyeball and retinal vessels enter

C. CHOKED DISK: a swelling of the optic disk due to increased intracranial pressure

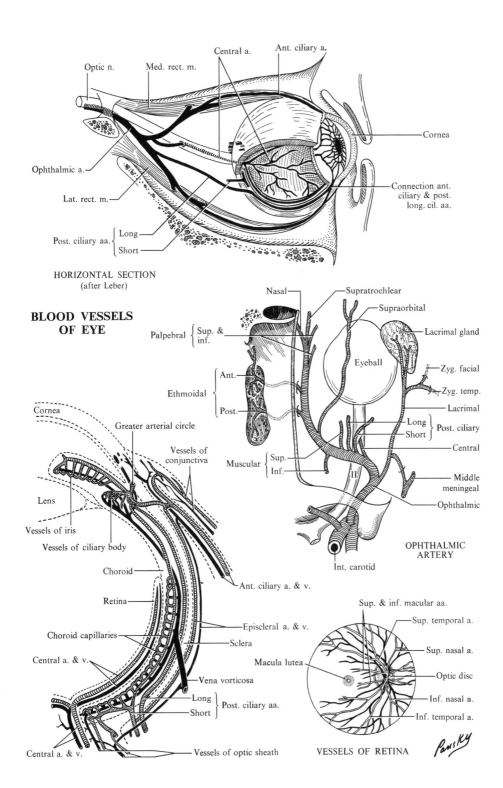

Central a.

Ant. ciliary a.

Optic n.

Med. rect. m.

Cornea

Ophthalmic a.

Connection ant.
ciliary & post.
long. cil. aa.

Lat. rect. m.

Post. ciliary aa. { Long Short

HORIZONTAL SECTION
(after Leber)

**BLOOD VESSELS
OF EYE**

Nasal

Supratrochlear

Supraorbital

Palpebral { Sup. & inf.

Lacrimal gland

Eyeball

Zyg. facial

Ethmoidal { Ant. Post.

Zyg. temp.

Lacrimal

Cornea

Greater arterial circle

Long } Post. ciliary
Short

Vessels of
conjunctiva

Central

Lens

Muscular { Sup. Inf.

II

Vessels of iris

Middle
meningeal

Vessels of ciliary body

Ophthalmic

Choroid

Retina

Int. carotid

**OPHTHALMIC
ARTERY**

Choroid capillaries

Ant. ciliary a. & v.

Sup. & inf. macular aa.

Central a. & v.

Episcleral a. & v.

Sup. temporal a.

Sclera

Macula lutea

Sup. nasal a.

Vena vorticosa

Optic disc

Long } Post. ciliary aa.
Short

Inf. nasal a.

Inf. temporal a.

Central a. & v.

Vessels of optic sheath

VESSELS OF RETINA

Pansky

54. NERVES OF THE ORBIT

I. Sensory
A. SPECIAL SENSE (see also p. 134)
  1. Optic: actually not a true nerve but a tract of central nervous system; enters orbit through optic foramen and joins eyeball just medial to posterior pole
B. GENERAL SENSE
  1. Ophthalmic nerve
    a. Origin: cell bodies in trigeminal (semilunar) ganglion
    b. Course and distribution: along lateral wall of cavernous sinus, below oculomotor and trochlear nerves, enters orbit through superior orbital fissure. Branches just before entering orbit:
      i. Lacrimal (smallest branch): along upper border of lateral rectus muscle to lacrimal gland, then rostral to conjunctiva and skin of upper lid
      ii. Frontal (largest branch) runs rostrally, above levator palpebrae muscle to divide into *supraorbital,* which leaves orbit through supraorbital notch, supplying upper lid, forehead, and scalp; *supratrochlear,* passes over trochlea of superior oblique muscle to conjunctiva of upper lid and forehead
      iii. Nasociliary crosses optic nerve, passes obliquely below superior rectus and superior oblique muscles to medial wall of orbit. Branches:
        (1) Communication to ciliary ganglion
        (2) Long ciliary to eyeball, especially iris and cornea
        (3) Infratrochlear, to medial angle of eye for conjunctiva, lacrimal sac, skin of lid, and side of nose
        (4) Anterior and posterior ethmoidals to sinuses (frontal, ethmoid, sphenoid) and nasal cavity

II. Motor (for visceral motor, see p. 114)
A. GENERAL MOTOR
  1. Oculomotor (III). For divisions and distribution, see page 106
    a. Origin: nucleus in midbrain tegmentum. The nerve appears in interpeduncular fossa, lies in most cephalic, lateral wall of cavernous sinus to enter orbit through superior orbital fissure
  2. Trochlear (IV). For distribution, see page 106
    a. Origin: nucleus in midbrain tegmentum. Fibers leave central nervous system through anterior medullary velum dorsally, cross to opposite side, pass rostrally and caudally, run in lateral wall of cavernous sinus between III and ophthalmic V, cross III, and pass through superior orbital fissure above other nerves
  3. Abducens (VI). For distribution, see page 106
    a. Origin: nucleus in tegmentum of pons. Fibers leave central nervous system in ventral groove between medulla and pons, pass through cavernous sinus between internal carotid artery and ophthalmic (V) to enter orbit through superior orbital fissure

III. Clinical considerations
A. PATHOLOGY IN THE CAVERNOUS SINUS: tumors or aneurysms of the carotid artery in this location may encroach upon all the nerves passing through the sinus, such as III, IV, and VI, thus causing complete paralysis of all the ocular muscles
B. THE LARGEST NERVE in the orbit, and the central structure in the muscle cone, is *the optic nerve.* This leaves the back of the eyeball and in the orbit is surrounded by a heavy sheath, *the external sheath* of the nerve, which is continuous with the dura and arachnoid in the skull; on the surface of the nerve is a very thin layer of pia mater, *the internal sheath.* Between the 2 layers is *the intervaginal space,* which is a continuation of the subarachnoid space

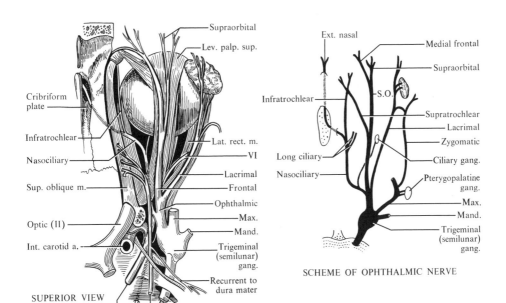

Supraorbital

Lev. palp. sup.

Cribriform plate

Infratrochlear

Nasociliary

Sup. oblique m.

Optic (II)

Int. carotid a.

SUPERIOR VIEW

Lat. rect. m.

VI

Lacrimal

Frontal

Ophthalmic

Max.

Mand.

Trigeminal (semilunar) gang.

Recurrent to dura mater

III VI IV

Sensory root
Motor root } (V)

Ext. nasal

Infratrochlear

Long ciliary

Nasociliary

Medial frontal

Supraorbital

S.O.

Supratrochlear

Lacrimal

Zygomatic

Ciliary gang.

Pterygopalatine gang.

Max.

Mand.

Trigeminal (semilunar) gang.

SCHEME OF OPHTHALMIC NERVE

NERVES OF ORBIT

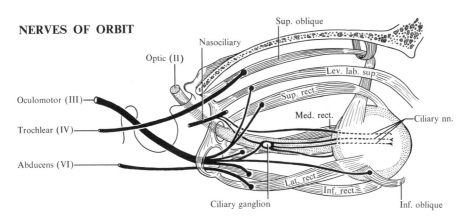

Sup. oblique

Nasociliary

Optic (II)

Oculomotor (III)

Trochlear (IV)

Abducens (VI)

Lev. lab. sup.

Sup. rect.

Med. rect.

Ciliary nn.

Ciliary ganglion

Lat. rect.

Inf. rect.

Inf. oblique

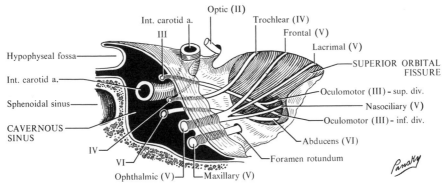

Optic (II)

Int. carotid a.

III

Trochlear (IV)

Frontal (V)

Lacrimal (V)

Hypophyseal fossa

Int. carotid a.

Sphenoidal sinus

CAVERNOUS SINUS

IV

VI

Ophthalmic (V)

Maxillary (V)

SUPERIOR ORBITAL FISSURE

Oculomotor (III) - sup. div.

Nasociliary (V)

Oculomotor (III) - inf. div.

Abducens (VI)

Foramen rotundum

Pansky

55. AUTONOMIC NERVES OF THE ORBIT

I. Lacrimal gland
A. PARASYMPATHETIC (CRANIOSACRAL)
1. Preganglionic fibers arise from cells in superior salivatory nucleus to run in nervus intermedius (with facial) then to greater petrosal nerve, which becomes part of nerve of pterygoid canal, to enter and synapse in pterygopalatine ganglion
2. Postganglionic fibers arise from cells of pterygopalatine ganglion, pass through pterygopalatine nerves to maxillary nerve, and then through the zygomatic branch of the latter. These fibers enter the zygomaticotemporal branch of the zygomatic nerve in the orbit and communicate with the lacrimal branch of ophthalmic nerve to reach the gland
B. SYMPATHETIC (THORACOLUMBAR)
1. Preganglionics arise in intermediolateral gray column of upper thoracic cord, enter the sympathetic trunk, ascend in cervical sympathetic trunk to end by synapses in superior cervical ganglion
2. Postganglionics arise in cells of superior cervical ganglion, pass through carotid plexus, continue rostrally in the deep petrosal nerve, which becomes part of the nerve of the pterygoid canal, pass through pterygopalatine ganglion, without synapse, and are distributed as above (A2)

II. Muscles of iris (sphincter and dilator pupillae)
A. PARASYMPATHETICS TO SPHINCTER MUSCLE (CONSTRICTS PUPIL)
1. Preganglionics arise in Edinger-Westphal nucleus of midbrain, travel through oculomotor (III) nerve, run into its inferior division and through short root to end by synapse in ciliary ganglion
2. Postganglionics arise from cells of ciliary ganglion, leave by short ciliary nerves, which pierce sclera posteriorly and run to iris
B. SYMPATHETIC TO DILATOR PUPILLAE MUSCLE (DILATES PUPIL)
1. Preganglionics: as IB
2. Postganglionics arise in cells of superior cervical ganglion, ascend through carotid and cavernous plexuses. Some fibers join ophthalmic (V) nerve to continue in its nasociliary branch and are carried to the eye with the long ciliary branches of this nerve; others from cavernous plexus enter sympathetic "root" of ciliary ganglion, pass through without synapse, and run with short ciliary nerves

III. Muscles of ciliary body: constriction causes thickening of lens as in accommodation to near vision
A. PARASYMPATHETICS: same as IIA
B. SYMPATHETICS: none

IV. Clinical considerations
A. HORNER'S SYNDROME: characterized by constriction of pupil, partial drooping of upper eyelid, sinking in of eyeball, dilation of vessels of face and conjunctiva, and lack of sweating. This can result from any lesion in the sympathetic pathways described above
B. ARGYLL ROBERTSON PUPIL: in this condition, the pupil does not constrict for light but will constrict for near vision. Usually caused by a lesion in the pretectal zone of the midbrain, the center for light reflexes

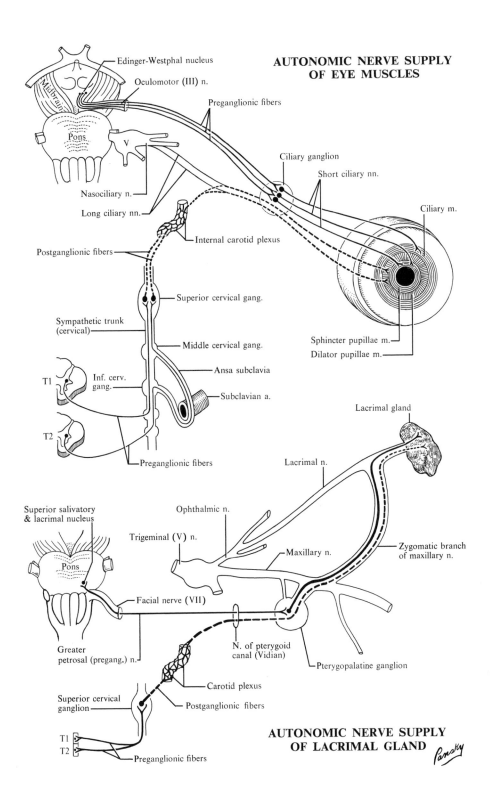

AUTONOMIC NERVE SUPPLY
OF EYE MUSCLES

Edinger-Westphal nucleus

Oculomotor (III) n.

Preganglionic fibers

Ciliary ganglion

Short ciliary nn.

Ciliary m.

Nasociliary n.

Long ciliary nn.

Internal carotid plexus

Postganglionic fibers

Superior cervical gang.

Sympathetic trunk
(cervical)

Middle cervical gang.

Sphincter pupillae m.

Dilator pupillae m.

Ansa subclavia

Inf. cerv.
gang.

T1

T2

Subclavian a.

Preganglionic fibers

Lacrimal gland

Lacrimal n.

Superior salivatory
& lacrimal nucleus

Ophthalmic n.

Trigeminal (V) n.

Maxillary n.

Zygomatic branch
of maxillary n.

Facial nerve (VII)

Greater
petrosal (pregang.) n.

N. of pterygoid
canal (Vidian)

Pterygopalatine ganglion

Carotid plexus

Superior cervical
ganglion

Postganglionic fibers

T1
T2

Preganglionic fibers

AUTONOMIC NERVE SUPPLY
OF LACRIMAL GLAND

-115-

56. THE EXTERNAL EAR

I. Ear consists of three parts: external, middle, internal

II. Structure of the external ear: 2 parts

A. AURICLE (PINNA): projects from side of head
1. Lateral surface: irregular and concave with crests and grooves
   a. Helix (rim): with small auricular tubercle
   b. Anthelix: elevation rostral and parallel to helix
      i. Splits into 2 crura, which bound triangular fossa
   c. Scapha: groove between a and b above
   d. Concha: deep concavity rostral to and bordered by anthelix
   e. Tragus: a dorsal projection, rostral to concha and over meatus
   f. Antitragus: a dorsal tubercle opposite tragus
   g. Intertragic notch lies between e and f, above
   h. Lobule: caudal part of pinna, below antitragus and notch
2. Medial (cranial) surface shows 2 eminences, corresponding to depressions of lateral surface
3. Internal structure: basic framework of elastic cartilage and dense fibrous tissue to which thin skin is firmly attached
4. Auricular muscles (see p. 22)

B. EXTERNAL AUDITORY MEATUS
1. Extent: from bottom of concha to tympanic membrane, about 2.4 cm
2. Course: S-shaped, at first being convex upward and convex backward
3. Form: cylindrical, with 2 constrictions, one near outer end, the other, the *isthmus,* located deeper
4. Tympanic membrane, at medial end, set obliquely, so that floor and rostral wall are longer than roof and dorsal wall
5. Parts
   a. Cartilaginous: about 8.0 mm long and continuous with cartilage of pinna. Is firmly attached to auricular process of temporal bone
   b. Osseous: about 16 mm long, narrower than cartilaginous part, inner end narrower than outer
      i. Tympanic sulcus: a narrow groove to which periphery of tympanic membrane is attached

III. Relations of external ear

A. ROSTRALLY: condyle of mandible and parotid gland
B. DORSALLY: mastoid air cells

IV. Arterial supply: branches of posterior auricular of external carotid artery, anterior auricular from superficial temporal artery, and deep auricular branch from maxillary artery

V. Nerves: auricular branch of vagus and auriculotemporal of mandibular (V) nerve

VI. Clinical considerations

A. LESIONS IN EXTERNAL AUDITORY CANAL: infections or boils in canal may cause nausea and vomiting because the general somatic afferent fibers (pain, etc.) from this area are carried in the vagus nerve (X), which also carries all the parasympathetic fibers to the upper gastrointestinal tract

EXTERNAL EAR

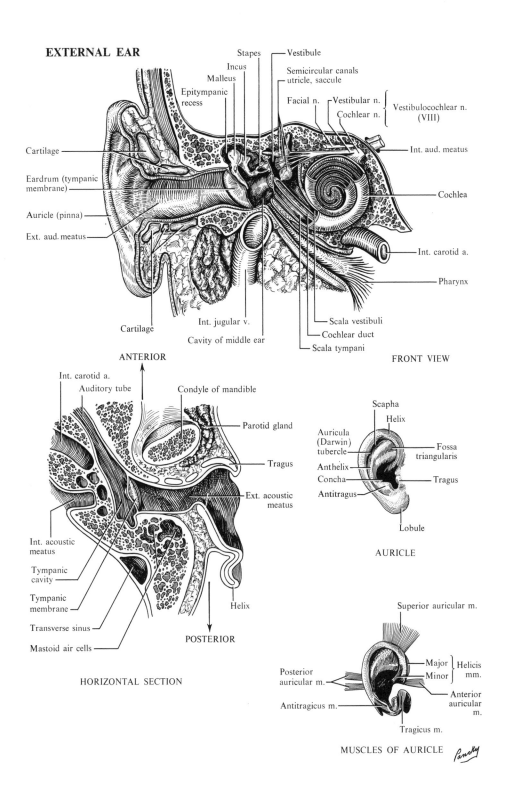

Stapes
Incus
Malleus
Epitympanic recess
Vestibule
Semicircular canals
utricle, saccule
Facial n.
Vestibular n.
Cochlear n.
Vestibulocochlear n. (VIII)
Int. aud. meatus
Cartilage
Eardrum (tympanic membrane)
Auricle (pinna)
Ext. aud. meatus
Cochlea
Int. carotid a.
Pharynx
Cartilage
Int. jugular v.
Cavity of middle ear
Scala vestibuli
Cochlear duct
Scala tympani

ANTERIOR

FRONT VIEW

Int. carotid a.
Auditory tube
Condyle of mandible
Parotid gland
Tragus
Ext. acoustic meatus
Int. acoustic meatus
Tympanic cavity
Tympanic membrane
Transverse sinus
Mastoid air cells
Helix

POSTERIOR

HORIZONTAL SECTION

Scapha
Helix
Auricula (Darwin) tubercle
Anthelix
Concha
Antitragus
Fossa triangularis
Tragus
Lobule

AURICLE

Superior auricular m.
Posterior auricular m.
Antitragicus m.
Major
Minor
Helicis mm.
Anterior auricular m.
Tragicus m.

MUSCLES OF AURICLE

57. THE MIDDLE EAR

I. Middle ear: 2 parts
A. TYMPANIC CAVITY PROPER: opposite tympanic membrane
B. EPITYMPANIC RECESS: above level of membrane

II. Form: a box with a roof, floor, lateral, medial, posterior, and anterior walls
A. ROOF (TEGMENTAL WALL): bony plate, between cranial and tympanic cavities
B. FLOOR (JUGULAR WALL): plate of bone between cavity and jugular fossa
C. LATERAL (MEMBRANOUS WALL): formed by tympanic membrane
  1. Structure and relations
    a. Set obliquely, forming angle of 55° with floor of external meatus
    b. Periphery, where attached to bone in tympanic sulcus, is fibrocartilaginous (anulus)
    c. Where tympanic sulcus is lacking at *tympanic notch (of Rivinus)* the 2 bands, *anterior and posterior malleolar folds,* extend from notch to lateral process of malleus
      i. Triangular area between folds is *pars flaccida* of membrane
    d. Manubrium of malleus is firmly fixed to membrane at its middle
      i. Umbo: caudal end of manubrium, where membrane is concave
    e. Chorda tympani nerve runs between handle of malleus and incus
  2. Special features
    a. Notch of Rivinus: deficiency in bony rim
    b. Iter chordae anterius and posterius
    c. Petrotympanic fissure
D. MEDIAL (LABYRINTHIC WALL): bone between middle and inner ear
  1. Vestibular window (oval) with long axis horizontal
  2. Cochlear window (round) contains secondary tympanic membrane
  3. Promontory: rounded eminence between 1 and 2
  4. Facial canal: above oval window, then curves downward along posterior wall
  5. Prominence of lateral semicircular canal
E. POSTERIOR (MASTOID WALL)
  1. Entrance to *tympanic antrum,* actually from epitympanic recess to mastoid
  2. Pyramidal eminence
  3. Fossa of incus
F. ANTERIOR (CAROTID WALL): bone separating cavity from carotid artery
  1. Perforated by tympanic branch of internal carotid artery and sympathetic nerves
  2. Auditory tube and canal for tensor tympani

III. Auditory ossicles, which transfer sound waves across middle ear
A. MALLEUS with head, neck, anterior process, and lateral process
B. INCUS with body, short crus, and long crus
C. STAPES with head, neck, 2 crura, and foot plate

IV. Intrinsic muscles
A. TENSOR TYMPANI arises from cartilaginous part of auditory tube and great wing of sphenoid. Inserts on malleus. Action: tenses membrane. Nerve: mandibular (V)
B. STAPEDIUS arises from inner wall of pyramidal eminence. Inserts on neck of stapes. Nerve: branch of facial (VII). Action: draws stapes back

V. Clinical considerations
A. MIDDLE EAR INFECTIONS: quite prevalent and may become extensive due to connection to both the mastoid cells and the nasopharynx by way of the eustachian tube
B. OTOSCOPIC EXAMINATION: when reflected light is used, the cone of light seen has its apex at the umbo and expands downward and forward. This cone changes in disease

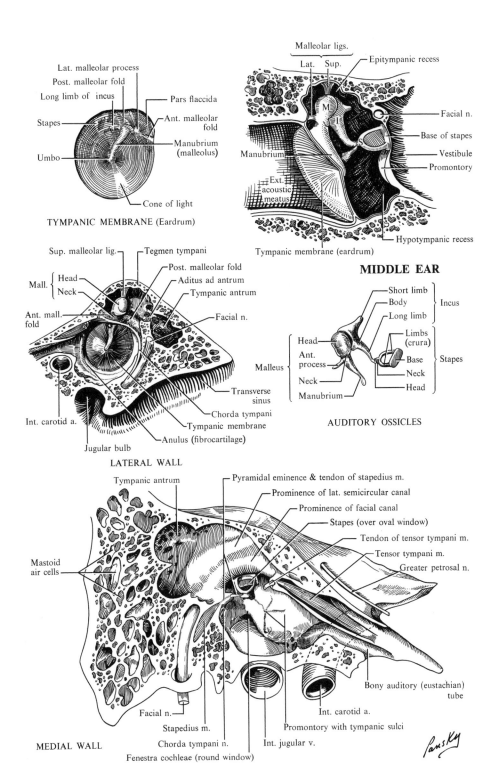

TYMPANIC MEMBRANE (Eardrum)

Lat. malleolar process
Post. malleolar fold
Long limb of incus
Stapes
Umbo
Pars flaccida
Ant. malleolar fold
Manubrium (malleolus)
Cone of light

Malleolar ligs.
Lat.   Sup.
Epitympanic recess
Facial n.
Base of stapes
Vestibule
Promontory
Manubrium
Ext. acoustic meatus
Hypotympanic recess
Tympanic membrane (eardrum)

MIDDLE EAR

Sup. malleolar lig.
Mall. { Head
       Neck
Ant. mall. fold
Tegmen tympani
Post. malleolar fold
Aditus ad antrum
Tympanic antrum
Facial n.
Transverse sinus
Chorda tympani
Tympanic membrane
Anulus (fibrocartilage)
Int. carotid a.
Jugular bulb

LATERAL WALL

Short limb
Body
Long limb } Incus

Head
Ant. process
Neck } Malleus
Manubrium

Limbs (crura)
Base
Neck
Head } Stapes

AUDITORY OSSICLES

Tympanic antrum
Pyramidal eminence & tendon of stapedius m.
Prominence of lat. semicircular canal
Prominence of facial canal
Stapes (over oval window)
Tendon of tensor tympani m.
Tensor tympani m.
Greater petrosal n.
Mastoid air cells
Bony auditory (eustachian) tube
Facial n.
Stapedius m.
Chorda tympani n.
Int. jugular v.
Int. carotid a.
Promontory with tympanic sulci
Fenestra cochleae (round window)

MEDIAL WALL

Pansky

58. THE INNER EAR

I. Inner ear (labyrinth) consists of 2 parts
A. OSSEOUS: cavities in the petrous bone
B. MEMBRANOUS: made up of interconnected sacs and tubes (ducts)

II. Osseous labyrinth has 3 parts, lined with periosteum. Filled with *perilymph*
A. VESTIBULE: central part of labyrinth
1. In lateral wall: oval window
2. In medial wall: perforations for nerves and opening for endolymphatic duct (vestibular aqueduct)
3. Dorsally: 5 openings for semicanals
4. Rostrally: opening for cochlea
B. SEMICIRCULAR CANALS: 3 in number
1. Anterior: lateral end has a dilatation, the *ampulla,* which opens into vestibule; medial end joins posterior canal forming the *common crus*
2. Posterior: caudal end has an *ampulla,* which opens into vestibule
3. Lateral has *ampulla,* which opens into vestibule at lateral end; other end opens into dorsal part of vestibule
C. COCHLEA (SNAIL SHELL): has apex and base. Apex directed rostrally and laterally; base faces dorsomedially and lies at bottom of internal acoustic meatus
1. Modiolus: the bony, conical axis
2. Bony canal: makes $2\frac{1}{2}$ turns around modiolus; lower end diverging from modiolus has openings for *round window*
3. Osseous spiral lamina: a bony shelf projecting from modiolus. With basilar membrane it divides an upper scala vestibuli and lower scala tympani

III. Membranous labyrinth: connective tissue and epithelium, filled with *endolymph*
A. UTRICLE: in bony vestibule. Has 5 openings for semicircular canals and *ductus utriculosaccularis*
B. SACCULE: in vestibule. Has opening into endolymphatic duct and cochlear duct (*ductus reuniens*)
C. SEMICIRCULAR CANALS open into utricle, conform to bony labyrinth
D. COCHLEAR DUCT: that part of labyrinth separated from scala tympani by the basilar membrane and from scala vestibuli by the vestibular membrane

IV. Sense organs
A. FOR EQUILIBRIUM: macula utriculi; macula sacculi; and cristae ampullaris
B. FOR HEARING: the *spiral organ (of Corti)* in floor of cochlear duct

V. Nerves
A. VESTIBULAR PORTION OF VIII for equilibrium
1. Ganglion: vestibular, located in acoustic meatus
2. Branches of nerve: *superior,* to macula utriculi and cristae of superior and lateral canals; *inferior,* to macula sacculi; and *posterior,* to crista of posterior canal
B. AUDITORY PORTION OF VIII for hearing
1. Ganglion; spiral, in spiral canal at center of bony modiolus
2. Distribution: fibers pass through osseous spiral lamina to the organ of Corti

VI. Clinical considerations
A. DEAFNESS CAN RESULT from damage to the organ of Corti, cochlear duct, or acoustic (VIII) nerve and its ganglion
B. MENIÈRE'S DISEASE is characterized by a loss of balance and ringing in the ears, which is due to edema of the labyrinth or inflammation of the vestibular nerve

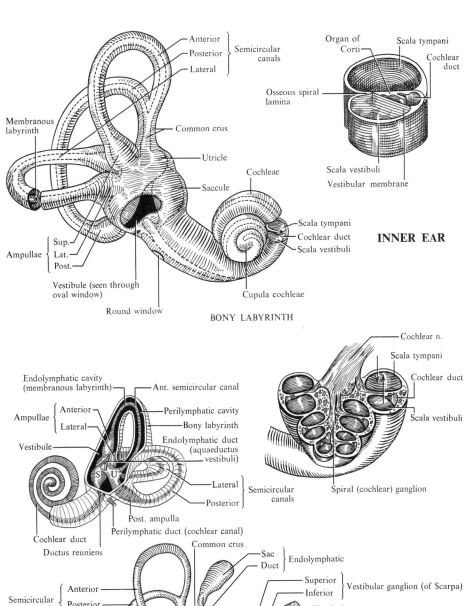

Anterior ⎤
Posterior ⎬ Semicircular canals
Lateral ⎦

Membranous labyrinth

Common crus

Utricle

Saccule

Cochleae

Ampullae { Sup. Lat. Post.

Vestibule (seen through oval window)

Round window

Scala tympani
Cochlear duct
Scala vestibuli

Cupula cochleae

BONY LABYRINTH

Organ of Corti

Osseous spiral lamina

Scala tympani

Cochlear duct

Scala vestibuli

Vestibular membrane

INNER EAR

Endolymphatic cavity (membranous labyrinth)

Ampullae { Anterior Lateral

Vestibule

Cochlear duct

Ductus reuniens

Ant. semicircular canal

Perilymphatic cavity

Bony labyrinth

Endolymphatic duct (aquaeductus vestibuli)

Lateral ⎤
Posterior ⎦ Semicircular canals

Post. ampulla

Perilymphatic duct (cochlear canal)

Cochlear n.

Scala tympani

Cochlear duct

Scala vestibuli

Spiral (cochlear) ganglion

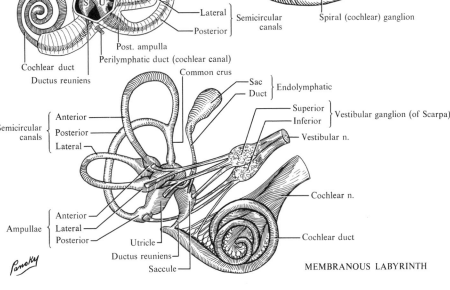

Common crus

Sac ⎤
Duct ⎦ Endolymphatic

Superior ⎤
Inferior ⎦ Vestibular ganglion (of Scarpa)

Vestibular n.

Semicircular canals { Anterior Posterior Lateral

Ampullae { Anterior Lateral Posterior

Utricle

Ductus reuniens

Saccule

Cochlear n.

Cochlear duct

MEMBRANOUS LABYRINTH

Pansky

59. THE NOSE—PART I

I. External nose
A. PARTS: root, dorsum or bridge, tip, alae, base, with nares separated by septum
B. NERVES (see also p. 132)
   1. Motor: facial (VII) to muscles (see p. 22)
   2. Sensory: infratrochlear from nasociliary branches of ophthalmic (V), to skin of root, alae, and nostrils; infraorbital branch of maxillary (V) to sides of nose
C. VESSELS (see p. 34)
   1. Arteries: facial to sides, alae, and septum; dorsal nasal branch of ophthalmic to root and dorsum; infraorbital branch of maxillary to sides
   2. Veins drain into the ophthalmic and anterior facial veins
D. FRAMEWORK: hyaline cartilage and bone
   1. Cartilage
      a. Septal: partition, between right and left nasal cavities
         i. Attachments: perpendicular plate of ethmoid, vomer, maxilla, nasal, septal process of lower nasal cartilage
      b. Lateral nasal (upper lateral)
         i. Attachments: nasal bone, frontal process of maxilla, lower nasal cartilage becomes continuous with septal cartilage
      c. Greater alar (lower lateral, alar)
         i. U-shaped, open posteriorly; has medial and lateral crura
         ii. Medial crura attach with each other and upper nasal cartilage
      d. Lesser alar
   2. Bones: 2 nasal bones and frontal process of maxillae

II. Nasal cavities: right and left separated by nasal septum
A. ANTERIOR APERTURE OR NARES (NOSTRILS) open into:
   1. Vestibule, a dilated area bounded by ala and crus of nasal cartilage
B. LATERAL WALL: divisions created by the nasal conchae
   1. Sphenoethmoidal recess: above superior concha
   2. Superior meatus: below superior concha, above middle concha
   3. Middle meatus: below and lateral to middle concha, above inferior concha
   4. Inferior meatus: below and lateral to inferior concha, above palate
C. OPENINGS INTO THE MEATUSES OR RECESSES
   1. Sphenoethmoidal recess: sphenoidal sinus
   2. Superior meatus: posterior ethmoidal air cells
   3. Middle meatus: frontonasal duct anteriorly (50 per cent of cases), accessory ostium for maxillary sinus below posterior end of middle concha, and semilunar hiatus which receives:
      a. Anteriorly: frontonasal duct (50 per cent); anterior ethmoid air cells
      b. Posteriorly: ostium from maxillary sinus
      c. On or above bulla: middle ethmoidal air cells
   4. Inferior meatus: nasolacrimal duct
D. MEDIAL WALL: recesses, spaces, or hiatuses
   1. Nasopalatine recess: depression in septum over incisive canal
   2. Vomeronasal organ (see p. 125)
E. POSTERIOR APERTURE: 2 *choanae* opening into nasopharynx

NASAL CAVITY

Nasal bone

Septal cartilage

Great alar (lower) cartilage

Lateral (upper) nasal cartilage

Lesser alar cartilages

Fibro-fatty tissue

CARTILAGES OF NOSE

Frontal sinus

Crista galli

Superior concha

Sphenoethmoidal recess

Middle concha

Great alar (lower) cartilage

Septal cartilage

Fibro-fatty tissue

Atrium

CARTILAGES OF NOSE
FROM BELOW

Vestibule

Hard palate

Inferior meatus

Inferior concha

Middle meatus

Soft palate

Semilunar hiatus

Bulla ethmoidalis

Maxillary sinus opening

Frontal sinus & opening

Middle ethmoidal air cells (sinus) opening

Sphenoidal sinus & opening

Posterior ethmoidal air cells (sinus) opening

Superior concha

Middle concha (cut away)

Auditory (eustachian) tube opening

Opening of nasolacrimal duct

Inferior nasal concha (cut away)

LATERAL WALL

DETAILS OF
MIDDLE MEATUS

Infundibular opening (frontal)

Infundibulum

Hiatus semilunaris

Bulla ethmoidalis

Middle ethmoidal air cells (sinus) opening

Superior concha

Middle concha (cut)

Anterior ethmoidal sinus opening

Uncinate process (cut)

Maxillary sinus opening

Superior concha

Middle concha (cut)

Pansky

60. THE NOSE—PART II

I. Nasal cavity
A. ANTERIOR APERTURE: pear-shaped, bounded by nasal bone and anterior border of maxillae
B. FRAMEWORK
  1. Medial (septal) wall
    a. Septal cartilage
    b. Perpendicular plate of ethmoid
    c. Vomer
    d. Projection of other bones which join those listed above
      i. Palatine
      ii. Maxillary
      iii. Frontal
      iv. Nasal
      v. Sphenoid
  2. Lateral wall
    a. Superior and middle conchae of ethmoid bone
    b. Inferior concha
    c. Nasal bone
    d. Frontal process and nasal surface of maxilla
    e. Lacrimal bone
    f. Perpendicular plate of the palatine
    g. Medial pterygoid plate and body of sphenoid
  3. Roof
    a. Upper nasal cartilage
    b. Nasal bone
    c. Spine of frontal bone
    d. Cribriform plate of ethmoid bone
    e. Body of sphenoid bone
  4. Floor
    a. Palatine process of maxilla
    b. Horizontal plate of palatine bone
C. POSTERIOR APERTURE (CHOANA) opens into the nasopharynx

II. Clinical considerations
A. DEVIATED SEPTUM: a deflection of the septum from the midline can block or partly block the nasal passageways on the side toward which the deviation occurs
B. THE CLOSE RELATIONSHIP between the roof of the nasal cavity and the anterior cranial cavity as well as the orbit should be recalled when dealing with trauma, disease, or surgery of the nose
C. RHINITIS is an inflammation of the nasal cavity
D. THE NASAL POLYP is a focal submucosal thickening due to edema, which is pinkish-gray and edematous and may attain a remarkably large size

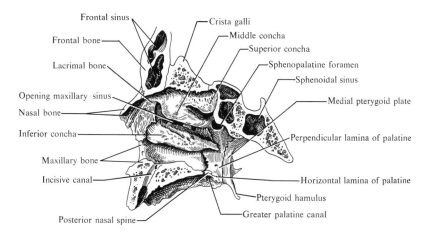

Frontal sinus
Frontal bone
Lacrimal bone
Opening maxillary sinus
Nasal bone
Inferior concha
Maxillary bone
Incisive canal
Posterior nasal spine

Crista galli
Middle concha
Superior concha
Sphenopalatine foramen
Sphenoidal sinus
Medial pterygoid plate
Perpendicular lamina of palatine
Horizontal lamina of palatine
Pterygoid hamulus
Greater palatine canal

BONES OF LATERAL NASAL WALL

NASAL CAVITY

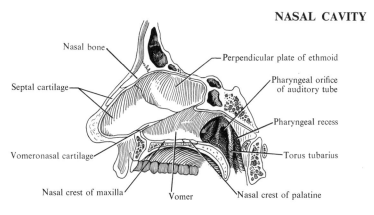

Nasal bone
Septal cartilage
Vomeronasal cartilage
Nasal crest of maxilla
Vomer

Perpendicular plate of ethmoid
Pharyngeal orifice of auditory tube
Pharyngeal recess
Torus tubarius
Nasal crest of palatine

NASAL SEPTUM

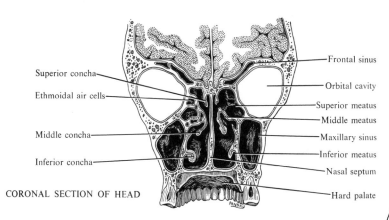

Superior concha
Ethmoidal air cells
Middle concha
Inferior concha

CORONAL SECTION OF HEAD

Frontal sinus
Orbital cavity
Superior meatus
Middle meatus
Maxillary sinus
Inferior meatus
Nasal septum
Hard palate

Pansky

61. MUCOUS MEMBRANE OF THE NOSE AND SINUSES

I. Mucous membrane
A. DIVISIONS: vestibule, respiratory area, olfactory area
B. STRUCTURE
    1. Vestibule: skin turned in at nares with coarse hairs and sebaceous glands
    2. Olfactory: over superior concha, roof, and upper third of septum; thick epithelial layer with supporting and olfactory cells. Proximal ends of latter are part of the olfactory nerve
    3. Respiratory: covers remainder of nasal cavity, continues into sinuses. Epithelium, pseudostratified columnar ciliated, with goblet cells

II. Sinuses (see also p. 125)
A. FRONTAL
    1. Location: behind supraciliary ridges of frontal bone, often extends dorsally into orbital roof, usually divided by a septum
    2. Drainage: frontonasal duct, through rostral ethmoidal cells to middle meatus
B. ETHMOIDAL
    1. Location: aggregations of thin-walled spaces (cells) in ethmoidal labyrinth between orbit and nasal cavities, arranged in 3 sets
    2. Drainage
        a. Anterior, into infundibulum of middle meatus
        b. Middle, on or above ethmoid bulla in middle meatus
        c. Posterior, into superior meatus, with interconnections to sphenoidal sinus
C. SPHENOIDAL
    1. Location: in body of sphenoid, usually asymmetrically divided by a septum
    2. Drainage: into sphenoethmoidal recess
D. MAXILLARY (largest)
    1. Location: in body of maxilla; roofed by orbit, wall of nasal cavity is medial; the alveolar process is lateral, and the sinus extends into the zygoma
    2. Drainage: into semilunar hiatus of middle meatus; accessory opening frequently found dorsal to hiatus

III. Clinical considerations
A. STRUCTURE OF MUCOUS MEMBRANE, with its motile cilia, glands, and rich blood supply is adapted to purifying, moistening, and warming air in order to protect lungs
B. SINUSES EFFECTIVE IN MAKING HEAD LIGHTER but, in bipeds, such as man, frequently cause difficulty because:
    1. Numerous connections to nasal cavities lead to easy infection
    2. When sinuses are infected, swelling of nasal mucosa around orifices slows drainage
    3. In maxillary and sphenoid, the lowest portions of the sinus lie below opening into nose, thus making complete drainage difficult
    4. Because of proximity of roots of upper teeth to maxillary sinus, it is sometimes difficult to differentiate between toothache and sinusitis
    5. Due to the thin bone between many of the sinuses and meninges there is a chance of infecting the latter
C. SINUSITIS is an inflammation of one or more accessory nasal sinuses and may be acute or chronic

NASAL SINUSES

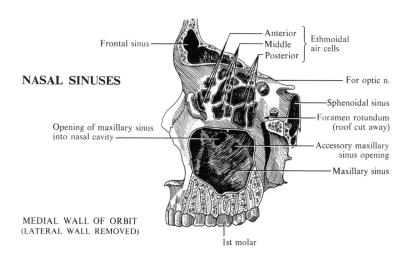

Frontal sinus

Anterior ⎱
Middle ⎰ Ethmoidal air cells
Posterior ⎱

For optic n.

Sphenoidal sinus

Foramen rotundum
(roof cut away)

Opening of maxillary sinus
into nasal cavity

Accessory maxillary
sinus opening

Maxillary sinus

MEDIAL WALL OF ORBIT
(LATERAL WALL REMOVED)

1st molar

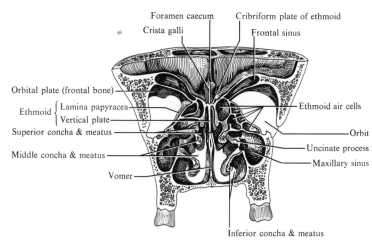

Foramen caecum   Cribriform plate of ethmoid

Crista galli   Frontal sinus

Orbital plate (frontal bone)

Ethmoid { Lamina papyracea
          Vertical plate

Ethmoid air cells

Orbit

Superior concha & meatus

Uncinate process

Middle concha & meatus

Maxillary sinus

Vomer

Inferior concha & meatus

FRONTAL SECTION THRU NASAL CAVITY

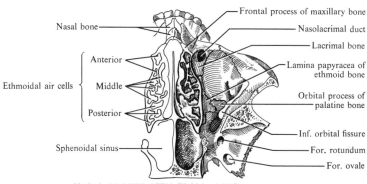

Nasal bone

Frontal process of maxillary bone

Nasolacrimal duct

Lacrimal bone

Anterior

Ethmoidal air cells {

Middle

Lamina papyracea of
ethmoid bone

Orbital process of
palatine bone

Posterior

Inf. orbital fissure

Sphenoidal sinus

For. rotundum

For. ovale

NASAL SINUSES SEEN FROM ABOVE (ALSO FLOOR OF ORBIT)

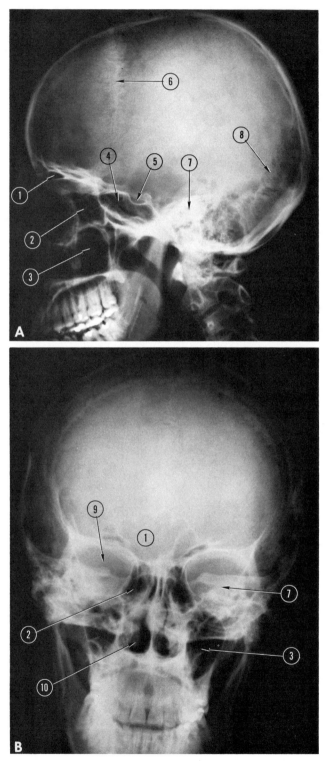

FIGURE 5. **Normal skull. A, Lateral view; B, anterior view.** *1,* Frontal sinus; *2,* ethmoid sinus; *3,* maxillary sinus; *4,* sphenoid sinus; *5,* sella turcica; *6,* coronal suture; *7,* petrous temporal bone; *8,* lambdoidal suture; *9,* orbit; *10,* nasal cavity.

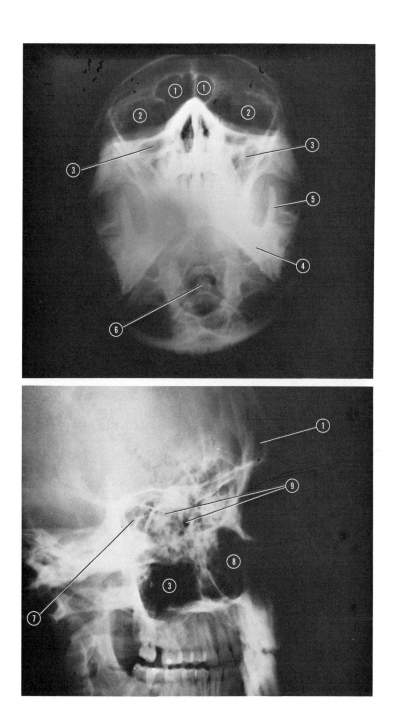

FIGURE 6. **Paranasal sinuses.** *1,* Frontal sinus; *2,* orbits; *3,* maxillary sinuses; *4,* mandible; *5,* coronoid process of mandible; *6,* dens of axis; *7,* sphenoid sinus; *8,* nasal cavity; *9,* ethmoidal sinuses.

62. ARTERIES OF THE NOSE

I. Anterior ethmoid
A. ORIGIN: ophthalmic
B. COURSE: from orbit through anterior ethmoid canal with nasal branches that pass through groove near crista galli
C. DISTRIBUTION: to both lateral wall and septum; also external branch to dorsolateral external surface of nose

II. Posterior ethmoid
A. ORIGIN: ophthalmic
B. COURSE: from orbit through posterior ethmoid canal with nasal branches through cribriform plate of ethmoid
C. DISTRIBUTION: anastomoses with branches of sphenopalatine artery and distributed with them

III. Sphenopalatine
A. ORIGIN: maxillary, third part
B. COURSE: leaves pterygopalatine fossa and enters nose through sphenopalatine foramen
C. DISTRIBUTION
   1. Posterior lateral nasal branches: conchae, meatuses
   2. Posterior septal branch: to nasal septum

IV. Greater palatine
A. ORIGIN: maxillary, third part
B. COURSE: leaves pterygopalatine fossa through pterygopalatine canal, runs in roof of mouth forward to incisive canal where terminal branch ascends to septum
C. DISTRIBUTION: to septum with septal branch of III, above

V. Septal and alar branches
A. ORIGIN: superior labial branch of facial
B. COURSE: cephalically from upper lip
C. DISTRIBUTION: to rostral septum and vestibule

VI. Clinical considerations
A. VASCULAR BED OF NASAL MUCOSA: vascularity is extremely great. Thus:
   1. Injuries to the membrane lead to profuse bleeding, especially where the arteries are large as they enter the nose. When severe, as at the posterior end of the middle concha, ligation of the external carotid may be necessary. When in the roof of the nose, intraorbital ligation of the ethmoidal arteries may be needed
   2. Irritating substances, either infectious or allergenic agents, which could lead to either engorgement or increased capillary permeability, will cause great swelling and occlusion of the air passageways
B. THE ARTERIES OF THE NOSE run in general with the larger nerves.
C. IN THE ANTERIOR LOWER PORTION of the nasal septum there are broad anastomoses between the major arteries of the nose, and this area often is involved when there is a nosebleed
D. BLEEDING close to the back end of the middle concha is usually from the sphenopalatine artery and can often be checked by packs placed here (if severe, the artery is ligated); bleeding from the roof of the nasal cavity originates from one of the ethmoidal arteries

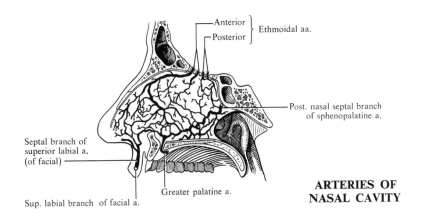

Anterior ⎫ Ethmoidal aa.
Posterior ⎬

Post. nasal septal branch
of sphenopalatine a.

Septal branch of
superior labial a.
(of facial)

**ARTERIES OF
NASAL CAVITY**

Greater palatine a.

Sup. labial branch of facial a.

ARTERIES OF NASAL SEPTUM

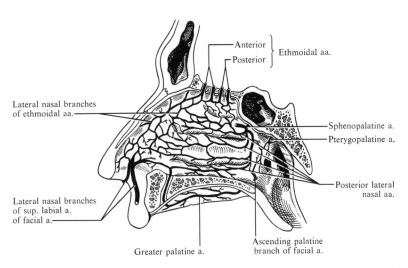

Anterior ⎫ Ethmoidal aa.
Posterior ⎬

Lateral nasal branches
of ethmoidal aa.

Sphenopalatine a.

Pterygopalatine a.

Posterior lateral
nasal aa.

Lateral nasal branches
of sup. labial a.
of facial a.

Greater palatine a.

Ascending palatine
branch of facial a.

ARTERIES OF LATERAL NASAL WALL

Pansky

63. NERVES OF THE NASAL CAVITY

I. Special sensory
A. OLFACTORY (I): nerve of smell
  1. Receptors: the neuroepithelial cells of the olfactory mucosa lie on the upper third of the nasal septum and on the superior nasal concha. Are a part of nerve
  2. Course: the fibers are gathered into bundles that pass through the cribriform plate of the ethmoid to enter the olfactory bulb

II. General sensory
A. OPHTHALMIC DIVISION OF THE TRIGEMINAL (V) NERVE, with its cell bodies in the trigeminal (semilunar) ganglion
  1. Nasociliary branch gives off an anterior ethmoid nerve, which leaves the orbit through the anterior ethmoid canal and sends nasal branches through the cleft near the crista galli to the anterior part of the nasal septum and lateral wall
B. MAXILLARY DIVISION OF THE TRIGEMINAL (V) NERVE, with its cell bodies in the trigeminal (semilunar) ganglion
  1. Pterygopalatine nerves
    a. Greater palatine nerve gives off posterior inferior nasal branches as this nerve lies in the pterygopalatine canal. Distributed to inferior concha, inferior and middle meatuses
    b. Posterior superior nasal branches enter nose through the sphenopalatine foramen to the superior and middle conchae
      i. Nasopalatine nerve runs over septum to the incisive canal
  2. Anterior superior alveolar nerve runs to the rostral part of the inferior meatus and the floor of the nasal cavity

III. Motor: autonomic system distributed to the glands and blood vessels of the mucous membrane
A. CRANIOSACRAL DIVISION (PARASYMPATHETIC)
  1. Preganglionic fibers, arising in the superior salivatory nucleus of the pons via the nervus intermedius and greater petrosal nerve to the pterygopalatine ganglion where they terminate
  2. Postganglionic fibers arise from cells in the pterygopalatine ganglion and are distributed with the pterygopalatine nerves
B. THORACOLUMBAR (SYMPATHETIC)
  1. Preganglionic fibers arise in the intermediolateral gray column of the upper thoracic cord and ascend in the cervical sympathetic trunk to terminate in the superior cervical ganglion
  2. Postganglionic fibers arise from cells in the superior cervical ganglion, ascend to the carotid plexus, and form the deep petrosal nerve, which joins the greater petrosal nerve to form the nerve of the pterygoid canal (vidian). The sympathetic fibers of this nerve pass through the pterygopalatine ganglion without synapse and are distributed as given above under IIIA, 2. It is thought that the effect of activity of this system is due to vasoconstriction of the arteries

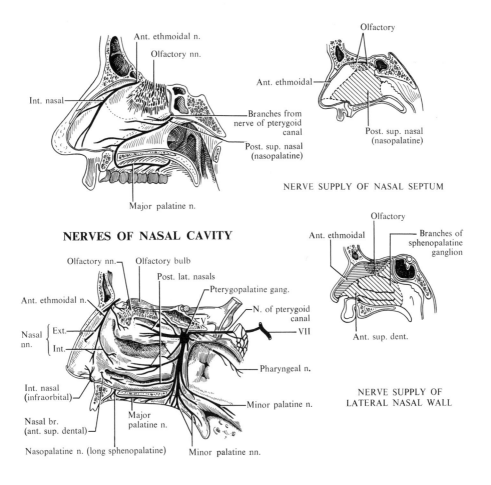

Ant. ethmoidal n.

Olfactory nn.

Int. nasal

Branches from
nerve of pterygoid
canal

Post. sup. nasal
(nasopalatine)

Major palatine n.

Olfactory

Ant. ethmoidal

Post. sup. nasal
(nasopalatine)

NERVE SUPPLY OF NASAL SEPTUM

NERVES OF NASAL CAVITY

Olfactory nn.

Olfactory bulb

Post. lat. nasals

Pterygopalatine gang.

Ant. ethmoidal n.

N. of pterygoid
canal

Nasal { Ext.
nn. { Int.

VII

Int. nasal
(infraorbital)

Pharyngeal n.

Nasal br.
(ant. sup. dental)

Minor palatine n.

Major
palatine n.

Nasopalatine n. (long sphenopalatine)

Minor palatine nn.

Olfactory

Ant. ethmoidal

Branches of
sphenopalatine
ganglion

Ant. sup. dent.

NERVE SUPPLY OF
LATERAL NASAL WALL

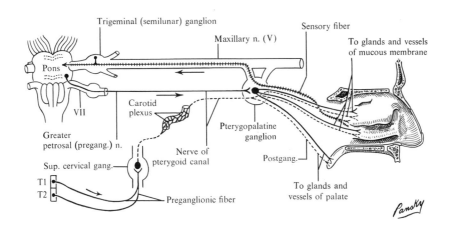

Trigeminal (semilunar) ganglion

Maxillary n. (V)

Sensory fiber

To glands and vessels
of mucous membrane

Pons

VII

Carotid
plexus

Pterygopalatine
ganglion

Greater
petrosal (pregang.) n.

Nerve of
pterygoid canal

Postgang.

Sup. cervical gang.

T1
T2

Preganglionic fiber

To glands and
vessels of palate

Pansky

- 133 -

64. SUMMARY OF THE CRANIAL NERVES

Name	Nuclei of Origin and Termination	Distribution	Function
I. Olfactory (sensory)	Central or deep process of olfactory bulb	Nasal mucous membranes	Sense of smell
II. Optic (sensory)	Ganglionic cells of retina	Retina of eye	Sense of sight
III. Oculomotor (motor)	Nucleus in floor of cerebral aqueduct	Sup., inf., and med. recti; inf. oblique; ciliaris; sphincter pupillae mm.	Motion
IV. Trochlear (motor)	Nucleus in floor of cerebral aqueduct	Superior oblique of eye	Motion
V. Trigeminal (mixed)	Fibers of sensory root arise from semilunar ganglion	1. *Ophthalmic* to cornea; ciliary body; iris; lacrimal gl.; conjunct.; mucous membrane of nasal cavity; skin of forehead, eyelid, eyebrow, and nose	Sensation
		2. *Maxillary* to dura, forehead, lower eyelid, lat. angles of orbit, upper lip, gums and teeth of upper jaw, mucous membrane, and skin of cheek and nose	Sensation
	Fibers of motor root arise from nucleus in pons	3. *Mandibular* to temple, auricle of ear, lower lip, lower part of face, teeth and gums of mandible; muscles of mastication; to mucous membrane of ant. part of tongue	Sensation and motion (some of VII reach tongue via lingual for taste)
VI. Abducens (motor)	Nucleus beneath floor of fourth ventricle	Lateral rectus muscle of eye	Motion

Name	Nuclei of Origin and Termination	Distribution	Function
VII. Facial (mixed)	Sensory from geniculate ganglion on nerve	Ant. 2/3 tongue (taste) via *chorda tympani* Middle ear	Taste
	Motor from nucleus in lower part of pons Sup. salivatory nucleus	To muscles of face, scalp, auricle, and superfic. neck. Stimulatory or excitatory to subman. and sublingual gls.	Gen. sense Secretion
VIII. Acoustic (sensory) 2 sets	Cochlear from bipolar cells in spiral gang. of cochlea	To organ of Corti	Sense of hearing
	Vestibular from bipolar cells in vestibular gang.	To semicircular canals	Sense of equilib.
IX. Glossopharyngeal (mixed)	Sensory fibers from sup. and inf. gang. on trunk of nerve Nucleus ambiguus and inf. salivatory nucleus	To mucous memb. of fauces, tonsils, pharynx, and post. 1/3 tongue To mm. of pharynx and secretory fibers to parotid gland	Sense of taste and gen. sense Motion Secretion
X. Vagus (mixed)	Sensory fibers from sup. gang. and inf. gang. (nodosum) on nerve trunk	To mucous memb. of larynx, trachea, lungs, esophagus, stomach, intestines, and gallbl.	Sensation
	Motor fibers from nucleus ambiguus in medulla and dorsal motor nucleus	To larynx, esophagus, stomach, sml. intestine, and part of lg. intest. Excitatory fibers to gastric and pancreatic glands	Motion Secretion
XI. Accessory (2 parts: cranial, spinal)	Cranial: from n. ambiguus	To pharyngeal and laryngeal brs. of vagus to pharynx and larynx	Motion
	Spinal: from spinal cord as low as C5	To sternocleidomastoid and trapezius mm.	Motion
XII. Hypoglossal (motor)	From hypoglossal nucleus in medulla	To muscles of tongue	Motion

65. HEAD, VISCERAL ARCHES, AND DETAILED NECK ANATOMY

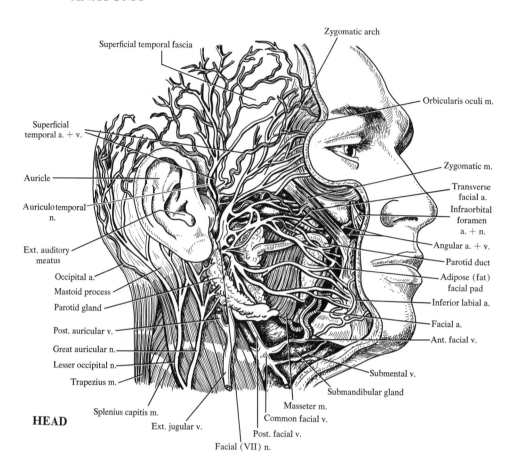

Zygomatic arch

Superficial temporal fascia

Orbicularis oculi m.

Superficial temporal a. + v.

Zygomatic m.

Auricle

Transverse facial a.

Infraorbital foramen a. + n.

Auriculotemporal n.

Angular a. + v.

Parotid duct

Ext. auditory meatus

Adipose (fat) facial pad

Occipital a.

Inferior labial a.

Mastoid process

Parotid gland

Facial a.

Post. auricular v.

Ant. facial v.

Great auricular n.

Lesser occipital n.

Trapezius m.

Submental v.

Submandibular gland

Masseter m.

HEAD

Splenius capitis m.

Common facial v.

Ext. jugular v.

Post. facial v.

Facial (VII) n.

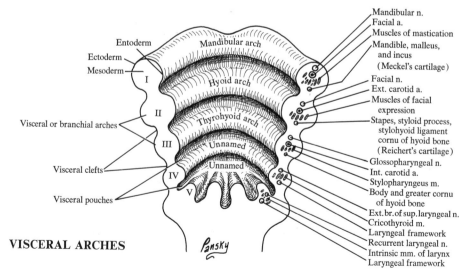

Mandibular n.

Facial a.

Entoderm

Mandibular arch

Muscles of mastication

Ectoderm

Mandible, malleus, and incus (Meckel's cartilage)

Mesoderm

I

Hyoid arch

Facial n.

Ext. carotid a.

Muscles of facial expression

II

Thyrohyoid arch

Visceral or branchial arches

Stapes, styloid process, stylohyoid ligament cornu of hyoid bone (Reichert's cartilage)

III

Unnamed

Glossopharyngeal n.

Unnamed

Int. carotid a.

Visceral clefts

Stylopharyngeus m.

IV

Body and greater cornu of hyoid bone

Visceral pouches

V

Ext. br. of sup. laryngeal n.

Cricothyroid m.

Laryngeal framework

VISCERAL ARCHES

Pansky

Recurrent laryngeal n.

Intrinsic mm. of larynx

Laryngeal framework

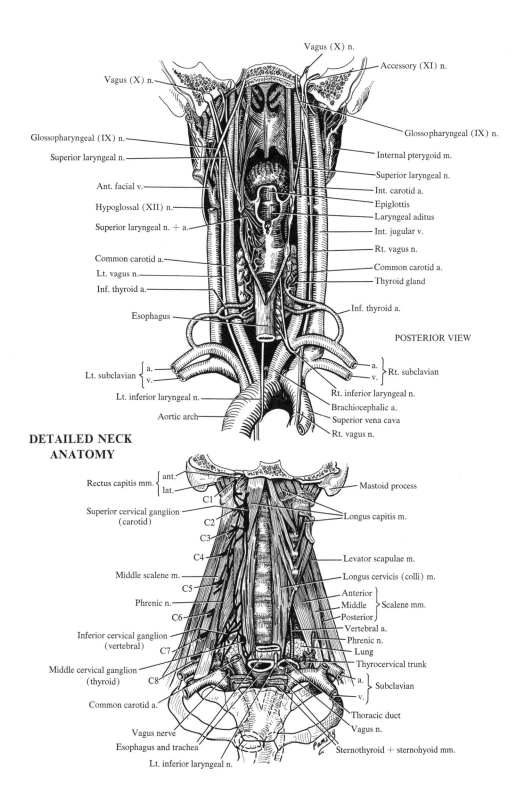

Vagus (X) n.

Accessory (XI) n.

Vagus (X) n.

Glossopharyngeal (IX) n.

Glossopharyngeal (IX) n.

Superior laryngeal n.

Internal pterygoid m.

Superior laryngeal n.

Ant. facial v.

Int. carotid a.

Hypoglossal (XII) n.

Epiglottis

Laryngeal aditus

Superior laryngeal n. + a.

Int. jugular v.

Rt. vagus n.

Common carotid a.

Common carotid a.

Lt. vagus n.

Thyroid gland

Inf. thyroid a.

Inf. thyroid a.

Esophagus

POSTERIOR VIEW

Lt. subclavian { a. v. }

{ a. v. } Rt. subclavian

Lt. inferior laryngeal n.

Rt. inferior laryngeal n.

Aortic arch

Brachiocephalic a.

Superior vena cava

Rt. vagus n.

**DETAILED NECK
ANATOMY**

Rectus capitis mm. { ant. lat. }

Mastoid process

C1

Superior cervical ganglion
(carotid)

Longus capitis m.

C2

C3

C4

Levator scapulae m.

Middle scalene m.

Longus cervicis (colli) m.

C5

Anterior

Phrenic n.

Middle } Scalene mm.

Posterior

C6

Vertebral a.

Phrenic n.

Inferior cervical ganglion
(vertebral)

Lung

C7

Thyrocervical trunk

Middle cervical ganglion
(thyroid)

C8

{ a. v. } Subclavian

Common carotid a.

Thoracic duct

Vagus nerve

Vagus n.

Esophagus and trachea

Sternothyroid + sternohyoid mm.

Lt. inferior laryngeal n.

- 137 -

UNIT TWO

Back

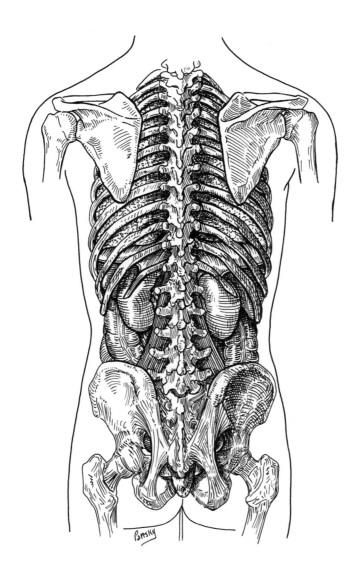

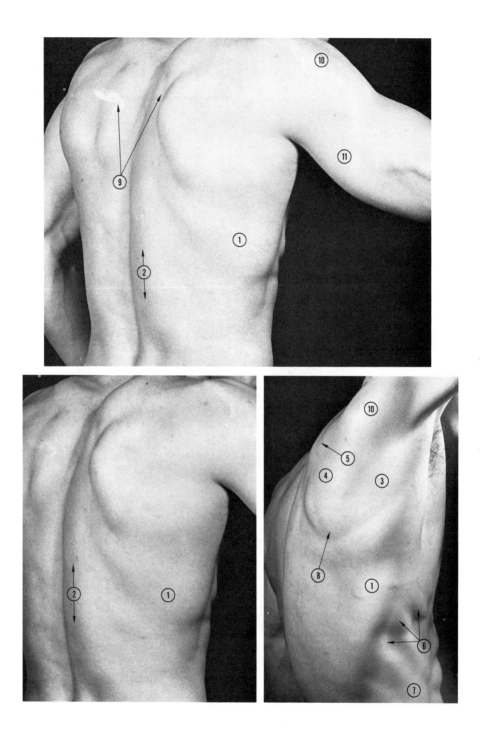

FIGURE 7. **Surface anatomy of back.** *1*, Latissimus dorsi; *2*, erector spinae; *3*, teres major m.; *4*, infraspinatus m.; *5*, supraspinatus m.; *6*, serratus anterior m.; *7*, external oblique m.; *8*, inferior angle of scapula; *9*, trapezius m.; *10*, deltoid m.; *11*, triceps brachii m.

– 140 –

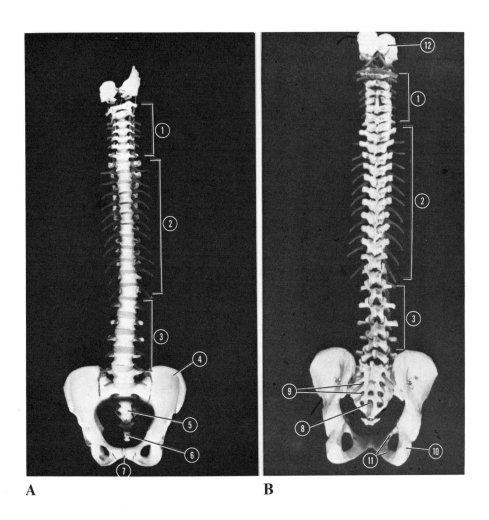

A   B

FIGURE 8. **Vertebral column. A, Anterior view; B, posterior view.** *1,* Cervical vertebra; *2,* thoracic vertebrae; *3,* lumbar vertebrae; *4,* ilium; *5,* sacral vertebrae; *6,* coccyx; *7,* pubic symphysis; *8,* sacral hiatus; *9,* posterior sacral foramina; *10,* ischium; *11,* pubis.

66. SPINAL NERVES AND CUTANEOUS NERVES OF THE BACK

I. **Spinal nerves:** arranged in 31 pairs grouped regionally: 8 cervical, 12 thoracic, 5 lumbar, 5 sacral, and 1 coccygeal
A. ATTACHMENTS TO SPINAL CORD
1. Posterior root: numerous rootlets along posterolateral sulcus
a. Has an oval swelling composed of nerve cell bodies—*posterior root ganglion*
2. Anterior root: rootlets from anterolateral sulcus
a. White ramus: preganglionic sympathetic fibers from root to sympathetic trunk
B. BEYOND GANGLION roots join to form nerve trunk
C. BRANCHES OF SPINAL NERVE TRUNK
1. Gray ramus: on all spinal nerves; contains postganglionic sympathetic fibers
2. Terminal branches: posterior and anterior primary divisions

II. **Distribution of primary divisions**
A. POSTERIOR PRIMARY DIVISION: all, except C1, S4, S5, and coccygeal have lateral and medial branches
1. C1: entire dorsal division supplies suboccipital muscles
2. C2: medial branch large, *great occipital nerve:* lateral branch, muscular
3. C3: medial branch to lower part of back of head; lateral branch, muscular
4. C4–C8: medial branches of C4 and 5 muscular and cutaneous; lateral branch is muscular; below C5 both divisions muscular
5. T1–T12: medial branches of upper 6 nerves to skin; lateral branches muscular. Medial branches of lower 6 nerves muscular; lateral branches cutaneous
6. L1–L5: medial branches, all muscular; lateral branches, L1–L3 to buttock as superior cluneal nerves; other lateral branches muscular
7. S1–S3: medial branches all muscular; lateral branches to skin on buttock
8. S4, S5, and coccygeal: posterior division to skin over coccyx
B. ANTERIOR PRIMARY DIVISIONS
1. C1–C4, C5–T1, L1–L4, and L4–S3 form plexuses
2. Thoracic nerves: T1–T11 *intercostal,* T12 *subcostal*
a. T1 has larger and smaller parts; larger to brachial plexus; small, first intercostal nerve
b. T2–T6: in intercostal spaces are thoracic intercostal nerves. Have muscular and cutaneous branches
i. Lateral cutaneous: to lateral part of chest and back. From T2: to arm, as *intercostobrachial nerve*
ii. Anterior cutaneous: termination of intercostal nerves. Supply medial chest
c. T6–T11 continue to abdomen as *thoracoabdominal nerves.* Have same arrangement as T2–T6
d. T12: below twelfth rib, the subcostal nerve
i. Lateral branch to anterolateral gluteal region

III. **Special features**
A. NOT ALL THE DORSAL RAMI OF SPINAL NERVES have cutaneous branches, and there is some variation in distribution
1. The first cervical nerve usually has no cutaneous branch
2. The cutaneous branches of the 2nd and 3rd cervical nerves are distributed upward to the scalp
3. The medial branches of the 4th and 5th cervical nerves typically reach skin on the back of the neck, while the next three typically have no cutaneous branches
4. All the posterior rami of the thoracic nerves typically have cutaneous branches

CUTANEOUS NERVES AND DERMATOMES

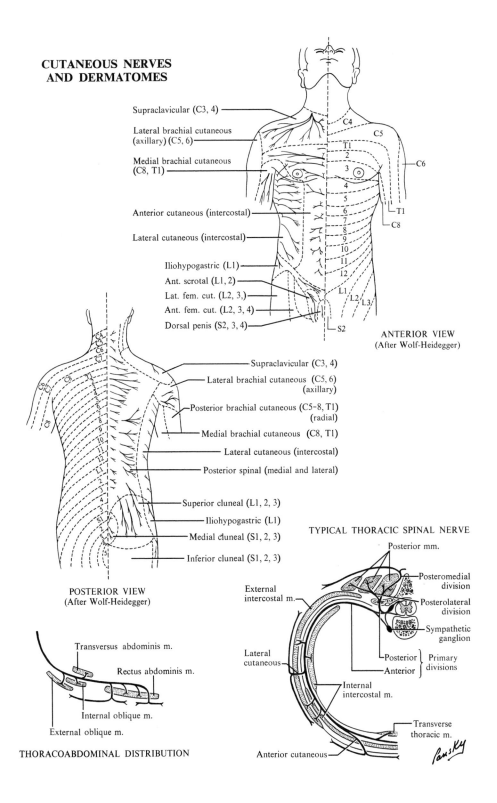

Supraclavicular (C3, 4)

Lateral brachial cutaneous (axillary) (C5, 6)

Medial brachial cutaneous (C8, T1)

Anterior cutaneous (intercostal)

Lateral cutaneous (intercostal)

Iliohypogastric (L1)

Ant. scrotal (L1, 2)

Lat. fem. cut. (L2, 3,)

Ant. fem. cut. (L2, 3, 4)

Dorsal penis (S2, 3, 4)

C4
C5
T1
2
3
4
5
6
7
8
9
10
11
12
L1
L2
L3
C6
T1
C8
S2

ANTERIOR VIEW
(After Wolf-Heidegger)

Supraclavicular (C3, 4)

Lateral brachial cutaneous (C5, 6) (axillary)

Posterior brachial cutaneous (C5-8, T1) (radial)

Medial brachial cutaneous (C8, T1)

Lateral cutaneous (intercostal)

Posterior spinal (medial and lateral)

Superior cluneal (L1, 2, 3)

Iliohypogastric (L1)

Medial cluneal (S1, 2, 3)

Inferior cluneal (S1, 2, 3)

POSTERIOR VIEW
(After Wolf-Heidegger)

Transversus abdominis m.

Rectus abdominis m.

Internal oblique m.

External oblique m.

THORACOABDOMINAL DISTRIBUTION

TYPICAL THORACIC SPINAL NERVE

Posterior mm.

Posteromedial division

Posterolateral division

Sympathetic ganglion

Posterior } Primary
Anterior } divisions

External intercostal m.

Lateral cutaneous

Internal intercostal m.

Transverse thoracic m.

Anterior cutaneous

- 143 -

67. THE VERTEBRAL COLUMN

I. Composition: 32–34 vertebrae; 7 cervical, 12 thoracic, 5 lumbar, 5 sacral, 3–5 coccygeal

II. Length
A. IN MALE: 71 cm
B. IN FEMALE: 61 cm

III. Curvatures
A. CERVICAL: CONVEX ANTERIORLY, from apex of odontoid process to T2
B. THORACIC: CONCAVE ANTERIORLY, from middle T2 to middle T12
C. LUMBAR: CONVEX ANTERIORLY, from middle T12 to sacrovertebral articulation; convexity greatest at lower 3 segments
D. SACRAL: CONCAVE ANTEROCAUDALLY
E. AT BIRTH only the thoracic and sacral curves (primary) are present. The secondary curves develop after birth: cervical from elevation and extension of head in infancy; lumbar from assumption of erect position when child begins to walk

IV. Surfaces
A. ANTERIOR: in general, width gradually broadens with widest point at base of sacrum, then tapers sharply to apex at tip of coccyx
B. POSTERIOR: spinous processes in midline
   1. Spinous processes
      a. Cervical, except for 2 and 7, are short and horizontal, have bifid extremity
      b. Thoracic: long, upper spines directed obliquely caudally with some separation; middle spines, almost vertical in direction with overlapping; lower spines, shorter and directed almost posteriorly with separation between them
      c. Lumbar: short, thick, and separated by interval
   2. Vertebral groove, on either side of spinous processes
      a. Formed by lamina and transverse processes; shallow in cervical and lumbar areas; deep in thoracic area
   3. The articular processes are lateral to the groove
   4. The transverse processes are the most lateral
C. LATERAL: separated from posterior surface by articular processes in cervical and lumbar areas, transverse processes in thoracic region
   1. Anteriorly: sides of bodies of vertebrae show articular facets for ribs in thorax
   2. Intervertebral foramina formed by the opposition of notches in adjoining vertebrae
      a. Increase in size from above downward

V. Vertebral canal consists of the foramina of adjoining vertebrae

VI. Special features
A. IN THE MIDTHORACIC REGION the tips of the spines are below the level of the bodies of the corresponding vertebrae
B. THE TRANSVERSE PROCESSES are in front of the articular processes in the cervical region and in line with the intervertebral foramina, in the thoracic region they are behind both, and in the lumbar region they are behind the foramina but in front of the articular processes

VII. Clinical considerations
A. ABNORMAL CURVATURE
   1. Kyphosis: increased backward curvature of spine (mostly in thorax)
   2. Lordosis: an exaggerated forward curvature of the lumbar segment
   3. Scoliosis: commonest. Lateral deviation most frequent in thorax

VERTEBRAL COLUMN

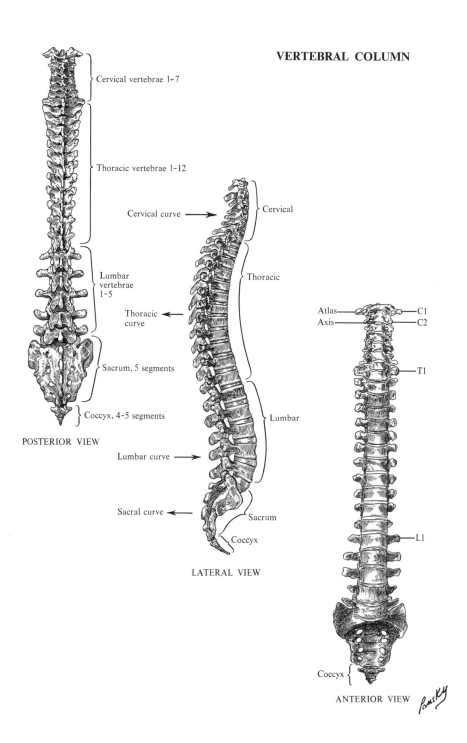

Cervical vertebrae 1-7

Thoracic vertebrae 1-12

Lumbar vertebrae 1-5

Sacrum, 5 segments

Coccyx, 4-5 segments

POSTERIOR VIEW

Cervical curve → Cervical

Thoracic

Thoracic curve ←

Lumbar

Lumbar curve →

Sacral curve ← Sacrum

Coccyx

LATERAL VIEW

Atlas — C1
Axis — C2

T1

L1

Coccyx

ANTERIOR VIEW

Pansky

-145-

68. TYPICAL AND CERVICAL VERTEBRAE

I. **Vertebrae made up of 2 parts:** a body and vertebral (neural) arch. The *vertebral foramen* is surrounded by parts of both
A. Body is largest and heaviest part
B. Vertebral arch consists of 2 *pedicles* and 2 *laminae* and has 7 *processes:* 1 *spinous,* 4 *articular,* and 2 *transverse*
    1. Pedicle: joins arch to posterolateral body
       a. Concavities in upper and lower surfaces: *vertebral notches*
    2. Laminae: plates extending posteriorly and medially from pedicle
    3. Spinous processes directed posteriorly and caudally from union of laminae
    4. Articular processes extend upward and downward from point where pedicles and laminae join
       a. 2 superior, articular surfaces face posteriorly
       b. 2 inferior, articular surfaces face anteriorly
    5. Transverse processes project laterally between the superior and inferior articular processes

II. **Typical cervical vertebrae:** smallest of true vertebrae
A. Body, small
B. Vertebral foramen, large and triangular
C. Spinous process, short and bifid
D. Transverse process contains a foramen—*transverse foramen*
    1. Anterior and posterior tubercles on process

III. **Atypical cervical vertebrae** (only points of difference will be given)
A. First—the atlas
    1. Has no body, is more or less circular
    2. Anterior arch has an anterior tubercle, posterior to which is an oval articular facet
    3. Posterior arch has a posterior tubercle
    4. Superior articular facets are very large concave ovals, facing upward
    5. Inferior articular facets are circular
    6. Transverse processes are large, anterior and posterior tubercles fused
B. Second—the axis (or epistropheus)
    1. Body has a long, pointed projection directed cranially—the *dens* (odontoid process)
       a. Process has an oval articular facet on anterior surface
    2. Pedicles are strong, are fused with sides of body and *dens* (odontoid process), and their upper surface forms the superior articular facet
    3. Transverse processes are small, end in single tubercle
       a. Foramen transversarium set obliquely
C. Seventh—vertebra prominens
    1. Spinous process: thick, directed almost straight posteriorly; is not bifid, but ends in tubercle
    2. Transverse processes are large, tubercles not clear

IV. **Special features**
A. The cervical vertebrae are distinguished by their small size
B. The spinous processes are typically short, but those of the 6th and 7th are much longer; the 3rd through the 6th are usually bifid in white persons, but not in black persons; and there is no spinous process on the 1st cervical vertebra

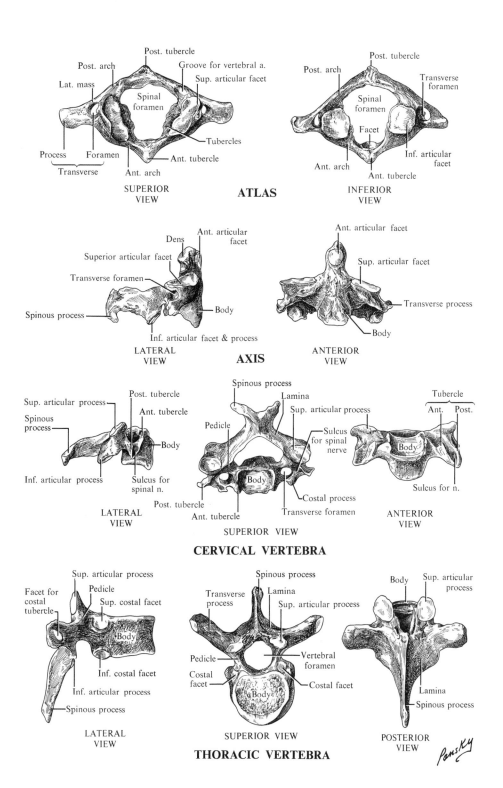

Post. tubercle
Post. arch
Lat. mass
Spinal foramen
Groove for vertebral a.
Sup. articular facet
Process
Foramen
Transverse
Ant. tubercle
Ant. arch
Tubercles

SUPERIOR VIEW

Post. tubercle
Post. arch
Spinal foramen
Transverse foramen
Facet
Ant. arch
Ant. tubercle
Inf. articular facet

INFERIOR VIEW

ATLAS

Dens
Ant. articular facet
Superior articular facet
Transverse foramen
Spinous process
Body
Inf. articular facet & process

LATERAL VIEW

Ant. articular facet
Sup. articular facet
Transverse process
Body

ANTERIOR VIEW

AXIS

Sup. articular process
Spinous process
Post. tubercle
Ant. tubercle
Body
Inf. articular process
Sulcus for spinal n.

LATERAL VIEW

Spinous process
Lamina
Sup. articular process
Pedicle
Sulcus for spinal nerve
Body
Post. tubercle
Ant. tubercle
Costal process
Transverse foramen

SUPERIOR VIEW

Tubercle
Ant.   Post.
Body
Sulcus for n.

ANTERIOR VIEW

CERVICAL VERTEBRA

Sup. articular process
Facet for costal tubercle
Pedicle
Sup. costal facet
Body
Inf. costal facet
Inf. articular process
Spinous process

LATERAL VIEW

Spinous process
Transverse process
Lamina
Sup. articular process
Pedicle
Vertebral foramen
Costal facet
Body
Costal facet

SUPERIOR VIEW

Body
Sup. articular process
Lamina
Spinous process

POSTERIOR VIEW

THORACIC VERTEBRA

Pansky

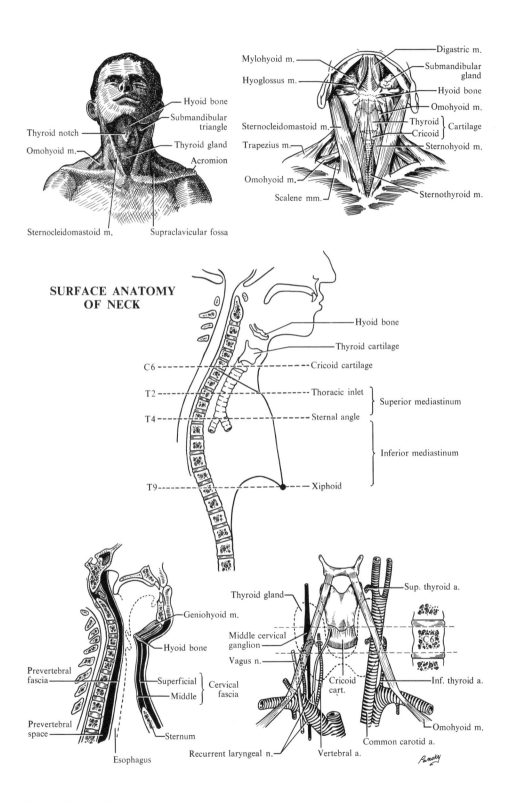

SURFACE ANATOMY
OF NECK

FIGURE 9. **Relations of cervical region.**

-148-

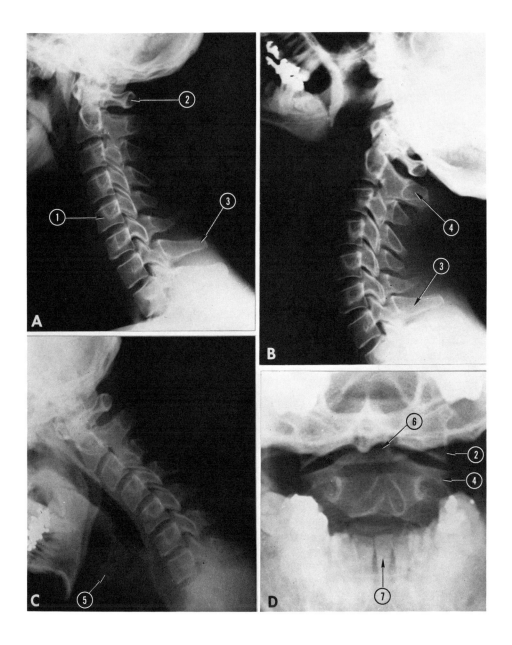

FIGURE 10. **Cervical vertebrae.** **A, Head erect; B, head extended; C, head flexed; D, view through open mouth showing second cervical vertebra.** *1,* Body, C5; *2,* atlas, C1; *3,* spine C7; *4,* axis, C2; *5,* hyoid bone; *6,* dens of axis; *7,* teeth of lower jaw.

69. THE THORACIC AND LUMBAR VERTEBRAE—SACRUM AND COCCYX

I. Thoracic vertebrae (see p. 147)

A. GENERAL CHARACTERISTICS: body increases in size from above downward, has facets or demifacets for rib articulation; laminae are broad and thick; spinous processes are long and directed obliquely caudally; the superior articular processes are thin with facets directed posteriorly; the inferior articular processes are short, and their facets are directed anteriorly; the transverse processes are thick, strong, and have articular facets for rib tubercles

B. SPECIAL FEATURES
   1. T1 has one entire facet on each side of body for first rib
   2. T10 has a single articular facet on each side
   3. T11: large body, large articular facet; spinous process is short and almost horizontal; transverse process is short with no articular facet
   4. T12 resembles both T11 and L1; inferior articular facet is directed laterally; transverse process has superior, inferior, and lateral tubercles

II. Lumbar vertebrae (largest)

A. GENERAL CHARACTERISTICS: body is large, wide, and thick; pedicles are strong and directed posteriorly; laminae are broad and strong; spinous processes are thick, broad, and directed posteriorly, superior articular processes are directed medially and posteriorly; inferior articular processes are directed anteriorly and laterally; transverse processes are long, slender, and have upper tubercle at junction with superior articular process called *mammillary process* and inferior tubercle at base of process called *accessory process*

B. SPECIAL FEATURE: L5 has a heavy body, small spinous process, and thick transverse process

III. Sacrum: a fusion of 5 segments, triangular in shape

A. PELVIC SURFACE: concave, crossed by 4 transverse ridges; *pelvic (anterior) sacral foramina* are seen at ends of the ridges

B. POSTERIOR SURFACE: convex, *median sacral crest* at midline, *sacral groove* on either side of median crest, *sacral articular (intermediate) crest* lateral to groove, which terminate as the *sacral cornua,* row of *posterior sacral foramina* lateral to articular crest; *lateral crests* lie lateral to foramina

C. LATERAL SURFACE: upper half, *auricular surface* with *sacral tuberosity* just behind this; inferolateral angle is at lower end of this surface

D. BASE: directed upward with a large, oval articular surface in middle of body just behind which is the *sacral canal*

E. SUPERIOR SURFACE exhibits a projecting anterior border—the *sacral promontory; superior articular processes* are supported by short, heavy pedicles and laminae which enclose sacral canal

F. ALA: on either side of body of sacrum, formed of costal and transverse processes

G. APEX: directed caudally, has oval articular facet at end

IV. Coccyx: formed by fusion of 3–5 segments

A. ANTERIOR SURFACE: slightly convex with transverse ridges

B. POSTERIOR SURFACE: convex, with transverse ridges; has articular crest (as sacrum), the cephalic end of which projects upward as *coccygeal cornua*

C. BASE: oval articular facet

D. APEX: caudally directed, rounded, but may be bifid

3RD LUMBAR VERTEBRA

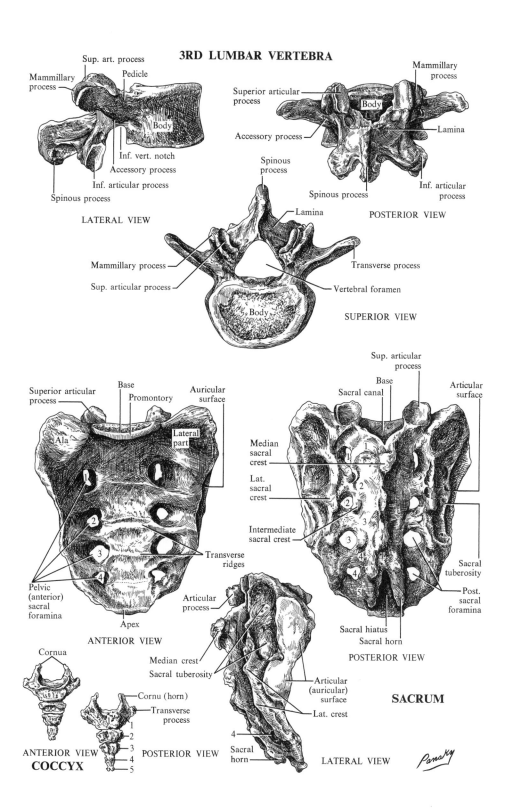

Sup. art. process
Pedicle
Mammillary process
Body
Inf. vert. notch
Accessory process
Inf. articular process
Spinous process

LATERAL VIEW

Mammillary process
Superior articular process
Body
Lamina
Accessory process
Spinous process
Inf. articular process

POSTERIOR VIEW

Spinous process
Lamina
Mammillary process
Sup. articular process
Transverse process
Vertebral foramen
Body

SUPERIOR VIEW

Superior articular process
Base
Promontory
Auricular surface
Ala
Lateral part
1
2
3
4
Pelvic (anterior) sacral foramina
Transverse ridges
Apex

ANTERIOR VIEW

Sup. articular process
Base
Sacral canal
Articular surface
Median sacral crest
Lat. sacral crest
Intermediate sacral crest
1
2
3
4
5
Sacral tuberosity
Post. sacral foramina
Sacral hiatus
Sacral horn

POSTERIOR VIEW

SACRUM

Cornua

ANTERIOR VIEW
COCCYX

Cornu (horn)
Transverse process
1
2
3
4
5

POSTERIOR VIEW

Articular process
Median crest
Sacral tuberosity
Articular (auricular) surface
Lat. crest
4
Sacral horn

LATERAL VIEW

Pansky

70. SUPERFICIAL BACK MUSCLES

I.

Name	Origin	Insertion	Action	Nerve
Trapezius	External occipital protuberance Medial part superior nuchal line of occipital bone Ligamentum nuchae Spinous proc. C7, T1–T12 vertebrae Supraspinous ligament	Post. border lat. clavicle Med. margin acromion Post. border spine of scapula Tubercle on spine of scapula	Rotates scapula to raise point of shoulder Adducts scapula Upper part raises scapula Lower part lowers, and pulls scapula down Upper part draws head to same side and turns face to opposite side Two sides together draw head back	Spinal accessory (XI) C3 C4
Latissimus dorsi	Through lumbar aponeurosis to spines T6–12 vertebrae Spines of lumbar and sacral vertebrae and supraspinous lig. Post. iliac crest Directly from iliac crest Lower 4 ribs	Bottom of intertubercular groove of humerus	Extends, adducts, and rotates arm medially Draws shoulder downward and backward Helps in climbing	Thoraco-dorsal (long subscapular)

II. Special features

A. THE APONEUROSIS in the lower fibers of the trapezius as they converge near scapula glides over a smooth area at the medial end of scapular spine

B. THERE ARE INTERDIGITATIONS of the costal origin of the latissimus dorsi with the external abdominal oblique muscle

C. THERE IS A TWISTING of the fibers of the latissimus dorsi as they converge toward insertion on humerus

D. THE UPPER PART OF THE TRAPEZIUS, in rotating the scapula, helps the serratus anterior (see p. 196) in making possible abduction of more than 90°

E. WHEN THE HANDS ARE FIXED by gripping an object above the head, the latissimus dorsi helps the pectoralis major (see p. 196) in drawing the body upward

F. TRIANGLE OF AUSCULTATION: medially, the trapezius; laterally, the scapula; below, the latissimus dorsi. The triangle can be enlarged by flexing the trunk with the arms across the chest

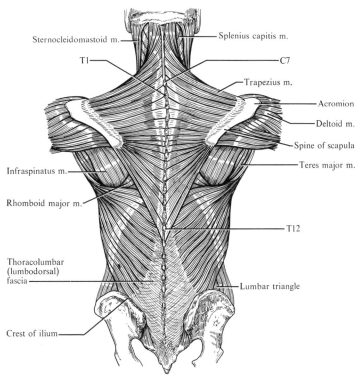

Sternocleidomastoid m.
Splenius capitis m.
T1
C7
Trapezius m.
Acromion
Deltoid m.
Spine of scapula
Teres major m.
Infraspinatus m.
Rhomboid major m.
T12
Thoracolumbar (lumbodorsal) fascia
Lumbar triangle
Crest of ilium

SUPERFICIAL POSTERIOR (DORSAL) MUSCLES

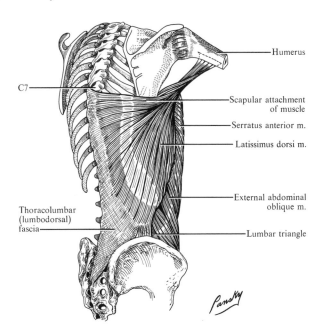

Humerus
C7
Scapular attachment of muscle
Serratus anterior m.
Latissimus dorsi m.
External abdominal oblique m.
Thoracolumbar (lumbodorsal) fascia
Lumbar triangle

Pansky

71. MUSCLES OF THE BACK, THE RHOMBOID LAYER, AND THE THORACOLUMBAR (LUMBODORSAL) FASCIA

I. Muscles

Name	Origin	Insertion	Action	Nerve
Levator scapulae	Posterior tubercles, transverse process C1–4	Medial border, scapula above spiné	Elevates scapula Rotates scapula Scapula fixed, extends & lat. bends neck	C3, C4; C5, through dorsal scapular
Rhomboid minor	Lig. nuchae Spines C7 through T1	Medial border, scapula at root of spine	Draws scapula medially Holds scapula to chest Depresses shoulder	Dorsal scapular
Rhomboid major	Spines, T2 through T5 Supraspinous lig.	Medial border, scapula below spine	As above	Dorsal scapular

II. **Nuchal and thoracolumbar (lumbodorsal) fascia:** muscles of back and dorsum of neck are enclosed by fascia. This fascial sheath attaches medially to the ligamentum nuchae, tips of spinous processes, and supraspinous ligaments of entire column and to the medial crest of sacrum. In the cervical and lumbar regions it is attached to the transverse processes of the vertebrae. In the thoracic region it joins the angles of the ribs (lateral to the iliocostalis muscle and to the intercostal fascia)

  A. IN THE THORACIC REGION, it is thin and transparent; in the lumbar region, it is dense and very strong

  B. IN THE LUMBAR REGION, it extends lateral to the transverse processes, investing the sacrospinalis muscle to become continuous with the aponeurosis of origin of the transversus abdominis muscle. Thus, it has 2 layers

    1. Posterior layer (lumbar aponeurosis) passes over the sacrospinalis muscle

    2. Anterior layer lies deep to the sacrospinalis. Medially it is attached to the transverse processes of the lumbar vertebrae

      a. The posterior lumbocostal ligament is a thickening of this layer between twelfth rib and the transverse process of L1

  C. BELOW THE LUMBAR REGION (see p. 153) it attaches to the iliac crest and lateral crest of sacrum

III. Clinical considerations

  A. NODOSE LUMBAGO (RHEUMATISM): a form of rheumatism characterized by nodule formation, often in areas where the posterior layer of the thoracolumbar fascia attaches to the bones (especially the iliac and sacral crests). This may lead to severe and incapacitating pain

  B. RHEUMATOID SPONDYLITIS (Marie-Strümpell disease) is characterized by inflammation of the cartilaginous joints between the vertebral bodies and of the gliding joints between the vertebral arches of the spine. It begins in the lumbar area and extends upward

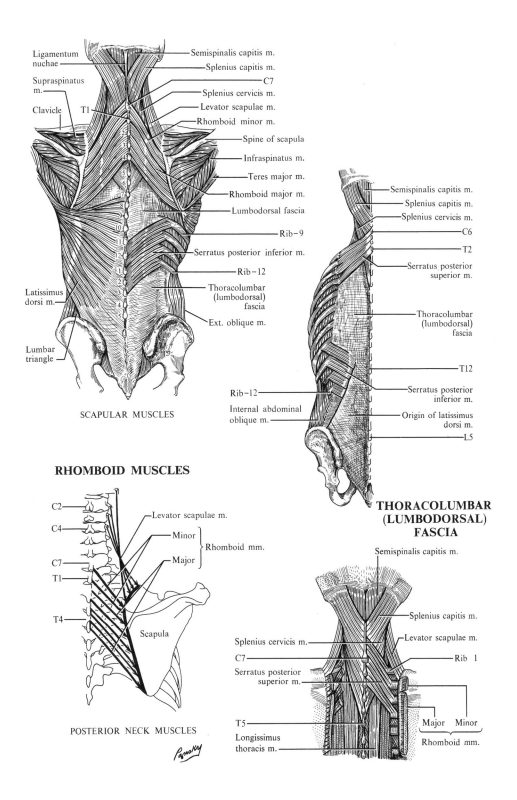

Ligamentum nuchae

Supraspinatus m.

Clavicle

T1

Latissimus dorsi m.

Lumbar triangle

Semispinalis capitis m.

Splenius capitis m.

C7

Splenius cervicis m.

Levator scapulae m.

Rhomboid minor m.

Spine of scapula

Infraspinatus m.

Teres major m.

Rhomboid major m.

Lumbodorsal fascia

Rib-9

Serratus posterior inferior m.

Rib-12

Thoracolumbar (lumbodorsal) fascia

Ext. oblique m.

SCAPULAR MUSCLES

Semispinalis capitis m.

Splenius capitis m.

Splenius cervicis m.

C6

T2

Serratus posterior superior m.

Thoracolumbar (lumbodorsal) fascia

T12

Rib-12

Serratus posterior inferior m.

Internal abdominal oblique m.

Origin of latissimus dorsi m.

L5

THORACOLUMBAR (LUMBODORSAL) FASCIA

RHOMBOID MUSCLES

C2

C4

C7

T1

T4

Levator scapulae m.

Minor

Major

Rhomboid mm.

Scapula

POSTERIOR NECK MUSCLES

Semispinalis capitis m.

Splenius capitis m.

Splenius cervicis m.

Levator scapulae m.

C7

Rib 1

Serratus posterior superior m.

T5

Longissimus thoracis m.

Major Minor

Rhomboid mm.

-155-

72. DEEP MUSCLES OF THE BACK—PART I

I. Superficial layer. Transversocostal group (see p. 159)

Name	Origin	Insertion
Splenius capitis	Lig. nuchae Spines C7, T1–3	Lat. part of occipital bone Mastoid of temporal bone
Splenius cervicis	Spines T3–6	Trans. proc. C1–3
Erector spinae Iliocostalis lumborum	Mid. crest, sacrum Spines T11–L5 Post. iliac crest Lat. crest of sacrum	Angles, lower 6 ribs
Iliocostalis thoracis	Upper borders of angles of lower 6 ribs	Upper borders, ribs 1–6 Trans. proc. C7
Iliocostalis cervicis	Angles, ribs 3–6	Trans. proc. C4–6
Longissimus thoracis	See Erector spinae above	Trans. proc. lumbar and thoracic vertebrae Lower 10 ribs
Longissimus cervicis	Trans. proc. T1–5	Trans. proc. C2–6
Longissimus capitis	Trans. proc. T1–5 Artic. proc. C5–7	Post. margin, mastoid proc.
Spinalis thoracis	Spines T11–L2	Spines of T1–8
Spinalis cervicis	Spine C7 (T1–2)	Spine C2 (C2–4)

II. Deep layer. Transversospinal group (see p. 159)

Name	Origin	Insertion
Semispinalis thoracis	Trans. proc. T6–10	Spines C6–T4
Semispinalis cervicis	Trans. proc. T1–6	Spines C2–5
Semispinalis capitis	Trans. proc. C7–T7 Artic. proc. C4–6	Planum nuchale of occip. bone
Spinalis capitis	With semispinalis capitis	
Multifidus Sacral part	Post. sacrum Aponeurosis of sacrospinalis Post. sup. iliac spine Post. sacroiliac lig.	Each part of the muscle crosses 1 to 4 vertebrae to reach spines of vertebrae from C2–L5
Lumbar part	From all mammillary proc.	
Thoracic part	From all transverse proc.	
Cervical part	Artic. proc. C4–7	

Rotatores
  Longi
  Breves
	
	Trans. proc. of 1 vertebra
	Trans. proc. of 1 vertebra
		
		Spine, 2 vertebrae above
		Spine, next vertebrae above

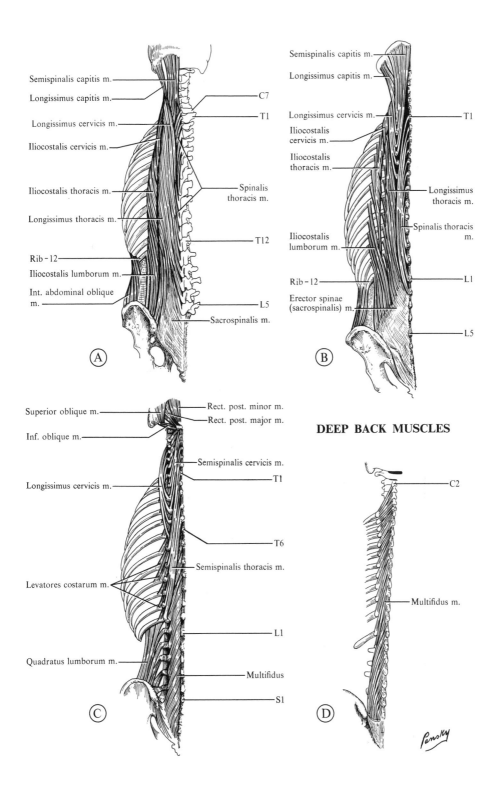

A

Semispinalis capitis m.
Longissimus capitis m.
Longissimus cervicis m.
Iliocostalis cervicis m.
Iliocostalis thoracis m.
Longissimus thoracis m.
Rib - 12
Iliocostalis lumborum m.
Int. abdominal oblique m.

C7
T1
Spinalis thoracis m.
T12
L5
Sacrospinalis m.

B

Semispinalis capitis m.
Longissimus capitis m.
Longissimus cervicis m.
Iliocostalis cervicis m.
Iliocostalis thoracis m.
Iliocostalis lumborum m.
Rib - 12
Erector spinae (sacrospinalis) m.

T1
Longissimus thoracis m.
Spinalis thoracis m.
L1
L5

C

Superior oblique m.
Inf. oblique m.
Longissimus cervicis m.
Levatores costarum m.
Quadratus lumborum m.

Rect. post. minor m.
Rect. post. major m.
Semispinalis cervicis m.
T1
T6
Semispinalis thoracis m.
L1
Multifidus
S1

DEEP BACK MUSCLES

D

C2
Multifidus m.

Pansky

73. MUSCLES OF THE BACK—PART II

II. Deep layer (continued)

Name	Location and Attachments
Interspinalis	Connect the apices of spinous processes of adjoining vertebrae from C2 to C3, C7 to T2, T11 to T12, L1 to L5
Intertransverse	
Anterior	Interconnect the ant. tubercles of trans. processes from C1 to T1, T10 to L1
Lateral	Between adjoining trans. processes of lumbar vertebrae
Medial	Between accessory process and mammillary processes of adjoining vertebrae of lumbar region
Posterior	Interconnect post. tubercles of trans. processes of lumbar vertebrae

III. Nerve supply: all by posterior primary divisions of spinal nerves

IV. Action: in general, to extend vertebral column
  A. Specific, actions in addition to the above:
  1. Splenius draws head back, bends head laterally and rotates face to the same side
  2. Iliocostalis bend vertebral column to side; the lumborum group depress ribs
  3. Longissimus thoracis and cervicis bend column to one side, depress ribs
  4. Longissimus capitis extends head, bends head to side, rotates face to same side
  5. Semispinalis thoracis and cervicis rotate column to opposite side
  6. Semispinalis capitis extends head, rotates head to opposite side
  7. Multifidis rotates column to opposite side
  8. Rotatores rotate column to opposite side
  9. Intertransverse bends column to same side

V. Suboccipital muscles

A. Name	Origin	Insertion	Action
Rectus capitis post. major	Spinous process, axis	Inf. nuchal line and bone below	Extends head Rotates to same side
Rectus capitis post. minor	Tubercle on post. arch of atlas	Inf. nuchal line and bone medial to it	Extends head
Obliquus capit. inferior	Apex, spine of axis	Trans. proc. atlas	Turns head, same side
Obliquus capit. superior	Trans. proc. of atlas	Occip. bone between sup. and inf. nuchal lines	Extends head and bends it to same side

  B. SPECIAL FEATURES
  1. Suboccipital triangle (see p. 55): medially and above, by rectus capitis posterior major; above and laterally, by obliquus capitis superior; below and laterally, obliquus capitis inferior; roof, fascia under semispinalis capitis; floor, atlanto-occipital membrane and posterior arch of atlas. The vertebral artery and the first cervical nerve lie on the upper surface of the arch of atlas

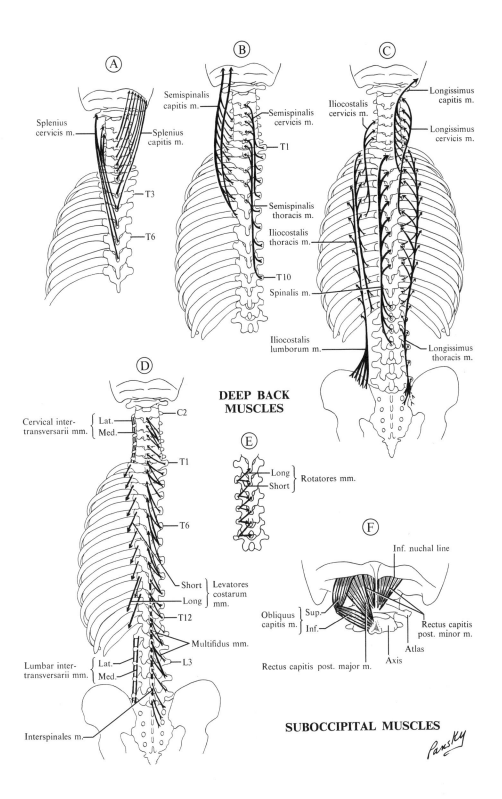

(A)

Splenius
cervicis m.

Splenius
capitis m.

T3

T6

(B)

Semispinalis
capitis m.

Semispinalis
cervicis m.

T1

Semispinalis
thoracis m.

Iliocostalis
thoracis m.

T10

Spinalis m.

(C)

Iliocostalis
cervicis m.

Longissimus
capitis m.

Longissimus
cervicis m.

Iliocostalis
lumborum m.

Longissimus
thoracis m.

**DEEP BACK
MUSCLES**

(D)

C2

Cervical inter-
transversarii mm. { Lat.
Med.

T1

T6

Short } Levatores
costarum
Long } mm.

T12

Multifidus mm.

Lumbar inter-
transversarii mm. { Lat.
Med.

L3

Interspinales m.

(E)

Long }
Short } Rotatores mm.

(F)

Inf. nuchal line

Obliquus
capitis m. { Sup.
Inf.

Rectus capitis
post. minor m.

Atlas

Axis

Rectus capitis post. major m.

SUBOCCIPITAL MUSCLES

Pansky

- 159 -

74. ARTERIAL BLOOD SUPPLY OF THE SPINAL CORD

I. Posterior spinal arteries: 1 on each side

A. ORIGIN: from vertebral artery, lateral to medulla oblongata

B. COURSE: descends through foramen magnum to spinal cord, where it runs downward anterior to dorsal roots of spinal nerves. Extends to lower end of cord and onto the cauda equina

C. BRANCHES: forms free anastomoses around dorsal roots with communications to:
1. Opposite side
2. Posterior funiculus
3. Posterior gray columns
4. Lateral funiculus (in part)

II. Anterior spinal arteries

A. ORIGIN: medial branch from each vertebral artery

B. COURSE: 2 arteries course downward and fuse into a single channel, anterior to medulla at level of the foramen magnum. Single *anterior spinal artery* runs caudally along anterior side of spinal cord, just in front of the anterior median fissure, and extends as a fine vessel along filum terminale

C. DISTRIBUTION
1. Through anterior median fissure to gray commissure
   a. To medial side of anterior gray column
2. To anterior roots and lateral side of anterior gray columns
3. To anterior funiculus

III. Reinforcing system: since both vessels in I and II, above, are small, they are joined in their caudal extent by small vessels, entering the vertebral column through intervertebral or anterior sacral foramina, which arise from sources indicated below. Each of these reinforcing vessels divides and follows the posterior and anterior spinal arteries

A. SOURCES
1. Vertebral, ascending cervical, and inferior thyroid in cervical region
2. Intercostals in thoracic region
3. Lumbar in abdomen
4. Lateral sacral in pelvis

B. DISPOSITION
1. The reinforcing vessels have ascending and descending branches, thus continuing the longitudinal course of the anterior and posterior spinal arteries
2. The main channels have medial and lateral branches. These anastomose posteriorly, anteriorly, and laterally, thus forming a circular arterial complex around periphery of cord

IV. Clinical considerations

A. THERE ARE SAID TO BE ARTERIAL WEAKNESSES anteriorly at the level of T4 and L1. Posteriorly, the weakness is said to be between T1 and T3.

B. BOTH THE ARTERIES AND VEINS of the cord are primarily longitudinally running vessels that communicate above with cranial vessels, but are reinforced at irregular intervals by segmental radicular vessels that come in along the nerve roots. These are important for the cord since the anterior and posterior spinal arteries are small even at their origins and can supply by themselves only a short upper portion of the cord

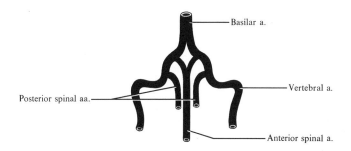

ORIGIN OF SPINAL ARTERIES (SCHEMATIC)

**ARTERIES OF
SPINAL CORD**

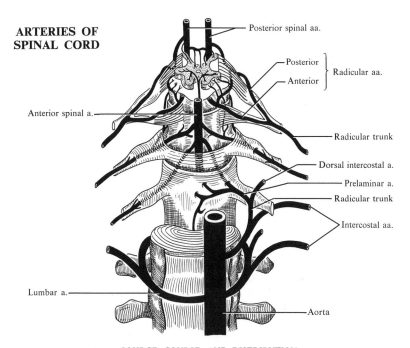

SOURCE, COURSE, AND DISTRIBUTION

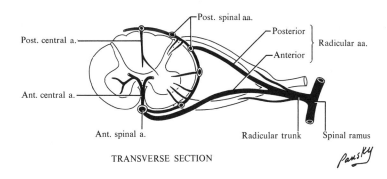

TRANSVERSE SECTION

75. VEINS OF CORD AND VERTEBRAL COLUMN

I. **The veins of the vertebral column** are arranged in a plexiform manner along the length of the vertebral column, both inside and outside the vertebral canal
   A. EXTERNAL VENOUS PLEXUS: composed of 2 freely anastomosing parts
      1. Anterior in front of vertebrae
         a. Communicate with *basivertebral* and *intervertebral* veins
         b. Receive tributaries from bodies of vertebrae
      2. Posterior on posterior surfaces of vertebral arches and processes
         a. Anastomose with vertebral, occipital, and deep cervical veins
   B. INTERNAL VENOUS PLEXUS lies in vertebral canal, between dura and bone. Tends to run longitudinally, receiving drainage from bone and cord. Four main channels:
      1. Anterior (2): lie on posterior surface of bodies of the vertebrae, 1 on each side of posterior longitudinal ligament, interconnected by transverse branches that receive basivertebral veins
      2. Posterior (2): on either side of midline, anterior to vertebral arches and ligamenta flava
         a. Communications
            i. Anastomose with posterior external plexus by veins piercing ligament
            ii. Venous rings at each vertebral level connect with anterior internal plexus
         b. Terminations
            i. Vertebral veins
            ii. Occipital sinus
            iii. Basilar plexus
            iv. Condyloid emissary veins
   C. BASIVERTEBRAL VEINS lie in large network of channels in bodies of vertebrae
      1. Communications: anteriorly with anterior external plexus; posteriorly converge into 1 or 2 large channels which open into transverse veins between anterior internal veins
   D. INTERVERTEBRAL VEINS run through intervertebral foramina
      1. Drain spinal cord and both internal and external plexuses
      2. Terminate in vertebral, intercostal, lumbar, and lateral sacral veins

II. **Veins of cord** lie in fine plexuses in the pia, with 6 longitudinal channels: posterior and anterior median, related to the posterior median sulcus and anterior median fissure; a vein just posterior to each of the posterior roots; and a vein just posterior to each of the anterior roots
   A. LATERALLY, send communications to the intervertebral veins
   B. CEPHALICALLY, coalesce into 2 or 3 vessels that communicate with the vertebral veins and terminate in the inferior petrosal sinus

III. **Clinical considerations**
   A. Since veins from all parts of the body interconnect with those around the spinal cord, these channels act as collateral venous pathways. Thus, metastases from cancer in the pelvis could involve the spinal cord

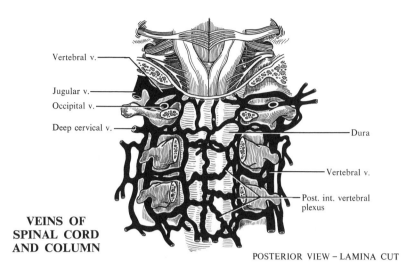

Vertebral v.

Jugular v.

Occipital v.

Deep cervical v.

Dura

Vertebral v.

Post. int. vertebral
plexus

**VEINS OF
SPINAL CORD
AND COLUMN**

POSTERIOR VIEW – LAMINA CUT

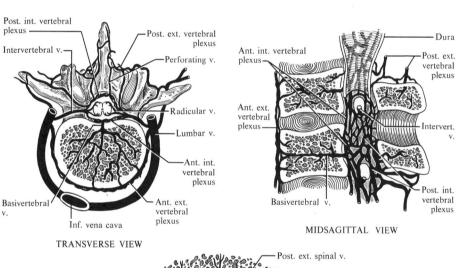

Post. int. vertebral
plexus

Intervertebral v.

Post. ext. vertebral
plexus

Perforating v.

Radicular v.

Lumbar v.

Ant. int.
vertebral
plexus

Basivertebral
v.

Inf. vena cava

Ant. ext.
vertebral
plexus

TRANSVERSE VIEW

Ant. int. vertebral
plexus

Ant. ext.
vertebral
plexus

Dura

Post. ext.
vertebral
plexus

Intervert.
v.

Post. int.
vertebral
plexus

Basivertebral v.

MIDSAGITTAL VIEW

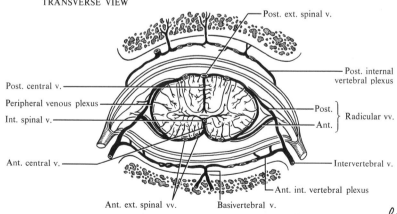

Post. ext. spinal v.

Post. internal
vertebral plexus

Post. central v.

Peripheral venous plexus

Int. spinal v.

Post.
Ant.

} Radicular vv.

Ant. central v.

Intervertebral v.

Ant. ext. spinal vv.

Basivertebral v.

Ant. int. vertebral plexus

TRANSVERSE SECTION

Pansky

76. THE ATLANTO-OCCIPITAL, ATLANTOEPISTROPHIC, AND INTERVERTEBRAL ARTICULATIONS

I. **Articulation between atlas and axis** consists of a medial and 2 lateral joints. Since the lateral are somewhat similar to those between arches of other vertebrae, these will not be included here
   A. TYPE: trochoid (pivot)
   B. MOVEMENTS: rotation
   C. BONES: dens (odontoid process) of axis with anterior arch and transverse ligament of atlas
   D. LIGAMENTS
      1. Anterior atlantoaxial: from body of axis to border of anterior arch of atlas
      2. Posterior atlantoaxial: lamina of axis to posterior arch of atlas
      3. Transverse: across arch of atlas to hold dens against anterior arch of atlas
         a. Has an upward prolongation to occipital bone and a caudal extension to the posterior surface of body of axis, forming *cruciform ligament of atlas*

II. **Articulation between atlas and occipital bone**
   A. TYPE: condyloid
   B. MOVEMENTS: flexion, extension, and lateral motion (abduction)
   C. BONES: condyles of occipital bone and superior facets of atlas
   D. LIGAMENTS OR MEMBRANES
      1. Articular capsules: surround condyles and superior articular processes of atlas
      2. Anterior atlanto-occipital: membrane between foramen magnum and anterior arch of atlas
      3. Posterior atlanto-occipital: membrane between margin of foramen magnum and posterior arch of atlas
      4. Lateral: transverse processes of atlas to jugular processes of occipital

III. **Union of axis with occipital bone**
   A. TECTORIAL MEMBRANE: inside vertebral canal and is a cephalic extension of posterior longitudinal ligament extending from body of axis to occipital bone
   B. ALAR (2): from sides of dens to condyles of occipital bone
   C. APICAL ODONTOID: from apex of dens to foramen magnum

IV. **Muscles acting between atlas and axis**
   A. ROTATION OF HEAD
      1. To opposite side: sternocleidomastoid muscle
      2. To same side: semispinalis capitis, longus capitis, splenius capitis, longissimus capitis, rectus capitis posterior major, oblique capitis superior and inferior muscles

V. **Muscles acting on the atlanto-occipital joint**

Flexion	Extension	Lateral Bending
Longus capitis	Rectus capitis post. major	Trapezius of same side
Rectus capitis anterior	Rectus capitis post. minor	Splenius capitis of same side
	Superior oblique	Sternocleidomastoid of same side
	Semispinalis capitis	
	Splenius capitis	
	Sternocleidomastoid	
	Upper trapezius	

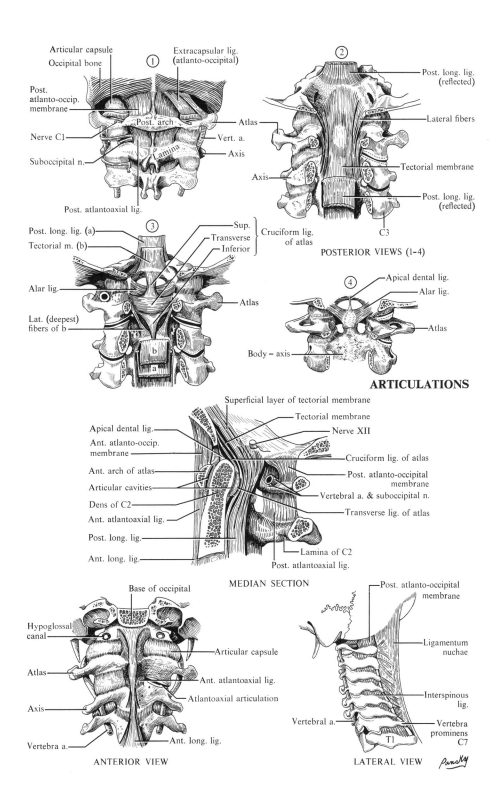

Figure 1

Articular capsule
Occipital bone
① Extracapsular lig. (atlanto-occipital)
Post. atlanto-occip. membrane
Nerve C1
Post. arch
Suboccipital n.
Lamina
Atlas
Vert. a.
Axis
Post. atlantoaxial lig.

Figure 2

②
Post. long. lig. (reflected)
Lateral fibers
Axis
Tectorial membrane
Post. long. lig. (reflected)
C3

POSTERIOR VIEWS (1-4)

Figure 3

Post. long. lig. (a)
Tectorial m. (b)
③
Sup.
Transverse
Inferior
Cruciform lig. of atlas
Alar lig.
Atlas
Lat. (deepest) fibers of b
b
a

Figure 4

④
Apical dental lig.
Alar lig.
Atlas
Body – axis

ARTICULATIONS

Superficial layer of tectorial membrane
Tectorial membrane
Apical dental lig.
Nerve XII
Ant. atlanto-occip. membrane
Cruciform lig. of atlas
Ant. arch of atlas
Post. atlanto-occipital membrane
Articular cavities
Vertebral a. & suboccipital n.
Dens of C2
Transverse lig. of atlas
Ant. atlantoaxial lig.
Post. long. lig.
Ant. long. lig.
Lamina of C2
Post. atlantoaxial lig.

MEDIAN SECTION

Base of occipital
Hypoglossal canal
Atlas
Articular capsule
Ant. atlantoaxial lig.
Atlantoaxial articulation
Axis
Vertebra a.
Ant. long. lig.

ANTERIOR VIEW

Post. atlanto-occipital membrane
Ligamentum nuchae
Interspinous lig.
Vertebral a.
Vertebra prominens C7
T1

LATERAL VIEW

Pansky

77. INTERVERTEBRAL AND COSTOVERTEBRAL ARTICULATIONS—PART I

I. Intervertebral

A. ARTICULATION BETWEEN BODIES OF VERTEBRAE

1. Type: amphiarthrodial
2. Movement: slight
3. Ligaments
   a. Anterior longitudinal covers ventral surfaces of vertebrae from C2 to sacrum
      i. Closely attached to disks and margins of vertebral bodies
   b. Posterior longitudinal in vertebral canal over posterior surface of bodies, extending from C2 to sacrum
      i. Closely adherent to disks and margins of bodies; more loosely attached over concavities and permits transverse veins to cross
   c. Intervertebral disks: form chief connections between bodies from C2 to sacrum. Variable in size, with size of adjacent bodies
      i. In cervical and lumbar regions, it is thicker in front than behind, thus helping to form the convex curvatures in these regions
      ii. One-fourth length of column due to disks
      iii. Where disk adjoins bone, cartilage is hyaline; in center is fibrocartilage
      iv. Nucleus pulposus: soft, pulpy, yellowish elastic material lying in center of disk (embryonic notochord)
      v. Fibrous ring (anulus): concentric ring of fibrous tissue and fibrocartilage
      vi. Not only do disks join bones, but they absorb shock

B. ARTICULATIONS BETWEEN ARCHES OF VERTEBRAE

1. Type: arthrodial
2. Bones: superior articular process with inferior articular process
3. Movements: gliding
4. Ligaments
   a. Articular capsule attached around articular facets of articular processes
   b. Ligamenta flava interconnect laminae of vertebrae
   c. Supraspinal ligament joins tips of vertebrae from C7 to sacrum
   d. Ligamentum nuchae corresponds to B4c, above, in cervical region. Extends longitudinally from external occipital protuberance and median nuchal line to spine of C7. Anteriorly, it is attached to the posterior tubercle of atlas and spines of C2–C6
   e. Interspinal interconnects spines of vertebrae, extending from root to apex of spinous process
   f. Intertransverse joins transverse processes of vertebrae

II. Movements of vertebral column as a whole

A. FLEXION: bending forward. Anterior parts of disks flattened; posterior parts expanded; ligamenta flava, posterior longitudinal, supraspinous, and interspinal ligaments all stretch and help to check action. Strongest check is tension of extensor muscles

B. EXTENSION: bending backward. Here anterior longitudinal ligament is stretched and helps check action. Contact of spinous processes also checks

C. LATERAL FLEXION: bending to either side. Stretches opposite ligamentum flavum and part of anterior longitudinal ligament and intertransverse ligaments, which help to check action. The opposing extensor muscles also check

D. ROTATION at any one articulation is slight, but total rotation is a summation of movement at several joints

E. FOR MUSCLES PRODUCING MOVEMENTS at *all* intervertebral joints (see p. 176)

ARTICULATIONS

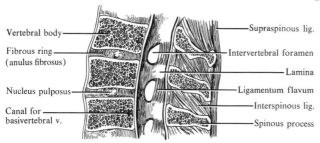

Vertebral body

Fibrous ring
(anulus fibrosus)

Nucleus pulposus

Canal for
basivertebral v.

Supraspinous lig.

Intervertebral foramen

Lamina

Ligamentum flavum

Interspinous lig.

Spinous process

MEDIAN SECTION – LUMBAR REGION

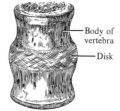

Body of
vertebra

Disk

INTERVERTEBRAL DISK – ANTERIOR

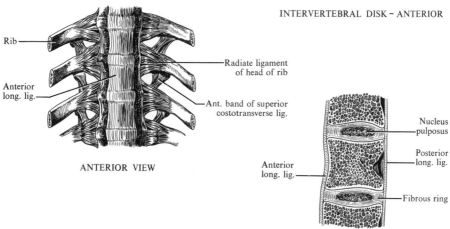

Rib

Anterior
long. lig.

Radiate ligament
of head of rib

Ant. band of superior
costotransverse lig.

ANTERIOR VIEW

Anterior
long. lig.

Nucleus
pulposus

Posterior
long. lig.

Fibrous ring

SAGITTAL SECTION – LUMBAR REGION

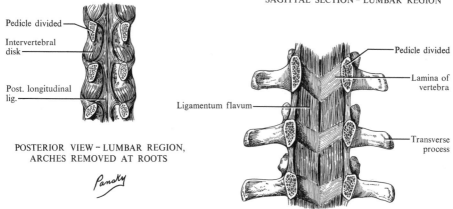

Pedicle divided

Intervertebral
disk

Post. longitudinal
lig.

POSTERIOR VIEW – LUMBAR REGION,
ARCHES REMOVED AT ROOTS

Panaky

Ligamentum flavum

Pedicle divided

Lamina of
vertebra

Transverse
process

FRONT VIEW – BODIES OF VERTEBRAE REMOVED

78. INTERVERTEBRAL AND COSTOVERTEBRAL ARTICULATIONS—PART II

III. Articulations between ribs and vertebrae
A. COSTOVERTEBRAL at head of rib
1. Type: arthrodial
2. Bones: head of rib and facet on side of vertebral body
3. Movements: gliding
4. Ligaments
    a. Articular capsule surrounds joint
    b. Radiate: from anterior head of rib to bodies and intervertebral disk
    c. Intra-articular: from interarticular crest of rib to fibrocartilage. Located within joint and divides it into 2 parts
B. COSTOTRANSVERSE between rib and transverse process
1. Type: arthrodial
2. Bones: tubercle (and neck) of rib with transverse process
3. Movements: gliding
4. Ligaments
    a. Articular capsule attached to edges of articular facets
    b. Superior costotransverse from cephalic border of neck of rib to transverse process above
    c. Posterior costotransverse from neck of rib to transverse process and inferior articular process of vertebra above
    d. Ligament of neck of rib (costotransverse ligament) from back of neck of rib to ventral surface of adjoining transverse process
    e. Ligament of tubercle of rib (lateral costrotransverse ligament) from tubercle of rib to apex of transverse process

IV. Movement of ribs: total effect is a rotation of head of rib in its own axis so that ribs are raised or lowered
A. MUSCLES INVOLVED IN PROCESS are those used in respiration (see p. 306)

V. Clinical considerations
A. HERNIATED (SLIPPED) DISK: after unusual strain to the vertebral column, the disk itself or its center, the nucleus pulposus, may be extruded beyond its normal limits and fail to return to normal position. Slight posterior protrusion may cause nerve root pain because of pressure on spinal nerve roots. The protrusion may occur at any level but the most common location is low in the lumbar region and the most frequent symptom is sciatic pain
B. MOST DISK RUPTURES occur in the third and fourth decades, which represent the most active period of adult life
C. DISK HERNIATION THAT DOES NOT INVOLVE A NERVE ROOT may result in back pain without sciatica, but from a clinical standpoint diagnosis of ruptured disk is made only when nerve root symptoms are present
D. HERNIATION OF NUCLEAR MATERIAL usually occurs through a small tear or rent in the anulus in a posterolateral direction, lateral to the posterior longitudinal ligament
E. THE NERVE THAT EXITS THROUGH THE FORAMEN BELOW the ruptured disk is usually the one affected
F. THE DISKS contribute about 25% of the length of the vertebral column above the sacrum. Their high water content means they are subject to dehydration, accounting for as much as 1.875 cm (0.75 in.) in loss of height by a man, and 1.25 cm (0.5 in.) by a woman during the course of a day. This dehydration is usually made up by reabsorption of water while we lie down, but over a period of years reabsorption does not equal water loss and the disks gradually thin. This accounts for part of the loss in height between young adulthood and old age

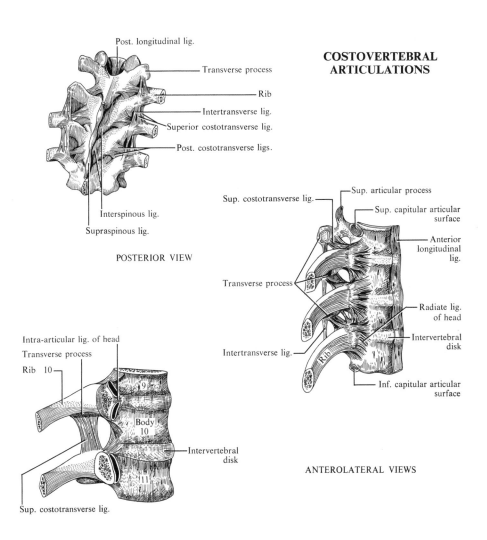

COSTOVERTEBRAL
ARTICULATIONS

POSTERIOR VIEW

Post. longitudinal lig.
Transverse process
Rib
Intertransverse lig.
Superior costotransverse lig.
Post. costotransverse ligs.
Interspinous lig.
Supraspinous lig.

Sup. costotransverse lig.
Sup. articular process
Sup. capitular articular surface
Anterior longitudinal lig.
Transverse process
Radiate lig. of head
Intertransverse lig.
Rib
Intervertebral disk
Inf. capitular articular surface

ANTEROLATERAL VIEWS

Intra-articular lig. of head
Transverse process
Rib 10
9
Body 10
11
Intervertebral disk
Sup. costotransverse lig.

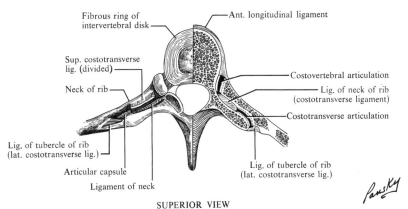

Fibrous ring of intervertebral disk
Ant. longitudinal ligament
Sup. costotransverse lig. (divided)
Neck of rib
Costovertebral articulation
Lig. of neck of rib (costotransverse ligament)
Costotransverse articulation
Lig. of tubercle of rib (lat. costotransverse lig.)
Articular capsule
Lig. of tubercle of rib (lat. costotransverse lig.)
Ligament of neck

SUPERIOR VIEW

79. SPINAL MENINGES

I. Dura, outer covering: dense and tough

A. CORRESPONDS WITH MENINGEAL LAYER of cranial dura; vertebrae have their own periosteal layer

B. EPIDURAL SPACE filled with fat, loose connective tissue, and veins that lie between dura and bony canal

C. EXTENT

    1. Cephalic end: foramen magnum to which it is attached (also attached to vertebrae C2 and C3)

    2. Extends downward as sheath larger than cord. Narrows down and fuses with filum terminale at level of second sacral segment

    3. From second sacral segment continues to be attached to back of coccyx

    4. Continues with spinal nerves through intervertebral foramina to fuse with their sheaths (epineurium)

II. Arachnoid, middle covering: thin and delicate

A. CONTINUOUS WITH CRANIAL ARACHNOID

B. OUTER SURFACE covered with a mesothelium, separated from dura by *subdural space*

C. INNER SURFACE connected across the subarachnoid space by delicate connective tissue, the *arachnoid trabeculae*

D. SUBARACHNOID SPACE continuous with that of cranium and extends to second sacral segment and outward to intervertebral foramina

    1. Longitudinal subarachnoid septum (septum posticum) incompletely subdivides space. Septum joins pia to arachnoid along post. median sulcus of cord

III. Pia, inner covering: delicate connective tissue, highly vascular, closely applied to cord

A. COMPOSED OF 2 LAYERS

    1. Outer layer consists of longitudinal connective tissue fibers

        a. Forms a median, longitudinal, anterior band, the *linea splendens*

        b. Along sides of cord, longitudinal fibers are concentrated as *denticulate* ligaments

            i. Extend full length of spinal cord, lying between attachments of posterior and anterior roots

            ii. At 21 points, beginning at foramen magnum and ending at conus medullaris, and are attached (on each side) to dura

    2. Inner layer, intimately adherent to cord, sends a septum into anterior median fissure

B. At apex of caudal end of cord the pia continues along filum terminale and, with addition of dura below S2, forms *central ligament of spinal cord,* which is attached to posterior of coccyx to help anchor the cord

IV. Clinical considerations

A. INFECTION, HEMORRHAGE, AND TUMOR FORMATION: the meninges may be the seat of any of these conditions. Infection of the meninges is called meningitis; it may involve the dura (pachymeningitis) or the pia-arachnoid (leptomeningitis). Headache; neck stiffness; flexion at ankle, knee, and hip when the neck is bent; inability to extend the leg completely when in the sitting position or when lying with the thigh flexed upon the abdomen; fever; mental confusion—all are indications of meningeal infection

B. ANESTHETIC AGENTS introduced into the subarachnoid space mix with cerebrospinal fluid and bathe the spinal cord and nerve roots, producing spinal anesthesia

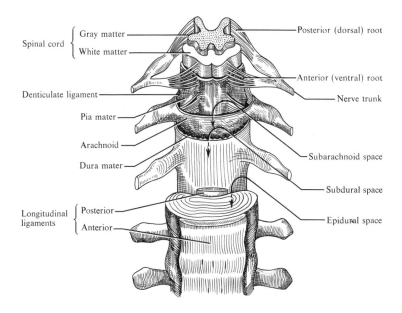

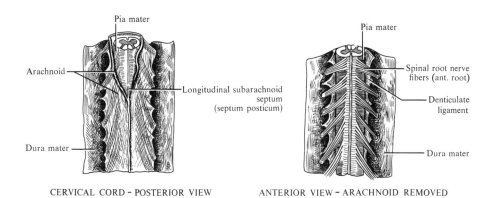

SPINAL MENINGES AND THEIR SPACES

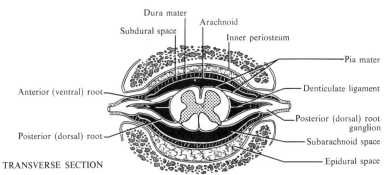

CERVICAL CORD - POSTERIOR VIEW   ANTERIOR VIEW - ARACHNOID REMOVED

TRANSVERSE SECTION

80. SPINAL CORD

I. Extent and size
A. BEGINS: cephalic border of atlas; ends in filum terminale at lower first or upper second lumbar vertebrae
B. LENGTH: 42 to 45 cm
C. DIAMETER varies at different levels

II. External appearance
A. ENLARGEMENTS
   1. Cervical: from C3 to T2 vertebral levels
   2. Lumbar, beginning at end of ninth thoracic, reaching largest diameter at twelfth thoracic
      a. Tapers down rapidly as conus medullaris, ending as the filum terminale
B. FISSURES AND SULCI
   1. Posterior median sulcus and anterior median fissure divide cord into halves
   2. Posterior lateral and anterior lateral sulci extend longitudinally along length of cord, represent line for attachment of posterior and anterior roots
   3. Posterior intermediate sulcus between posteromedial and posterolateral sulci in upper thoracic and cervical regions

III. Relationship between vertebrae and pairs of spinal nerves
A. BECAUSE OF GROWTH INEQUALITIES between vertebral column and spinal cord, the nerves, especially at lower levels, do not emerge from vertebral canal at level of attachment to cord
   1. In third fetal month, cord fills canal; at birth, end of cord lies at the lower border of third lumbar vertebra
   2. Relationship in adult (see following table)

Spinal Nerve and Cord Segment	Vertebral Segment	Exit
C1	C1	Above C1
C8	C7	Between C7–T1
T6	T5	Between T6–T7
T12	T8	Between T12–L1
L2	T10	Between L1–L2
L5	T11	Below L5
S3	T12	Third sacral foramen

IV. Internal structure
A. COMPOSITION
   1. Centrally located, H-shaped gray matter containing cell bodies of efferent and internuncial neurons
      a. Two, narrow posterior projections—posterior columns (horns)
      b. Two, usually thick anterior projections—anterior columns (horns)
      c. Cross bar—the gray commissure (intermediate substance)
   2. Peripherally located white matter made up of nerve fibers and glia
      a. The posterior lateral and anterior lateral sulci divide cord into posterior, lateral, and anterior funiculi
         i. The posterior funiculus of cervical cord is subdivided into fasciculus gracilis and fasciculus cuneatus by the posterior intermediate sulcus

SPINAL CORD

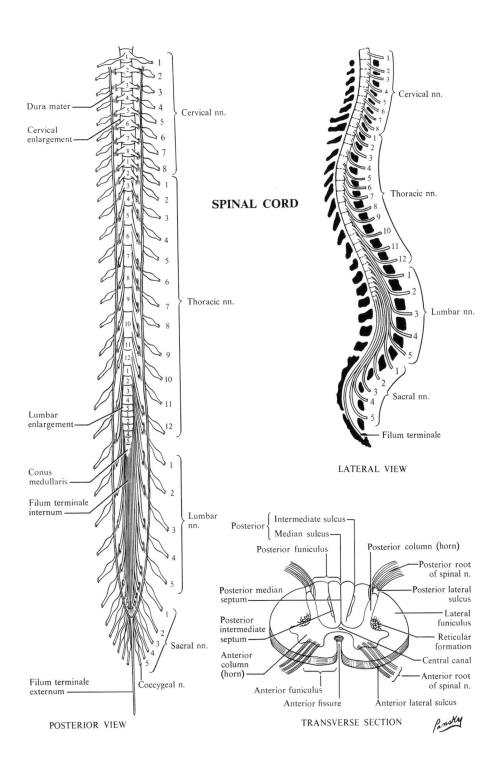

Dura mater

Cervical enlargement

Lumbar enlargement

Conus medullaris

Filum terminale internum

Filum terminale externum

Cervical nn.

Thoracic nn.

Lumbar nn.

Sacral nn.

Coccygeal n.

POSTERIOR VIEW

Cervical nn.

Thoracic nn.

Lumbar nn.

Sacral nn.

Filum terminale

LATERAL VIEW

Posterior { Intermediate sulcus
          { Median sulcus

Posterior funiculus

Posterior median septum

Posterior intermediate septum

Anterior column (horn)

Anterior funiculus

Anterior fissure

Posterior column (horn)

Posterior root of spinal n.

Posterior lateral sulcus

Lateral funiculus

Reticular formation

Central canal

Anterior root of spinal n.

Anterior lateral sulcus

TRANSVERSE SECTION

81. CAUDAL END OF SPINAL CORD AND ITS MENINGES

I. Caudal extent of spinal cord
A. IN ADULT, tip of conus medullaris lies at lower border of first lumbar or upper border of second lumbar vertebrae (see p. 173)

II. Filum terminale: a filamentous continuation of caudal tip of cord (conus medullaris) with a covering of pia
A. LIES IN CENTER OF THE DURAL SHEATH, which continues at same size to level of second sacral segment where dura moves down to join filum, thus forming the *central ligament* of the cord
B. WITHIN DURAL SHEATH, filum is surrounded by cauda equina, which is composed of the lumbar and sacral nerves arising from cord to descend to the appropriate intervertebral or sacral foramen for exit from the vertebral canal

III. Clinical considerations
A. SPINAL PUNCTURE: for withdrawal of fluid for diagnostic purposes; treatment, as antibiotics in meningitis; anesthesia, for operations below the thorax
   1. Site: below L2, since spinal cord ends at or above this level. Most usually at L4, since it is easily located by a line drawn across level of highest point on iliac crests. This line passes through spinous process of L4
   2. In the newborn, the spinal cord terminates at the lower border of L3
B. EXTRADURAL INJECTIONS: to produce anesthesia over a more limited area and without the dangers inherent in spinal puncture
   1. Caudal, usually for obstetric, gynecologic, and urologic surgery
      a. Site: *sacral hiatus,* located between sacral and coccygeal cornua
   2. Transsacral: to inject anesthetic material by way of posterior sacral foramina
      a. Site: using posterior superior iliac spine as landmark: 1.0 cm directly medial and slightly above locates first sacral foramen. Line drawn 0.5 cm downward and 1.0 cm medially from first foramen should locate second sacral foramen; one fingerbreadth down and slightly medial is point for the third foramen while the fourth foramen is a fingerbreadth below and medial to the third. It usually is not necessary to locate the fifth, but, when present, is a small fingerbreadth below and slightly medial to the fourth
C. PILONIDAL SINUS is variously explained as a persistence of the neurenteric canal in the coccygeal region, an epidermal inclusion, or a failure of fusion of the skin of the 2 halves of the body in this region. The sinus opens onto the skin over the coccyx
D. SPINA BIFIDA is a failure of development of the dorsal arches of 1 or more vertebrae usually in the lumbosacral region. A *tight filum terminale* may be associated with spina bifida occulta (no protrusion above skin). If there is protrusion, a sac is formed (spina bifida aperta). Its contents may be of 3 types: (1) *meningocele,* a leptomeningeal protrusion through a bony and dural defect in the skull or vertebral column (usually sacral); (2) *meningomyelocele,* protrusion of both leptomeninges and cord through a bony and dural defect in the vertebrae; or (3) *syringomyelocele,* a rare defect in which the central canal of the cord in myelocele is distended with C.S.F. with resultant pressure atrophy of the spinal cord

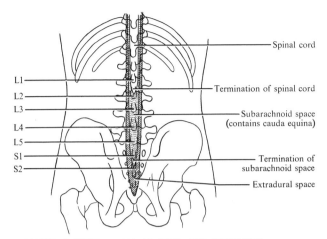

Spinal cord

L1 — Termination of spinal cord
L2
L3 — Subarachnoid space
(contains cauda equina)
L4
L5
S1 — Termination of
S2 subarachnoid space
— Extradural space

VERTEBRAL SPINE LEVELS OF CORD AND MENINGES

SPINAL CORD

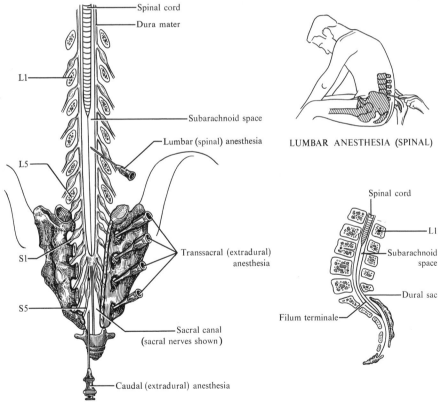

Spinal cord
Dura mater

L1

Subarachnoid space

Lumbar (spinal) anesthesia

LUMBAR ANESTHESIA (SPINAL)

L5

Spinal cord

L1

Subarachnoid
space

S1

Transsacral (extradural)
anesthesia

S5

Dural sac

Filum terminale

Sacral canal
(sacral nerves shown)

Caudal (extradural) anesthesia

SPINAL CORD WITH RELATION TO ANESTHESIA

82. MUSCLES PRODUCING MOVEMENTS ON THE VERTEBRAL COLUMN

Flexion*	Extension*	Lateral Bending†	Rotation†
Sternocleidomastoid	Erector spinae	Same muscles listed under flexion and extension plus rectus capitis lateralis (except for the interspinous)	All the muscles listed under flexion and extension rotate column to the same side except sternocleidomastoid, ext. abd. oblique, semispinalis cervicis and capitis, multifidus, and the rotatores, which rotate to opposite side (the longus coli and scalenes give only a slight rotation)
Longus capitis	Semispinalis		
Rectus capitis ant.	Splenius		
Longus coli	Multifidus		
Scaleni	Rotatores		
Psoas major	Interspinous		
Ext. abdominal oblique			
Int. abdominal oblique	Intertransverse		
Rectus abdominis	All posterior sub-occipital muscles		

*Flexion and extension are accomplished when the bilateral muscles act together.
†Lateral bending and rotation are accomplished when the muscles on one side contract unilaterally.

I. General function
A. THE POSTURE of the vertebral column is maintained by the intrinsic back muscles, but these muscles do not act simultaneously. They come into action to maintain balance but are quiescent most of the time. They are most completely relaxed in extreme flexion since the ligaments of the column then take over the load
B. IN WALKING, the pelvis is prevented from falling, to the side where the foot is raised, by the contraction of the vertebral muscles on that same side
C. IN GENERAL, all the extensor muscles on the dorsum of the back can be considered to be antigravity muscles enabling us to maintain the erect position

II. Special features
A. THE FLEXIBILITY OF THE COLUMN is controlled by groups of muscles that are vertically disposed (therefore they are parallel to the column), as well as by obliquely disposed muscles swathed around the torso
B. THE MUSCLES OF THE BACK are the principal extensors; the two rectus abdominis muscles are powerful flexors; the oblique fibers of the abdominal muscles act as rotators, and their vertical fibers form part of the side flexor group, along with the appropriate rectus abdominis, quadratus lumborum, and erector spinae
C. GRAVITY must be taken into account, and thus the muscle groups mentioned above have the specific actions when they act against a resistance, which may, indeed, be gravity

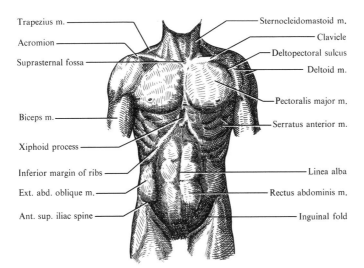

Trapezius m.
Acromion
Suprasternal fossa
Biceps m.
Xiphoid process
Inferior margin of ribs
Ext. abd. oblique m.
Ant. sup. iliac spine

Sternocleidomastoid m.
Clavicle
Deltopectoral sulcus
Deltoid m.
Pectoralis major m.
Serratus anterior m.
Linea alba
Rectus abdominis m.
Inguinal fold

SURFACE ANATOMY – THORAX AND DORSUM

POSTERIOR VIEW

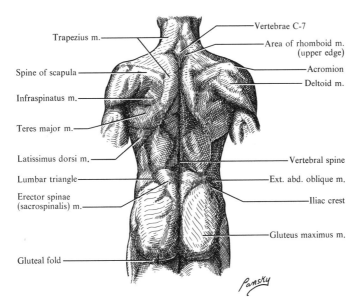

Trapezius m.
Spine of scapula
Infraspinatus m.
Teres major m.
Latissimus dorsi m.
Lumbar triangle
Erector spinae
(sacrospinalis) m.
Gluteal fold

Vertebrae C-7
Area of rhomboid m.
(upper edge)
Acromion
Deltoid m.
Vertebral spine
Ext. abd. oblique m.
Iliac crest
Gluteus maximus m.

83. SYMPATHETIC SYSTEM, SUBOCCIPITAL AND BACK DETAIL

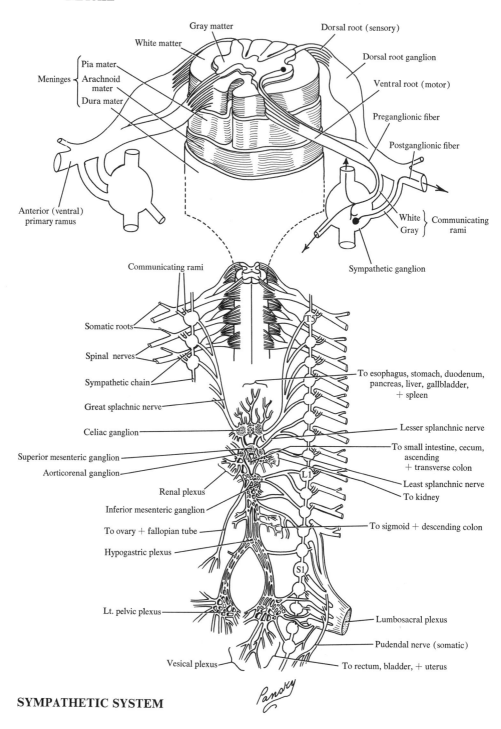

Gray matter

White matter

Dorsal root (sensory)

Dorsal root ganglion

Ventral root (motor)

Pia mater

Meninges { Arachnoid mater

Dura mater

Preganglionic fiber

Postganglionic fiber

Anterior (ventral) primary ramus

White
Gray } Communicating rami

Sympathetic ganglion

Communicating rami

Somatic roots

Spinal nerves

Sympathetic chain

Great splachnic nerve

Celiac ganglion

Superior mesenteric ganglion

Aorticorenal ganglion

Renal plexus

Inferior mesenteric ganglion

To ovary + fallopian tube

Hypogastric plexus

Lt. pelvic plexus

Vesical plexus

To esophagus, stomach, duodenum, pancreas, liver, gallbladder, + spleen

Lesser splanchnic nerve

To small intestine, cecum, ascending + transverse colon

Least splanchnic nerve

To kidney

To sigmoid + descending colon

Lumbosacral plexus

Pudendal nerve (somatic)

To rectum, bladder, + uterus

T5

L1

S1

SYMPATHETIC SYSTEM

Pansky

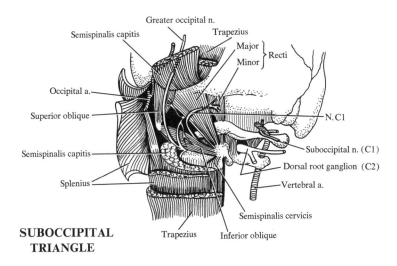

Greater occipital n.

Semispinalis capitis

Trapezius

Major
}Recti
Minor

Occipital a.

Superior oblique

Semispinalis capitis

Splenius

N.C1

Suboccipital n. (C1)

Dorsal root ganglion (C2)

Vertebral a.

Semispinalis cervicis

**SUBOCCIPITAL
TRIANGLE**

Trapezius

Inferior oblique

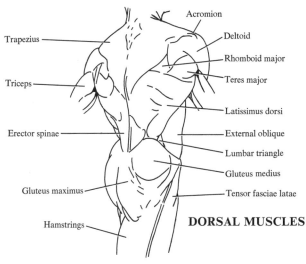

Acromion

Trapezius

Deltoid

Rhomboid major

Teres major

Triceps

Latissimus dorsi

Erector spinae

External oblique

Lumbar triangle

Gluteus medius

Gluteus maximus

Tensor fasciae latae

Hamstrings

DORSAL MUSCLES

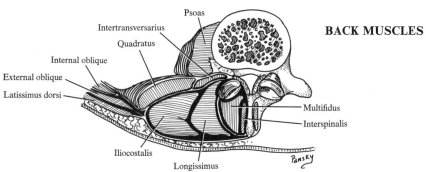

Psoas

Intertransversarius

BACK MUSCLES

Quadratus

Internal oblique

External oblique

Latissimus dorsi

Multifidus

Interspinalis

Iliocostalis

Longissimus

PANSKY

UNIT THREE

Upper
Extremity

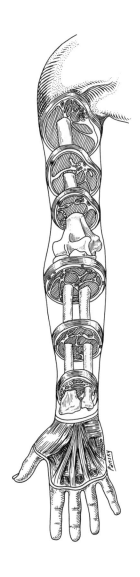

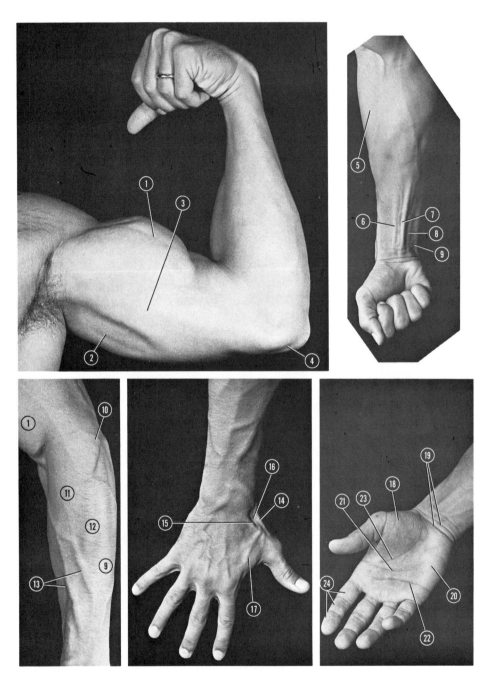

FIGURE 11. **Surface anatomy of upper extremity.** *1*, Biceps m.; *2*, long head of triceps m.; *3*, medial head of triceps m.; *4*, medial epicondyle; *5*, brachioradialis m.; *6*, flexor carpi radialis m.; *7*, palmaris longus tendon; *8*, flexor digitorum superficialis m.; *9*, flexor carpi ulnaris m.; *10*, epicondyle; *11*, extensor digitorum m.; *12*, extensor carpi ulnaris m.; *13*, superficial veins; *14*, extensor pollicis brevis m.; *15*, extensor pollicis longus m.; *16*, anatomical snuffbox; *17*, first dorsal interosseous m.; *18*, thenar eminence; *19*, wrist creases; *20*, hypothenar eminence; *21*, proximal transverse crease; *22*, distal transverse crease; *23*, transverse crease; *24*, phalanges.

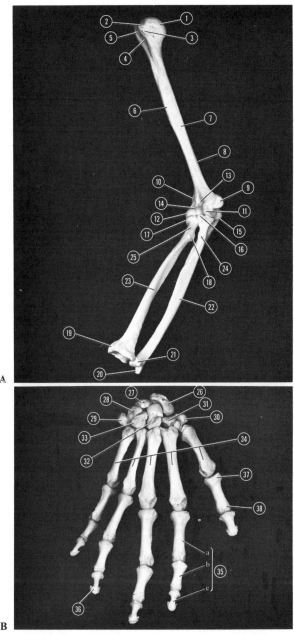

A

B

FIGURE 12. **A, Right humerus, ulna, and radius.** *1,* Head of humerus; *2,* anatomical neck; *3,* lesser tubercle; *4,* intertubercular groove; *5,* greater tubercle; *6,* deltoid tuberosity; *7,* shaft of humerus; *8,* medial supracondylar ridge; *9,* medial epicondyle; *10,* lateral supracondylar ridge; *11,* trochlea; *12,* lateral epicondyle; *13,* coronoid fossa; *14,* radial fossa; *15,* olecranon; *16,* coronoid process; *17,* head of radius; *18,* tuberosity of radius; *19,* styloid process of radius; *20,* styloid process of ulna; *21,* head of ulna; *22,* shaft of ulna; *23,* shaft of radius; *24,* tuberosity of ulna; *25,* neck of radius. **B, Posterior view of hand.** *26,* Scaphoid; *27,* lunate; *28,* triquetrum; *29,* pisiform; *30,* trapezium; *31,* trapezoid; *32,* capitate; *33,* hamate; *34,* metacarpal bones; *35,* phalanges (*a,* proximal; *b,* middle; *c,* distal); *36,* tuberosity; *37,* base of phalanx; *38,* head of phalanx.

84. CUTANEOUS NERVES AND DERMATOMES OF THE UPPER EXTREMITY

I. Sources of cutaneous nerves: cervical plexus, brachial plexus, posterior primary divisions of upper thoracic nerves, and lateral cutaneous branches of second and third intercostals

II. Distribution

A. CERVICAL PLEXUS (C3 and C4) as medial, intermediate, and lateral *supraclavicular nerves* to upper pectoral, deltoid, and outer trapezius areas

B. POSTERIOR DIVISIONS of upper 3 thoracic nerves over trapezius area to spine of scapula

C. BRACHIAL PLEXUS (see p. 198)

1. Superior lateral brachial cutaneous nerve and other cutaneous twigs from axillary over deltoid and upper arm
2. Medial brachial cutaneous to medial side of arm as far as elbow
3. Posterior brachial cutaneous of radial over the posterior arm, nearly to the olecranon
4. Posterior antebrachial cutaneous from radial to lateral side of volar surface of arm, most of the posterior side of the lower arm, and middle posterior of the forearm
5. Lateral antebrachial cutaneous from musculocutaneous to lateral third of posterior of forearm and lateral half of anterior side of forearm
6. Medial antebrachial cutaneous from medial cord to middle third of anterior side of arm, medial half of anterior forearm, and medial third of posterior forearm
7. Palmar branch of median nerve to midpalm and thenar eminence
8. First, second, and third common palmar digitals to palmar surfaces of thumb, first and second fingers and lateral side of ring finger, and posterior surface of terminal phalanges of these same fingers
9. Palmar cutaneous branch of ulnar to medial palm
10. Dorsal digital and metacarpal nerves from dorsal branch of ulnar to medial side of dorsum of hand, both sides of little finger, and medial side of ring finger to the tip
11. Lateral and medial branches of the superficial radial nerve to lateral part of dorsum of hand, dorsum of thumb, first 2 fingers, and lateral side of ring finger as far as terminal phalanx

D. LATERAL CUTANEOUS BRANCHES

1. Intercostobrachial from second intercostal to medial arm, usually with medial brachial cutaneous (another branch from the third sometimes occurs)

III. The dermatome is a specific skin area supplied by a specific spinal nerve regardless of the cutaneous nerve within which the spinal segmental fibers lie. Knowledge of the dermatomes is helpful in localizing certain lesions of the nervous system. However, it is not enough to know merely the cord segment involved. When surgery is contemplated, it is also necessary to know at which level of the vertebral column a particular cord segment lies. For example, a patient showing numbness on the fifth digit and the medial side of the hand has a lesion in the eighth spinal cord segment, and this lies at the level of the seventh cervical vertebra

IV. Special features

A. THE 3RD AND 4TH CERVICAL NERVES supply a limited area of skin over the upper parts of the pectoral region and shoulder, and the 2nd thoracic nerve usually sends a branch to skin of the medial and upper part of the arm; otherwise, the skin of the upper limb is supplied by branches from the brachial plexus

B. IN THE LIMBS, the overlap between peripheral nerves is less extensive than in the trunk, so that complete interruption of a single peripheral nerve typically produces changes in sensation that are appreciated by the patient

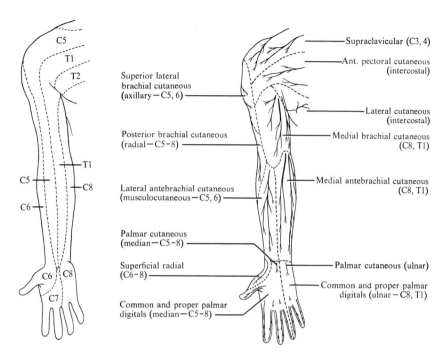

C5
T1
T2

Superior lateral
brachial cutaneous
(axillary—C5, 6)

Posterior brachial cutaneous
(radial—C5-8)

Lateral antebrachial cutaneous
(musculocutaneous—C5, 6)

Palmar cutaneous
(median—C5-8)

Superficial radial
(C6-8)

Common and proper palmar
digitals (median—C5-8)

T1
C5
C6
C8

C6 C8
C7

Supraclavicular (C3, 4)
Ant. pectoral cutaneous
(intercostal)

Lateral cutaneous
(intercostal)
Medial brachial cutaneous
(C8, T1)

Medial antebrachial cutaneous
(C8, T1)

Palmar cutaneous (ulnar)
Common and proper palmar
digitals (ulnar—C8, T1)

DERMATOMES

CUTANEOUS NERVES

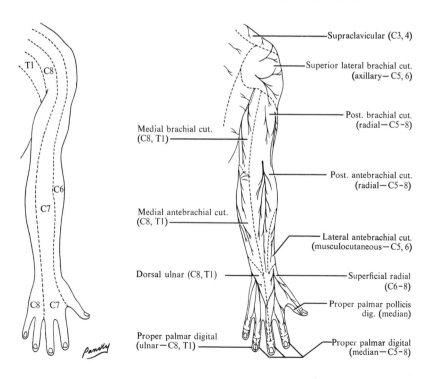

T1 C8

Medial brachial cut.
(C8, T1)

Medial antebrachial cut.
(C8, T1)

Dorsal ulnar (C8, T1)

Proper palmar digital
(ulnar—C8, T1)

C6
C7

C8 C7

Supraclavicular (C3, 4)
Superior lateral brachial cut.
(axillary—C5, 6)

Post. brachial cut.
(radial—C5-8)

Post. antebrachial cut.
(radial—C5-8)

Lateral antebrachial cut.
(musculocutaneous—C5, 6)
Superficial radial
(C6-8)
Proper palmar pollicis
dig. (median)
Proper palmar digital
(median—C5-8)

-185-

85. SUPERFICIAL VEINS OF THE UPPER EXTREMITY

I. General: lie between 2 layers of superficial fascia

II. Specific veins

A. CEPHALIC VEIN
   1. Origin: lateral side of dorsal venous rete on hand
   2. Course: from dorsum curves around to lateral anterior side of forearm, ascends in front of elbow in groove between biceps and brachialis, then along lateral side of biceps to lie in deltopectoral triangle (groove), and pierces clavipectoral fascia to reach axillary vein
   3. Termination: axillary vein, below clavicle
   4. Tributaries from:
      a. Lateral side of palmar venous plexus
      b. Posterior and anterior veins on radial side of forearm
      c. Lower lateral arm
      d. Entire upper arm and shoulder region
   5. Communications
      a. Median cubital vein, running obliquely across antecubital area to join basilic vein

B. BASILIC VEIN
   1. Origin: medial side of dorsal venous rete
   2. Course: posterior side of medial forearm; below elbow crosses to volar side, ascends in groove between biceps and pronator teres muscles, continues along medial border of biceps, pierces deep fascia just below midarm, and runs along medial side of brachial artery to lower border of teres major, where it joins brachial vein
   3. Termination: joins brachial to form axillary vein
   4. Tributaries from:
      a. Medial side of palmar network of hand
      b. Posterior and medial side of forearm and lower arm
      c. Median cubital vein, from cephalic vein

C. ACCESSORY CEPHALIC VEIN: often small, arising either from dorsal network on medial hand or a network on dorsal forearm. It ends in cephalic vein, below elbow

D. MEDIAN ANTEBRACHIAL VEIN: arises in palmar venous network, ascends on ulnar side of forearm, and ends in either basilic or median cubital veins

III. Clinical considerations

A. VEINS OF UPPER EXTREMITY are most often used for either drawing of blood samples or administration of drugs, fluids, or nutriments. The median cubital or the basilic veins, because of the strong support given by the bicipital aponeurosis, are most frequent sites

B. DUE TO THE PROXIMITY of the lateral antebrachial cutaneous nerve to the cephalic vein and the medial antebrachial cutaneous nerve to the basilic vein, care must be used to avoid these if venipuncture of these veins is attempted

SUPERFICIAL VEINS

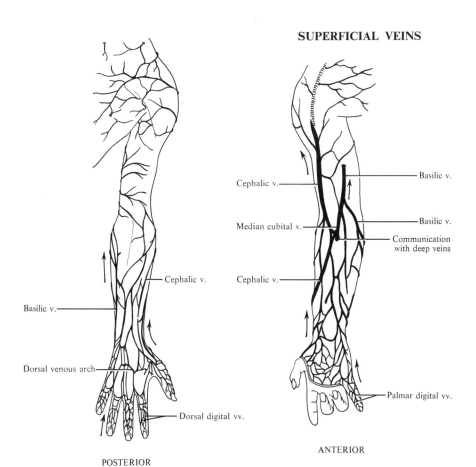

Cephalic v.

Basilic v.

Dorsal venous arch

Dorsal digital vv.

POSTERIOR

Cephalic v.

Median cubital v.

Cephalic v.

Basilic v.

Basilic v.

Communication
with deep veins

Palmar digital vv.

ANTERIOR

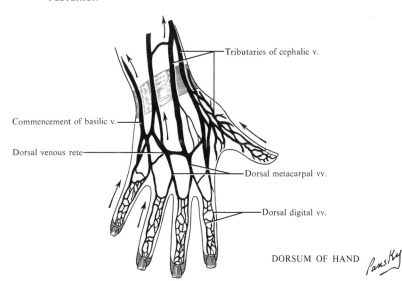

Tributaries of cephalic v.

Commencement of basilic v.

Dorsal venous rete

Dorsal metacarpal vv.

Dorsal digital vv.

DORSUM OF HAND

86. LYMPHATICS OF THE UPPER EXTREMITY

I. Superficial vessels

A. BEGIN IN NETWORK ON HAND. The plexuses of the palm and palmar surface of the fingers are more numerous than on dorsum

B. FROM FINGERS, vessels on either side of digit go toward dorsum of hand

C. FROM PALM may go toward fingers, wrist, or either side of hand

D. RUN BOTH IN FRONT OF AND BEHIND WRIST forming radial, median, and ulnar groups of vessels, which travel with the cephalic, median antebrachial, and basilic veins
1. Some of the ulnar vessels end in epitrochlear nodes
2. Some of the radial vessels go to deltopectoral nodes
3. Most go to lateral axillary nodes

II. Deep vessels run with deep arteries and veins

A. IN FOREARM, 4 sets: along the radial, ulnar, dorsal, and volar interosseous vessels
1. Some terminate in nodes along brachial artery
2. Most end in lateral axillary nodes

III. Lymph nodes of upper extremity

A. SUPERFICIAL
1. Supratrochlear above medial epicondyle, medial to basilic vein: receives afferents from medial fingers, palm, and forearm. Efferents go to deep nodes
2. Deltopectoral in deltopectoral groove, along cephalic vein: receive afferents from vessels of radial side that follow cephalic vein. Efferents go to either subclavian or inferior cervical nodes

B. Deep
1. A few are sometimes found in forearm along radial, ulnar, and interosseous vessels and in arm along medial side of brachial artery
2. Axillary
   a. Lateral located on medial and posterior aspects of axillary artery. Receive afferents from all upper extremity except those nodes around cephalic vein. Efferents go to central and subclavicular nodes
   b. Pectoral located at lower border of pectoralis minor muscle. Receive afferents from anterior and lateral thorax, central and lateral mammary gland. Efferents go to central and subclavicular nodes
   c. Subscapular located along course of subscapular artery. Receive afferents from lower back of neck and posterior wall of thorax. Efferents go to central nodes
   d. Central located toward apex of axilla. Receive afferents from all the above nodes. Efferents go to subclavicular
   e. Apical (subclavicular) located behind and above pectoralis minor muscle. Receive afferents from deltopectoral nodes; a direct connection with upper mammary gland and all other axillary nodes. Efferents form the subclavian trunk, which enters: jugular trunk, subclavian vein, or thoracic duct (a few go to inferior cervical nodes)

IV. Clinical considerations

A. AXILLARY LYMPH NODES: a patient with an infection or malignancy in the upper extremity may complain of tenderness and swelling in the axilla. This is due to involvement of the axillary lymph nodes, especially the lateral group along the axillary artery, since these filter lymph from much of the upper limb

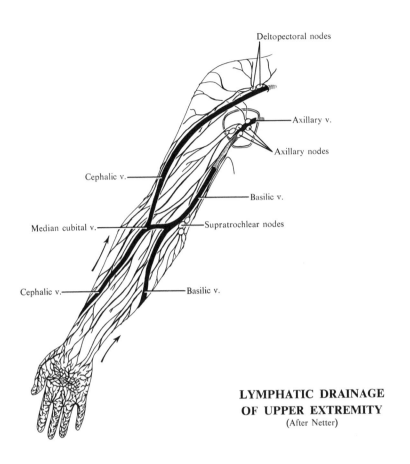

Deltopectoral nodes

Axillary v.

Axillary nodes

Cephalic v.

Basilic v.

Median cubital v.

Supratrochlear nodes

Cephalic v.

Basilic v.

**LYMPHATIC DRAINAGE
OF UPPER EXTREMITY**
(After Netter)

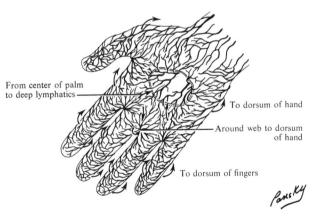

From center of palm
to deep lymphatics

To dorsum of hand

Around web to dorsum
of hand

To dorsum of fingers

Pansky

87. BONES OF THE SHOULDER GIRDLE

I. Clavicle (collar bone)
A. PARTS: sternal extremity, acromial extremity, and body
   1. Sternal extremity: articulates with sternum and has an articular facet for the first rib
   2. Acromial extremity: flat and rough, has area for articulation with acromion on undersurface
   3. Body
      a. Medial part is rounded; lateral part is flattened
      b. Double curvature: convex anteriorly at medial end and concave anteriorly at lateral end
      c. Lateral part has an upper surface and undersurface, the latter showing the *conoid tubercle* and *trapezoid line.* It also has an anterior and posterior border
      d. Medial part has anterior, superior, and posterior borders; anterior, posterior, and inferior surfaces. The impression for the costoclavicular ligament and groove for the subclavius muscle appear on the inferior surface

II. Scapula
A. FLAT AND TRIANGULAR with 2 surfaces, 3 borders, and 3 angles
   1. Costal (anterior) surface concave—subscapular fossa
   2. Posterior surface divided by *spine* into *supra-* and *infraspinous fossae*
   3. Spine of scapula ends in the flattened *acromion*
      a. *Great notch of scapula* joins the infra- and supraspinous fossae
   4. The *scapular notch* is in the superior border
   5. At the lateral angle is located the *glenoid cavity,* with its *supra-* and *infraglenoid tuberosities.* Medial to the cavity is the *coracoid process*
      a. The *neck* is a constriction just medial to the glenoid cavity

III. Ossification

Location	When Appears	When Closes
A. CLAVICLE, from 3 centers (ossified before any other bone in body)		
Medial (in body)	5th–6th fetal week	
Lateral (in body)	5th–6th fetal week	Close by 25th year
In sternal end	18th–20th year	
B. SCAPULA, from seven or more centers		
Body	8th fetal week	15th year
Middle coracoid	15th–18th month	15th year
Upper glenoid	10th year	16th–18th year
Root of coracoid	14th–20th year	By 25th year
Base of acromion	14th–20th year	By 25th year
Inferior angle	14th–20th year	By 25th year
End of acromion	14th–20th year	By 25th year
Vertebral border	14th–20th year	By 25th year

IV. Clinical considerations
A. FRACTURES OF THE CLAVICLE usually occur in the medial part, with medial and downward displacement of the distal fragment due to the weight of the shoulder and the pull of the pectoralis major muscle

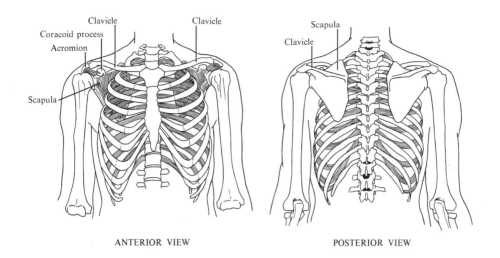

ANTERIOR VIEW   POSTERIOR VIEW

RT. CLAVICLE

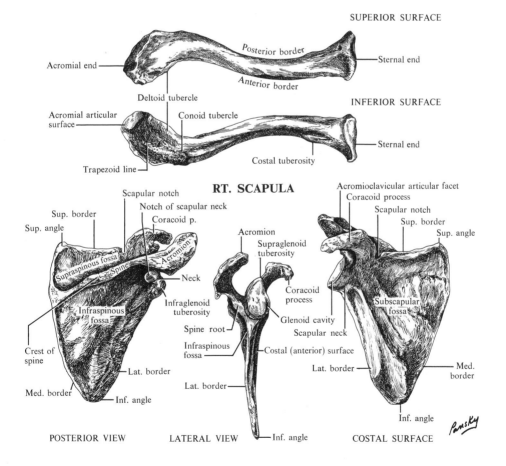

SUPERIOR SURFACE

Acromial end — Posterior border — Sternal end

Anterior border

Deltoid tubercle

INFERIOR SURFACE

Acromial articular surface — Conoid tubercle

Trapezoid line — Costal tuberosity — Sternal end

RT. SCAPULA

Scapular notch
Sup. border   Notch of scapular neck   Acromioclavicular articular facet
Sup. angle   Coracoid p.   Coracoid process
                               Acromion   Scapular notch
Supraspinous fossa   Spine   Sup. border
                               Neck   Acromion   Sup. angle
Infraspinous fossa   Infraglenoid   Supraglenoid   Subscapular fossa
                               tuberosity   tuberosity
Crest of spine   Coracoid process
Med. border   Spine root   Glenoid cavity
                     Infraspinous   Scapular neck Lat. border   Med. border
Lat. border   fossa   Costal (anterior) surface
Inf. angle   Lat. border   Inf. angle

POSTERIOR VIEW   LATERAL VIEW — Inf. angle   COSTAL SURFACE

Pansky

88. ARTICULATIONS OF THE SHOULDER GIRDLE

I. Definition: bones through which upper extremity is attached to trunk

II. Bones (see p. 190)

III. Articulations
- A. ACROMIOCLAVICULAR
  1. Type: arthrodial
  2. Bones: acromion of scapula with clavicle
  3. Movement: gliding, rotation of scapula on clavicle
  4. Ligaments
     - a. Articular capsule completely surrounds articular areas
     - b. Superior acromioclavicular covers joint cephalically
     - c. Inferior acromioclavicular covers joint caudally
     - d. Articular disk frequently absent or incomplete
     - e. Coracoclavicular not part of joint but keeps clavicle against acromion
       - i. Trapezoid from upper coracoid process to oblique line on caudal surface of clavicle. Limits ventral rotation of scapula
       - ii. Conoid from base of coracoid process to coracoid tuberosity on caudal surface of clavicle. Limits dorsal rotation of scapula
- B. STERNOCLAVICULAR (see p. 268)
  1. Type: arthrodial
  2. Bones: clavicle with manubrium of sternum and cartilage of 1st rib
  3. Movements: gliding, with some motion in almost any direction
  4. Ligaments
     - a. Articular capsule surrounds articulation
     - b. Anterior sternoclavicular from clavicle to ventral surface of manubrium
     - c. Posterior sternoclavicular between dorsal surfaces of clavicle and manubrium
     - d. Interclavicular joins cephalic surface of the 2 clavicles and limits depression of shoulder
     - e. Costoclavicular from first costal cartilage to costal tuberosity on caudal surface of clavicle. Limits elevation of shoulder
     - f. Articular disk between sternum and clavicle and attached to clavicle and sternum. Helps limit depression of shoulder

IV. Muscles acting on shoulder girdle

A.	*Elevation*	*Depression*	*Protraction**	*Retraction*†
	Upper trapezius	Subclavius	Serratus anterior	Rhomboids
	Levator scapulae	Pect. minor	Pect. minor	Middle and lower
	Sternocleidomastoid	Lower trapezius		trapezius
	Rhomboids			

B.	*Rotation of Scapula*	*Up*	*Down*
		Upper and lower trapezius	Levator scapulae
		Serratus anterior	Rhomboids
			Pectoralis major and minor
			Latissimus dorsi

*Protraction: drawing forward or ventral.
†Retraction: drawing backward or dorsal.

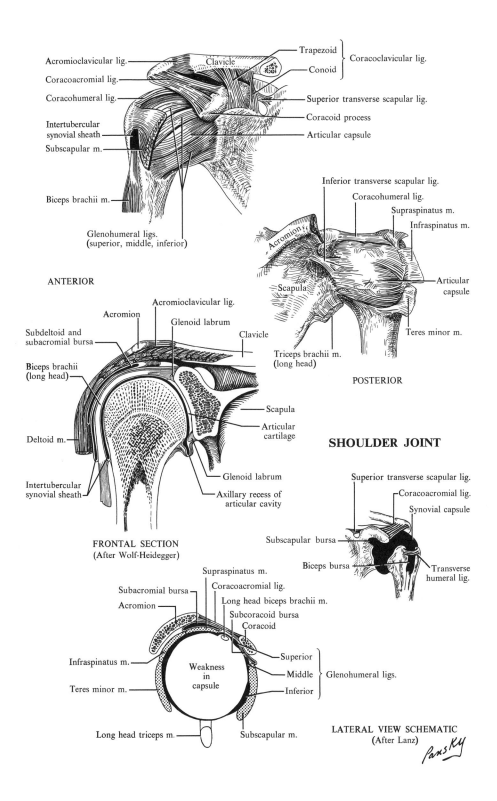

Acromioclavicular lig.
Coracoacromial lig.
Coracohumeral lig.
Intertubercular synovial sheath
Subscapular m.
Biceps brachii m.
Glenohumeral ligs. (superior, middle, inferior)
Clavicle
Trapezoid
Conoid
Coracoclavicular lig.
Superior transverse scapular lig.
Coracoid process
Articular capsule

ANTERIOR

Inferior transverse scapular lig.
Coracohumeral lig.
Supraspinatus m.
Infraspinatus m.
Acromion
Scapula
Articular capsule
Teres minor m.
Triceps brachii m. (long head)

POSTERIOR

Subdeltoid and subacromial bursa
Acromion
Acromioclavicular lig.
Glenoid labrum
Clavicle
Biceps brachii (long head)
Deltoid m.
Intertubercular synovial sheath
Scapula
Articular cartilage
Glenoid labrum
Axillary recess of articular cavity

FRONTAL SECTION
(After Wolf-Heidegger)

SHOULDER JOINT

Superior transverse scapular lig.
Coracoacromial lig.
Synovial capsule
Subscapular bursa
Biceps bursa
Transverse humeral lig.

Supraspinatus m.
Coracoacromial lig.
Long head biceps brachii m.
Subcoracoid bursa
Coracoid
Subacromial bursa
Acromion
Infraspinatus m.
Teres minor m.
Weakness in capsule
Superior
Middle
Inferior
Glenohumeral ligs.
Long head triceps m.
Subscapular m.

LATERAL VIEW SCHEMATIC
(After Lanz)

Pansky

- 193 -

89. HUMERUS

I. General: largest and longest bone of upper extremity

II. Parts: proximal extremity, shaft, and distal extremity

A. PROXIMAL: head, anatomical neck, greater tubercle, lesser tubercle, intertubercular groove (bicipital groove), and crests of the tubercles (are lips of groove)

B. SHAFT
1. Borders: lateral and medial
2. Surfaces: anterolateral, anteromedial, and posterior
3. Special features: radial sulcus
4. Lateral supracondylar ridge

C. DISTAL: medial and lateral epicondyles, capitulum, radial fossa, ulnar sulcus, trochlea, coronoid fossa, olecranon fossa, lateral and medial supracondylar ridges

III. Ossification of humerus: from 8 centers

Part	When Appears		When Closes
Body	8th week, fetal		As joined by the following
Head	1st year		
Greater tubercle	3rd year	All form 1 center in 6th year	Join body in 16–17th year
Lesser tubercle	5th year		
Capitulum	2nd year		16–17th year
Trochlea	12th year	All form 1 center and join body at same time	
Lateral epicondyle	13th year		
Medial epicondyle	5th year		18th year

IV. Clinical considerations

A. SURGICAL NECK—constriction just below the tubercles. Frequent site of fractures

B. ALTHOUGH TRAUMA often determines how bones will be displaced when broken, the displacement due to muscle insertion is also important, especially on the shorter of the two broken pieces. This is an important consideration in reduction of fractures

C. FRACTURES OF THE MIDDLE PART of shaft may injure radial nerve

D. TRAUMA AT MEDIAL EPICONDYLE may injure ulnar nerve

E. COMPLETE FRACTURES of the humerus usually show some overriding and other displacement of the ends as a result of muscular pull; rotatory and angular displacements of the upper fragment are characteristic of fractures of the surgical neck, and angular displacement is characteristic of lower fractures

F. EXPOSURE OF THE HUMERAL SHAFT, in surgery, is difficult because the bone is not subcutaneous anywhere, and important vessels and nerves are closely related to it

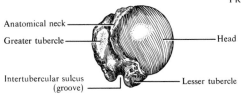

UPPER EXTREMITY
FROM ABOVE

Anatomical neck

Greater tubercle

Head

Intertubercular sulcus
(groove)

Lesser tubercle

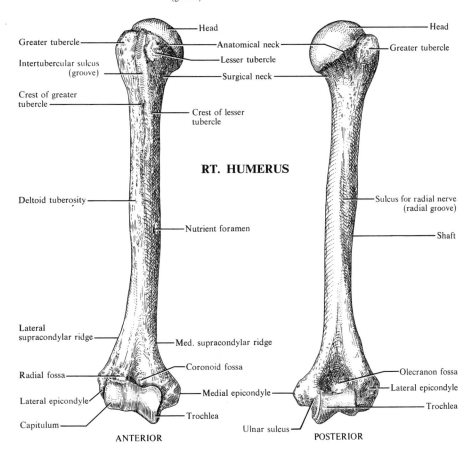

Head

Greater tubercle

Intertubercular sulcus
(groove)

Crest of greater
tubercle

Head

Anatomical neck

Lesser tubercle

Surgical neck

Crest of lesser
tubercle

Head

Greater tubercle

RT. HUMERUS

Deltoid tuberosity

Sulcus for radial nerve
(radial groove)

Nutrient foramen

Shaft

Lateral
supracondylar ridge

Med. supracondylar ridge

Radial fossa

Coronoid fossa

Olecranon fossa

Lateral epicondyle

Medial epicondyle

Lateral epicondyle

Capitulum

Trochlea

Trochlea

Ulnar sulcus

ANTERIOR

POSTERIOR

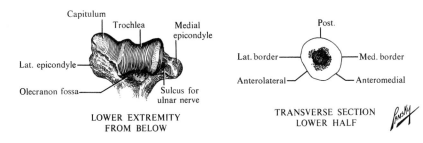

Capitulum

Trochlea

Medial
epicondyle

Post.

Lat. epicondyle

Lat. border

Med. border

Olecranon fossa

Anterolateral

Anteromedial

Sulcus for
ulnar nerve

LOWER EXTREMITY
FROM BELOW

TRANSVERSE SECTION
LOWER HALF

90. MUSCLES OF THE CHEST WALL

I.

Name	Origin	Insertion	Action	Nerve
Platysma (see p. 40)	Fascia over pect. maj. & deltoid	Mandible Skin of lower face	Draws lower lip down and back Opens jaws Pulls up skin from clavicle	Facial (VII)
Pectoralis major (see p. 208)	Medial clavicle Lat. sternum to 7th cost. cart. Costal carts. 2–6 Apon. of ext. abd. oblique	Crest of greater tubercle of humerus	Flexes, adducts, and rotates arm medially In climbing draws body upward	Med. and lat. ant. thoracic
Pectoralis minor (see p. 208)	Upper outer surface, ribs 3–5	Coracoid process of scapula	Draws scapula down and forward	Med. ant. thoracic
Subclavius	First rib cartilage	Subclavian groove of clavicle	Draws shoulder down and forward	N. to sub-clavius
Serratus anterior (see p. 208)	Outer surface and upper border of ribs 1–8	Med. angle, vertebral border, and inf. angle of scapula	Rotates scapula, raising point of shoulder Draws scapula forward	Long thoracic
Subscapularis (see p. 208)	Medial part, subscapular fossa	Lesser tubercle of humerus Capsule of shoulder joint	Rotates arm medially, helps in adduction, abduction, flexion, and extension	Upper and lower sub-scapular
Teres major (see p. 208)	Dorsal surface inf. angle of scapula	Crest of lesser tubercle of humerus	Adducts, extends, and rotates arm medially	Lower sub-scapular

II. Special features
  A. THE DIRECTION OF THE FIBERS of the pectoralis major should be noted: from clavicular origin, down and laterally; from sternum, horizontally; from lower ribs, upward and laterally
  B. THE ACTION OF THE SERRATUS ANTERIOR in the rotation of the scapula makes possible abduction of the arm of more than 90°
  C. WHEN THE HANDS ARE FIXED, as in gripping an object over the head, the pectoralis major helps to draw the body toward the hands
  D. THE APPARENT INCONGRUITY in the action of the subscapularis in performing opposing actions depends on position of the arm; if extended, the muscle flexes; if flexed, it helps to extend, etc.

MUSCLES OF CHEST WALL

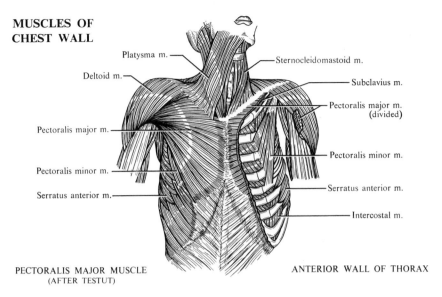

Platysma m.

Deltoid m.

Pectoralis major m.

Pectoralis minor m.

Serratus anterior m.

Sternocleidomastoid m.

Subclavius m.

Pectoralis major m. (divided)

Pectoralis minor m.

Serratus anterior m.

Intercostal m.

ANTERIOR WALL OF THORAX

PECTORALIS MAJOR MUSCLE
(AFTER TESTUT)

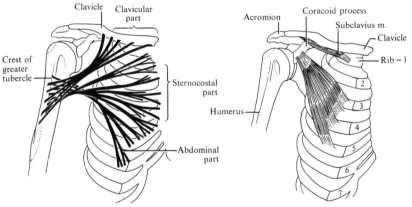

Clavicle

Clavicular part

Crest of greater tubercle

Sternocostal part

Abdominal part

Acromion

Coracoid process

Subclavius m.

Clavicle

Rib – 1

Humerus

2

3

4

5

6

7

PECTORALIS MINOR MUSCLE

SERRATUS ANTERIOR MUSCLE

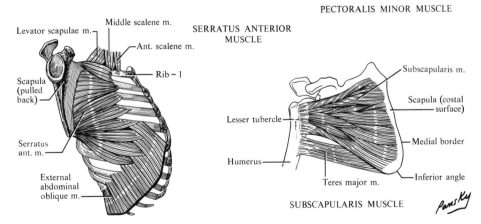

Levator scapulae m.

Middle scalene m.

Ant. scalene m.

Rib – 1

Scapula (pulled back)

Serratus ant. m.

External abdominal oblique m.

Subscapularis m.

Scapula (costal surface)

Lesser tubercle

Medial border

Humerus

Teres major m.

Inferior angle

SUBSCAPULARIS MUSCLE

Pansky

91. PARTS OF THE AXILLARY ARTERY AND THE BRACHIAL PLEXUS

I. Axillary artery
A. ORIGIN: subclavian artery becomes axillary artery at lateral border of first rib
B. TERMINATION: becomes brachial artery at lower border of tendon of teres major muscle
C. DIVISIONS
   1. First part from first rib to upper border of pectoralis minor muscle
   2. Second part lies behind pectoralis minor muscle
   3. Third part from lower border of pectoralis minor muscle to lower border of tendon of teres major muscle

II. Brachial plexus
A. ORIGIN: chiefly, anterior primary divisions of cervical nerves 5–8 plus first thoracic nerve
B. COMPONENTS: roots, trunks, and cords
C. MANNER OF FORMATION
   1. Roots (the anterior primary divisions) of C5 and C6 join to form *upper trunk;* C7 continues by itself as *middle trunk;* C8 and T1 join to form *lower trunk.* This occurs near lateral border of the scalenes in the posterior triangle of neck
   2. Trunks are short and split into anterior and posterior divisions. The anterior divisions of the upper and middle trunks form the *lateral cord;* the anterior part of the lower trunk becomes the *medial cord;* the posterior divisions of all trunks form the *posterior cord*
   3. The cords give rise to terminal branches
D. BRANCHES FROM

Part	Spinal Segment	Part	Spinal Segment
1. Roots		2. Trunks	
a. Dorsal scapular	C5	a. N. to subclavius	C5, 6
b. Long thoracic	C5, 6, 7	b. Suprascapular	C5, 6
3. Cords		4. Terminal	
a. Pectorals (lat. and med.)	C5, 6, 7, 8, T1	a. Musculo-cutaneous (lateral cord)	C5, 6, 7
		b. Median (lateral and medial cord)	C6, 7, 8, T1
Post. { b. Subscapular (lower and upper)	C5, 6		
c. Thoracodorsal	C5, 6, 7	c. Ulnar (medial cord)	C8, T1
d. Axillary	C5, 6	d. Radial (posterior cord)	C5, 6, 7, 8, T1
Med. { e. Med. brach. cutan.	C8, T1		
f. Med ante-brach. cutan.	C8, T1		

III. Special features
A. TWO ROOTS of the brachial plexus lie above and two below the level of the cricoid cartilage
B. THE TRUNKS of the plexus lie between the scalene muscles. The muscles do not lie lateral to the midpoint of the clavicle, and the same may be said of the trunks
C. THE DIVISIONS of the plexus lie adjacent to the 3rd part of the subclavian artery
D. THE CORDS of the plexus extend from the midpoint of the clavicle to the inferomedial aspect of the coracoid process of the scapula

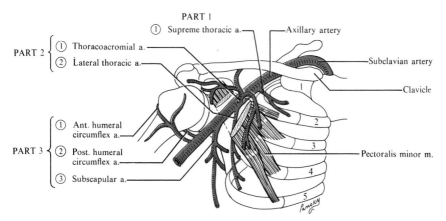

PART 1
① Supreme thoracic a.
Axillary artery

PART 2 { ① Thoracoacromial a.
② Lateral thoracic a.

Subclavian artery

Clavicle

1

2

3

PART 3 { ① Ant. humeral circumflex a.
② Post. humeral circumflex a.
③ Subscapular a.

Pectoralis minor m.

4

5

BRANCHES OF THE AXILLARY ARTERY

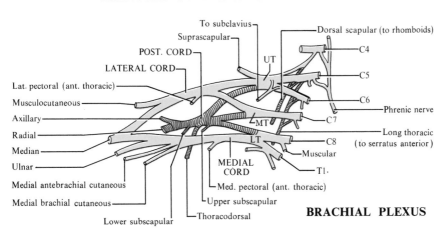

To subclavius
Suprascapular
POST. CORD
LATERAL CORD

Dorsal scapular (to rhomboids)
C4
UT
C5
C6

Lat. pectoral (ant. thoracic)
Musculocutaneous
Axillary
Radial
Median
Ulnar
Medial antebrachial cutaneous
Medial brachial cutaneous

MT
LT
MEDIAL CORD
Med. pectoral (ant. thoracic)
Upper subscapular
Lower subscapular
Thoracodorsal

Phrenic nerve
C7
Long thoracic (to serratus anterior)
C8
Muscular
T1

BRACHIAL PLEXUS

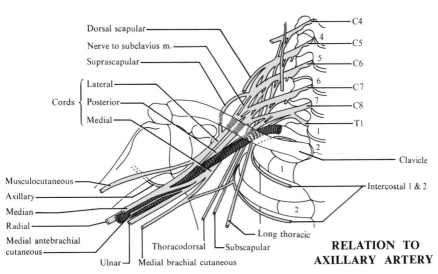

Dorsal scapular
Nerve to subclavius m.
Suprascapular

C4
4
C5
5
C6
6
C7
7
C8
T1

Cords { Lateral
Posterior
Medial

1
2

Clavicle

1

Musculocutaneous
Axillary
Median
Radial
Medial antebrachial cutaneous

Intercostal 1 & 2

2

Ulnar
Thoracodorsal
Subscapular
Medial brachial cutaneous
Long thoracic

RELATION TO AXILLARY ARTERY

92. THE AXILLARY REGION

I. The axilla
A. A SPACE, pyramidal in shape, between the arm and thoracic wall
B. BOUNDARIES
   1. Apex: directed upward and medialward, ending in the *cervicoaxillary canal,* which leads into the posterior triangle of neck
   2. Base: formed by axillary fascia and skin
   3. Anterior wall: pectoralis major and minor muscles, clavipectoral fascia
   4. Posterior wall: subscapularis, teres major, latissimus dorsi muscles
   5. Medial wall: first 4 ribs and intercostal muscles, upper part of serratus anterior muscle
   6. Lateral wall: humerus, coracobrachialis and biceps muscles
C. CONTENTS
   1. Axillary vessels and their branches
   2. Brachial plexus and its branches
   3. Lymph nodes, embedded in fat

II. The fascia
A. AXILLARY. Investing layer extending from the pectoralis major to latissimus dorsi muscles, arching inward to form the hollow of the armpit. It is continuous with the fasciae covering the muscles bounding the axilla
B. CLAVIPECTORAL (CORACOCLAVICULAR). Attached above in front and behind the subclavius muscle. At lower border of this muscle, these 2 sheets form a single layer, which extends downward and laterally to the border of the pectoralis minor muscle. Here, the fascia splits to surround that muscle and again forms a single sheet at its lower border, which is continuous with the axillary fascia as the suspensory lig.
   1. Costocoracoid ligament: that part of the above fascia, greatly strengthened, which lies in front of the subclavius muscle, extending from the first rib to the coracoid process
   2. Coracoclavicular fascia: that portion of the clavipectoral fascia lying between the pectoralis minor and subclavius muscles. Is pierced by thoracoacromial vessels, cephalic vein, and lateral anterior thoracic nerve
C. AXILLARY SHEATH. At apex of axilla, fascia of first two ribs in first interspace becomes continuous with scalene fascia. Here this fascia forms a tubular sheath for vessels and nerves entering the axilla. Under subclavius and pectoralis minor muscles it is adherent to the clavipectoral fascia

III. Special features
A. THE AXILLA houses the great vessels and nerves of the limb, which are closely grouped together and enclosed in a layer of fascia (*axillary fascia*)
B. MOST OF THE NERVES appearing in the axilla are the lower part of the brachial plexus and its major branches, and closely surround the axillary artery
C. THE AXILLARY ARTERY is conveniently thought of as the central structure of the axilla, with axillary veins closely related to it
D. LYMPH NODES of the axilla lie in the looser connective tissue that serves as padding about the vital structures and are particularly related to the blood vessels

AXILLA

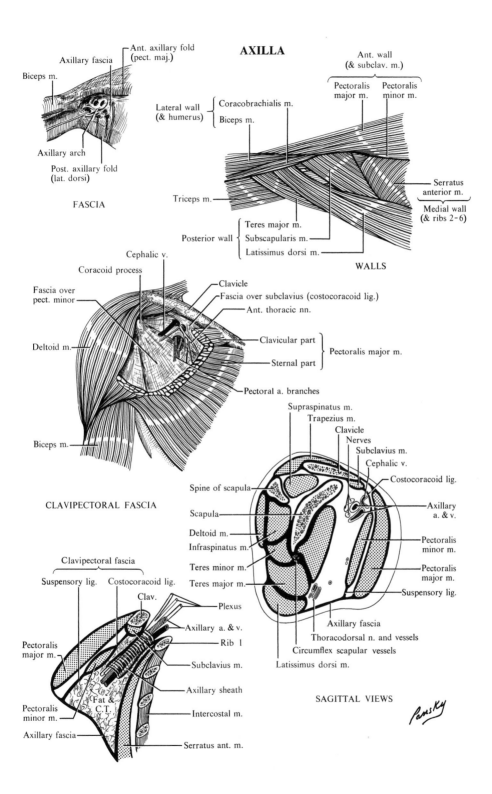

FASCIA

Ant. axillary fold (pect. maj.)
Axillary fascia
Biceps m.
Axillary arch
Post. axillary fold (lat. dorsi)

Lateral wall (& humerus) { Coracobrachialis m. / Biceps m.
Triceps m.

Ant. wall (& subclav. m.)
Pectoralis major m. / Pectoralis minor m.
Serratus anterior m.
Medial wall (& ribs 2-6)

Posterior wall { Teres major m. / Subscapularis m. / Latissimus dorsi m.

WALLS

CLAVIPECTORAL FASCIA

Cephalic v.
Coracoid process
Fascia over pect. minor
Deltoid m.
Biceps m.
Clavicle
Fascia over subclavius (costocoracoid lig.)
Ant. thoracic nn.
Clavicular part / Sternal part } Pectoralis major m.
Pectoral a. branches

Clavipectoral fascia
Suspensory lig.
Costocoracoid lig.
Clav.
Plexus
Axillary a. & v.
Rib 1
Subclavius m.
Axillary sheath
Intercostal m.
Serratus ant. m.
Pectoralis major m.
Pectoralis minor m.
Axillary fascia
Fat & C.T.

Supraspinatus m.
Trapezius m.
Clavicle
Nerves
Subclavius m.
Cephalic v.
Costocoracoid lig.
Spine of scapula
Scapula
Deltoid m.
Infraspinatus m.
Teres minor m.
Teres major m.
Axillary a. & v.
Pectoralis minor m.
Pectoralis major m.
Suspensory lig.
Axillary fascia
Thoracodorsal n. and vessels
Circumflex scapular vessels
Latissimus dorsi m.

SAGITTAL VIEWS

Pansky

93. VESSELS AND NERVES OF AXILLA

I. Axillary artery
A. RELATIONS
1. Part I

Anterior
Subclav. and pect. minor mm., clavipect. fascia, lat. ant. thoracic n., thoracoacromial vessels, cephalic v.

Lateral	**Axillary I**	Medial
Brachial plexus		Axillary v.

Posterior
First intercost. and serratus ant. mm., Long thoracic n.

2. Part II

Anterior
Pect. major and minor mm.

Lateral	**Axillary II**	Medial
Lat. cord of plexus		Med. cord brachial plexus, axillary v.

Posterior
Post. cord of plexus, subscapular m.

3. Part III

Anterior
Skin, fascia, pect. major m., medial head of median n.

Lateral	**Axillary III**	Medial
Coracobrach. m., median and musculocutaneous nn.		Med. antebrach. cutaneous and ulnar nn., axillary v.

Posterior
Subscap. m., tendons of teres maj. and latissimus dorsi mm., radial and axillary nn.

B. BRANCHES
1. Part I: supreme (superior) thoracic
2. Part II: thoracoacromial and lateral thoracic
3. Part III: subscapular; anterior and posterior circumflex humeral

II. Origin and distribution of some nerves from brachial plexus

Name	Distribution
Long thoracic	Serratus ant. m.
N. to subclavius	Subclavius m.
Lat. pectoral	Pect. major m.
Med. pectoral	Pect. major and minor mm.
Subscapular	Subscap. and teres major mm.
Thoracodorsal	Latissimus dorsi m.
Axillary	Deltoid, teres minor mm., and skin of arm
Musculocutaneous	Muscles of arm and skin of forearm

III. Axillary vein begins at union of basilic and brachial veins and terminates at first rib as subclavian vein. Lies medial to and partly overlaps axillary artery with the medial cord, median, ulnar, and medial anterior pectoral nerves between them. Receives tributaries corresponding to branches of artery plus the cephalic vein

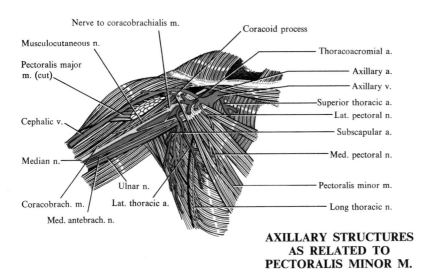

Nerve to coracobrachialis m.

Musculocutaneous n.

Pectoralis major m. (cut)

Cephalic v.

Median n.

Coracobrach. m.

Ulnar n.

Lat. thoracic a.

Med. antebrach. n.

Coracoid process

Thoracoacromial a.

Axillary a.

Axillary v.

Superior thoracic a.

Lat. pectoral n.

Subscapular a.

Med. pectoral n.

Pectoralis minor m.

Long thoracic n.

AXILLARY STRUCTURES AS RELATED TO PECTORALIS MINOR M.

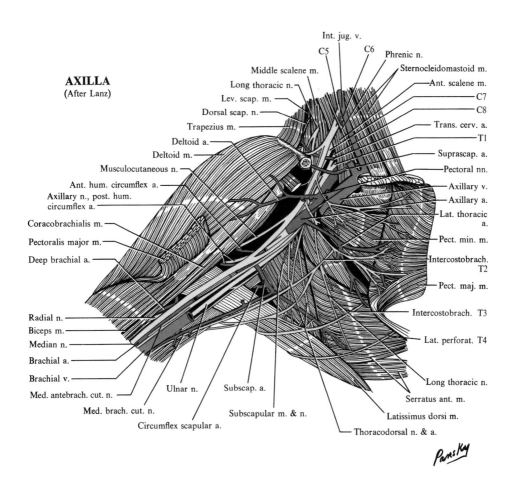

AXILLA
(After Lanz)

Int. jug. v.

C5   C6   Phrenic n.

Middle scalene m.

Sternocleidomastoid m.

Long thoracic n.

Ant. scalene m.

Lev. scap. m.

C7

Dorsal scap. n.

C8

Trapezius m.

Trans. cerv. a.

Deltoid a.

T1

Deltoid m.

Suprascap. a.

Musculocutaneous n.

Pectoral nn.

Ant. hum. circumflex a.

Axillary v.

Axillary n., post. hum. circumflex a.

Axillary a.

Coracobrachialis m.

Lat. thoracic a.

Pectoralis major m.

Pect. min. m.

Deep brachial a.

Intercostobrach. T2

Pect. maj. m.

Radial n.

Intercostobrach. T3

Biceps m.

Median n.

Lat. perforat. T4

Brachial a.

Brachial v.

Med. antebrach. cut. n.

Ulnar n.

Subscap. a.

Long thoracic n.

Med. brach. cut. n.

Serratus ant. m.

Subscapular m. & n.

Circumflex scapular a.

Latissimus dorsi m.

Thoracodorsal n. & a.

Pansky

-203-

94. MUSCLES OF THE SHOULDER

I.

Name	Origin	Insertion	Action	Nerve
Deltoid (see p. 208)	Ant. border and upper surface of clavicle Lat. margin, upper surface of acromion Lower border of spine of scapula	Deltoid tubercle of humerus	Abducts arm Ant. fibers flex and rotate arm medially Post. fibers extend and rotate arm laterally	Axillary
Supra-spinatus	Medial part of supraspinous fossa	Highest impression, great tubercle of humerus	Abducts arm	Supra-scapular
Infra-spinatus	Medial part of infraspinous fossa	Middle impression, great tubercle of humerus	Rotates arm laterally Upper fibers abduct Lower fibers adduct	Supra-scapular
Teres minor	Upper dorsal axillary border of scapula	Lowest impression, great tubercle of humerus United to joint capsule	Rotates arm laterally and adducts it	Axillary
Subscapularis (see p. 196) Teres major (see p. 196)				

II. Special features

A. THE DELTOID MUSCLE is responsible for the roundness of the shoulder

B. CERTAIN SPACES BETWEEN ADJOINING MUSCLES or parts of muscle and bone should be noted, for they permit passage of vessels and nerves
   1. Triangular space, a triangular hiatus whose base is lateral; bounded by the long head of the triceps, the teres minor, and teres major
   2. Quadrilateral space, formed by the teres minor, the teres major, the humerus, and the long head of the triceps
   3. Triangular interval, formed by the teres major, lateral head of the triceps, and long head of the triceps

C. TENDONS of the supraspinatus, infraspinatus, and teres major all strengthen the shoulder joint

D. THE SUPRASPINATUS, INFRASPINATUS, AND TERES MINOR MUSCLES are sometimes referred to as the great "sit" muscles of the shoulder: *s* for supra, *i* for infra, and *t* for teres minor, which all "sit" on the great tubercle of the humerus

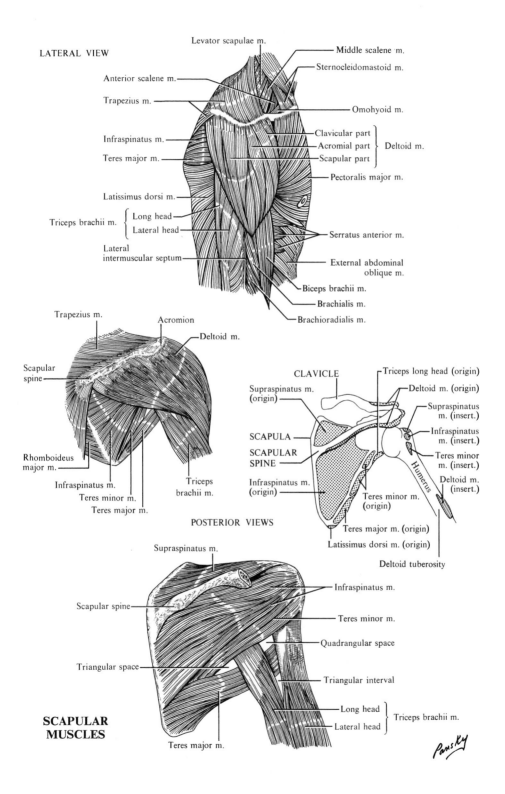

LATERAL VIEW

Levator scapulae m.

Middle scalene m.

Sternocleidomastoid m.

Anterior scalene m.

Trapezius m.

Omohyoid m.

Infraspinatus m.

Clavicular part

Acromial part ⎱ Deltoid m.

Teres major m.

Scapular part

Pectoralis major m.

Latissimus dorsi m.

Triceps brachii m. ⎰ Long head

Lateral head

Serratus anterior m.

Lateral
intermuscular septum

External abdominal
oblique m.

Biceps brachii m.

Brachialis m.

Brachioradialis m.

Trapezius m.

Acromion

Deltoid m.

Scapular
spine

CLAVICLE

Triceps long head (origin)

Supraspinatus m.
(origin)

Deltoid m. (origin)

Supraspinatus
m. (insert.)

SCAPULA

Infraspinatus
m. (insert.)

Rhomboideus
major m.

SCAPULAR
SPINE

Teres minor
m. (insert.)

Infraspinatus m.

Infraspinatus m.
(origin)

Deltoid m.
(insert.)

Teres minor m.

Triceps
brachii m.

Teres minor m.
(origin)

Teres major m.

POSTERIOR VIEWS

Teres major m. (origin)

Latissimus dorsi m. (origin)

Deltoid tuberosity

Supraspinatus m.

Scapular spine

Infraspinatus m.

Teres minor m.

Quadrangular space

Triangular space

Triangular interval

Long head

Lateral head

Triceps brachii m.

**SCAPULAR
MUSCLES**

Teres major m.

Pansky

- 205 -

95. COLLATERAL CIRCULATION AROUND THE SHOULDER

I. Collateral circulation for ligation of or obstruction in any artery is important and is accomplished through the anastomosing of arteries that arise from the main channel above the stoppage with those that arise below that point

II. Collateral circulation for the ligation of the axillary artery

Arteries Above Ligature	Arteries Below Ligature
A. IN AXILLARY 1, above the origin of the thoracoacromial artery	
1. Suprascapular (from thyrocervical trunk)	Circumflex scapular (from subscapular)
2. Descending (deep) ramus (from trans. cervical of thyrocervical trunk)	Circumflex scapular
3. Descending scapular (dorsal scapular) (from subclavian or thyrocervical trunk)	Circumflex scapular
4. Intercostals (from aorta)	Thoracodorsal (from axillary 2)
B. IN AXILLARY 2, between origins of thoracoacromial and lateral thoracic	
1. Suprascapular—as above	Circumflex scapular—as above
2. Descending ramus—as above	Circumflex scapular—as above
3. Descending scapular—as above	Circumflex scapular—as above
4. Intercostals (from aorta)	Lateral thoracic (from axillary 2) and thoracodorsal (from subscapular)
5. Pectoral branch (from thoracoacromial)	Lateral thoracic (from axillary 2)
C. IN AXILLARY 2, between lateral thoracic and subscapular	
1. Suprascapular—as above	As above
2. Descending ramus—as above	As above
3. Descending scapular—as above	As above
4. Lateral thoracic (from axillary 2)	Thoracodorsal (from subscapular)
5. Acromial branch (thoracoacromial)	Post. circumflex humeral (axillary 3)
D. IN AXILLARY 3, below the subscapular and the 2 circumflex humeral	
1. Ant. and post. circumflex humeral (from axillary 3)	Deep brachial (from brachial)

III. Sites where anastomoses occur
A. BETWEEN THE SUPRASCAPULAR AND CIRCUMFLEX SCAPULAR in infraspinous fossa
B. BETWEEN DESCENDING SCAPULAR AND CIRCUMFLEX SCAPULAR along medial border of scapula
C. BETWEEN PECTORAL BRANCHES, INTERCOSTALS, LATERAL THORACIC, OR THORACODORSAL on thoracic walls
D. BETWEEN ACROMIAL AND POSTERIOR CIRCUMFLEX HUMERAL around acromion
E. BETWEEN CIRCUMFLEX HUMERAL AND DEEP BRACHIAL: in or under triceps muscle

IV. Special features
A. THE CIRCUMFLEX SCAPULAR ARTERY leaves axilla through the triangular space
B. THE POSTERIOR CIRCUMFLEX HUMERAL ARTERY leaves the axilla via the quadrilateral space with the axillary nerve
C. SUPRASCAPULAR NERVE AND SUPRASCAPULAR ARTERY run together, but the nerve goes under and the artery passes over the transverse scapular ligament
D. THE DORSAL SCAPULAR NERVE descends with the descending branch of the transverse cervical artery

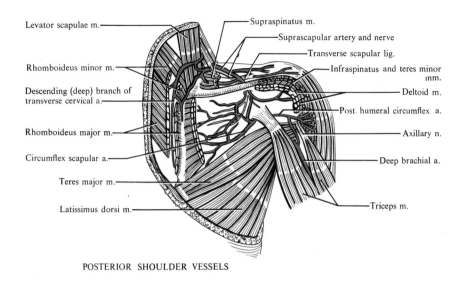

Levator scapulae m.
Supraspinatus m.
Suprascapular artery and nerve
Transverse scapular lig.
Rhomboideus minor m.
Infraspinatus and teres minor mm.
Descending (deep) branch of transverse cervical a.
Deltoid m.
Post. humeral circumflex a.
Rhomboideus major m.
Axillary n.
Circumflex scapular a.
Deep brachial a.
Teres major m.
Latissimus dorsi m.
Triceps m.

POSTERIOR SHOULDER VESSELS

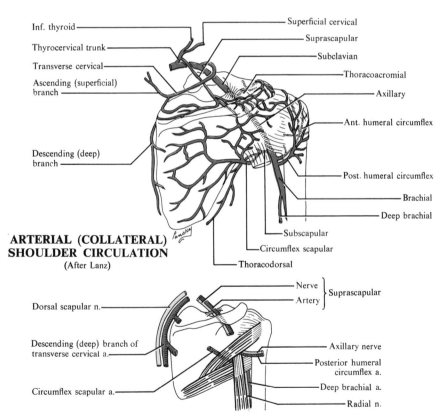

Inf. thyroid
Superficial cervical
Thyrocervical trunk
Suprascapular
Transverse cervical
Subclavian
Ascending (superficial) branch
Thoracoacromial
Axillary
Ant. humeral circumflex
Descending (deep) branch
Post. humeral circumflex
Brachial
Deep brachial
ARTERIAL (COLLATERAL) SHOULDER CIRCULATION
(After Lanz)
Subscapular
Circumflex scapular
Thoracodorsal

Nerve
Artery
} Suprascapular
Dorsal scapular n.
Descending (deep) branch of transverse cervical a.
Axillary nerve
Posterior humeral circumflex a.
Circumflex scapular a.
Deep brachial a.
Radial n.

96. SHOULDER JOINT

I. Type: enarthrodial (ball and socket)

II. Bones: spherical head of humerus and glenoid cavity of scapula

III. Movements: flexion, extension, abduction, adduction, medial rotation, lateral rotation, and circumduction

IV. Ligaments

A. ARTICULAR CAPSULE from edge of glenoid fossa and glenoidal labrum to anatomical neck of humerus. Strengthened: anteriorly, subscapular muscle; posteriorly, infraspinatus and teres minor muscle; above, supraspinatus muscle; below, long head of triceps

B. CORACOHUMERAL from coracoid to greater tubercle

C. GLENOHUMERAL: usually 3 bands
   1. From medial side of glenoid cavity to lower part of lesser tubercle
   2. From lower edge of glenoid cavity to lower anatomical neck
   3. From above glenoid rim near root of coracoid process along medial side of biceps tendon to a depression above lesser tubercle of humerus

D. TRANSVERSE HUMERAL bridges intertubercular sulcus (groove)

E. GLENOIDAL LABRUM: fibrocartilage attached to edges of glenoid fossa to deepen it and protect bone. Above it is continuous with biceps tendon (long head)

V. Synovial membrane: from glenoid cavity over labrum; lines inside of capsule and is reflected on anatomical neck to articular cartilage. Encloses tendon of long head of biceps muscle in a tubular sheath, intertubercular synovial sheath

VI. Muscles acting on joint

Flexion	Extension	Abduction	Adduction	Med. Rotation	Lat. Rotation
Pect. major (clav. head)	Latissimus dorsi	Deltoid (as whole)	Pect. major (as whole)	Pect. major (as whole)	Infraspinatus
Deltoid (ant. fibers)	Teres major	Supra-spinatus	Latissimus dorsi	Latissimus dorsi	Teres minor
Coraco-brach.	Deltoid (post. fibers)		Teres major	Teres major	Deltoid (post. fibers)
Biceps (long head)	Triceps (long head)		Subscap-ularis	Subscap-ularis	
			Triceps (long head)	Deltoid (ant. fibers)	

VII. Clinical considerations

A. MOST OF THE STRENGTH OF THE JOINT is provided by muscle tendons that run behind, above, and in front of the capsule (see p. 201)

B. WEAKEST PART IS BELOW, where dislocation is easiest, especially when arm is abducted. In this position the humeral head breaks through the capsule below the glenoid cavity

C. THE UPPER SURFACE of the musculotendinous cuff is separated from the overlying acromion, coracoacromial ligament, and the upper part of the deltoid muscle by *the subacromial bursa* (there may be a lateral *subdeltoid bursa,* but the two are usually fused). Calcium deposits in the tendinous floor of the bursa may cause disability of the shoulder, provoking a *bursitis,* which makes movements painful

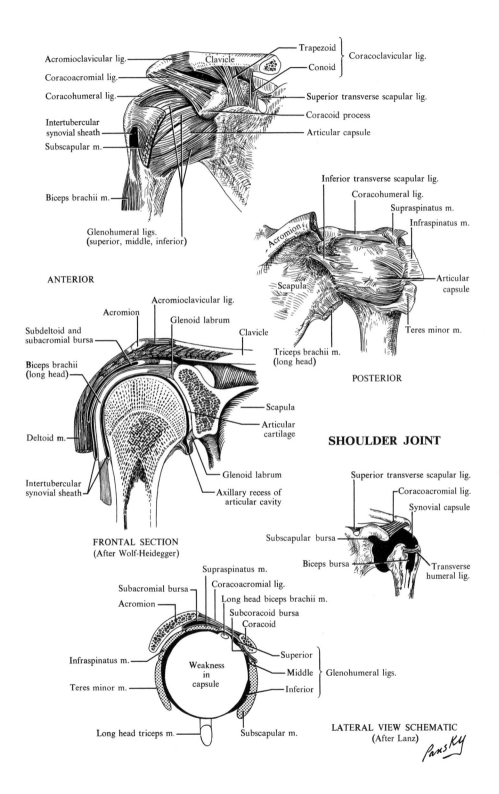

Acromioclavicular lig.
Coracoacromial lig.
Coracohumeral lig.
Intertubercular synovial sheath
Subscapular m.
Biceps brachii m.
Glenohumeral ligs. (superior, middle, inferior)

Trapezoid
Conoid
} Coracoclavicular lig.
Clavicle
Superior transverse scapular lig.
Coracoid process
Articular capsule

ANTERIOR

Inferior transverse scapular lig.
Coracohumeral lig.
Supraspinatus m.
Infraspinatus m.
Acromion
Scapula
Articular capsule
Teres minor m.
Triceps brachii m. (long head)

POSTERIOR

Acromioclavicular lig.
Acromion
Glenoid labrum
Clavicle
Subdeltoid and subacromial bursa
Biceps brachii (long head)
Deltoid m.
Intertubercular synovial sheath
Scapula
Articular cartilage
Glenoid labrum
Axillary recess of articular cavity

FRONTAL SECTION
(After Wolf-Heidegger)

SHOULDER JOINT

Superior transverse scapular lig.
Coracoacromial lig.
Synovial capsule
Subscapular bursa
Biceps bursa
Transverse humeral lig.

Supraspinatus m.
Coracoacromial lig.
Long head biceps brachii m.
Subcoracoid bursa
Coracoid
Subacromial bursa
Acromion
Superior
Middle } Glenohumeral ligs.
Inferior
Infraspinatus m.
Weakness in capsule
Teres minor m.
Long head triceps m.
Subscapular m.

LATERAL VIEW SCHEMATIC
(After Lanz)

Pansky

- 209 -

97. MUSCLES OF THE ARM

I.

Name	Origin	Insertion	Action	Nerves
Coraco-brachialis	Apex, coracoid proc. of scapula	Middle shaft of humerus	Flexes and adducts arm	Musculo-cutaneous
Biceps				
Long head	Supraglenoid tuberosity	Tuberosity of radius; bicipital aponeurosis to fascia of forearm	Flexes arm and forearm	Musculo-cutaneous
Short head	Apex, coracoid proc. of scapula		Supinates hand	Musculo-cutaneous
Brachialis	Lower half of front of humerus	Tuberosity and coronoid process of ulna	Flexes forearm	Musculo-cutaneous, radial, median
Triceps				
Long head	Infraglenoid tuberosity of scapula	Post. and upper olecranon Fascia of forearm	Extends and adducts arm (long head only)	Radial
Lat. head	Post. humerus above radial groove Lat. side of humerus	Post. and upper olecranon Fascia of forearm	Extends forearm	Radial
Med. head	Post. humerus below radial groove	Post. and upper olecranon Fascia of forearm	Extends forearm	Radial

II. Special features

A. TENDON OF ORIGIN OF LONG HEAD OF BICEPS runs over head of humerus, ensheathed by a layer of synovial membrane that follows the tendon as far as the surgical neck of humerus. The tendon is held in the intertubercular groove by the *transverse humeral ligament* and a prolongation of the tendon of the pectoralis major muscle

B. THE APONEUROSIS OF THE BICEPS overlies the brachial artery, thus protecting it. Further, it supports the median cubital vein, making it easier to introduce a needle

III. Clinical considerations

A. FRACTURES OF HUMERUS: as noted previously (p. 194), displacement of bone fragments after fracture may be determined by muscle attachments. For example, after fracture through the surgical neck: both medial and lateral rotators are attached to the proximal fragment, therefore this part is not rotated since they are balanced; the supraspinatus muscle is attached to the proximal piece and is unopposed, therefore the proximal fragment is abducted; the distal fragment has no lateral rotators attached to it but does have medial rotators, resulting in this part's being medially rotated; the distal part also has attached to it the large and powerful adductors—latissimus dorsi and pectoralis major—causing the distal piece to be adducted; and there will be overriding of the ends of the two pieces because of the pull of such muscles as the biceps and triceps

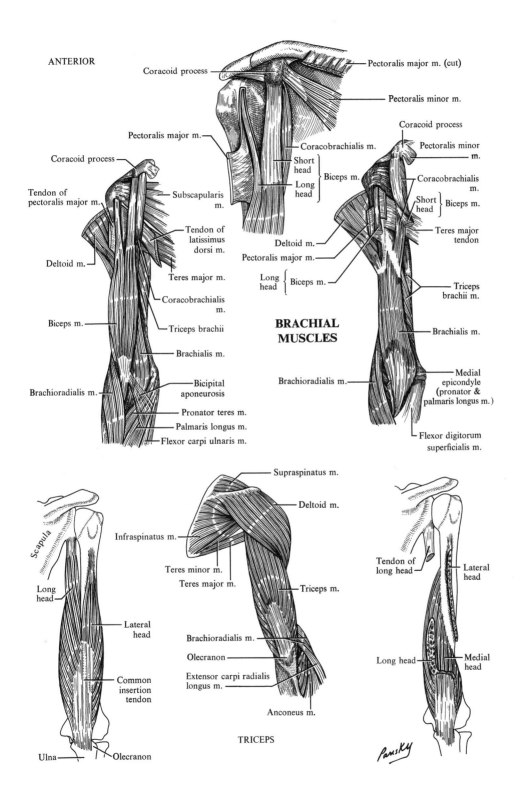

ANTERIOR

Coracoid process

Pectoralis major m. (cut)

Pectoralis minor m.

Coracoid process

Pectoralis major m.

Coracobrachialis m.

Pectoralis minor m.

Coracoid process

Tendon of pectoralis major m.

Short head } Biceps m.
Long head

Coracobrachialis m.

Subscapularis m.

Short head } Biceps m.

Tendon of latissimus dorsi m.

Deltoid m.

Teres major tendon

Deltoid m.

Pectoralis major m.

Teres major m.

Long head } Biceps m.

Triceps brachii m.

Biceps m.

Coracobrachialis m.

Triceps brachii

Brachialis m.

Brachialis m.

BRACHIAL MUSCLES

Brachioradialis m.

Medial epicondyle (pronator & palmaris longus m.)

Brachioradialis m.

Bicipital aponeurosis

Pronator teres m.

Palmaris longus m.

Flexor carpi ulnaris m.

Flexor digitorum superficialis m.

Supraspinatus m.

Deltoid m.

Infraspinatus m.

Scapula

Teres minor m.

Teres major m.

Triceps m.

Tendon of long head

Lateral head

Long head

Lateral head

Brachioradialis m.

Olecranon

Long head

Medial head

Common insertion tendon

Extensor carpi radialis longus m.

Anconeus m.

Ulna

Olecranon

TRICEPS

Pansky

98. VESSELS AND NERVES OF THE ARM

I. **Brachial fascia:** continuous with axillary, pectoral, and deltoid fasciae above, is attached to the humerus and ulna, and is continuous with the antebrachial fascia below

II. **Intermuscular septa:** separate posterior and anterior muscle groups
A. LATERAL from brachial fascia to lateral side of humerus
B. MEDIAL from brachial fascia to medial side of humerus

III. **Brachial artery** is a direct continuation of the axillary artery and terminates as the radial and ulnar arteries

A. RELATIONS

	Anterior	
	Skin, fascia, bicipital aponeurosis, median n., med. cubital v.	
Lateral		*Medial*
Coracobrachialis and biceps mm., median n.	**Brachial artery**	Median., med. antebrach. cutaneous, and ulnar nn.; basilic v.
	Posterior	
	Triceps and brachialis mm., radial n., deep brachial a.	

B. BRANCHES (see also p. 227)
1. Deep brachial enters radial sulcus (groove) behind humerus and terminates as radial and middle collateral arteries
2. Superior ulnar collateral runs with ulnar nerve behind medial intermuscular septum
3. Inferior ulnar collateral descends to back of elbow

IV. **Nerves** (see also p. 199)
A. MEDIAN first lateral to, then crosses, and finally is medial to brachial artery. No branches in arm
B. MUSCULOCUTANEOUS pierces coracobrachialis muscle, sending branches to it, the biceps, and brachialis muscles. Continues as lateral antebrachial cutaneous nerve
C. RADIAL passes in groove on back of humerus with deep brachial artery, pierces lateral intermuscular septum, and divides into superficial and deep branches in front of lateral epicondyle. In arm, gives branches to triceps and to skin (posterior brachial, inferior lateral brachial, and posterior antebrachial cutaneous nerves)
D. ULNAR medial to axillary and brachial arteries to the middle of arm, pierces medial intermuscular septum and runs with superior ulnar collateral artery to groove behind medial epicondyle of humerus. No branches in arm

V. **Special features**
A. THE BRACHIAL ARTERY lies successively on 3 muscles, gives 3 main branches, is in contact with 3 important nerves, and is associated with 3 veins
1. The 3 muscles that constitute the floor on which the artery runs are (from above downward): the long head of the triceps, the coracobrachialis, and the brachialis
2. The 3 main branches of the artery are: the profunda (deep) brachii, the superior ulnar collateral, and the inferior ulnar collateral
3. The 3 important nerves associated with the artery are: radial, ulnar, and median
4. The 3 veins associated with the brachial artery are: its two venae comitantes (brachial veins) and the basilic
B. THE BIFURCATION of the brachial artery is not constant and may take place at a high level in the arm

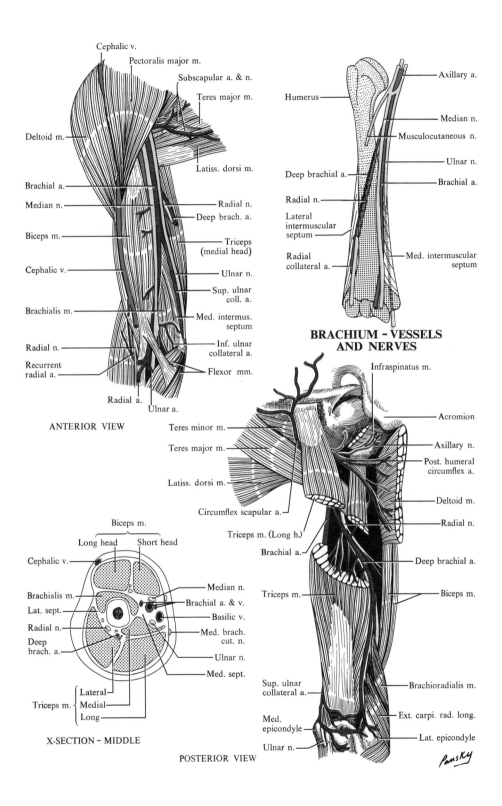

Cephalic v.

Pectoralis major m.

Subscapular a. & n.

Teres major m.

Deltoid m.

Latiss. dorsi m.

Brachial a.

Median n.

Radial n.

Deep brach. a.

Biceps m.

Triceps
(medial head)

Cephalic v.

Ulnar n.

Sup. ulnar
coll. a.

Brachialis m.

Med. intermus.
septum

Radial n.

Inf. ulnar
collateral a.

Recurrent
radial a.

Flexor mm.

Radial a.

Ulnar a.

ANTERIOR VIEW

Humerus

Axillary a.

Median n.

Musculocutaneous n.

Ulnar n.

Deep brachial a.

Brachial a.

Radial n.

Lateral
intermuscular
septum

Radial
collateral a.

Med. intermuscular
septum

**BRACHIUM - VESSELS
AND NERVES**

Infraspinatus m.

Teres minor m.

Acromion

Teres major m.

Axillary n.

Post. humeral
circumflex a.

Latiss. dorsi m.

Deltoid m.

Circumflex scapular a.

Radial n.

Triceps m. (Long h.)

Brachial a.

Deep brachial a.

Biceps m.

Triceps m.

Biceps m.

Long head

Short head

Cephalic v.

Median n.

Brachialis m.

Brachial a. & v.

Lat. sept.

Basilic v.

Radial n.

Med. brach.
cut. n.

Deep
brach. a.

Ulnar n.

Med. sept.

Lateral

Medial

Triceps m.

Long

Sup. ulnar
collateral a.

Brachioradialis m.

Med.
epicondyle

Ext. carpi. rad. long.

Lat. epicondyle

Ulnar n.

X-SECTION - MIDDLE

POSTERIOR VIEW

Pansky

99. BONES OF THE FOREARM

I. Ulna: medial bone of forearm

A. PARTS: proximal extremity, shaft, and distal extremity
   1. Proximal: olecranon, trochlear notch with smooth articular surface, and the coronoid process with a tuberosity medially and a radial notch laterally
   2. Shaft. Three borders: anterior, posterior, and lateral (interosseous). Three surfaces: anterior, medial, and posterior
   3. Distal: head, with styloid process and articular surface

II. Radius: lateral bone of forearm, parallel to above

A. PARTS: proximal extremity, shaft, and distal extremity
   1. Proximal: head, neck, and tuberosity; and shallow depression (fovea) on upper head
   2. Shaft. Three borders: anterior, posterior, and medial (interosseous). Three surfaces: anterior, posterior, and lateral
   3. Distal: large carpal articular surface with ulnar notch on medial side, 3 distinct grooves on dorsal side for tendons, and styloid process

III. Ossification

Location	When Appears	When Closes
A. RADIUS from 3 centers		
Shaft	8th fetal week	——
Distal	2nd year	20th year
Proximal	5th year	17th–18th years
B. ULNA from 3 centers		
Shaft	8th fetal week	——
Head	4th year	20th year
Olecranon	10th year	16th year

IV. Clinical considerations

A. ULNA IS MORE SUBJECT TO FRACTURE DUE TO TRAUMA AT ELBOW; radius is more subject to fracture from falls on hands
B. COLLES' FRACTURE is a fracture of the lower end of the radius with displacement of the hand backward and outward
C. MALUNITED FRACTURE OF THE HEAD OF THE RADIUS often causes severe disability, since rotation of the forearm is restricted and extension of the elbow may be impaired
D. MALUNION OR EVEN NONUNION OF FRACTURES OF THE OLECRANON often results in little or no limitation of elbow function
E. THE CORRECT POSITION FOR A RADIUS FRACTURED ABOVE the insertion of the pronator teres is with the elbow flexed and the hand supinated. In fractures below the pronator teres, the thumb-up (midprone) position is used with flexion at the elbow. The important rule in all fractures of the radius above the position of a Colles' fracture is to keep the elbow flexed; otherwise it will be impossible to maintain the forearm in any given position of rotation
F. A REVERSE COLLES' FRACTURE is called *Smith's fracture* and is usually produced by a fall on the back of the hand with the wrist flexed
G. EPIPHYSEAL SEPARATION may be confused with a Colles' fracture, is common in children, and may occur at any time up to the 18th or the 20th year

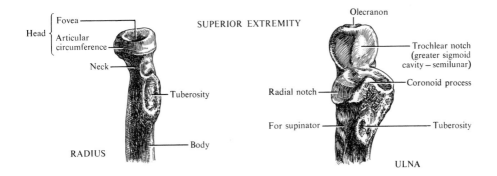

Head { Fovea
Articular circumference
Neck

SUPERIOR EXTREMITY

Tuberosity

Body

RADIUS

Olecranon

Trochlear notch
(greater sigmoid
cavity — semilunar)

Radial notch

Coronoid process

For supinator

Tuberosity

ULNA

RT. RADIUS AND ULNA

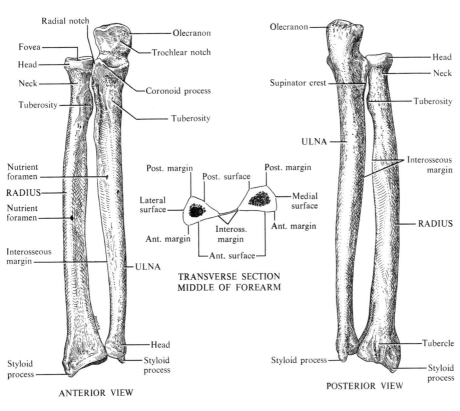

Radial notch

Fovea

Head

Neck

Tuberosity

Olecranon

Trochlear notch

Coronoid process

Tuberosity

Nutrient foramen

RADIUS

Nutrient foramen

Interosseous margin

ULNA

Post. margin

Post. surface

Post. margin

Lateral surface

Medial surface

Ant. margin

Inteross. margin

Ant. margin

Ant. surface

TRANSVERSE SECTION
MIDDLE OF FOREARM

Head

Styloid process

Styloid process

ANTERIOR VIEW

Olecranon

Head

Neck

Supinator crest

Tuberosity

ULNA

Interosseous margin

RADIUS

Tubercle

Styloid process

Styloid process

POSTERIOR VIEW

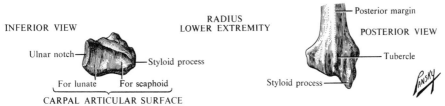

INFERIOR VIEW

RADIUS
LOWER EXTREMITY

POSTERIOR VIEW

Ulnar notch

Styloid process

For lunate

For scaphoid

CARPAL ARTICULAR SURFACE

Posterior margin

Tubercle

Styloid process

100. ANTECUBITAL REGION

I. Cubital fossa: a triangular region at front of elbow

A. BOUNDARIES: base, a line drawn between the 2 humeral epicondyles; lateral side, the brachioradialis muscle; medial side, the pronator teres muscle; roof, the deep fascia strengthened by the aponeurosis of the biceps muscle (see p. 221); and floor, the brachialis and supinator muscles

II. Superficial structures: overlying roof of fossa

A. VEINS

1. Cephalic runs at lateral edge of fossa along brachialis muscle
2. Median cubital, from cephalic vein below elbow, upward and medialward to join basilic vein. Crosses superficial to aponeurosis of biceps
3. Basilic, runs along medial edge of fossa, overlying aponeurosis of biceps muscle as it merges with the antebrachial fascia
4. Median basilic and median cephalic veins are often present (see opposite page)

B. NERVES

1. Medial antebrachial cutaneous sends branches both deep and superficial to the basilic vein at the aponeurosis of biceps
2. Lateral antebrachial cutaneous lies deep to the cephalic vein at elbow and here splits into posterior and anterior branches

III. Deep structures in fossa

A. ARTERIES

1. Brachial runs through middle of fossa, dividing into its terminal branches, *radial and ulnar,* opposite neck of radius
   a. Relations: behind, brachialis muscle; in front, skin and fascia, separated from median cubital vein by the aponeurosis of the biceps muscle; medially, above is median nerve, below is ulnar head of pronator teres muscle; and laterally, the tendon of biceps muscle
2. Radial lies on tendon of biceps, supinator, and pronator teres muscles
3. Ulnar passes almost at once beneath pronator teres muscle and leaves fossa

B. NERVES

1. Median in upper fossa lies close to medial side of brachial artery. Lower down, it passes between heads of pronator teres muscle and is separated from the ulnar artery by the medial head of the muscle
2. Radial usually lies beneath the brachioradialis muscle and is not in the fossa. It splits beneath this muscle in front of the lateral epicondyle into superficial and deep branches
   a. Superficial branch continues down forearm under brachioradialis muscle
   b. Deep branch curves around lateral side of radius between layers of the supinator muscle to the dorsum of forearm
3. Ulnar stays behind medial intermuscular septum of arm and passes behind medial epicondyle. Is never related to fossa

IV. Clinical considerations

A. VENIPUNCTURE: this is an extremely common procedure performed when a large blood sample is needed for transfusions, for intravenous feeding, and for intravenous anesthetics. Anatomically, the median cubital vein should be used because it overlies the bicipital aponeurosis which offers some support and also some protection for underlying parts, and it is not accompanied by sizable cutaneous nerves

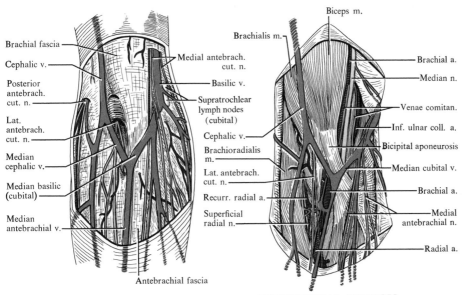

SUPERFICIAL VEINS

Brachial fascia
Cephalic v.
Posterior antebrach. cut. n.
Lat. antebrach. cut. n.
Median cephalic v.
Median basilic (cubital)
Median antebrachial v.

Medial antebrach. cut. n.
Basilic v.
Supratrochlear lymph nodes (cubital)
Cephalic v.
Brachioradialis m.
Lat. antebrach. cut. n.
Recurr. radial a.
Superficial radial n.

Antebrachial fascia

Biceps m.
Brachialis m.
Brachial a.
Median n.
Venae comitan.
Inf. ulnar coll. a.
Bicipital aponeurosis
Median cubital v.
Brachial a.
Medial antebrachial n.
Radial a.

ANTECUBITAL REGION

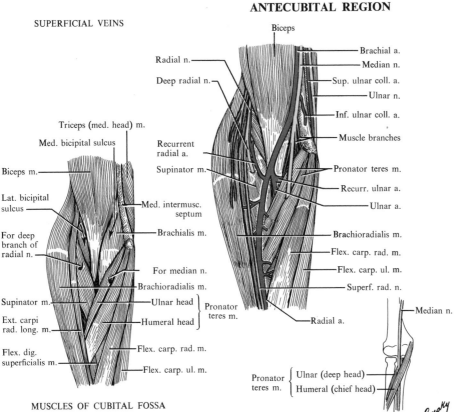

Triceps (med. head) m.
Med. bicipital sulcus
Biceps m.
Lat. bicipital sulcus
For deep branch of radial n.
Supinator m.
Ext. carpi rad. long. m.
Flex. dig. superficialis m.

Recurrent radial a.
Supinator m.
Med. intermusc. septum
Brachialis m.
For median n.
Brachioradialis m.
Ulnar head
Humeral head
Flex. carp. rad. m.
Flex. carp. ul. m.

Pronator teres m.

MUSCLES OF CUBITAL FOSSA

Biceps
Radial n.
Deep radial n.
Brachial a.
Median n.
Sup. ulnar coll. a.
Ulnar n.
Inf. ulnar coll. a.
Muscle branches
Pronator teres m.
Recurr. ulnar a.
Ulnar a.
Brachioradialis m.
Flex. carp. rad. m.
Flex. carp. ul. m.
Superf. rad. n.
Radial a.

Median n.

Pronator teres m. { Ulnar (deep head)
Humeral (chief head)

101. MUSCLES OF THE VOLAR FOREARM

Name	Origin	Insertion	Action	Nerve
I. Superficial group				
Pronator teres	Above med. epicond. Med. side of coronoid process of ulna	Lat. side of radius	Pronates hand	Median
Flex. carpi radialis	Med. epicond.	Bases of second and third metacarpals	Flexes forearm Flexes and abducts hand	Median
Palmaris longus	Med. epicond.	Trans. carpal lig. Palmar aponeurosis	Flexes hand and forearm	Median
Flex. carpi ulnaris	Med. epicond. Med. olecranon Post. border ulna	Pisiform, hamate, and fifth metacarpal	Flexes and adducts hand Flexes forearm	Ulnar
Flex. digit superficialis	Med. epicond. Med. coronoid proc. of ulna Oblique line, radius	Sides of second phalanx of 4 fingers	Flexes first and second phalanges Flexes hand Flexes forearm	Median
II. Deep group				
Flex. digit. profundus	Upper anterior and med. ulna Med. coronoid proc. and upper posterior ulna Interos. memb.	Bases of terminal phalanges of 4 fingers	Flexes all phalanges Flexes hand	Anterior interosseous of median; ulnar
Flex. pol. longus	Anterior shaft, radius Interos. memb.	Base distal phalanx of thumb	Flexes thumb Flexes and adducts first metacarpal	Anterior interos. of median
Pronator quadratus	Lower anterior shaft, ulna Med. anterior surface, distal ulna	Lower lat. border and lower anterior surface, shaft of radius	Pronates hand	Anterior interos. of median

I. Special features

A. THE FLEXOR MUSCLES of the forearm are concerned with pronation of the forearm, with flexion and abduction at the wrist, and with flexion of the digits

B. THE MOST COMMON VARIATION AMONG THIS GROUP OF MUSCLES is the absence of the palmaris longus muscle (about 12%)

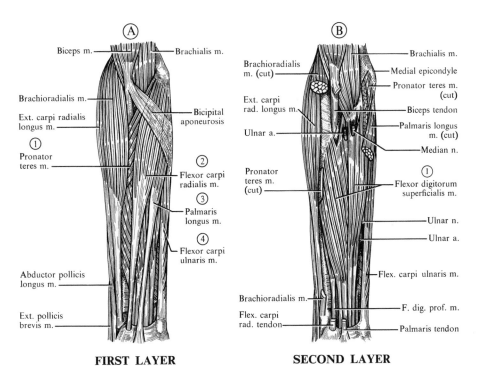

A

Biceps m.

Brachialis m.

Brachioradialis m.

Ext. carpi radialis longus m.

① Pronator teres m.

Bicipital aponeurosis

② Flexor carpi radialis m.

③ Palmaris longus m.

④ Flexor carpi ulnaris m.

Abductor pollicis longus m.

Ext. pollicis brevis m.

FIRST LAYER

B

Brachioradialis m. (cut)

Brachialis m.

Medial epicondyle

Pronator teres m. (cut)

Ext. carpi rad. longus m.

Biceps tendon

Palmaris longus m. (cut)

Ulnar a.

Median n.

Pronator teres m. (cut)

① Flexor digitorum superficialis m.

Ulnar n.

Ulnar a.

Flex. carpi ulnaris m.

Brachioradialis m.

Flex. carpi rad. tendon

F. dig. prof. m.

Palmaris tendon

SECOND LAYER

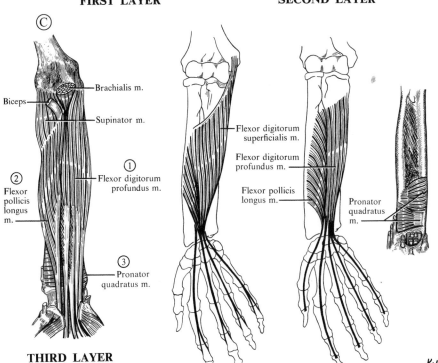

C

Brachialis m.

Biceps

Supinator m.

② Flexor pollicis longus m.

① Flexor digitorum profundus m.

③ Pronator quadratus m.

THIRD LAYER

Flexor digitorum superficialis m.

Flexor digitorum profundus m.

Flexor pollicis longus m.

Pronator quadratus m.

Pansky

102. DEEP VESSELS AND NERVES OF THE VOLAR FOREARM

I. **Radial artery** extends from neck of radius to medial side of radial styloid process

A. RELATIONS

Anterior
Skin and fascia, brachioradialis

Lateral	**Radial**	*Medial*
Brachioradialis m., superfic. radial n.	**Artery**	Pronator teres and flex. carp. radialis mm.

Posterior
Tendon biceps, supinator, flex. digit. superfic., pronator teres,
flex. pol. long., and pronat. quadrat. mm., radius

B. BRANCHES: radial recurrent, muscular, palmar carpal, and superficial palmar

II. **Ulnar artery** extends from neck of radius to flexor retinaculum at wrist

A. RELATIONS

Anterior
Skin and fascia, superfic. flexor mm., median n.

Lateral	**Ulnar**	*Medial*
Flex. digit. superfic. m.	**Artery**	Flex. carpi ulnaris m., ulnar n.

Posterior
Brachialis and flex. digit. profundus mm.

B. BRANCHES: anterior and posterior ulnar recurrent, common interosseous, and muscular

III. **Common interosseous artery:** trunk arising from the ulnar artery below radial tuberosity to the upper interosseous membrane, where it divides into:
A. POSTERIOR INTEROSSEOUS (see p. 224)
B. ANTERIOR INTEROSSEOUS: down forearm on interosseous membrane to upper border of the pronator quadratus muscle. Here it sends a branch through the interosseous membrane to dorsum to join the posterior interosseous artery, and a small branch continues under the pronator quadratus muscle to the palmar carpal net. Gives muscular and nutrient (radius and ulna) branches and a branch to median nerve

IV. **Radial nerve** divides into superficial and deep branches
A. SUPERFICIAL runs beneath brachioradialis muscle just lateral to radial artery. In lower third of forearm, crosses to dorsum. Cutaneous to dorsum of hand and thumb
B. DEEP (see p. 224)

V. **Median nerve** passes between heads of pronator teres muscle; lies between flexor digitorum superficialis and profundus muscles; near wrist is more superficial, between tendons of flexor digitorum superficialis and flexor carpi radialis muscles; and then passes deep and medial to the tendon of the palmaris longus muscle
A. BRANCHES to all superficial muscles except flexor carpi ulnaris. The anterior interosseous branch accompanies the anterior interosseous artery on the interosseous membrane to supply the flexor digitorum profundus, flexor pollicis longus and pronator quadratus muscles. The palmar branch goes to skin of palm

VI. **Ulnar nerve** from behind medial epicondyle enters forearm through flexor carpi ulnaris muscle, continues between this and flexor digitorum profundus, and in the lower half of the forearm the ulnar artery lies close to its medial side. Both artery and nerve are covered by skin and fascia lateral to the flexor carpi ulnaris muscle
A. BRANCHES to flexor carpi ulnaris and medial half of the flexor digitorum profundus muscles; palmar cutaneous to medial side of palm and a posterior branch (see p. 248)

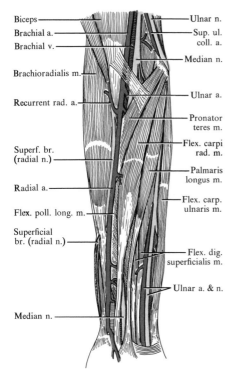

Biceps
Brachial a.
Brachial v.
Brachioradialis m.
Recurrent rad. a.
Superf. br. (radial n.)
Radial a.
Flex. poll. long. m.
Superficial br. (radial n.)
Median n.

Ulnar n.
Sup. ul. coll. a.
Median n.
Ulnar a.
Pronator teres m.
Flex. carpi rad. m.
Palmaris longus m.
Flex. carp. ulnaris m.
Flex. dig. superficialis m.
Ulnar a. & n.

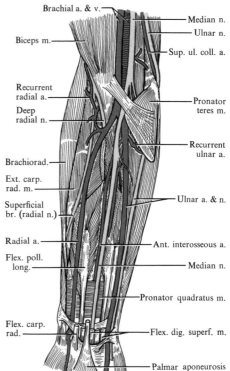

Brachial a. & v.
Biceps m.
Recurrent radial a.
Deep radial n.
Brachiorad.
Ext. carp. rad. m.
Superficial br. (radial n.)
Radial a.
Flex. poll. long.
Flex. carp. rad.

Median n.
Ulnar n.
Sup. ul. coll. a.
Pronator teres m.
Recurrent ulnar a.
Ulnar a. & n.
Ant. interosseous a.
Median n.
Pronator quadratus m.
Flex. dig. superf. m.
Palmar aponeurosis

VOLAR FOREARM
DEEP VESSELS AND NERVES

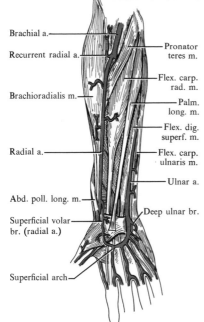

Brachial a.
Recurrent radial a.
Brachioradialis m.
Radial a.
Abd. poll. long. m.
Superficial volar br. (radial a.)
Superficial arch

Pronator teres m.
Flex. carp. rad. m.
Palm. long. m.
Flex. dig. superf. m.
Flex. carp. ulnaris m.
Ulnar a.
Deep ulnar br.

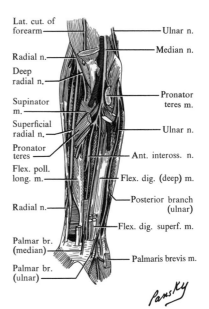

Lat. cut. of forearm
Radial n.
Deep radial n.
Supinator m.
Superficial radial n.
Pronator teres
Flex. poll. long. m.
Radial n.
Palmar br. (median)
Palmar br. (ulnar)

Ulnar n.
Median n.
Pronator teres m.
Ulnar n.
Ant. inteross. n.
Flex. dig. (deep) m.
Posterior branch (ulnar)
Flex. dig. superf. m.
Palmaris brevis m.

Pansky

103. MUSCLES OF THE POSTERIOR FOREARM

Name	Origin	Insertion	Action	Nerve
I. Superficial group				
Brachio-radialis	Lat. supracondylar ridge of humerus	Lat. side of base, radial styloid	Flexes forearm	Radial
Ext. carpi radialis long.	Lat. supracondylar ridge of humerus	Post. base of second metacarpal	Extends and abducts hand	Radial
Ext. carpi radialis brev.	Lat. epicondyle of humerus	Post. base of third metacarpal	Extends and abducts hand	Radial
Ext. digit.	Lat. epicondyle of humerus	Mid. base of second and third phalanges of 4 fingers	Extends fingers Extends hand	Deep radial
Ext. digit. minimi	Common extensor tendon	Aponeurosis; post., first phalanx of little finger	Extends little finger	Deep radial
Ext. carpi ulnaris	Lat. epicondyle Post. border of ulna	Tubercle at base of fifth metacarpal	Extends and adducts hand	Deep radial
Anconeus	Back of lat. epicondyle	Olecranon, upper posterior ulna	Extends forearm	Radial
II. Deep group				
Supinator	Lat. epicondyle Ulna below radial notch	Radial tubercle Oblique line of radius	Supinates hand	Deep radial
Abductor pollicis longus	Lat. post. ulna Interos. memb. Post. radius	Lat. side of base of first metacarpal	Abducts thumb and hand	Deep radial
Ext. pollicis brev.	Post. radius Interos. memb.	Base of first phalanx of thumb	Extends last phalanx of thumb Abducts hand	Deep radial
Ext. pollicis long.	Lat. side, post. surface of ulna	Base of last phalanx of thumb	Extends first phalanx of thumb Abducts hand	Deep radial
Ext. indicis	Post. shaft of ulna Interos. memb.	Tendon of ext. dig. index finger	Extends and adducts index finger	Deep radial

III. Special features
  A. INSERTION OF THE TENDONS of the lumbricales and interossei is in a common extensor aponeurosis on the dorsum of each first phalanx
  B. OPPOSITE THE FIRST INTERPHALANGEAL JOINT is a splitting of the extensor aponeurosis into 3 slips with the middle going to the base of the second phalanx and the 2 lateral going to the base of the third phalanx
  C. THE DEEP RADIAL NERVE passes between the 2 layers of the supinator muscle

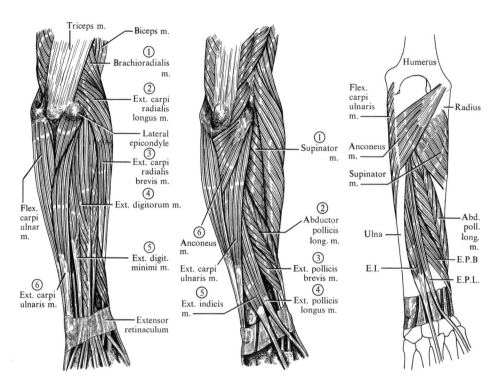

Triceps m.

Biceps m.

① Brachioradialis m.

② Ext. carpi radialis longus m.

Lateral epicondyle

③ Ext. carpi radialis brevis m.

④ Ext. digitorum m.

⑤ Ext. digit. minimi m.

⑥ Ext. carpi ulnaris m.

Flex. carpi ulnar m.

Extensor retinaculum

⑥ Anconeus m.

Ext. carpi ulnaris m.

⑤ Ext. indicis m.

① Supinator m.

② Abductor pollicis long. m.

③ Ext. pollicis brevis m.

④ Ext. pollicis longus m.

Humerus

Flex. carpi ulnaris m.

Radius

Anconeus m.

Supinator m.

Ulna

E.I.

Abd. poll. long. m.

E.P.B

E.P.L.

SUPERFICIAL GROUP OF MUSCLES

DEEP GROUP OF MUSCLES

DORSAL MUSCLES

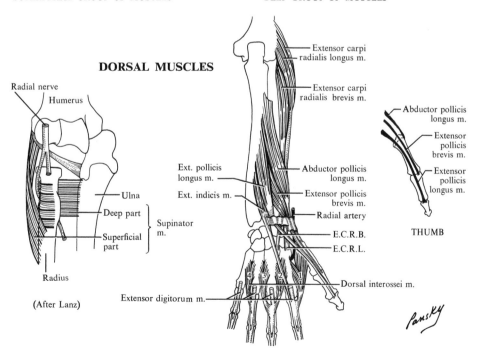

Radial nerve

Humerus

Ulna

Deep part

Supinator m.

Superficial part

Radius

(After Lanz)

Extensor carpi radialis longus m.

Extensor carpi radialis brevis m.

Ext. pollicis longus m.

Ext. indicis m.

Abductor pollicis longus m.

Extensor pollicis brevis m.

Radial artery

E.C.R.B.

E.C.R.L.

Dorsal interossei m.

Extensor digitorum m.

Abductor pollicis longus m.

Extensor pollicis brevis m.

Extensor pollicis longus m.

THUMB

Pansky

-223-

104. COMPARTMENTS OF FOREARM, DEEP VESSELS AND NERVES OF POSTERIOR ASPECT

I. **Forearm is divided into posterior and anterior compartments** by medial and lateral intermuscular septa of the antebrachial fascia, the radius, the ulna, and the interosseus membrane

A. ANTEBRACHIAL FASCIA is fused to the ulna throughout its length

B. IN LOWER FOREARM fascia splits into 2 layers: posterior and anterior to the palmaris longus, flexor carpi radialis, and flexor carpi ulnaris muscles

II. **Vessels**

A. POSTERIOR INTEROSSEOUS ARTERY arises from common interosseous artery at upper border of interosseous membrane over which it passes, comes through between the borders of supinator and abductor pollicis longus muscles, and descends through forearm between superficial and deep layers. In lower forearm is joined by one of the terminal branches of the anterior interosseous artery and enters the dorsal carpal network

1. Branches: muscular; interosseous recurrent artery, which ascends between olecranon and lateral epicondyle under anconeus muscle to anastomose with the middle collateral artery, inferior ulnar collateral artery, and posterior ulnar recurrent artery

III. **Nerves**

A. RADIAL NERVE pierces lateral intermuscular septum of arm, runs between brachialis and brachioradialis muscles, and in front of the lateral epicondyle divides into:

1. Superficial branch (see p. 248)
2. Deep: curves around lateral border of radius between layers of supinator muscle, continues between superficial and deep layers of muscles to the middle of forearm, and then, as the *posterior interosseous nerve,* it lies between the interosseous membrane and extensor pollicis longus muscle. It continues to the posterior of the carpus, giving branches to all extensor muscles

IV. **Clinical considerations**

A. DAMAGE TO THE RADIAL NERVE, especially above the elbow, leads to "wrist drop," the inability to extend the hand at the wrist. Extension of the fingers is also impossible. Sensation on the lateral side of the back of the hand is also lost

1. Because of the shortness of the extensor muscles of the digits, however, a person with wrist drop can passively extend his wrist by tightly clenching his fingers

B. LESIONS OF THE DEEP BRANCH OF THE RADIAL may have varying effects, but complete ones abolish extension at the metacarpophalangeal joints of the digits while allowing extension of the wrist through the radial extensors

C. ALL THE EXTENSOR MUSCLES ARE INNERVATED through the radial nerve. The lateral ones (brachioradialis and radial extensors) receive their innervation from the nerve while it lies on the front of the arm or forearm; the posterior ones (extensor digitorum, digiti minimi, and carpi ulnaris) receive their innervation on their deep surfaces from the deep branch of the radial nerve

D. IT IS USUALLY STATED THAT: the brachioradialis receives fibers from the 5th and 6th cervical nerves; the radial extensors from about C6 and C7; the extensor digitorum and extensor digiti minimi from C6, C7, and C8; and the ulnar extensor from these or C7 and C8 only

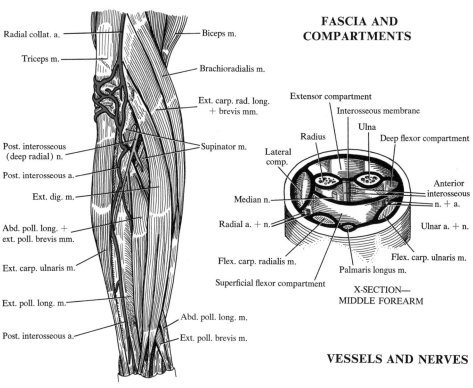

Radial collat. a.

Triceps m.

Biceps m.

Brachioradialis m.

Ext. carp. rad. long. + brevis mm.

Post. interosseous (deep radial) n.

Supinator m.

Post. interosseous a.

Ext. dig. m.

Abd. poll. long. + ext. poll. brevis mm.

Ext. carp. ulnaris m.

Ext. poll. long. m.

Post. interosseous a.

Abd. poll. long. m.

Ext. poll. brevis m.

FASCIA AND COMPARTMENTS

Extensor compartment

Interosseous membrane

Ulna

Deep flexor compartment

Radius

Lateral comp.

Anterior interosseous n. + a.

Median n.

Radial a. + n.

Ulnar a. + n.

Flex. carp. ulnaris m.

Flex. carp. radialis m.

Palmaris longus m.

Superficial flexor compartment

X-SECTION— MIDDLE FOREARM

VESSELS AND NERVES

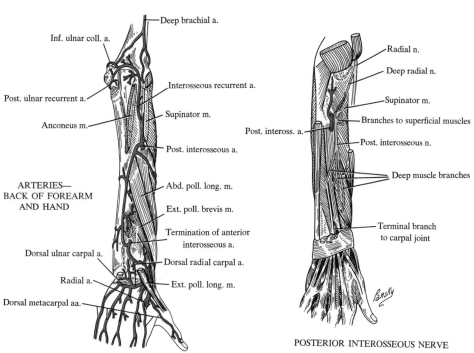

Deep brachial a.

Inf. ulnar coll. a.

Interosseous recurrent a.

Post. ulnar recurrent a.

Supinator m.

Anconeus m.

Post. interosseous a.

Radial n.

Deep radial n.

Supinator m.

Branches to superficial muscles

Post. inteross. a.

Post. interosseous n.

Deep muscle branches

Abd. poll. long. m.

Ext. poll. brevis m.

ARTERIES— BACK OF FOREARM AND HAND

Termination of anterior interosseous a.

Terminal branch to carpal joint

Dorsal ulnar carpal a.

Dorsal radial carpal a.

Radial a.

Ext. poll. long. m.

Dorsal metacarpal aa.

POSTERIOR INTEROSSEOUS NERVE

105. COLLATERAL CIRCULATION INVOLVING THE ANASTOMOSIS AROUND THE ELBOW

I. Collateral circulation for ligation of the brachial artery
A. LIGATION BETWEEN ORIGINS OF THE SUPERIOR ULNAR COLLATERAL AND THE PROFUNDA ARTERIES

Arteries Above Ligature	Arteries Below Ligature
1. Radial collateral (from deep brachial)	Radial recurrent (from radial)
2. Medial collateral (from deep brachial)	Interosseous recurrent (from posterior interosseous)

B. LIGATION BETWEEN ORIGINS OF SUPERIOR AND INFERIOR ULNAR COLLATERALS

1. Radial collateral, as above	
2. Medial collateral, as above	
3. Sup. ulnar collateral (from brachial)	Inf. ulnar collateral (from brachial)
	Posterior ulnar recurrent (from ulnar)

C. LIGATION BETWEEN ORIGIN OF INFERIOR ULNAR COLLATERAL AND END OF BRACHIAL ARTERY

1. Radial collateral, as above	
2. Medial collateral, as above	Interosseous recurrent (from posterior
3. Sup. ulnar collateral, as above	interosseous)
4. Inf. ulnar collateral (from brachial)	Anterior ulnar recurrent (from ulnar)
	Posterior ulnar recurrent (from ulnar)

II. Sites where anastomoses occur
A. BETWEEN ANTERIOR ULNAR RECURRENT AND INFERIOR ULNAR COLLATERAL in front of the medial epicondyle
B. POSTERIOR ULNAR RECURRENT AND POSTERIOR BRANCHES OF INFERIOR ULNAR COLLATERAL behind medial epicondyle
C. BETWEEN INTEROSSEOUS RECURRENT AND MEDIAL COLLATERAL in interval between lateral epicondyle of humerus and olecranon
D. BETWEEN RADIAL RECURRENT AND RADIAL COLLATERAL in front of the lateral epicondyle

III. Clinical considerations
A. THE ABOVE-MENTIONED ANASTOMOSES, in theory, may preserve the limb distal to the interruption of the main vessel. However, the small anastomotic channels, in reality, may not open in time, especially when the major blood supply is cut off suddenly

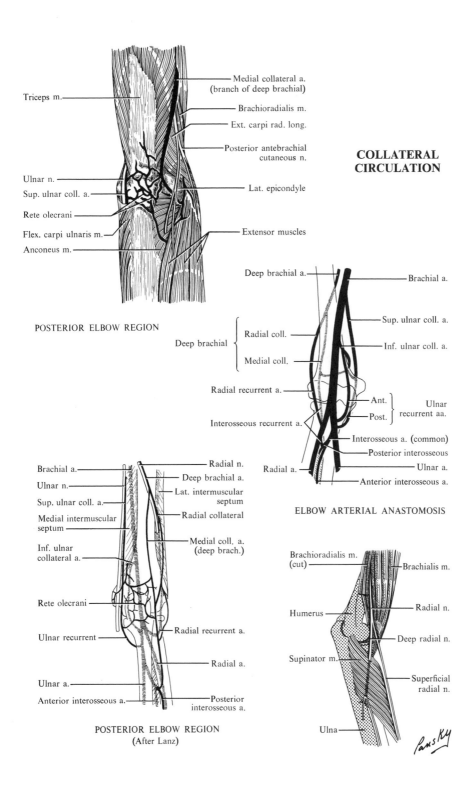

Triceps m.

Medial collateral a.
(branch of deep brachial)

Brachioradialis m.

Ext. carpi rad. long.

Posterior antebrachial
cutaneous n.

**COLLATERAL
CIRCULATION**

Ulnar n.

Sup. ulnar coll. a.

Rete olecrani

Flex. carpi ulnaris m.

Anconeus m.

Lat. epicondyle

Extensor muscles

POSTERIOR ELBOW REGION

Deep brachial a.

Brachial a.

Sup. ulnar coll. a.

Inf. ulnar coll. a.

Deep brachial { Radial coll.

Medial coll.

Radial recurrent a.

Ant.
Post. } Ulnar
recurrent aa.

Interosseous recurrent a.

Interosseous a. (common)

Posterior interosseous

Radial a.

Ulnar a.

Anterior interosseous a.

ELBOW ARTERIAL ANASTOMOSIS

Brachial a.

Radial n.

Ulnar n.

Deep brachial a.

Sup. ulnar coll. a.

Lat. intermuscular
septum

Medial intermuscular
septum

Radial collateral

Inf. ulnar
collateral a.

Medial coll. a.
(deep brach.)

Brachioradialis m.
(cut)

Brachialis m.

Humerus

Radial n.

Rete olecrani

Deep radial n.

Ulnar recurrent

Radial recurrent a.

Supinator m.

Radial a.

Superficial
radial n.

Ulnar a.

Anterior interosseous a.

Posterior
interosseous a.

POSTERIOR ELBOW REGION
(After Lanz)

Ulna

Pansky

-227-

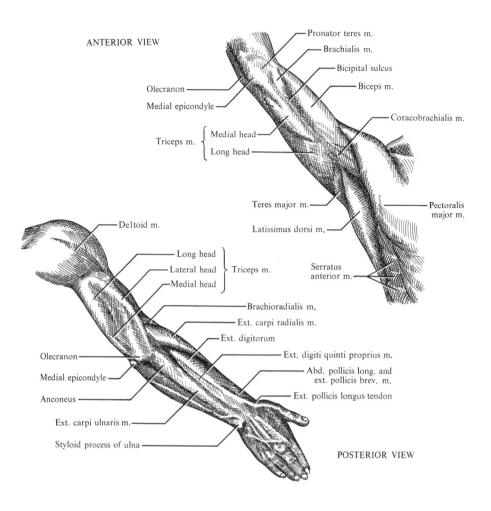

ANTERIOR VIEW

Pronator teres m.
Brachialis m.
Bicipital sulcus
Biceps m.
Olecranon
Medial epicondyle
Coracobrachialis m.
Triceps m. { Medial head
Long head
Teres major m.
Latissimus dorsi m.
Pectoralis major m.
Serratus anterior m.

Deltoid m.
Long head
Lateral head } Triceps m.
Medial head
Brachioradialis m.
Ext. carpi radialis m.
Ext. digitorum
Olecranon
Ext. digiti quinti proprius m.
Medial epicondyle
Abd. pollicis long. and ext. pollicis brev. m.
Anconeus
Ext. pollicis longus tendon
Ext. carpi ulnaris m.
Styloid process of ulna

POSTERIOR VIEW

SURFACE ANATOMY

ANTERIOR VIEW

Biceps m.
Deltoid m.
Brachioradialis m.
Cephalic v.
Thenar mm.
Lateral ext. mm.
Pectoralis major m.
Hypothenar mm.
Flexor mm.
Flexor carpi ulnaris m.
Cubital fossa
Flexor dig. sublimis tend.
Medial epicondyle
Palmaris longus tend.
Latissimus dorsi m.
Basilic v.
Teres major m.
Flexor carpi radialis tend.
Bicipital sulcus
Coracobrachialis m.

FIGURE 13. **Surface anatomy of upper extremity.**

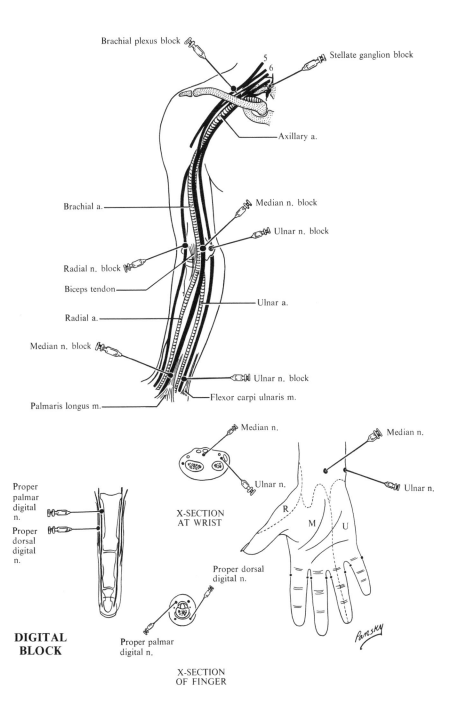

FIGURE 14. **Nerve blocks of upper extremity.**

106. THE ELBOW JOINT

I. Humeroulnar
A. TYPE: ginglymus (hinge)
B. BONES: trochlea of humerus with trochlear notch of ulna, capitulum of humerus with fovea of radial head
C. MOVEMENTS: flexion and extension
D. LIGAMENTS
1. Articular capsule in front, extends from medial epicondyle and front of humerus above coronoid and radial fossae to the anterior coronoid process, anular ligament, and the collateral ligaments. Behind, it extends from the humerus (behind capitulum) and medial side of trochlea to the margins of the olecranon, posterior part of ulna, and anular ligament
2. Ulnar collateral extends from front and back of the medial epicondyle to the medial side of the coronoid process and olecranon
3. Radial collateral extends from below the lateral epicondyle to the anular ligament and lateral margin of the ulna

E. SYNOVIAL MEMBRANE: very extensive. Extends from margins of articular areas of humerus. Lines the 3 fossae of the humerus and the inside of the capsule. It forms a sac between the head of radius, the radial notch, and the anular ligament. Three fat pads exist between the synovia and capsule: over the olecranon fossa, over the coronoid fossa, and over the radial fossa. The former is pressed by the triceps muscle into the fossa during flexion. The other 2 are pressed by the brachialis muscle into their fossae in extension of the forearm
F. MUSCLES ACTING ON THE JOINT

Flexion	Extension
Biceps	Triceps
Brachialis	Anconeus
Brachioradialis	Extensor carpi radialis longus and brevis
Pronator teres	Extensor digitorum
Flexor carpi radialis and ulnaris	Extensor digiti minimi
Palmaris longus	Extensor carpi ulnaris
Flexor digitorum superficialis	Supinator

II. Radioulnar
A. PROXIMAL
1. Type: trochoid (pivot)
2. Bones: head of radius and radial notch of ulna
3. Movements: rotation (pronation, supination)
4. Ligaments
a. Anular encircles radial head, attached to ends of radial notch
b. Quadrate between neck of radius and lower part of radial notch of ulna
5. Muscles acting on proximal radioulnar joint

Supination	Pronation
Biceps and supinator	Pronator teres
Extensors of thumb	Pronator quadratus

III. Clinical considerations
A. SINCE THE ULNA CANNOT BE ANTERIORLY DISLOCATED without a concomitant fracture, most dislocations at the elbow are posterior; the ulnar nerve is frequently injured
B. DISLOCATION OF THE HEAD OF THE RADIUS ALONE is usually anterior

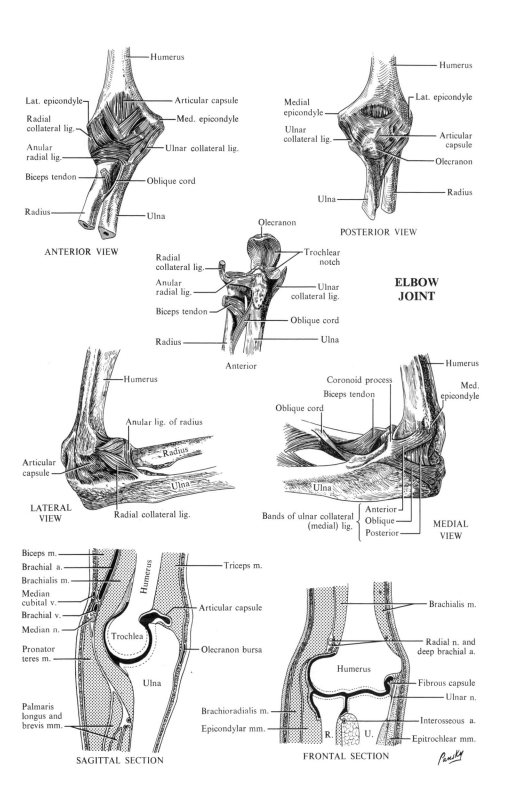

Humerus

Lat. epicondyle

Radial collateral lig.

Anular radial lig.

Biceps tendon

Radius

Articular capsule

Med. epicondyle

Ulnar collateral lig.

Oblique cord

Ulna

ANTERIOR VIEW

Medial epicondyle

Ulnar collateral lig.

Ulna

Humerus

Lat. epicondyle

Articular capsule

Olecranon

Radius

POSTERIOR VIEW

Olecranon

Radial collateral lig.

Anular radial lig.

Biceps tendon

Radius

Trochlear notch

Ulnar collateral lig.

Oblique cord

Ulna

Anterior

ELBOW JOINT

Humerus

Anular lig. of radius

Radius

Ulna

Articular capsule

LATERAL VIEW

Radial collateral lig.

Coronoid process

Biceps tendon

Oblique cord

Ulna

Humerus

Med. epicondyle

Bands of ulnar collateral (medial) lig.

Anterior

Oblique

Posterior

MEDIAL VIEW

Biceps m.

Brachial a.

Brachialis m.

Median cubital v.

Brachial v.

Median n.

Pronator teres m.

Palmaris longus and brevis mm.

Humerus

Trochlea

Ulna

Triceps m.

Articular capsule

Olecranon bursa

SAGITTAL SECTION

Humerus

R.   U.

Brachialis m.

Radial n. and deep brachial a.

Fibrous capsule

Ulnar n.

Interosseous a.

Epitrochlear mm.

Brachioradialis m.

Epicondylar mm.

FRONTAL SECTION

Pansky

– 231 –

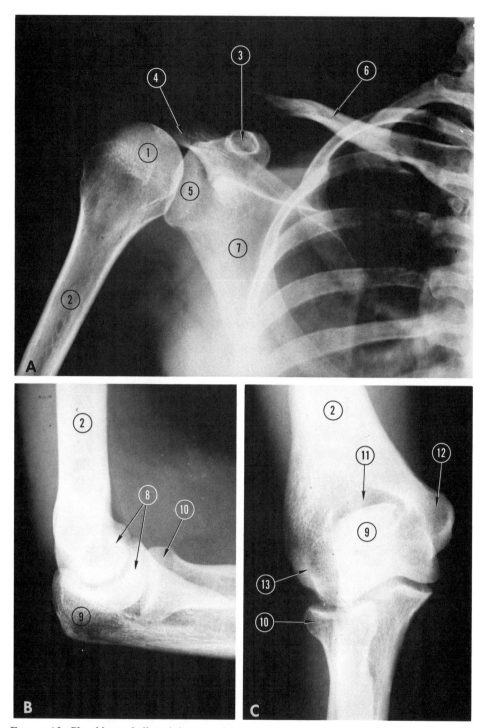

FIGURE 15. **Shoulder and elbow joints. A, Shoulder joint; B, elbow, lateral view; C, elbow AP view.** *1,* Head of humerus; *2,* shaft of humerus; *3,* coracoid process of scapula; *4,* acromion of scapula; *5,* glenoid fossa of scapula; *6,* clavicle; *7,* body of scapula; *8,* capitulum and trochlea of humerus; *9,* olecranon; *10,* head of radius; *11,* olecranon fossa; *12,* medial epicondyle of humerus; *13,* lateral epicondyle of humerus.

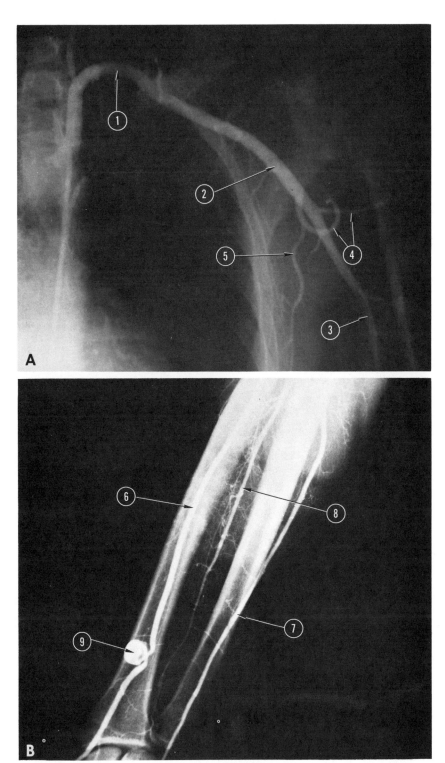

FIGURE 16. **Arteriogram of upper extremity.** **A, Subclavian, axillary, and brachial arteries; B, forearm.** *1,* Subclavian; *2,* axillary; *3,* brachial; *4,* circumflex humeral; *5,* subscapular; *6,* radial; *7,* ulnar; *8,* anterior interosseous; *9,* aneurysm of radial artery.

107. BONES OF THE WRIST AND HAND

I. Carpals (wrist bones): 8 in number arranged in proximal and distal rows
A. SCAPHOID (NAVICULAR) (boat-shaped): largest of proximal row, superior and inferior surfaces concave, has a tubercle
B. LUNATE (SEMILUNAR): crescent-shaped, superior surface convex, inferior concave
C. TRIQUETRAL: pyramidal in shape, has an oval articular facet on its palmar surface
D. PISIFORM: small, pear-shaped, oval articular facet in dorsal surface
E. TRAPEZIUM (GREATER MULTANGULAR): palmar surface shows an oblique deep groove
F. TRAPEZOID (LESSER MULTANGULAR): smallest of distal row, wedge-shaped
G. CAPITATE: largest of carpals, at center of wrist and the first to ossify
H. HAMATE: hook-shaped, has a curved, hooklike process (hamulus) from medial side of palmar surface

II. Metacarpals: 5, numbered from lateral to medial
A. COMMON CHARACTERISTICS. Each has a base proximally, a shaft, and a head distally
B. OUTSTANDING CHARACTERISTICS OF INDIVIDUAL BONES
  1. First: shorter and stouter, has tubercle on lateral side of base
  2. Second: longest with largest base
  3. Third: slightly smaller than second; styloid process on base
  4. Fifth: smallest, has tubercle on medial side of base

III. Phalanges: total of 14: 2 for thumb and 3 for each finger. Numbered or named on each finger from proximal to distal: 1 (proximal), 2 (middle), 3 (distal)
A. COMMON CHARACTERISTICS. Each has proximal base, tapering body, and distal extremity
  1. First phalanges have oval, concave articular surfaces; second and third show double concavities separated by a ridge at base
  2. First and second phalanges also show 2 condyles separated by a groove

IV. Ossification
A. CARPALS: from 1 center in each, which appear as follows: capitate and hamate, first year; triquetral, third year; lunate and trapezium, fifth year; scaphoid, sixth year; trapezoid, eighth year; pisiform, twelfth year
B. METACARPALS: from 2 centers, body and distal end, which appear and close as follows: body, eighth and ninth fetal week; extremity, third year. They join in twentieth year. First has center in body and base, which for body appears in the eighth to ninth fetal week; base appears in third year and they join in the twentieth year
C. PHALANGES: from 2 centers, 1 for body, 1 for base. Appear in body in eighth week, in base at 3 to 4 years (in proximal row), and later in distal row. Unite in twentieth year

V. Clinical considerations
A. OSSIFICATION CENTERS: knowledge of the appearance and closure of ossification centers is important both medicolegally and in the radiologic examination of the young
B. FRACTURES: 70–75% of fractures of the carpus include the scaphoid bone. In 10–15%, due to the lack of blood supply to the proximal part, the bone may not heal and necrosis of the scaphoid takes place

WRIST AND HAND BONES

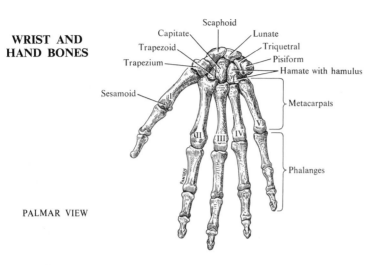

Scaphoid
Capitate
Trapezoid
Trapezium
Sesamoid
Lunate
Triquetral
Pisiform
Hamate with hamulus
Metacarpals
II III IV V
Phalanges

PALMAR VIEW

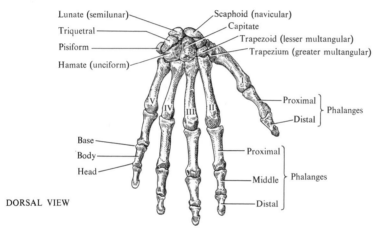

Lunate (semilunar)
Triquetral
Pisiform
Hamate (unciform)
Scaphoid (navicular)
Capitate
Trapezoid (lesser multangular)
Trapezium (greater multangular)
I
Proximal
Distal
Phalanges
V IV III II
Base
Body
Head
Proximal
Middle
Distal
Phalanges

DORSAL VIEW

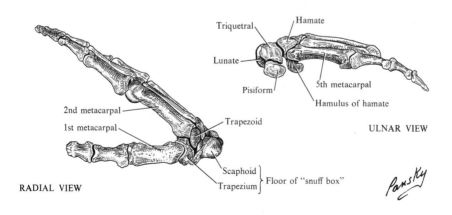

Triquetral
Lunate
Pisiform
Hamate
5th metacarpal
Hamulus of hamate

ULNAR VIEW

2nd metacarpal
1st metacarpal
Trapezoid
Scaphoid
Trapezium
Floor of "snuff box"

RADIAL VIEW

Pansky

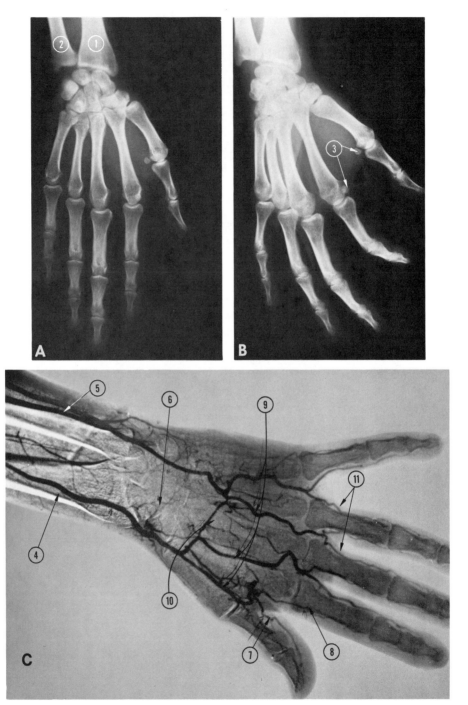

FIGURE 17. **Right hand, adult. A, AP view; B, oblique view; C, vascular anatomy, subtraction film.** *1,* Radius; *2,* ulna; *3,* sesamoid bones; *4,* radial artery; *5,* ulnar artery; *6,* carpal branches; *7,* principal artery to thumb; *8,* radial artery to index finger; *9,* common palmar digital arteries; *10,* superficial palmar arch; *11,* proper digital arteries.

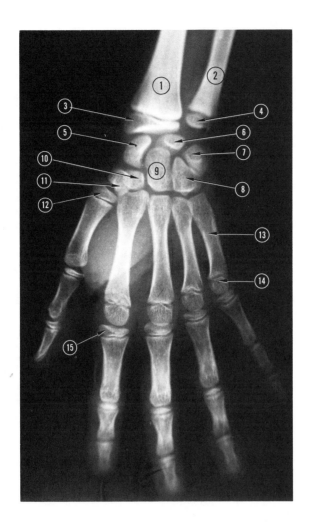

FIGURE 18. **Left hand, child (about 11 years).** *1*, Radius; *2*, ulna; *3*, distal epiphysis of radius; *4*, distal epiphysis of ulna; *5*, scaphoid; *6*, lunate; *7*, triquetral; *8*, hamate; *9*, capitate; *10*, trapezoid; *11*, trapezium; *12*, proximal epiphysis of 1st metacarpal; *13*, 5th metacarpal; *14*, head of 5th metacarpal; *15*, proximal epiphysis of 1st phalanx of index finger.

108. ANTERIOR AND LATERAL ASPECTS OF THE WRIST

I. **Antebrachial fascia**
A. PALMAR CARPAL LIGAMENT (volar carpal ligament) fascia strengthened by transverse fibers extending between the radial and ulnar styloid processes

II. **Vessels and nerves**
A. ULNAR ARTERY with ulnar nerve on its medial side is covered by carpal ligament and skin. The flexor carpi ulnaris is just medial to these
B. RADIAL ARTERY is also quite superficial, with the flexor carpi radialis tendon just medial to it. The superficial radial nerve lies close to its lateral side
C. MEDIAN NERVE is overlapped by the tendon of the palmaris longus

III. **Tendons**
A. TENDON OF PALMARIS LONGUS is superficial and crosses wrist near its middle, with the tendon of the flexor carpi radialis lying close to its lateral side
B. TENDONS OF THE FLEXOR DIGITORUM SUPERFICIALIS are arranged so that those to the middle and ring fingers are superficial to those to the index and little finger
C. TENDONS OF FLEXOR DIGITORUM PROFUNDUS are arranged in order from side to side, deep to the superficial flexor tendons
D. THE TENDON OF THE FLEXOR POLLICIS LONGUS lies in the same plane as the deep flexors, just to radial side of the tendon to the index finger

IV. **Special features**
A. The tendons of the extensor pollicis brevis and abductor pollicis longus run from the dorsum of the forearm, cross the tendons of the extensor carpi radialis longus and brevis to reach their insertion on the radial side of thumb. The extensor of the thumb also crosses the radial carpal extensors, but is separated from the short extensor of the thumb by a triangular space, the "anatomical snuffbox"
B. THE RADIAL ARTERY winds to the dorsum of the hand by passing under these tendons across the "snuffbox" to reach the base of the first interosseous space. Branches of the superficial radial nerve, especially the medial branch with its first dorsal digital, are related to this area superficially
C. THE PALMAR CARPAL LIGAMENT crosses superficial to all the superficial flexor muscles as well as the ulnar nerve and vessels

V. **Clinical considerations**
A. THE SUPERFICIAL POSITION OF THE VESSELS, NERVES, AND TENDONS at the wrist makes them exceedingly vulnerable to damage and injury
B. THE SUPERFICIAL POSITION OF THE RADIAL ARTERY to the radius makes it possible to take the pulse at the wrist
C. THE FREE ANASTOMOSES, both from side to side and from posterior to palmar surface, will preserve the hand even if a major vessel is cut. However, both ends of the severed artery usually must be clamped

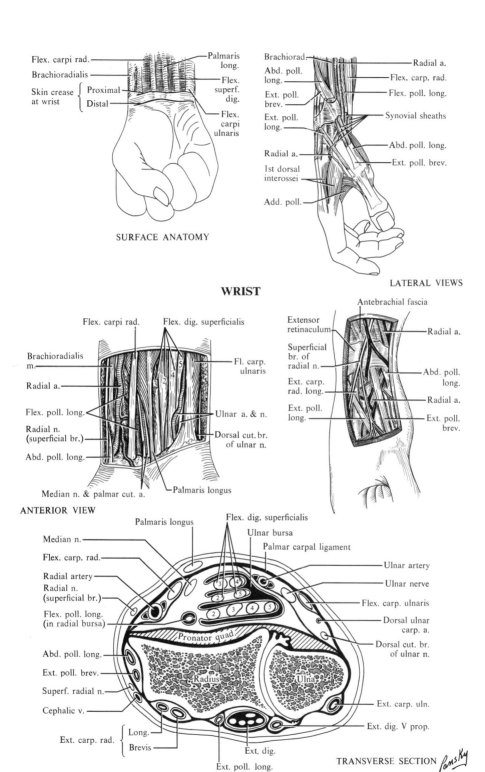

SURFACE ANATOMY

Flex. carpi rad.
Brachioradialis
Skin crease { Proximal
at wrist { Distal
Palmaris long.
Flex. superf. dig.
Flex. carpi ulnaris

LATERAL VIEWS

Brachiorad.
Abd. poll. long.
Ext. poll. brev.
Ext. poll. long.
Radial a.
1st dorsal interossei
Add. poll.
Radial a.
Flex. carp. rad.
Flex. poll. long.
Synovial sheaths
Abd. poll. long.
Ext. poll. brev.

WRIST

ANTERIOR VIEW

Flex. carpi rad.
Flex. dig. superficialis
Brachioradialis m.
Radial a.
Flex. poll. long.
Radial n. (superficial br.)
Abd. poll. long.
Fl. carp. ulnaris
Ulnar a. & n.
Dorsal cut. br. of ulnar n.
Median n. & palmar cut. a.
Palmaris longus

Antebrachial fascia
Extensor retinaculum
Superficial br. of radial n.
Ext. carp. rad. long.
Ext. poll. long.
Radial a.
Abd. poll. long.
Radial a.
Ext. poll. brev.

TRANSVERSE SECTION

Palmaris longus
Flex. dig. superficialis
Ulnar bursa
Palmar carpal ligament
Median n.
Flex. carp. rad.
Radial artery
Radial n. (superficial br.)
Flex. poll. long. (in radial bursa)
Pronator quad.
Abd. poll. long.
Ext. poll. brev.
Superf. radial n.
Cephalic v.
Radius
Ulna
Ext. carp. rad. { Long. Brevis
Ext. poll. long.
Ext. dig.
Ulnar artery
Ulnar nerve
Flex. carp. ulnaris
Dorsal ulnar carp. a.
Dorsal cut. br. of ulnar n.
Ext. carp. uln.
Ext. dig. V prop.

Pansky

- 239 -

109. SUPERFICIAL PALMAR HAND

I. Skin
A. SKIN OF PALM: thick, coarse, and more vascular than on dorsum
   1. Contains no hair or sebaceous glands, but sweat glands are numerous
   2. Creases: proximal, for movement of index finger (marks convexity of superficial palmar arch); distal, for movement of medial 3 digits (marks metacarpal heads); and vertical, limits thenar eminence, for movement of opposition of thumb
   3. Bound to palmar aponeurosis by fibrous septa between which is granular fat

II. Arteries
A. ULNAR. Passes deep to palmar carpal ligament, but superficial to the flexor retinaculum, and lies lateral to the pisiform bone. Branches:
   1. Palmar carpal joins palmar interosseous and radial palmar carpal, forming palmar carpal net
   2. Dorsal carpal (see p. 248)
   3. Deep palmar (see p. 247)
   4. Proper palmar digital to medial side of little finger
   5. Superficial palmar arch: running beneath palmar aponeurosis, but superficial to the long flexor tendons, lumbricales, ulnar, and median nerves. The arch is completed by a superficial palmar branch of the radial artery and gives rise to 3 common palmar digital arteries
      a. Common palmar digital arteries receive communications from the palmar meta-carpal arteries, and each divides into a *proper palmar digital* to adjacent sides of index, middle, and ring fingers
B. RADIAL (see p. 221). Its superficial branch helps make up the superficial palmar arch

III. Nerves
A. RADIAL
   1. Superficial: to lateral side of thumb
   2. Deep: does not enter hand
B. ULNAR
   1. Superficial
      a. Palmar cutaneous: medial palm
      b. Proper palmar digital: medial side of little finger
      c. Common palmar digital: adjacent sides of ring and little finger
   2. Deep: with deep palmar arch to supply all interossei and adductor pollicis
C. MEDIAN
   1. Palmar branch to lateral palm
   2. Common digitals (3): giving proper digitals to the adjacent sides of the other fingers

IV. Clinical considerations
A. KNOWLEDGE OF POSITION OF PROPER DIGITAL NERVES makes possible adequate nerve blocks for surgery on digits
B. IN CASE OF WOUNDS OF HAND INVOLVING ARTERIES, the extensive anastomoses make necessary the ligation of both ends of severed vessels
C. A DISEASE OF THE PALMAR APONEUROSIS, of unknown origin, typically produces abnormal bands of tissue that extend from the aponeurosis to the phalanges, and pull one or more digits into marked flexion at the metacarpophalangeal joint. This is called *Dupuytren's contracture*

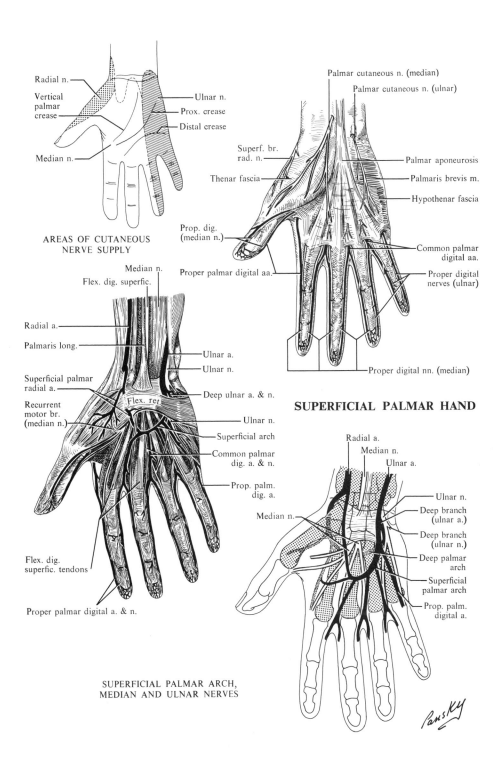

Radial n.

Vertical palmar crease

Median n.

Ulnar n.

Prox. crease

Distal crease

AREAS OF CUTANEOUS NERVE SUPPLY

Median n.

Flex. dig. superfic.

Radial a.

Palmaris long.

Superficial palmar radial a.

Recurrent motor br. (median n.)

Flex. ret.

Ulnar a.

Ulnar n.

Deep ulnar a. & n.

Ulnar n.

Superficial arch

Common palmar dig. a. & n.

Prop. palm. dig. a.

Flex. dig. superfic. tendons

Proper palmar digital a. & n.

SUPERFICIAL PALMAR ARCH, MEDIAN AND ULNAR NERVES

Palmar cutaneous n. (median)

Palmar cutaneous n. (ulnar)

Superf. br. rad. n.

Thenar fascia

Palmar aponeurosis

Palmaris brevis m.

Hypothenar fascia

Prop. dig. (median n.)

Proper palmar digital aa.

Common palmar digital aa.

Proper digital nerves (ulnar)

Proper digital nn. (median)

SUPERFICIAL PALMAR HAND

Radial a.

Median n.

Ulnar a.

Median n.

Ulnar n.

Deep branch (ulnar a.)

Deep branch (ulnar n.)

Deep palmar arch

Superficial palmar arch

Prop. palm. digital a.

I. **Palmar aponeurosis.** Made up of 2 layers: longitudinal and transverse

A. LONGITUDINAL is more superficial. Represents an extension of the tendon of the palmaris longus to fingers. Forms bands distally over the long flexor tendons

B. TRANSVERSE, deeper. Is attached to the flexor retinaculum. These fibers are especially prominent between the diverging longitudinal bands and extend as far distally as heads of metacarpals to form the *superficial transverse metacarpal ligament*

II. **Flexor retinaculum (transverse carpal ligament):** dense band over the groove formed on the palmar aspect of carpals. Medially attached to pisiform and hamulus of hamate; laterally, to tuberosity of scaphoid and ridge of trapezium. Attached superficially to palmar aponeurosis

III. **Muscles of hand** arranged in 3 groups: thenar, hypothenar, and intermediate

Name	Origin	Insertion	Action	Nerve
A. THENAR (to complete, see p. 244)				
Abductor pol. brev.	Flex. retinac. Tuberos. of scaphoid Trapezium	Lat. side, base of first phalanx of thumb	Abducts thumb	Median
B. HYPOTHENAR				
Palmaris brev.	Flex. retinac. Palmar aponeurosis	Skin, med. border of palm	Builds up hypothenar eminence	Ulnar
Abduct. dig. minimi	Pisiform Tendon of flex. carp. ulnaris	Med. side, base of first phalanx of small finger	Abducts small finger Flexes first phalanx	Ulnar
Flex. digit. minimi brev.	Hamulus of hamate Flex. retinac.	Med. side, base of first phalanx of little finger	Flexes first phalanx	Ulnar
Opponens digit. minimi	Hamulus of hamate Flex. retinac.	Med. side, shaft of fifth metacarpal	Opposes the fifth finger	Ulnar
C. INTERMEDIATE (for interossei, see p. 244)				
Lumbricales 1 and 2	Radial side deep flexor tendons to index and middle finger	Tendon, ext. digit. communis, on dorsum of first phalanx, index, and middle fingers	Flex first phalanx, extend second and third phalanges	Median
Lumbricales 3 and 4	Adjoining sides, deep flexor tendons to middle and ring fingers	Tendon, ext. digitorum, on dorsum of first phalanx, ring, and little fingers	Flex first phalanx, and extend distal phalanges, ring and little fingers	Deep ulnar

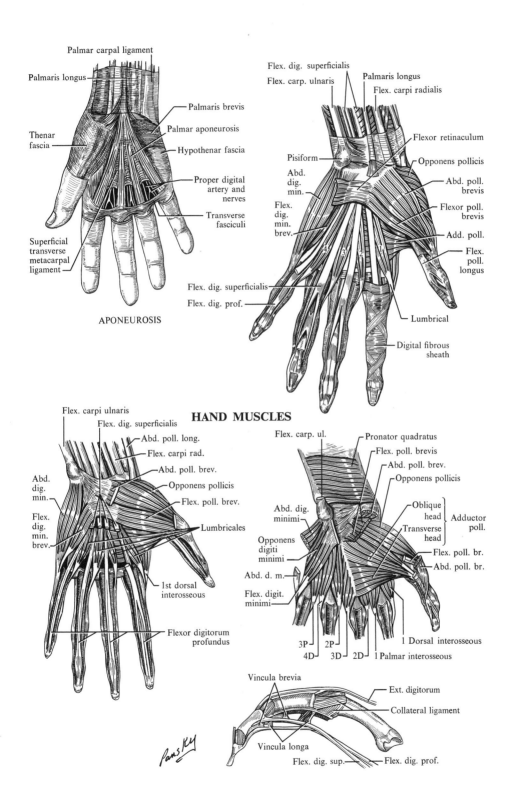

Palmar carpal ligament

Palmaris longus

Thenar fascia

Superficial transverse metacarpal ligament

Palmaris brevis

Palmar aponeurosis

Hypothenar fascia

Proper digital artery and nerves

Transverse fasciculi

APONEUROSIS

Flex. dig. superficialis

Flex. carp. ulnaris

Palmaris longus

Flex. carpi radialis

Pisiform

Abd. dig. min.

Flex. dig. min. brev.

Flexor retinaculum

Opponens pollicis

Abd. poll. brevis

Flexor poll. brevis

Add. poll.

Flex. poll. longus

Flex. dig. superficialis

Flex. dig. prof.

Lumbrical

Digital fibrous sheath

HAND MUSCLES

Flex. carpi ulnaris

Flex. dig. superficialis

Abd. poll. long.

Flex. carpi rad.

Abd. poll. brev.

Opponens pollicis

Flex. poll. brev.

Abd. dig. min.

Flex. dig. min. brev.

Lumbricales

1st dorsal interosseous

Flexor digitorum profundus

Flex. carp. ul.

Pronator quadratus

Flex. poll. brevis

Abd. poll. brev.

Opponens pollicis

Abd. dig. minimi

Opponens digiti minimi

Abd. d. m.

Flex. digit. minimi

Oblique head
Transverse head
} Adductor poll.

Flex. poll. br.

Abd. poll. br.

3P 2P

4D 3D 2D

1 Dorsal interosseous

1 Palmar interosseous

Vincula brevia

Ext. digitorum

Collateral ligament

Vincula longa

Flex. dig. sup.

Flex. dig. prof.

111. MUSCLES OF THE HAND—PART II

Name	Origin	Insertion	Action	Nerve
I. Muscles of the thumb (see also p. 242)				
Opponens pollicis	Flex. retinac. Trapezium	Radial side, first metacarpal	Abducts, flexes, rotates first metacarpal	Median
Flex. pol. brev.	Flex. retinac. Trapezium	Lat. side of base of first phalanx of thumb Med. side of base of first phalanx	Flexes and adducts thumb	Median and deep ulnar
Adductor pollicis				
Obliq.	Capitate Bases, second and third metacarpals	Med. side, base, first phalanx of thumb	Adducts thumb	Deep ulnar
Trans.	Shaft, third metacarpal	Med. side, base, first phalanx of thumb	Adducts thumb	Deep ulnar
II. Intermediate muscles (see also p. 242)				
Palmar interossei	1. Med. side, second meta-carpal	Med. side, base first phalanx, index finger Tendon ext. digitorum	All 3 adduct fingers, flex first phalanx, extend second and third	Deep ulnar
	2. Lat. side, fourth meta-carpal	Same as above for ring finger		
	3. Lat. side, fifth metacarpal	Same as above for little finger		
Dorsal interossei	1. Adjacent sides, second and first meta-carpals	Lat. side, base first phalanx, index finger Ext. digit.	All 4 abduct fingers, flex first phalanx, extend second and third	Deep ulnar
	2. Adjacent sides, second and third meta-carpals	Lat. side, base first phalanx, middle finger Ext. digit.		
	3. Adjacent sides, third and fourth meta-carpals	Med. side, base first phalanx, middle finger Ext. digit.		
	4. Adjacent sides, fourth and fifth meta-carpals	Med. side, base first phalanx, ring finger Tendon ext. digit. com.		

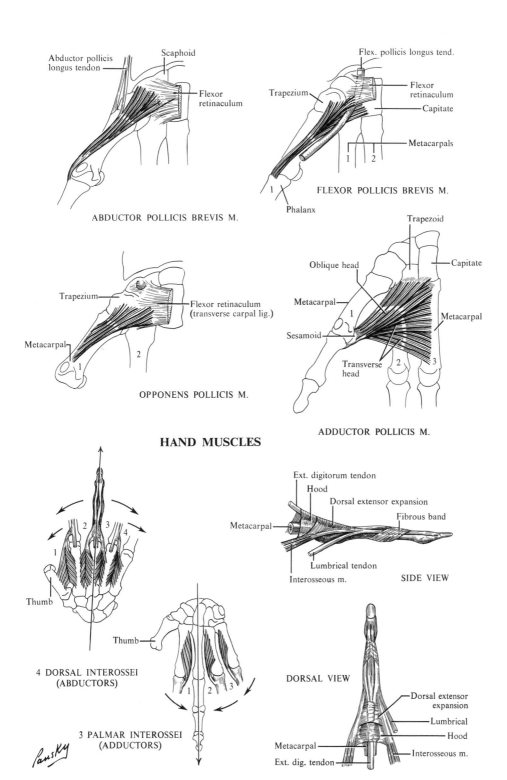

ABDUCTOR POLLICIS BREVIS M.

- Abductor pollicis longus tendon
- Scaphoid
- Flexor retinaculum

FLEXOR POLLICIS BREVIS M.

- Flex. pollicis longus tend.
- Trapezium
- Flexor retinaculum
- Capitate
- Metacarpals
- 1
- 2
- Phalanx
- 1

OPPONENS POLLICIS M.

- Trapezium
- Flexor retinaculum (transverse carpal lig.)
- Metacarpal
- 1
- 2

ADDUCTOR POLLICIS M.

- Trapezoid
- Oblique head
- Capitate
- Metacarpal
- 1
- Metacarpal
- Sesamoid
- Transverse head
- 2
- 3

HAND MUSCLES

4 DORSAL INTEROSSEI
(ABDUCTORS)

- 1
- 2
- 3
- 4
- Thumb

3 PALMAR INTEROSSEI
(ADDUCTORS)

- Thumb
- 1
- 2
- 3

SIDE VIEW

- Ext. digitorum tendon
- Hood
- Dorsal extensor expansion
- Fibrous band
- Metacarpal
- Lumbrical tendon
- Interosseous m.

DORSAL VIEW

- Dorsal extensor expansion
- Lumbrical
- Hood
- Metacarpal
- Interosseous m.
- Ext. dig. tendon

Pansky

112. THE SYNOVIAL BURSAE, COMPARTMENTS AND SPACES IN THE HAND

I. Bursae: synovial compartments for flexor tendons

A. RADIAL, smaller, begins proximal to flexor retinaculum, encloses tendon of flexor pollicis longus, and extends to base of terminal phalanx of thumb

B. ULNAR, large, begins proximal to flexor retinaculum, encloses tendons of flexor digitorum superficialis and profundus, and continues to terminal phalanx of little finger. For other fingers, ends near middle of metacarpals
  1. Arrangement of tendons
     a. Tendons of superficial flexors are paired. Those for the third and fourth fingers lie superficial to those of the second and fifth fingers. Each pair is covered medially, dorsally, and ventrally by a synovial membrane, the pairs separated by a space. Distally, in hand, paired arrangement disappears—tendons lie side by side
     b. The tendons of the deep flexors are enclosed in the same sheet, separated by a space from the deeper pair of superficial tendons

II. Fascial clefts (spaces): lines of separation between fascial membranes

A. MIDDLE PALMAR between deep surface of flexor tendons and fascia covering the interossei. Closed laterally by septum attached to third metacarpal, medially by hypothenar septum to fifth metacarpal
  1. Lumbrical canals: fascial tubes leading out of middle palmar cleft. The canals prolong the midpalmar cleft to the lateral side of fingers. They contain the lumbrical tendons and the proper digital vessels and nerves

B. THENAR overlies fascia of adductor pollicis muscle. Medial boundary is same as lateral septum for middle cleft; lateral is thenar septum and first metacarpal. Proximally it ends at flexor retinaculum, distally at transverse head of the adductor

III. Fascial compartments: areas enclosed by fascial membranes

A. THENAR (thenar eminence), closed proximally by attachments of thenar muscles to flexor retinaculum and carpals, distally by insertion of muscles to first phalanx of thumb, dorsally by subcutaneous border of first metacarpal, medially by attachment of thenar septum to adductors. Contains short thumb muscles, except adductor; tendon of flexor pollicis longus; palmar branch of radial artery

B. HYPOTHENAR (hypothenar eminence); proximally and distally closed as A above; laterally is the hypothenar septum, which ends between flexor digiti minimi muscles and third palmar interosseous
  Contents: all short muscles of small finger

C. CENTRAL between the thenar and hypothenar septa. Ventrally is the palmar aponeurosis, dorsally is the fascia on the dorsal surface of the long flexor tendons, distally the fascia fuses with the tissue in the webs of the fingers, and proximally the fascia is closed, but its contents extend to the forearm
  Contents: lumbrical muscles, superficial volar arch, palmar branch of median nerve, superficial branch of ulnar nerve, and all the flexor tendons

D. INTEROSSEUS-ADDUCTOR enclosed by the dorsal and palmar interosseus fascia and closed proximal and distally by the origins and insertions of the interosseus muscles
  Contents: palmar and dorsal interossei, adductor pollicis muscles, metacarpals 2–4, deep volar arch, and deep branch of the ulnar nerve

IV. Clinical considerations

A. TENOSYNOVITIS is an inflammation of the tendon sheaths contained within the bursae

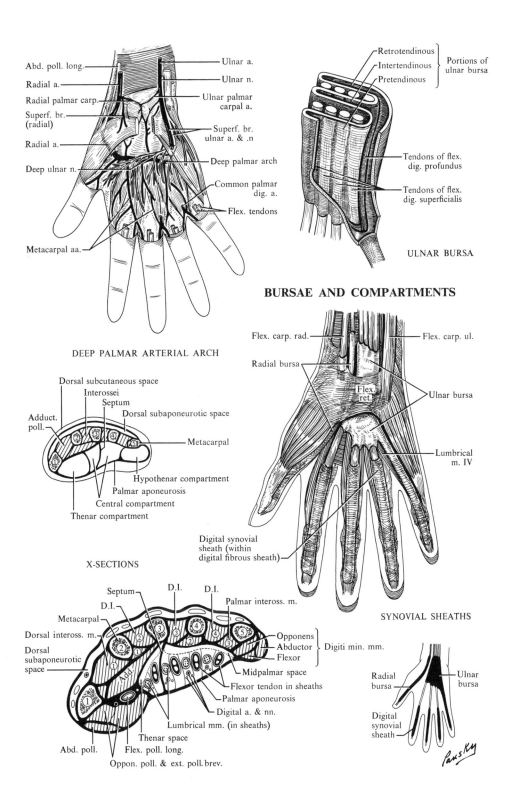

Abd. poll. long.
Radial a.
Radial palmar carp.
Superf. br. (radial)
Radial a.
Deep ulnar n.

Ulnar a.
Ulnar n.
Ulnar palmar carpal a.
Superf. br. ulnar a. & .n
Deep palmar arch
Common palmar dig. a.
Flex. tendons

Metacarpal aa.

DEEP PALMAR ARTERIAL ARCH

Retrotendinous
Intertendinous
Pretendinous
} Portions of ulnar bursa

Tendons of flex. dig. profundus
Tendons of flex. dig. superficialis

ULNAR BURSA

BURSAE AND COMPARTMENTS

Dorsal subcutaneous space
Interossei
Septum
Dorsal subaponeurotic space
Adduct. poll.
Metacarpal
Hypothenar compartment
Palmar aponeurosis
Central compartment
Thenar compartment

X-SECTIONS

Flex. carp. rad.
Radial bursa
Flex. ret.
Flex. carp. ul.
Ulnar bursa
Lumbrical m. IV

Digital synovial sheath (within digital fibrous sheath)

SYNOVIAL SHEATHS

Septum
D.I.
D.I.
D.I.
Palmar inteross. m.
Metacarpal
Dorsal inteross. m.
Dorsal subaponeurotic space
Opponens
Abductor
Flexor
} Digiti min. mm.
Midpalmar space
Flexor tendon in sheaths
Palmar aponeurosis
Digital a. & nn.
Lumbrical mm. (in sheaths)
Thenar space
Abd. poll.
Flex. poll. long.
Oppon. poll. & ext. poll. brev.

Radial bursa
Ulnar bursa
Digital synovial sheath

Pansky

113. RELATIONS AT DORSUM OF WRIST WITH VESSELS AND NERVES TO DORSUM OF HAND

I. The ligaments
A. EXTENSOR RETINACULUM: antebrachial fascia strengthened by circular fibers. Attached to styloid of ulna, the triquetral and the pisiform bones, and the radius. Between the medial and lateral attachments of the retinaculum, osseofibrous canals are formed for the tendons of the dorsal antebrachial muscles
B. CANALS FOR TENDONS: 6 in number, from lateral to medial
  1. On lateral side of radial styloid for abductor pollicis longus and extensor pollicis brevis
  2. Dorsal to styloid for extensor carpi radialis longus and brevis
  3. On middorsum of radius for extensor pollicis longus
  4. Medial to 3, above, for extensor digitorum and extensor indicis
  5. Between radius and ulna for extensor digiti minimi
  6. Between ulnar head and styloid for extensor carpi ulnaris
C. SYNOVIAL SHEATHS for all extensor tendons begin just above the retinaculum. They terminate either proximal to the heads of the metacarpals or at the metacarpophalangeal joints

II. Nerves
A. DORSAL OF ULNAR gives 2 dorsal digital branches and a metacarpal communicating branch
  1. Most medial digital branch to medial side of little finger; the other supplies adjacent sides of ring and little fingers
  2. Metacarpal branch to skin of dorsal hand on ulnar side
B. SUPERFICIAL BRANCH of radial gives a lateral and medial branch
  1. Lateral to radial side and ball of thumb
  2. Medial has 4 dorsal digital branches: first to medial thumb, second to lateral side of index finger, third to adjacent sides of index and middle fingers, and fourth to adjacent side of middle and ring fingers
C. POSTERIOR ANTEBRACHIAL CUTANEOUS (OF RADIAL NERVE) may extend to the middle part of the proximal hand
D. DORSAL BRANCH OF ULNAR AND LATERAL ANTEBRACHIAL CUTANEOUS (OF MUSCULO-CUTANEOUS) reach the wrist but do not extend onto hand

III. Arteries
A. DORSAL CARPAL NET formed by dorsal carpal branch of radial, dorsal carpal branch of ulnar, and posterior interosseous. This gives:
  1. Three dorsal metacarpals, which run on second, third, and fourth dorsal interossei. Each divides into dorsal digital branches to adjacent sides of index, middle, ring, and small fingers
    a. Superior perforating from dorsal metacarpals to deep palmar arch
    b. Inferior perforating from dorsal metacarpals to common palmar digitals
B. FIRST DORSAL METACARPAL, from radial, divides to supply the adjacent sides of thumb and index finger
C. FIFTH DORSAL METACARPAL, from ulnar, to medial side of little finger

DORSUM OF HAND AND WRIST

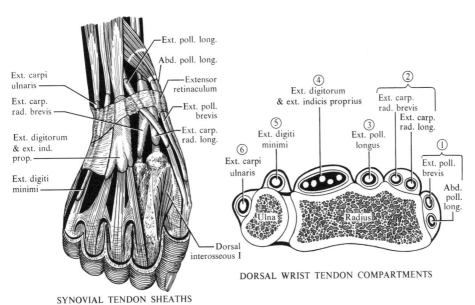

Ext. carpi ulnaris

Ext. carp. rad. brevis

Ext. digitorum & ext. ind. prop.

Ext. digiti minimi

Ext. poll. long.

Abd. poll. long.

Extensor retinaculum

Ext. poll. brevis

Ext. carp. rad. long.

Dorsal interosseous I

SYNOVIAL TENDON SHEATHS

④ Ext. digitorum & ext. indicis proprius

② Ext. carp. rad. brevis

⑤ Ext. digiti minimi

③ Ext. poll. longus

Ext. carp. rad. long.

⑥ Ext. carpi ulnaris

① Ext. poll. brevis

Abd. poll. long.

Ulna

Radius

DORSAL WRIST TENDON COMPARTMENTS

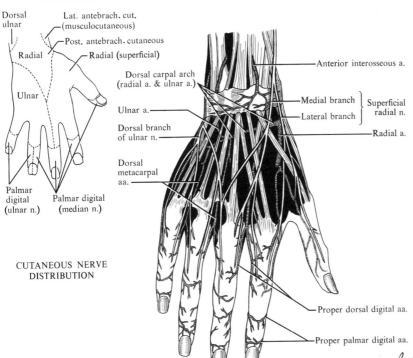

Dorsal ulnar

Lat. antebrach. cut. (musculocutaneous)

Post. antebrach. cutaneous

Radial

Radial (superficial)

Ulnar

Dorsal carpal arch (radial a. & ulnar a.)

Ulnar a.

Dorsal branch of ulnar n.

Dorsal metacarpal aa.

Palmar digital (ulnar n.)

Palmar digital (median n.)

Anterior interosseous a.

Medial branch

Lateral branch

Superficial radial n.

Radial a.

CUTANEOUS NERVE DISTRIBUTION

Proper dorsal digital aa.

Proper palmar digital aa.

Pansky

114. FASCIA, TENDON SHEATHS, NERVES, AND VESSELS OF THE FINGERS

I. **Digital tendon sheaths:** fibrous bands that form osseofibrous canals with palmar surface of phalanges to hold tendons of flexor digitorum superficialis and profundus (see p. 243)

   A. AT MIDDLE OF FIRST AND SECOND PHALANX, greatly strengthened by circular fibers— *digital vaginal ligaments*

   B. OVER JOINTS, sheaths thinner, fibers circular or obliquely crossed (*cruciate*)

II. **The vincula tendina**

   A. SHORT: 2 for each finger: (1) from superficial tendon to first interphalangeal joint and head of first phalanx, (2) from deep tendon to second interphalangeal joint and head of second phalanx

   B. LONG: 2 for each finger: (1) under side of profunda to the superficialis, (2) superficial tendon to proximal end of first phalanx

III. **Extensor aponeurosis** formed by tendons of extensor digitorum. At metacarpophalangeal joints, tendon is bound laterally with the collateral ligaments of this joint. Thus, the aponeurosis acts as dorsal ligaments, which are lacking, per se, in this joint. After crossing joint, tendon broadens to cover dorsal aspect of first phalanx, here reinforced by tendons of lumbricales and interossei. At first interphalangeal joint, splits into 3 slips, 2 lateral and 1 intermediate. The intermediate inserts in base of second phalanx. The 2 lateral reunite before they reach the second interphalangeal joint, to insert into base of third phalanx. This aponeurosis also serves as dorsal ligaments to interphalangeal joints

IV. **Synovial sheaths**

   A. EXTENSORS terminate in proximal third of hand and are not found in the fingers

   B. FLEXORS

      1. On thumb and little finger these sheaths continuous with the so-called *radial* and *ulnar bursae,* respectively. For fingers 2, 3, and 4 start at heads of their respective metacarpals. They continue distally to insertion of deep flexors

V. **Arteries**

   A. PROPER PALMAR DIGITAL ARTERIES extend on either side of finger, dorsal to the corresponding nerves. Cross anastomoses in fingertips and at interphalangeal joints occur, and anastomotic branches to dorsal digitals are present

   B. DORSAL DIGITAL ARTERIES run just ventral to their corresponding nerves but reach only to the first interphalangeal joint. The palmar digitals supply the second and third phalanges and nail bed

VI. **Nerves**

   A. PROPER PALMAR DIGITAL NERVES lie just ventral to arteries; they extend to end of finger supplying the skin of the ball, the nail bed, and the skin over the dorsum of the third phalanx

   B. DORSAL DIGITAL NERVES lie just above corresponding artery and extend to near the second interphalangeal joint

VII. **Special features**

   A. AS THE EXTENSOR TENDONS REACH THE DORSAL ASPECT of the metacarpophalangeal joints, they expand laterally over the sides of the joints in addition to their central thicker portion. This expansion of the extensor tendon is called *the extensor hood*

   B. THE FASCIA forms a pad (called "pulp") of considerable thickness on the distal phalanx; here the retinacula cutis are so arranged that they guide superficial infections to deep positions. Infection of the pulp of the terminal phalanx is called a *felon* or *Whitlow*

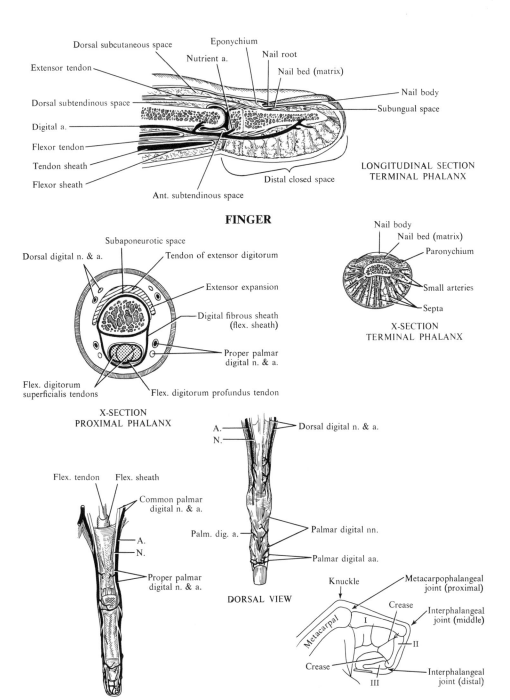

Dorsal subcutaneous space
Eponychium
Nutrient a.
Nail root
Extensor tendon
Nail bed (matrix)
Dorsal subtendinous space
Nail body
Subungual space
Digital a.
Flexor tendon
Tendon sheath
Flexor sheath
Distal closed space
Ant. subtendinous space

LONGITUDINAL SECTION
TERMINAL PHALANX

FINGER

Subaponeurotic space
Dorsal digital n. & a.
Tendon of extensor digitorum
Extensor expansion
Digital fibrous sheath
(flex. sheath)
Proper palmar
digital n. & a.
Flex. digitorum
superficialis tendons
Flex. digitorum profundus tendon

X-SECTION
PROXIMAL PHALANX

Nail body
Nail bed (matrix)
Paronychium
Small arteries
Septa

X-SECTION
TERMINAL PHALANX

A.
N.
Dorsal digital n. & a.

Flex. tendon
Flex. sheath
Common palmar
digital n. & a.
Palm. dig. a.
Palmar digital nn.
A.
N.
Palmar digital aa.
Proper palmar
digital n. & a.

DORSAL VIEW

Knuckle
Metacarpophalangeal
joint (proximal)
Crease
Interphalangeal
joint (middle)
Metacarpal
I
II
Crease
III
Interphalangeal
joint (distal)

PALMAR VIEW

JOINTS OF FINGER
AND SKIN CREASES

Pansky

-251-

115. JOINTS OF FOREARM, WRIST, AND HAND

I. Radioulnar
A. PROXIMAL (see p. 230)
B. MIDDLE considered a syndesmosis with very slight movement. Shafts of radius and ulna joined by oblique cord and interosseous membrane
C. DISTAL
    1. Type: trochoid (pivot)
    2. Bones: head of ulna and ulnar notch of radius
    3. Movements: rotation (pronation, supination)
    4. Ligaments
        a. Articular capsule with palmar and dorsal radioulnar ligaments
        b. Articular disk

II. Muscles acting on radioulnar joints (see p. 230)

III. Wrist joint (radiocarpal articulation)
A. TYPE: condyloid
B. BONES: distal radius and articular disk with proximal row of carpals
C. MOVEMENTS: flexion, extension, abduction, adduction
D. LIGAMENTS
    1. Palmar radiocarpal
    2. Dorsal radiocarpal
    3. Ulnar carpal collateral
    4. Radial carpal collateral

IV. Muscles acting on wrist joint

Flexion	Extension	Abduction	Adduction
Flex. carp. uln.	Ext. carp. rad. long.	Flex. carp. rad.	Flex. carp. uln.
Flex. carp. rad.	Ext. carp. rad. brev.	Ext. carp. rad. long.	Ext. carp. uln.
Palm. long.	Ext. carp. uln.	Ext. carp. rad. br.	Ext. indic. prop.
Flex. dig. superfic.	Ext. dig. communis	Ext. pol. long.	
Flex. dig. prof.		Ext. pol. brev.	

V. Intercarpal articulation
A. ARTICULATIONS OF CARPALS with each other: arthrodial joints with dorsal, palmar, and interosseous ligaments
B. ARTICULATIONS BETWEEN THE 2 ROWS OF CARPALS (*the midcarpal joint*): a hinge joint, acting with wrist. Has dorsal, palmar, and collateral ligaments

VI. Carpometacarpal
A. BETWEEN FIRST METACARPAL AND TRAPEZIUM: a saddle joint, permitting flexion, extension, abduction, adduction, and opposition
B. ALL OTHERS between distal row of carpals and bases of metacarpals: arthrodial, with dorsal, palmar, and interosseous ligaments

VII. Intermetacarpal among bases of metacarpals: dorsal, palmar, and interosseous ligaments
A. DEEP TRANSVERSE METACARPAL LIGAMENTS join heads of metacarpals

VIII. Metacarpophalangeal joints: condyloid, joined by 1 palmar and 2 collateral ligaments

IX. Interphalangeal joints: hinge joints, with 1 palmar and 2 collateral ligaments

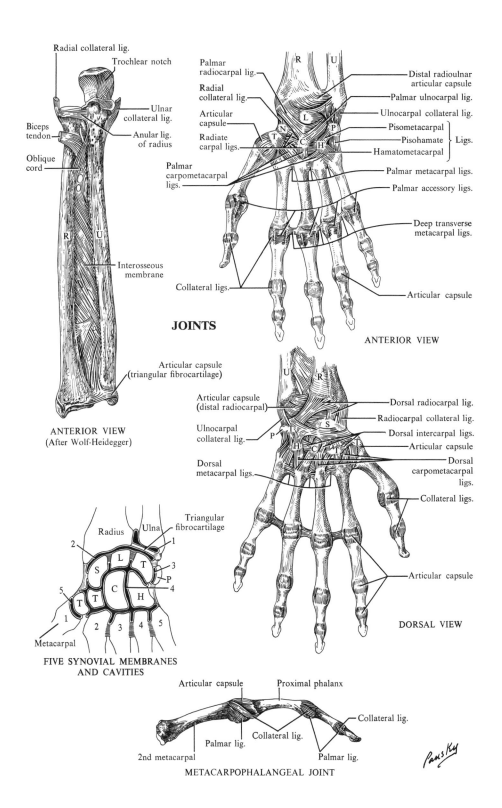

Radial collateral lig.
Trochlear notch
Biceps tendon
Ulnar collateral lig.
Anular lig. of radius
Oblique cord
R   U
Interosseous membrane

ANTERIOR VIEW
(After Wolf-Heidegger)

Palmar radiocarpal lig.
Radial collateral lig.
Articular capsule
Radiate carpal ligs.
Palmar carpometacarpal ligs.

R U
L
N
T   P
C   H

Distal radioulnar articular capsule
Palmar ulnocarpal lig.
Ulnocarpal collateral lig.
Pisometacarpal
Pisohamate   } Ligs.
Hamatometacarpal
Palmar metacarpal ligs.
Palmar accessory ligs.
Deep transverse metacarpal ligs.

Collateral ligs.

Articular capsule

JOINTS

ANTERIOR VIEW

Articular capsule (triangular fibrocartilage)

Articular capsule (distal radiocarpal)
Ulnocarpal collateral lig.
Dorsal metacarpal ligs.

U   R
S
P
H C

Dorsal radiocarpal lig.
Radiocarpal collateral lig.
Dorsal intercarpal ligs.
Articular capsule
Dorsal carpometacarpal ligs.
Collateral ligs.

Articular capsule

DORSAL VIEW

Triangular fibrocartilage
Radius   Ulna
2   1
S   L   T
T   3
P
C   4
T   T   H
1   2   3   4   5
Metacarpal
5

FIVE SYNOVIAL MEMBRANES
AND CAVITIES

Articular capsule   Proximal phalanx
Collateral lig.
2nd metacarpal
Palmar lig.   Collateral lig.
Palmar lig.

METACARPOPHALANGEAL JOINT

Pansky

- 253 -

116. SURFACE ANATOMY OF UPPER EXTREMITY

I. Arteries

A. AXILLARY: its medial end can be compressed against the 2nd rib by pressure applied below the middle of the clavicle just medial to the coracoid process of the scapula. The line extends laterally from here to the medial border of the biceps at level of the posterior axillary fold

B. BRACHIAL: its line extends downward along the medial side of the biceps (along bicipital groove) to a point in the cubital fossa midway between the humeral epicondyles and terminates about 2.5 cm below this point opposite the head of the radius

C. RADIAL: this can be indicated by a line beginning 1 in. below the center of the line joining the humeral epicondyles to a point just medial to the styloid process of the radius. From this point the vessel curves around the lateral side of the wrist to reach the proximal end of the first intermetacarpal space on the back of the hand

D. ULNAR: beginning at the terminal point of the brachial artery, the line runs obliquely downward to reach the ulnar side of the forearm about halfway between the elbow and wrist. Thence, the line continues downward to a point just lateral to the pisiform bone

E. SUPERFICIAL PALMAR ARCH: beginning just to the lateral side of the pisiform bone, the line arches distally and laterally across the palm. The convexity of the arch reaches a level of the medial border of the fully extended and abducted thumb

F. DEEP PALM ARCH: this is indicated by an almost directly transverse line about 1.0 cm proximal to the superficial arch

II. Nerves

A. MEDIAN: in the arm its line coincides with that for the brachial artery; in the lower cubital fossa it lies at the medial side of ulnar artery; just below the cubital fossa it crosses anterior to the ulnar artery and continues in the forearm on a line drawn down the center of the anterior surface of the forearm; it continues down the wrist along the middle of its anterior surface between the tendons of the flexor carpi radialis and palmaris longus

B. ULNAR: in the upper arm it follows the line of the brachial artery, thence it passes deep and runs in a groove just behind the medial epicondyle of the humerus; in the forearm it shifts anteriorly, and at the junction of the middle and upper thirds of the forearm it joins the ulnar artery and follows its line into the wrist

C. RADIAL: in upper part of arm it lies on the medial side of the humerus; at the junction of the upper and middle thirds of the humerus it passes behind the humerus in the radial sulcus (groove) on the posterior aspect of that bone; at the junction of the middle and lower thirds of arm it reaches the lateral side of the arm and continues downward across the lateral side of the elbow joint, lateral to the cubital fossa

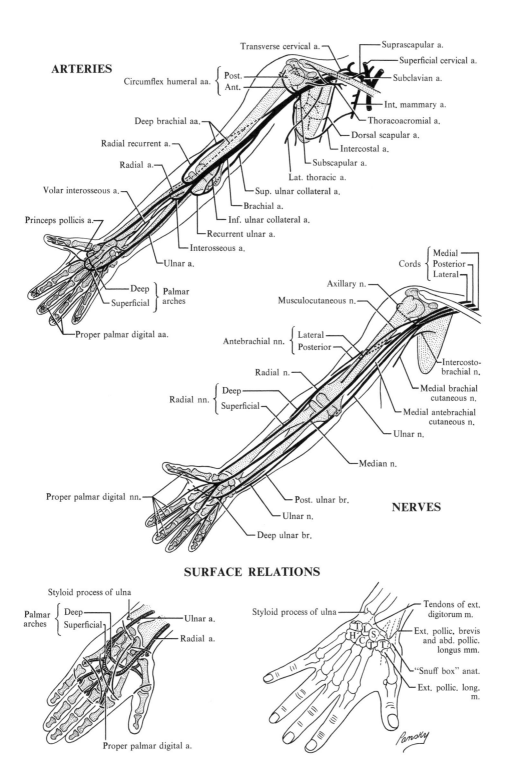

ARTERIES

Transverse cervical a.

Suprascapular a.

Superficial cervical a.

Circumflex humeral aa. { Post. Ant.

Subclavian a.

Int. mammary a.

Deep brachial aa.

Thoracoacromial a.

Radial recurrent a.

Dorsal scapular a.

Intercostal a.

Radial a.

Subscapular a.

Volar interosseous a.

Lat. thoracic a.

Sup. ulnar collateral a.

Princeps pollicis a.

Brachial a.

Inf. ulnar collateral a.

Recurrent ulnar a.

Interosseous a.

Ulnar a.

Deep } Palmar
Superficial } arches

Proper palmar digital aa.

Cords { Medial
Posterior
Lateral

Axillary n.

Musculocutaneous n.

Antebrachial nn. { Lateral
Posterior

Intercosto-
brachial n.

Radial n.

Medial brachial
cutaneous n.

Radial nn. { Deep
Superficial

Medial antebrachial
cutaneous n.

Ulnar n.

Median n.

Proper palmar digital nn.

Post. ulnar br.

NERVES

Ulnar n.

Deep ulnar br.

SURFACE RELATIONS

Styloid process of ulna

Palmar
arches { Deep
Superficial

Ulnar a.

Radial a.

Styloid process of ulna

Tendons of ext.
digitorum m.

Ext. pollic. brevis
and abd. pollic.
longus mm.

"Snuff box" anat.

Ext. pollic. long.
m.

Proper palmar digital a.

Pansky

117. MUSCLE-NERVE RELATIONSHIP: FUNCTIONAL

I. Nerve damage may be indicated in one of two ways
A. LOSS OF SENSATION in the area of cutaneous distribution
B. PARALYSIS OF VOLUNTARY MUSCLE SUPPLIED. This can be shown both by the inability to perform certain functions and by the atrophy of the muscle supplied, since muscle cannot long survive the loss of its nerve

II. Certain nerves are most vulnerable in specific regions, and the possibility of damage should always be considered when wounds, fractures, etc., occur in these regions

III. Specific cases of nerve injury
A. AXILLARY: loss of roundness of shoulder and loss of sensation over the deltoid area
B. RADIAL: if trauma occurred in the axilla, all cutaneous areas supplied (see p. 185) would be insensitive, the triceps would atrophy, and there would be wristdrop (when arm is extended, the hand passively falls and cannot be extended). If trauma occurred in midarm, the loss of sensation would be noted in the forearm and hand. Wristdrop would be present. The triceps (and its function) would be normal
C. ULNAR: loss of sensation on the medial side of the palm and the little finger, wasting of the hypothenar eminence, wasting of the interossei (causing the metacarpals to stand out); fingers cannot be abducted or adducted; the thumb cannot be adducted
D. MEDIAN: loss of sensation on radial palm and palmar surface of index and second fingers, wasting of the thenar eminence; the "hand of papal benediction" with the ring and little finger flexed while the index and second fingers are extended; loss of ability to oppose thumb (ape hand)
E. MUSCULOCUTANEOUS: no flexion at elbow, wasting of biceps, and loss of sensation on the lateral forearm
F. UPPER TRUNK (ERB'S) PALSY: involves C5 and C6 and their fibers, which travel in such branches of the brachial plexus as the axillary, dorsal scapular, subscapular, and long thoracic nerves. Symptoms: "winged scapula" due to weakness of the serratus anterior with other signs as noted under the axillary and musculocutaneous nerves

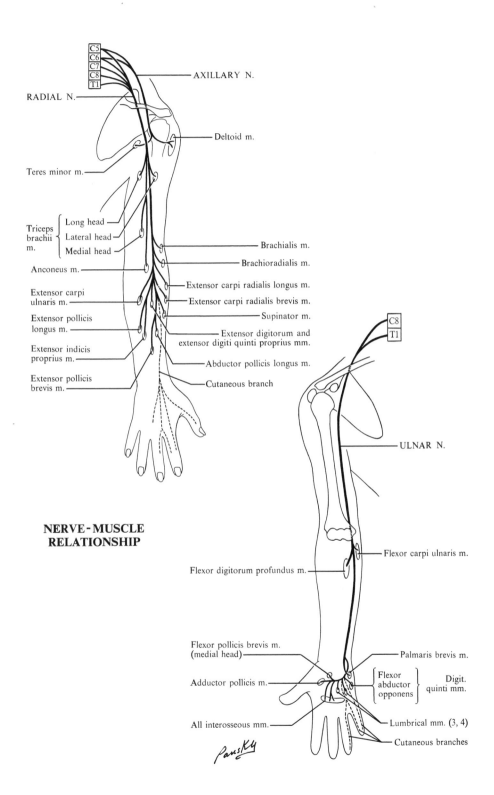

C5
C6
C7
C8
T1

RADIAL N.

AXILLARY N.

Deltoid m.

Teres minor m.

Triceps brachii m. { Long head — Lateral head — Medial head }

Anconeus m.

Extensor carpi ulnaris m.

Extensor pollicis longus m.

Extensor indicis proprius m.

Extensor pollicis brevis m.

Brachialis m.

Brachioradialis m.

Extensor carpi radialis longus m.

Extensor carpi radialis brevis m.

Supinator m.

Extensor digitorum and extensor digiti quinti proprius mm.

Abductor pollicis longus m.

Cutaneous branch

NERVE-MUSCLE RELATIONSHIP

C8
T1

ULNAR N.

Flexor carpi ulnaris m.

Flexor digitorum profundus m.

Flexor pollicis brevis m. (medial head)

Palmaris brevis m.

Adductor pollicis m.

{ Flexor abductor opponens } Digit. quinti mm.

All interosseous mm.

Lumbrical mm. (3, 4)

Cutaneous branches

Pansky

118. SUMMARY OF MUSCLE-NERVE RELATIONSHIPS IN UPPER EXTREMITY

Segment	Muscle	Named Nerve
C5–C6	Biceps brachii	Musculocutaneous
C5–C6	Brachialis	Musculocutaneous and radial
C5–C6	Brachioradialis	Radial
C5–C6	Coracobrachialis	Musculocutaneous
C5–C6	Deltoid	Axillary
C5–C6	Infraspinatus	Suprascapular
C5–C6	Subscapularis	Upper and lower subscapular
C5–C6	Supraspinatus	Suprascapular
C5–C6	Teres major	Lower subscapular
C5–C6	Teres minor	Axillary
C6–C7	Abd. pollicis longus	Deep radial
C6–C7	Ext. carpi radialis brevis	Deep radial
C6–C7	Ext. carpi radialis longus	Radial
C6–C7	Ext. pollicis brevis	Deep radial
C6–C7	Flex. carpi radialis	Median
C6–C7	Palmaris longus	Median
C6–C7	Pronator teres	Median
C6–C7	Supinator	Deep radial
C6–C7	Ext. carpi ulnaris	Deep radial
C6–C7	Ext. digitorum	Deep radial
C6–C7	Ext. indicis	Deep radial
C6–C7	Ext. pollicis longus	Deep radial
C6–C7	Triceps brachii	Radial
C7–C8	Anconeus	Radial
C7–C8	Ext. digiti minimi	Deep radial
C7–C8	Flex. digitorum profundus	Median and ulnar
C7–C8	Flex. digitorum superficialis	Median
C8–T1	Abd. digiti minimi	Deep ulnar
C8–T1	Abd. pollicis brevis	Median
C8–T1	Add. pollicis	Deep ulnar
C8–T1	Flex. carpi ulnaris	Ulnar
C8–T1	Flex. digiti minimi	Deep ulnar
C8–T1	Flex. pollicis brevis	Median
C8–T1	Flex. pollicis longus	Median
C8–T1	Interossei	Deep ulnar
C8–T1	Lumbricales	Deep ulnar and median
C8–T1	Opponens digiti minimi	Deep ulnar
C8–T1	Opponens pollicis	Median
C8–T1	Palmaris brevis	Superficial ulnar
C8–T1	Pronator quadratus	Median

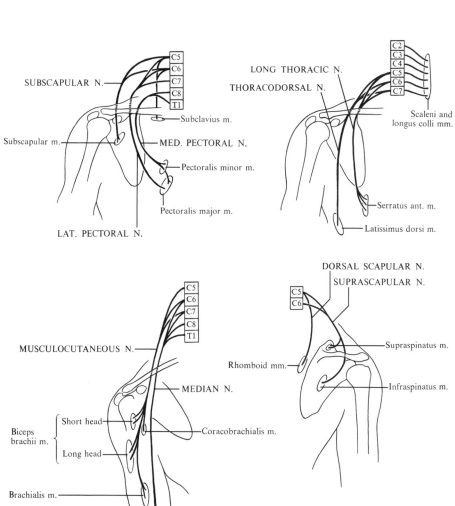

SUBSCAPULAR N.

C5
C6
C7
C8
T1

Subclavius m.

Subscapular m.

MED. PECTORAL N.

Pectoralis minor m.

Pectoralis major m.

LAT. PECTORAL N.

LONG THORACIC N.

THORACODORSAL N.

C2
C3
C4
C5
C6
C7

Scaleni and
longus colli mm.

Serratus ant. m.

Latissimus dorsi m.

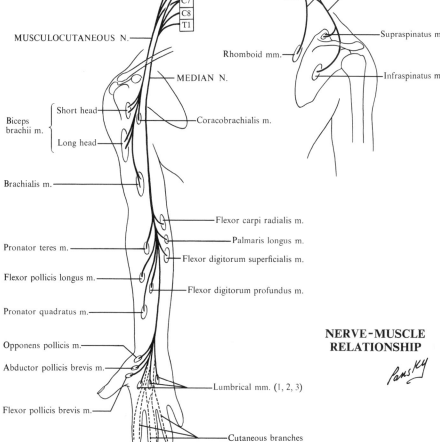

MUSCULOCUTANEOUS N.

C5
C6
C7
C8
T1

MEDIAN N.

Biceps
brachii m. {

Short head

Long head

Coracobrachialis m.

Brachialis m.

Flexor carpi radialis m.

Pronator teres m.

Palmaris longus m.

Flexor digitorum superficialis m.

Flexor pollicis longus m.

Flexor digitorum profundus m.

Pronator quadratus m.

Opponens pollicis m.

Abductor pollicis brevis m.

Lumbrical mm. (1, 2, 3)

Flexor pollicis brevis m.

Cutaneous branches

DORSAL SCAPULAR N.

SUPRASCAPULAR N.

C5
C6

Rhomboid mm.

Supraspinatus m.

Infraspinatus m.

**NERVE-MUSCLE
RELATIONSHIP**

Pansky

Thorax

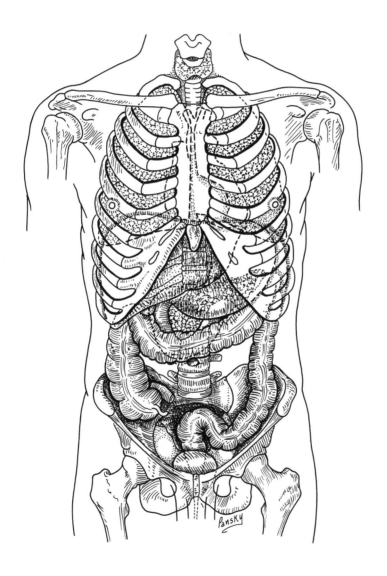

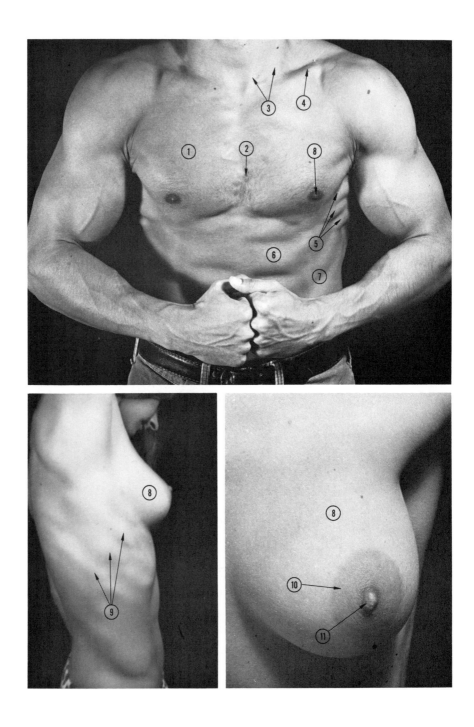

FIGURE 19. **Surface anatomy of thorax.** *1,* Pectoralis major m.; *2,* sternum; *3,* sternocleido-mastoid m.; *4,* clavicle; *5,* serratus anterior m.; *6,* rectus abdominis m.; *7,* external oblique m.; *8,* mammary gland; *9,* ribs; *10,* areola; *11,* nipple.

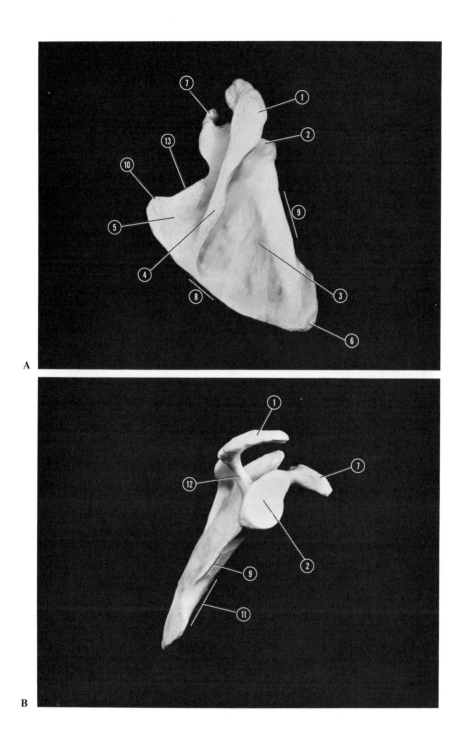

FIGURE 20. **Scapula. A, Posterior view; B, medial view.** *1,* Acromion; *2,* glenoid cavity; *3,* infraspinous fossa; *4,* spine; *5,* supraspinous fossa; *6,* inferior angle; *7,* coracoid process; *8,* medial border; *9,* lateral border; *10,* superior angle; *11,* subscapular fossa; *12,* spine root; *13,* superior border.

119. THE MAMMARY GLAND

I. **Location and extent:** vertically from second to sixth or seventh rib and horizontally from lateral border of sternum to beyond anterior axillary fold (into axilla). Mammary papilla (nipple) just below center of gland at the fourth interspace

II. **Structure**
A. EACH BREAST made up of 15 to 20 separate lobes, arranged like segments of a citrus fruit, each divided from its neighbor by connective tissue septa
B. TYPE OF GLAND: modified sweat gland, derived from skin
C. ENTIRE GLAND lies in the superficial fascia
   1. Deep surface is separated from muscle fascia by loose connective tissue
D. SUSPENSORY LIGAMENTS (LIGAMENTS OF COOPER) run from corium of skin, vertically through gland to deepest part of superficial fascia

III. **Ducts**
A. NEAR THE CENTRAL END of each lobe lactiferous (excretory) ducts are large. All converge toward nipple
B. AMPULLAE (lactiferous sinuses): dilatations of ducts beneath areola
C. AT BASE OF NIPPLE ducts narrow down, change direction from horizontal to vertical and run to summit of nipple

IV. **Areola:** pigmented area surrounding nipple for distance of 1.0 to 2.0 cm
A. AREOLAR GLANDS of Montgomery produce little irregularities in areolae

V. **Blood Supply**
A. ARTERIES
   1. From axillary artery: supreme thoracic, pectoral branch of thoracoacromial, and lateral thoracic arteries, which give rise to the *external mammary artery*
   2. From intercostal arteries: mammary branches especially from anterior rami in third, fourth, and fifth spaces
   3. From internal thoracic artery: perforating branches in second, third, and fourth interspaces
B. VEINS: from anastomotic circle around papilla. Branches pass peripherally from this to axillary and internal thoracic veins

VI. **Nerves:** mainly from anterior and lateral cutaneous branches of intercostal nerves in fourth, fifth, and sixth spaces

VII. **Lymphatic drainage**
A. PRINCIPAL: 2 trunks, which drain into pectoral group of axillary nodes
B. SECONDARY
   1. From medial side of gland to sternal nodes. Some cross to opposite gland
   2. From cephalic part through pectoralis major muscle to the apical axillary nodes. A few over clavicle to supraclavicular nodes
   3. From caudal portions, may be some few vessels to abdominal wall

VIII. **Clinical considerations**
A. BREAST CANCER: the most common form of cancer in women; spreads by way of all vascular and lymphatic channels as well as along the fibrous tissue of the gland to deeper structures. For this reason, radical mastectomy is often performed, the purpose of which is to remove the primary tumor in the breast, the underlying fascia, the pectoral muscles, and the axillary lymph nodes

BREAST

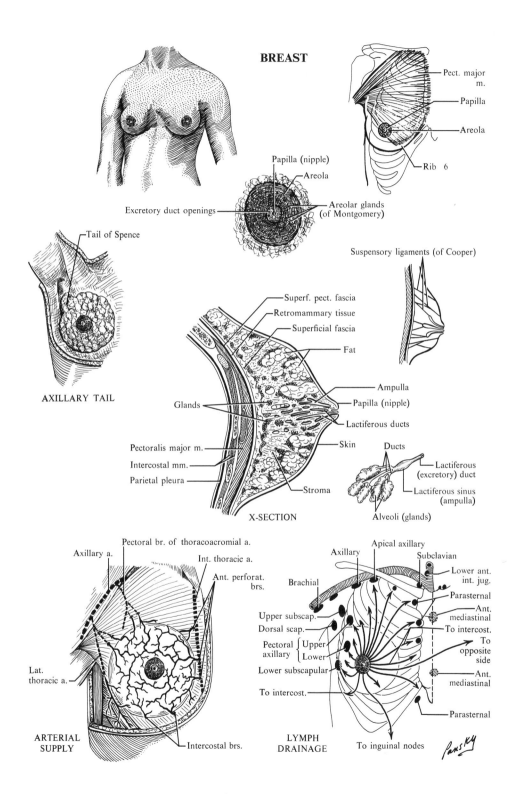

Papilla (nipple)

Areola

Pect. major m.

Papilla

Areola

Rib 6

Excretory duct openings

Areolar glands (of Montgomery)

Tail of Spence

Suspensory ligaments (of Cooper)

Superf. pect. fascia

Retromammary tissue

Superficial fascia

Fat

Ampulla

Papilla (nipple)

Lactiferous ducts

Glands

Skin

Ducts

Lactiferous (excretory) duct

AXILLARY TAIL

Pectoralis major m.

Intercostal mm.

Parietal pleura

Lactiferous sinus (ampulla)

Stroma

Alveoli (glands)

X-SECTION

Pectoral br. of thoracoacromial a.

Axillary a.

Int. thoracic a.

Ant. perforat. brs.

Apical axillary

Axillary

Subclavian

Brachial

Lower ant. int. jug.

Parasternal

Upper subscap.

Ant. mediastinal

Dorsal scap.

To intercost.

Pectoral axillary { Upper / Lower

To opposite side

Ant. mediastinal

Lat. thoracic a.

Lower subscapular

To intercost.

Parasternal

ARTERIAL SUPPLY

Intercostal brs.

LYMPH DRAINAGE

To inguinal nodes

Pansky

I. Sternum (breast bone)

A. PARTS

    1. Manubrium, most cephalic. Roughly quadrangular; has 2 surfaces (anterior and posterior) and 4 borders (superior, inferior, and 2 lateral)

        a. Both surfaces tend to be smooth and concave

        b. Jugular notch: in middle of superior border on either side of which is an oval articular surface for clavicle

        c. Inferior border: rough, normally covered with cartilage for articulation with body

        d. Near upper end of lateral border is distinct impression for first costal cartilage; at caudal extremity of this border is a small facet for articulation with part of second costal cartilage

    2. Body makes up most of bone. Long and narrow. Has 2 surfaces (anterior and posterior) and 4 borders (superior, inferior, 2 lateral)

        a. Anterior surface has 3 transverse ridges at level of articular depressions for third, fourth, and fifth costal cartilages

        b. Superior border, oval for articulation with manubrium. This articulation is the *sternal angle*

        c. On lateral border: at cephalic end, small depression for second costal cartilage. Below this, 4 depressions for cartilages of ribs 3–6. At caudal end of this border, a small notch which, with adjoining one in xiphoid, holds cartilage of seventh rib

        d. Inferior border: narrow, for articulation with xiphoid

    3. Xiphoid smallest and most caudal. May be bifid. Has 2 surfaces (anterior and posterior) and 3 borders (superior and 2 lateral)

        a. At cephalic end, has small depression for part of seventh costal cartilage

B. OSSIFICATION, from 6 centers: 1 manubrium, 4 body, 1 xiphoid

Name	When Appear	When Closed
Manubrium	6th fetal month	25th year
1st body	6th fetal month	25th year
2nd and 3rd body	7th fetal month	25th year
4th body	1st year postnatal	Puberty
Xiphoid	5th to 18th year	30 to after 40 years

II. Clinical considerations

A. SLIGHT DEFORMITIES AND ASYMMETRIES of the thorax are common, cause no disability, and often require no treatment (including flat chest associated with round back)

B. SERIOUS DEFORMITIES of the chest are almost predominantly congenital in origin and are always associated with overgrowth of the ribs

C. PIGEON BREAST (PECTUS CARINATUM): sternum projects forward and downward like the keel of a boat

D. FUNNEL CHEST (PECTUS EXCAVATUM): sternum pushed posteriorly by overgrowth of ribs

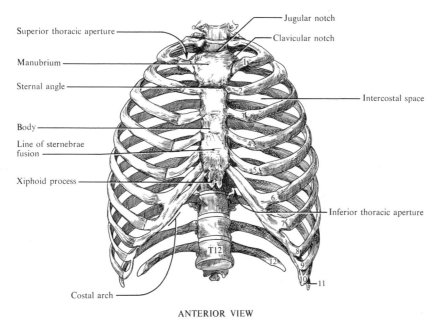

Jugular notch

Superior thoracic aperture

Clavicular notch

Manubrium

Sternal angle

Intercostal space

Body

Line of sternebrae fusion

Xiphoid process

Inferior thoracic aperture

T12

Costal arch

ANTERIOR VIEW

THORACIC CAGE

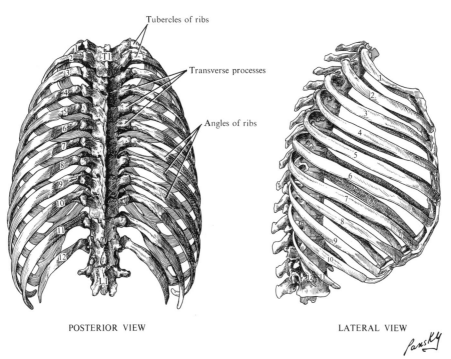

Tubercles of ribs

Transverse processes

Angles of ribs

POSTERIOR VIEW

LATERAL VIEW

121. STERNOCLAVICULAR, STERNOCOSTAL, AND INTRASTERNAL ARTICULATIONS

I. Sternoclavicular joint (see p. 192)

II. Sternocostal articulations

A. BETWEEN FIRST RIB AND STERNUM
   1. Type: synarthrosis (synchondrosis)
B. BETWEEN STERNUM AND RIBS 2–7
   1. Type: arthrodial
   2. Movements: gliding
   3. Ligaments
      a. Articular capsule
      b. Radiate sternocostal: from dorsal and ventral side of sternal ends of cartilages to dorsal and ventral side of sternum
      c. Intra-articular sternocostal: usually found with second rib only, where ligament runs from end of cartilage to the fibrocartilage between manubrium and body of sternum
      d. Costoxiphoid: join dorsal and ventral surfaces of seventh costal cartilage to dorsal and ventral sides of xiphoid process
C. INTERCHONDRAL between costal cartilages of ribs 6–8 (sometimes through 10) join together at adjacent articular facets
   1. Ligaments
      a. Articular capsules
      b. Interchondral, running from 1 cartilage to another
D. COSTOCHONDRAL between depression in medial end of rib and lateral end of costal cartilage. These are synarthroses, joined by periosteum

III. Intrasternal

A. BETWEEN MANUBRIUM AND BODY OF STERNUM
   1. Type: synarthrosis with fibrocartilage between the edges of bone, strengthened by fibrous tissue (probably periosteum). In some cases, a synovial cavity is found. In old age, ossification sometimes is found (a true synostosis)
B. BETWEEN XIPHOID AND BODY OF STERNUM
   1. Type: synarthrosis (synchondrosis)
   2. In general, by fifteenth year cartilage is replaced by bone in a synostosis

IV. Clinical considerations

A. DISLOCATION OF THE ACROMIOCLAVICULAR JOINT is a common injury (shoulder separation)
B. DISLOCATION OF THE STERNAL END OF THE CLAVICLE occurs much less frequently than acromioclavicular dislocation
C. STERNOCLAVICULAR DISLOCATION results from a violent fall or blow upon the shoulder: sternoclavicular ligaments are ruptured and the intra-articular fibrocartilage remains attached to the clavicle

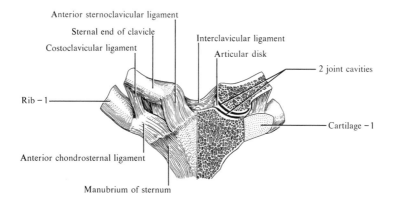

Anterior sternoclavicular ligament

Sternal end of clavicle

Costoclavicular ligament

Interclavicular ligament

Articular disk

2 joint cavities

Rib – 1

Cartilage – 1

Anterior chondrosternal ligament

Manubrium of sternum

STERNOCLAVICULAR

ARTICULATIONS – ANTERIOR VIEWS

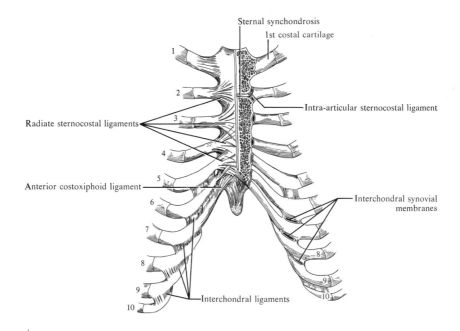

Sternal synchondrosis

1st costal cartilage

Radiate sternocostal ligaments

Intra-articular sternocostal ligament

Anterior costoxiphoid ligament

Interchondral synovial membranes

Interchondral ligaments

STERNOCOSTAL AND INTERCHONDRAL

122. THE RIBS

I. Classification
A. TRUE RIBS (VERTEBROCOSTAL) (1–7): articulate behind with the vertebrae and in front, through cartilages, with the sternum
B. FALSE RIBS
    1. Vertebrochondral (8–10): join the vertebrae behind, but ventrally join the costal cartilages of the ribs above
    2. Floating (11–12): the ventral ends of the ribs are free

II. General characteristics of a "typical" rib: each rib has a dorsal (*vertebral*) and a ventral (*sternal*) extremity with an intervening body (*shaft*)
A. DORSAL EXTREMITY
    1. Head: has an articular surface divided into 2 parts by a ridge
    2. Neck: about 1 in. in length, just lateral to the head
    3. Tubercle: at junction of neck and body. Has an articular facet
B. BODY (SHAFT): thin, flat, and arched with 2 surfaces (external and internal) and 2 borders (superior and inferior)
    1. Angle of rib: an oblique ridge just beyond the tubercle on the external surface
    2. Costal groove: along and just above the inferior border, on the internal surface
C. VENTRAL EXTREMITY ends in an oval concavity for costal cartilage

III. Atypical ribs
A. FIRST RIB has greatest curvature and is the shortest
    1. Flattened in a plane different from the others so that there are superior and inferior surfaces with lateral and medial borders
    2. Head has no division of its articular facet
    3. Tubercle is very thick and prominent
    4. There is no angle
    5. Upper surface shows 2 shallow grooves (for subclavian a. and v.), separated by the *scalene tubercle*
    6. No costal groove
    7. Anterior extremity is larger and thicker than other ribs
B. SECOND RIB: longer than first and intermediate in form between the first rib and the typical rib. Has the same curvature as the first rib, but is not quite as flattened. The angle is slight and close to the tubercle
C. TENTH RIB: like a typical rib except that it has a single articular facet on its head
D. ELEVENTH AND TWELFTH RIBS
    1. Have a single articular facet on head
    2. Have neither neck nor tubercle
    3. Ventral ends are pointed
    4. Eleventh has a very shallow costal groove; twelfth has none

IV. Ossification
A. RIBS 1–10 OSSIFY FROM 4 CENTERS: for body, for head, for articular part of tubercle, and for nonarticular part of tubercle. They appear in body near angle in the eighth fetal week (seen first in ribs 6 and 7). Centers for the head and tubercles appear between the sixteenth and twentieth years and unite with the body in the twenty-fifth year.

V. Special features
A. THE ARTICULATIONS of the ribs merit little description. Most have 2 *costovertebral joints*, one between the head of the rib and adjacent vertebral bodies, the other between the tubercle and the transverse process (*costotransverse joints*). See page 268 for description of the sternocostal joints

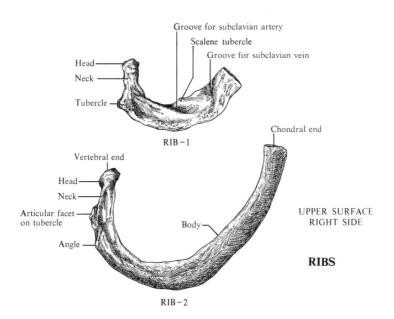

Groove for subclavian artery
Scalene tubercle
Groove for subclavian vein
Head
Neck
Tubercle

RIB-1

Vertebral end
Head
Neck
Articular facet
on tubercle
Angle
Body

Chondral end

UPPER SURFACE
RIGHT SIDE

RIBS

RIB-2

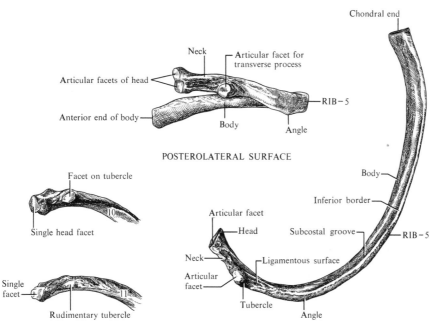

Neck
Articular facet for
transverse process
Articular facets of head
Anterior end of body
Body
Angle
RIB-5

POSTEROLATERAL SURFACE

Chondral end

Facet on tubercle
Single head facet
10

Single
facet
11
Rudimentary tubercle

Single
facet
12

SEEN FROM BELOW, RIGHT SIDE

Articular facet
Head
Neck
Articular
facet
Tubercle
Ligamentous surface
Angle

Body
Inferior border
Subcostal groove
RIB-5

UNDERSURFACE
RIGHT SIDE

123. MUSCLES, DEEP VESSELS, AND NERVES OF THE THORACIC WALL

I. **Muscles** (see p. 197)

A. INTERCOSTALS: all innervated by intercostal nerves. All draw ribs together, raise ribs

 1. External (11 pairs). Origin: lower border of rib; insertion: in upper border of rib below. Extend from tubercle of rib to costal cartilage. External intercostal membrane replaces muscle from costal cartilages to sternum

 2. Internal (11 pairs). Origin: from inner surface of rib; insertion: in upper border of rib below. Extend from sternum to angles of rib. Internal intercostal membrane replaces muscle between angle of rib and vertebrae

B. SUBCOSTAL: in lower thorax only. Origin: from inner surface near angle of rib; insertion: 2 or 3 ribs below. Action: when last rib is fixed by quadratus lumborum muscle, it lowers ribs

C. TRANSVERSE THORACIC: on inner chest wall. Origin: from posterior body and xiphoid of sternum and sternal ends of costal cartilages of ribs 4–6; insertion: on lower border and costal cartilages of ribs 2–6. Action: draws ribs down

D. LEVATORES COSTARUM (12 pairs): in posterior thorax. Origin: from ends and transverse processes of vertebrae C7–T11; insertion: into outer surface of rib immediately below origin. Action: raise ribs; extend, laterally flex, and rotate the vertebral column to opposite side

E. SERRATUS POSTERIOR SUPERIOR: on posterior of upper thorax (see p. 155). Origin: from ligamentum nuchae, supraspinal ligament, and spines of vertebrae C7–T3; insertion: on upper borders of ribs 2–5, lateral to angles. Action: raises ribs

F. SERRATUS POSTERIOR INFERIOR: on posterior of lower thorax (see p. 155). Origin: from supraspinous ligament and spines of vertebrae T11–L3; insertion: on lower border of ribs 9–12, lateral to angles. Action: lowers ribs

II. **Intercostal arteries**

A. AORTIC (POSTERIOR) INTERCOSTALS: 9 pairs, which run in lower 9 interspaces. Arise from dorsal side of aorta and divide into 2 rami

 1. Anterior ramus. In general, has vein above and nerve below. Branches: to muscles, lateral cutaneous and collateral intercostal, which run along border of rib below to join branches of the internal thoracic artery. The mammary branches arise in spaces 3–5

 2. Dorsal ramus. Runs backward between necks of ribs. Branches: to spinal cord, muscles, and skin of dorsum

B. INTERCOSTAL BRANCHES OF INTERNAL THORACIC ARTERY: 2 in each of upper 6 interspaces. Anastomose with anterior rami of aortic intercostals. Have perforating branches to muscle, skin, and mammary gland

C. SUPREME (HIGHEST) INTERCOSTAL ARTERY arises from costocervical trunk and gives intercostal branches to first 2 interspaces

III. **Intercostal nerves**

A. REPRESENTED BY INTERCOSTAL PART of first thoracic and anterior primary divisions of other thoracic nerves. Course similar to artery up to middle of rib; then it lies with the internal intercostal muscle as far as the costal cartilage; then it runs between internal intercostal muscle and pleura. Branches: anterior and lateral cutaneous and muscular

IV. **Clinical considerations**

A. BECAUSE THE CHIEF VESSELS AND NERVES are at the lower border of each rib, it is customary in passing a needle through the thoracic wall (*thoracentesis or paracentesis*) to place the needle close to the lower rather than the upper rib, and thus avoid injury to these nerves and vessels

DEEP THORACIC MUSCLES

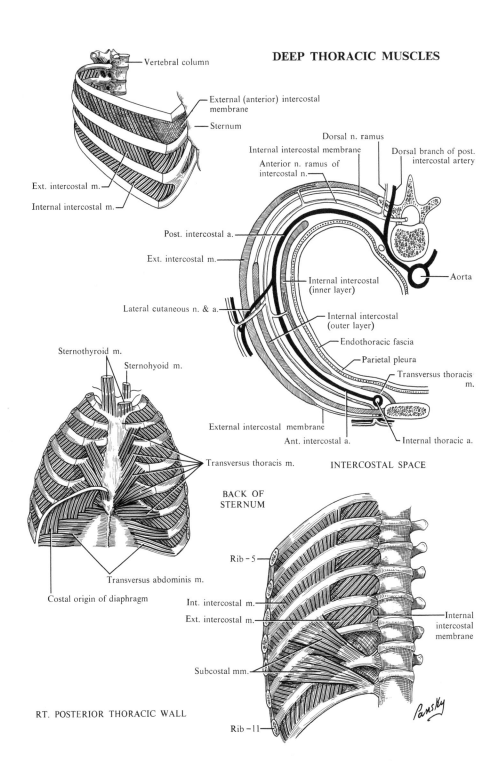

Vertebral column

External (anterior) intercostal membrane

Sternum

Ext. intercostal m.

Internal intercostal m.

Dorsal n. ramus

Internal intercostal membrane

Anterior n. ramus of intercostal n.

Dorsal branch of post. intercostal artery

Post. intercostal a.

Ext. intercostal m.

Internal intercostal (inner layer)

Aorta

Lateral cutaneous n. & a.

Internal intercostal (outer layer)

Endothoracic fascia

Parietal pleura

Transversus thoracis m.

External intercostal membrane

Ant. intercostal a.

Internal thoracic a.

INTERCOSTAL SPACE

Sternothyroid m.

Sternohyoid m.

Transversus thoracis m.

BACK OF STERNUM

Transversus abdominis m.

Costal origin of diaphragm

Rib – 5

Int. intercostal m.

Ext. intercostal m.

Internal intercostal membrane

Subcostal mm.

RT. POSTERIOR THORACIC WALL

Rib – 11

Pansky

-273-

124. SURFACE PROJECTIONS OF LUNGS AND PLEURA

I. Lungs

A. APEX: same for both lungs, about 2.5 cm above medial end of clavicle

B. MEDIAL MARGINS

1. Right descends in a slight curve from the sternoclavicular joint and crosses the midsternal line at the sternal angle, descends just to left of midline to sixth costosternal junction

2. Left descends in slight curve from sternoclavicular joint; stays to left of midline, parallel and close to the right as far as level of fourth costal cartilage; passes to the left along this cartilage to the parasternal line, descends across the fifth costal cartilage, and curves medially to upper sixth costal cartilage to the left of its costosternal union

C. INFERIOR MARGINS: same for both lungs. Extends along length of sixth cartilage, crosses sixth costochondral union, curves downward and lateralward to upper eighth rib in midaxillary line, continues to ninth or tenth rib in scapular line, then medially to eleventh costovertebral joint

D. POSTERIOR MARGINS: same for both lungs. Extends cephalically from eleventh costovertebral joint, on either side of vertebral spines to T2, then curves laterally to apex

E. FISSURES

1. Oblique (interlobar) starts 6 cm below apex, at level of third rib. With arm at side, this fissure crosses the back from third thoracic spine to scapular spine, then caudally around thorax to end of sixth rib

2. Horizontal (accessory) in right lung. Starting at the point where the oblique fissure crosses the midaxillary line, this fissure runs nearly horizontally to the ventral margin at level of fourth rib

II. The pleurae: except for the inferior limits, the pleurae and lungs are similar enough to make complete repetition unnecessary

A. RIGHT SIDE. From the xiphisternal junction downward across the sternocostal union, crosses the eighth costochondral articulation in the mammary line, the tenth rib in the midaxillary line, then runs posteriorly and medially to the spinous process of the twelfth thoracic vertebrae

B. LEFT SIDE. Same as right to level of fourth costal cartilage. Curves downward, just to the left of the sternal margin to sixth costal cartilage, then crosses the seventh costal cartilage, and from there is same as right. Although the divergence of the pleura to the left of the midline on the left side is more pronounced than on the right, it does not exhibit a cardiac notch as does the left lung

III. Clinical consideration

A. THE FACT THAT THE LUNGS do not go as low as the lowest limits of the pleurae creates a potential space, which makes possible the draining of blood or fluid from the pleural cavity without endangering the lungs (see p. 276). To drain the space (thoracentesis) one must not go below the 7th, 9th, or 11th ribs in the midclavicular, midaxillary, or posterior scapular lines, respectively, since these points are below the diaphragm. Insert needle at upper border of rib to avoid intercostal vessels; and direct upward to avoid puncturing diaphragm

B. THE BASE OF THE LUNG refers clinically not to the anatomic base, which is related to the diaphragm and, therefore, is inferior, but to the lower limits of the posterior surface of the lower lobe. To listen to the base of the lung, apply the stethoscope to the posterior chest wall about at a level with the 10th thoracic vertebra

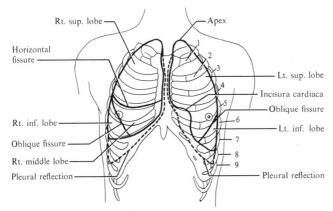

Rt. sup. lobe — Apex

Horizontal fissure —

1
2
3
4
— Lt. sup. lobe
— Incisura cardiaca
5
— Oblique fissure
6
Rt. inf. lobe — — Lt. inf. lobe
7
Oblique fissure — 8
Rt. middle lobe — 9
Pleural reflection — — Pleural reflection

ANTERIOR VIEW

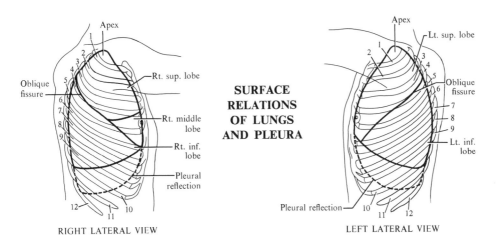

Apex
1
2
3
4
5
Oblique fissure — — Rt. sup. lobe
6
7
8
9
— Rt. middle lobe

— Rt. inf. lobe

— Pleural reflection

12   10
11

RIGHT LATERAL VIEW

**SURFACE
RELATIONS
OF LUNGS
AND PLEURA**

Apex
— Lt. sup. lobe
1
2
3
4
5
— Oblique fissure
6
7
8
9
— Lt. inf. lobe

Pleural reflection — 10
11   12

LEFT LATERAL VIEW

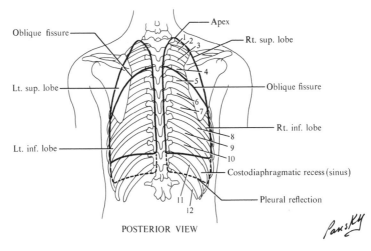

Oblique fissure — — Apex
1 2
3
— Rt. sup. lobe
4
5
— Oblique fissure
Lt. sup. lobe —
6
7
— Rt. inf. lobe
8
9
10
Lt. inf. lobe —
— Costodiaphragmatic recess (sinus)
11
12
— Pleural reflection

POSTERIOR VIEW

Pansky

125. LUNGS

I. Apex: rises 2.5 to 5 cm above first sternocostal articulation

II. Diaphragmatic surface (base): concave, where it rests on convexity of diaphragm

III. Surfaces and their impressions
A. COSTAL: convex to conform to thoracic wall
B. MEDIASTINAL
1. Cardiac impression for pericardium, deeper on left lung to form so-called *cardiac notch*
2. Hilum above cardiac impression
3. Above hilum, on right lung, is arched groove for azygos vein
4. Groove for superior vena cava and right brachiocephalic vein, some distance from apex toward ventral side of right lung
5. Groove for right subclavian artery lies behind this and nearer apex
6. Posterior to hilum is the groove for esophagus on right lung
7. Groove for arch of aorta is above hilum and descending groove for thoracic aorta behind hilum on left lung
8. Groove for subclavian artery runs upward and lateralward from groove for aortic arch on left lung
9. Anterior and below groove for subclavian artery is groove for the left brachiocephalic vein on left lung

IV. Borders
A. INFERIOR: sharp, where it separates base and costal surface; blunt at mediastinal border
B. POSTERIOR (VERTEBRAL): broad and round to fit in the groove lateral to the vertebrae
C. ANTERIOR: thin and sharp; overlaps pericardium on right almost straight; on left, cardiac notch

V. Fissures and lobes
A. LEFT divided into 2 lobes, superior and inferior, by oblique fissure
B. RIGHT divided into 3 lobes by 2 fissures. One fissure, similar to left lung, separates inferior from middle and superior lobes. The other, more horizontal, separates the superior and middle lobes

VI. Root structures at the hilum
A. RIGHT LUNG: pulmonary artery lies ventral to bronchus; 1 pulmonary vein lies just caudal to artery; the other vein is caudal to the bronchus
B. LEFT LUNG: pulmonary artery lies most cephalic; 1 vein lies ventral to bronchus, both being caudal to artery; the other pulmonary vein is caudal to bronchus

VII. Pleura
A. VISCERAL covers lungs and dips into fissures
B. PARIETAL lines thoracic walls. Named by location: costal, diaphragmatic, mediastinal, and cervical (cupula)
C. REFLECTIONS
1. Pulmonary ligament: a fold where the mediastinal pleura, reflected on dorsal and ventral sides of lung root, is continued downward toward diaphragm
2. Phrenicocostal recess (sinus): where costal pleura extends into groove between diaphragm and chest wall
3. Costomediastinal recess (sinus): where costal and mediastinal pleurae meet

LUNGS AND PLEURAE

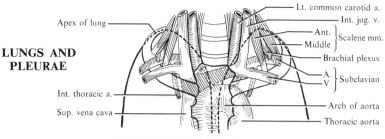

Apex of lung — Lt. common carotid a.

Int. jug. v.

Ant. } Scalene mm.
Middle }

Brachial plexus

A. } Subclavian
V. }

Int. thoracic a. —

Sup. vena cava —

Arch of aorta

Thoracic aorta

TOPOGRAPHIC RELATIONS OF LUNG APEX

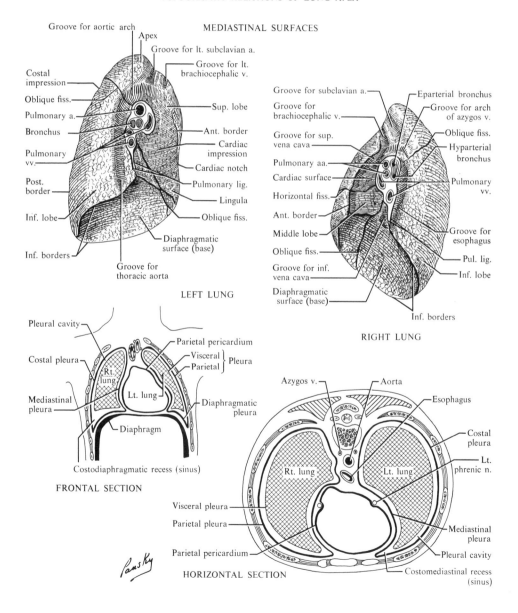

MEDIASTINAL SURFACES

LEFT LUNG

Groove for aortic arch
Apex
Groove for lt. subclavian a.
Groove for lt. brachiocephalic v.
Costal impression
Oblique fiss.
Pulmonary a.
Bronchus
Pulmonary vv.
Post. border
Inf. lobe
Inf. borders
Sup. lobe
Ant. border
Cardiac impression
Cardiac notch
Pulmonary lig.
Lingula
Oblique fiss.
Diaphragmatic surface (base)
Groove for thoracic aorta

RIGHT LUNG

Groove for subclavian a.
Groove for brachiocephalic v.
Groove for sup. vena cava
Pulmonary aa.
Cardiac surface
Horizontal fiss.
Ant. border
Middle lobe
Oblique fiss.
Groove for inf. vena cava
Diaphragmatic surface (base)
Eparterial bronchus
Groove for arch of azygos v.
Oblique fiss.
Hyparterial bronchus
Pulmonary vv.
Groove for esophagus
Pul. lig.
Inf. lobe
Inf. borders

FRONTAL SECTION

Pleural cavity
Costal pleura
Mediastinal pleura
Rt. lung
Lt. lung
Diaphragm
Parietal pericardium
Visceral } Pleura
Parietal }
Diaphragmatic pleura
Costodiaphragmatic recess (sinus)

HORIZONTAL SECTION

Azygos v.
Aorta
Esophagus
Costal pleura
Lt. phrenic n.
Rt. lung
Lt. lung
Visceral pleura
Parietal pleura
Parietal pericardium
Mediastinal pleura
Pleural cavity
Costomediastinal recess (sinus)

-277-

126. BRONCHOPULMONARY SEGMENTS

I. Primary bronchi (see p. 283)
A. RIGHT: 3 secondary bronchi come off here, 1 for each of the 3 lobes. The branch to the superior lobe lies above the pulmonary artery (*eparterial*). The branches to the middle and inferior lobes are below the artery (*hyparterial*). Each bronchus then branches according to its bronchopulmonary segments
   1. Superior lobe bronchus: 3 branches
   2. Middle lobe bronchus: 2 branches
   3. Inferior lobe bronchus: 5 branches
B. LEFT: 2 secondary bronchi come off here, 1 for each lobe. Both lie below the artery and are thus *both hyparterial*
   1. Superior lobe bronchus: 2 major branches, a superior and an inferior
      a. Superior division bronchus: 2 branches
      b. Inferior division bronchus: 2 branches
   2. Inferior lobe bronchus: 4 or 5 branches

II. Segments
A. RIGHT LUNG
   1. Superior lobe: apical, posterior, and anterior segments
   2. Middle lobe: lateral and medial segments
   3. Inferior lobe: superior; medial, anterior, lateral, and posterior basal segments
B. LEFT LUNG
   1. Superior lobe
      a. Superior division: apical-posterior and anterior segments
      b. Inferior division: superior and inferior (lingual) segments
   2. Inferior lobe: superior; anterior, medial, lateral, and posterior basal segments

III. Clinical considerations
A. SINCE THE BRONCHOPULMONARY SEGMENTAL TERRITORIES are relatively distinct, it is theoretically possible to separate or isolate the individual segments without having to remove an entire lobe of a lung. This has been accomplished in a number of surgical procedures
B. THE UPPER LOBE of the lung is related to more of the chest wall than meets the eye. Much of the upper lobe is related not only to the anterior chest between the clavicle and nipple but also to the lateral chest wall and can be examined here
C. THE MIDDLE LOBE is an anterior part of the right lung. It has no direct relationship to the posterior chest wall. It relates to the anterior chest wall between the 4th and 6th costal cartilages
D. POSTURAL DRAINAGE: there are many times in the practice of medicine when the assistance of gravity is sought in the care of the chest; for example, when it is necessary to promote drainage from all or part of the bronchial tree, the posture of the patient is adjusted to bring maximal influence of gravity to bear on the bronchus to be drained

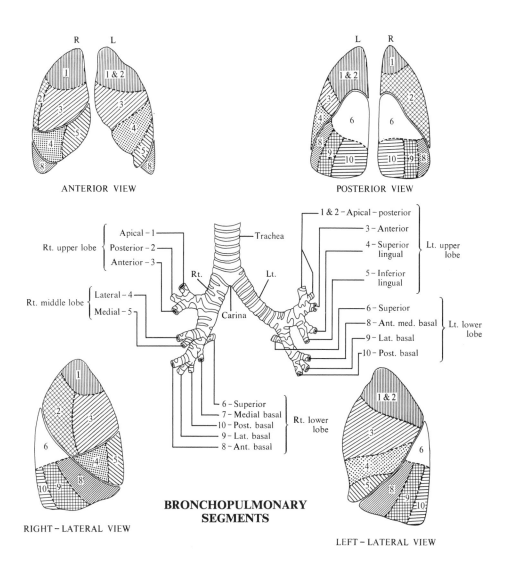

R   L   L   R

1   1
1 & 2   1 & 2
2   3   2
3   3   4   6   6
5   4   8
4   9 10   10 9 8
8   5
             8

ANTERIOR VIEW   POSTERIOR VIEW

                                        Trachea
                    Apical – 1   1 & 2 – Apical – posterior
Rt. upper lobe {   Posterior – 2   3 – Anterior
                    Anterior – 3   Rt.   Lt.   4 – Superior   } Lt. upper
                                                              lingual   lobe
                                        Carina   5 – Inferior
Rt. middle lobe {   Lateral – 4   lingual
                    Medial – 5   6 – Superior
                                                           8 – Ant. med. basal   } Lt. lower
                                                           9 – Lat. basal   lobe
                                                          10 – Post. basal

                              6 – Superior
                              7 – Medial basal
                             10 – Post. basal   } Rt. lower
                              9 – Lat. basal   lobe
                              8 – Ant. basal

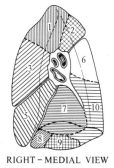

RIGHT – LATERAL VIEW

**BRONCHOPULMONARY
SEGMENTS**

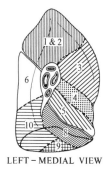

LEFT – LATERAL VIEW

RIGHT – MEDIAL VIEW   LEFT – MEDIAL VIEW

Pansky

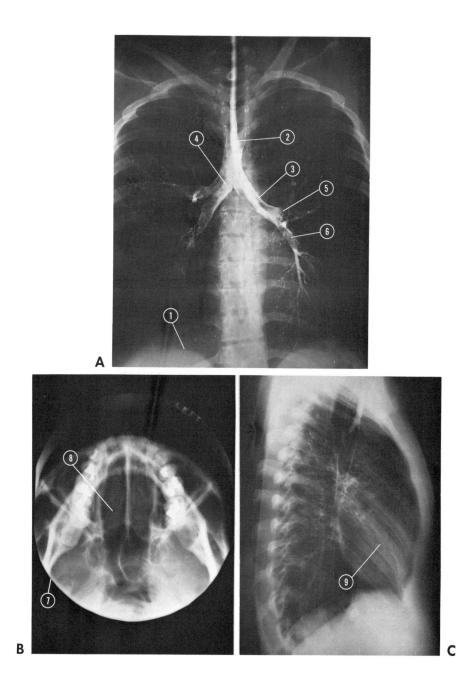

FIGURE 21. **Respiratory system and bronchography. A, Normal bronchogram; B, mouth; C, chest, lateral view.** *1*, Diaphragm; *2*, trachea; *3*, left primary bronchus; *4*, right primary bronchus; *5*, left upper secondary bronchus; *6*, left lower secondary bronchus; *7*, maxilla; *8*, hard palate; *9*, heart.

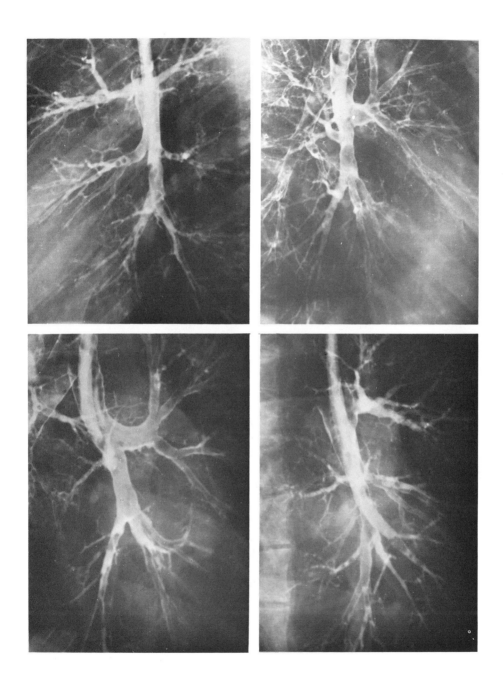

FIGURE 22. **Series of bronchograms from both the left and right sides.**

127. CIRCULATION AND INNERVATION OF THE LUNGS

I. Intrapulmonary bronchi
A. SECONDARY BRONCHI to lobes enter lung substance to become intrapulmonary. Immediately begin to subdivide, the smaller divisions being known as bronchioles
B. BRONCHIOLES
    1. Terminal (lobular) bronchioles: the last division that has semblance of typical bronchial structure, are not respiratory in function. Each divides into 2 or more respiratory bronchioles
    2. Respiratory bronchioles, which are somewhat similar to the terminal but with a variable number of respiratory units (single alveoli for respiration): these give rise to alveolar ducts
    3. Alveolar ducts: these are similar to 2, above, but with a greater number of respiratory units. These, in turn, lead into alveolar sacs
    4. Alveolar sacs: these are structures whose walls are composed of *pulmonary alveoli.* It is around these that most of respiration occurs. (NOTE: Small circular spaces, called *atria,* are sometimes spoken of as existing between the alveolar ducts and sacs)

II. Lobulation
A. PRIMARY LOBULE begins with the alveolar duct and includes all atria, alveolar sacs, and alveoli coming from it
B. SECONDARY LOBULE made up of several primaries

III. Blood supply
A. SYSTEMIC, through bronchial arteries, which arise from the aorta or upper intercostal arteries and accompany the bronchial tree as far as the respiratory bronchiole. Supplies walls of bronchi and bronchioles
B. PULMONARY, through the right and left pulmonary arteries from the pulmonary trunk. Follows the bronchial tree, behind the bronchi. Forms a capillary net around the alveoli

IV. Venous drainage
A. SYSTEMIC arises from the territory supplied by the bronchial arteries. Vessels converge to form 1 vein at the root of each lung. These end on the right in the azygos, on the left in the supreme intercostal vein
B. PULMONARY arises chiefly in the capillary net around the alveoli with a very small amount from the bronchial system. These veins pass through the lung substance independent of bronchi and pulmonary arteries. They converge to form 2 veins at the root of each lung. These terminate in the left atrium

V. Lymphatics: 2 plexuses.
Superficial, under the pleura, from tissue along the alveolar ducts and pleura. These curve around the borders of the lungs and terminate in nodes (bronchopulmonary) at the hilum. The *deep* vessels follow the bronchial tree, pass through the hilum to the *tracheobronchial nodes.* These nodes lie beside the trachea, under the primary bronchi, at the hilum, and on the large intrapulmonary bronchi. The afferents of these nodes arise from the lungs, bronchi, trachea, heart, and posterior mediastinum. The efferents join those of the internal thoracic and anterior mediastinum to form the *bronchomediastinal trunk*

VI. Nerves:
the innervation of the lungs is by both vagal and sympathetic fibers via the anterior and posterior pulmonary plexuses (see p. 297). The pulmonary arteries are apparently innervated by sympathetic fibers only, the smooth muscles of the bronchi by parasympathetic fibers, and the bronchial glands are innervated by sympathetic fibers. All afferent fibers from the lung are vagal fibers

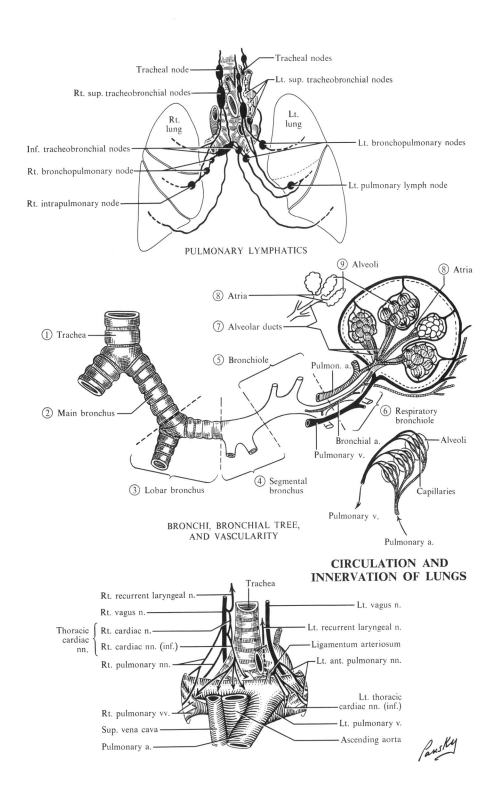

PULMONARY LYMPHATICS

Tracheal node
Rt. sup. tracheobronchial nodes
Rt. lung
Inf. tracheobronchial nodes
Rt. bronchopulmonary node
Rt. intrapulmonary node

Tracheal nodes
Lt. sup. tracheobronchial nodes
Lt. lung
Lt. bronchopulmonary nodes
Lt. pulmonary lymph node

BRONCHI, BRONCHIAL TREE, AND VASCULARITY

① Trachea
② Main bronchus
③ Lobar bronchus
④ Segmental bronchus
⑤ Bronchiole
⑥ Respiratory bronchiole
⑦ Alveolar ducts
⑧ Atria
⑨ Alveoli

Pulmon. a.
Bronchial a.
Pulmonary v.
Alveoli
Capillaries
Pulmonary v.
Pulmonary a.

CIRCULATION AND INNERVATION OF LUNGS

Trachea
Rt. recurrent laryngeal n.
Rt. vagus n.
Thoracic cardiac nn. { Rt. cardiac n.
Rt. cardiac nn. (inf.)
Rt. pulmonary nn.
Rt. pulmonary vv.
Sup. vena cava
Pulmonary a.

Lt. vagus n.
Lt. recurrent laryngeal n.
Ligamentum arteriosum
Lt. ant. pulmonary nn.
Lt. thoracic cardiac nn. (inf.)
Lt. pulmonary v.
Ascending aorta

Pansky

128. THE HEART, SURFACE PROJECTIONS, GREAT VESSELS, AND CORONARIES

I. Surface projections of the heart
A. APEX: in fifth interspace, approximately 8 cm from midsternal line
B. BASE: slightly oblique line at level of third costal cartilage projecting 2 cm to left and 1 cm to right of lateral border of sternum
C. INFERIOR (DIAPHRAGMATIC) BORDER: right end under sixth costosternal junction; line slopes down across xiphisternal junction to apex
D. RIGHT BORDER begins at right end of base, curves slightly to right, reaching 2.5 cm from sternal margin in fourth interspace, ends at right end of inferior border
E. LEFT BORDER curves upward and medialward from apex to left end of base line

II. Superficial features of heart
A. CORONARY SULCUS: groove separating atria from ventricles
B. ANTERIOR AND POSTERIOR INTERVENTRICULAR (LONGITUDINAL) SULCI divide ventricles into right and left
   1. Incisura apices cordis: notch near apex where the longitudinal sulci meet
C. APEX points down and to the left
D. BASE faces toward right, upward and backward
E. SURFACES: sternocostal beneath sternum and ribs, diaphragmatic against diaphragm
F. MARGINS: right (acute) runs from diaphragmatic to base; left (obtuse) runs from base to apex

III. Coronary vessels
A. ARTERIES
   1. Right originates in right aortic sinus, runs to right, beneath right auricle to coronary sulcus
      a. Posterior descending branch down posterior interventricular sulcus to apex
      b. Marginal branch follows right margin to apex
   2. Left originates in left aortic sinus, runs under left auricle and divides
      a. Anterior descending branch: runs in anterior interventricular sulcus to apex
      b. Circumflex: runs in left part of coronary sulcus, curving around to posterior interventricular sulcus
B. VEINS
   1. Coronary sinus: large vessel in posterior part of coronary sulcus, receives most veins of heart and ends in right atrium
      a. Great cardiac vein starts at apex, ascends in anterior interventricular sulcus, and ends in coronary sinus
      b. Small cardiac vein on right side of coronary sulcus, opens into end of sinus
         i. Right marginal: along right margin of heart
      c. Middle cardiac vein from apex ascends in posterior longitudinal sulcus to sinus
      d. Posterior vein of left ventricle on diaphragmatic surface, to sinus
      e. Oblique vein of left atrium descends on dorsum of atrium to sinus
   2. Veins opening directly into atrium (thebesian veins)
      a. Anterior (3 or 4) from right ventricular wall
      b. Smallest (very small) from cardiac muscle, usually ending in right atrium, but some terminate in ventricles

IV. Clinical considerations
A. OBSTRUCTION OF A CORONARY ARTERY can lead to anoxia of the heart area supplied, resulting in spasmodic contractions (heart attack), and eventually in death
B. 55–60% of cases, supplies SA node
C. 85–90% of cases, supplies AV node

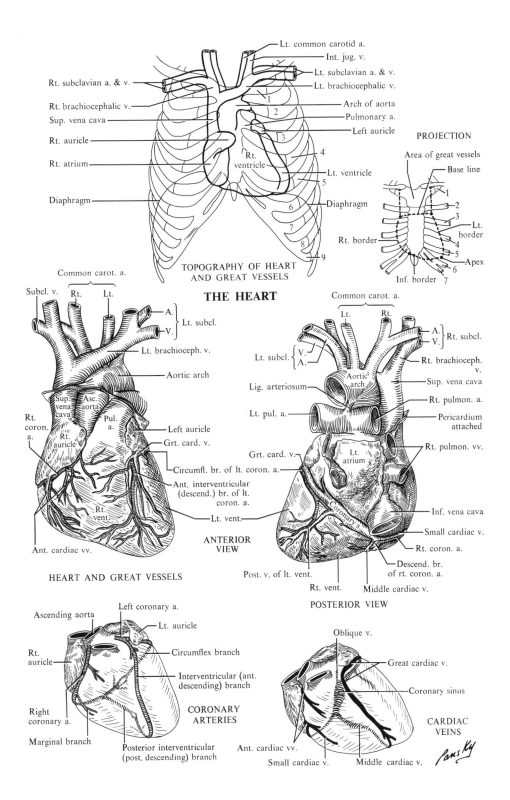

Lt. common carotid a.
Int. jug. v.
Rt. subclavian a. & v.
Lt. subclavian a. & v.
Lt. brachiocephalic v.
Rt. brachiocephalic v.
Arch of aorta
Sup. vena cava
Pulmonary a.
Rt. auricle
Left auricle
Rt. atrium
Rt. ventricle
Lt. ventricle
Diaphragm
Diaphragm

PROJECTION

Area of great vessels
Base line
Lt. border
Rt. border
Apex
Inf. border

TOPOGRAPHY OF HEART
AND GREAT VESSELS

THE HEART

Common carot. a.
Subcl. v.   Rt.   Lt.
A.
V.   } Lt. subcl.
Lt. brachioceph. v.
Aortic arch
Sup. vena cava
Asc. aorta
Rt. coron. a.
Rt. auricle
Pul. a.
Left auricle
Grt. card. v.
Circumfl. br. of lt. coron. a.
Ant. interventricular
(descend.) br. of lt. coron. a.
Rt. vent.
Lt. vent.
Ant. cardiac vv.

ANTERIOR VIEW

Common carot. a.
Lt.   Rt.
A.
V.   } Rt. subcl.
Lt. subcl. {
V.
A.
Rt. brachioceph. v.
Aortic arch
Lig. arteriosum
Sup. vena cava
Lt. pul. a.
Rt. pulmon. a.
Pericardium attached
Grt. card. v.
Lt. atrium
Rt. pulmon. vv.
Inf. vena cava
Small cardiac v.
Rt. coron. a.
Descend. br. of rt. coron. a.
Post. v. of lt. vent.
Rt. vent.
Middle cardiac v.

POSTERIOR VIEW

HEART AND GREAT VESSELS

Ascending aorta
Left coronary a.
Lt. auricle
Rt. auricle
Circumflex branch
Interventricular (ant. descending) branch
Right coronary a.
CORONARY ARTERIES
Marginal branch
Posterior interventricular (post. descending) branch

Oblique v.
Great cardiac v.
Coronary sinus
CARDIAC VEINS
Ant. cardiac vv.
Small cardiac v.
Middle cardiac v.

Pansky

-285-

129. THE PERICARDIUM

I. Components
A. FIBROUS: tough, fibrous sac, closed above by attachments to the great vessels. Outer layer is attached to:
  1. Manubrium and xiphoid process of sternum by the superior and inferior *sterno-pericardial ligaments*
  2. Vertebral column by the *vertebropericardial ligament*
  3. Diaphragm by a wide area of fibers attached to the central tendon of the diaphragm. The *phrenicopericardial ligament* is a thickened band near the inferior vena cava
  4. Pleura on either side where the fibrous pericardium and the mediastinal pleura meet. This union is loose and permits the phrenic nerve and vessels to run between the pleura and the pericardium
B. SEROUS: smooth membrane with a mesothelial layer that lines the fibrous sac and also covers the surface of the heart
  1. Visceral pericardium (epicardium) covers the entire surface of the heart and extends along its great vessels for 3 cm, where it is reflected onto the inner surface of the fibrous pericardium
  2. Parietal pericardium lines the inner surface of the fibrous pericardium

II. Pericardial cavity: the potential space between the parietal and visceral pericardium. The pericardial surfaces are in contact, covered with a watery fluid to allow for freedom of heart movement during its contractions

III. The mesocardia: reflections of the epicardium along the great vessels and onto the fibrous pericardial sac
A. ARTERIAL: tubular prolongation on aorta and pulmonary trunk
B. Venous: extension along the venae cavae and pulmonary veins
  1. Due to the arrangement of the veins, an inverted U-shaped pocket called the *oblique pericardial sinus* is formed
  2. The narrow, pericardial-lined groove between IIIA and B, above, is the *transverse pericardial sinus*

IV. Clinical considerations
A. IN INJURY OR DISEASE, fluid may accumulate in the pericardial cavity so that the parietal and visceral layers become greatly separated. This may be withdrawn by anterior thoracic approach, just above the 6th costal cartilage in the 5th interspace, one finger-breadth from the left side of the sternum; by the abdominal approach, passing needle upward and to the left between the xiphoid process and the left costal margin, directed toward the nipple; posterior thoracic approach, the needle being inserted over the 8th rib, midway between the inferior angle of scapula and midline, directing it toward the middle of the chest
B. IN SOME CASES OF CORONARY OCCLUSION, attempts have been made to create adhesions between the layers of the pericardium in the hope that the vessels in the pericardium could help supply the heart
C. THE TERM PERICARDIAL SAC refers to the fibrous plus parietal layer of serous pericardium. It does not have a very rich blood supply. It surrounds the heart loosely, and if the heart gradually enlarges so does the sac.
D. AS A RESULT OF SEVERE PERICARDITIS, the pericardium becomes greatly thickened and adherent to the heart (constrictive pericarditis)

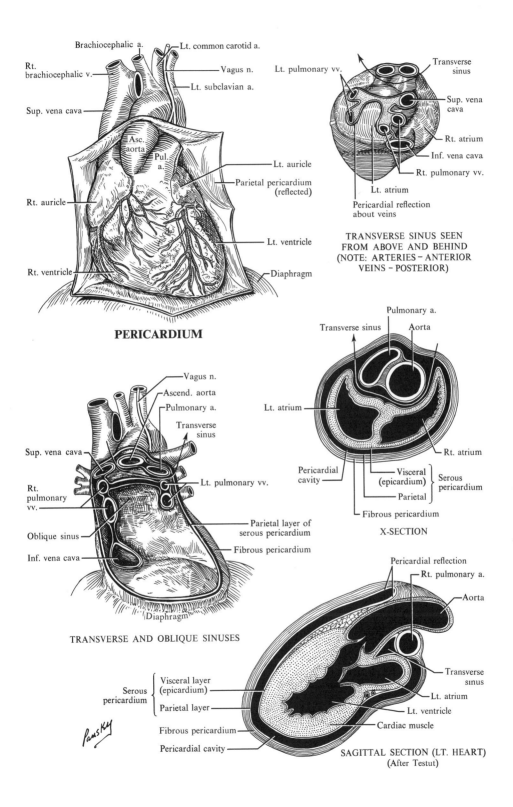

Brachiocephalic a.
Rt. brachiocephalic v.
Lt. common carotid a.
Vagus n.
Lt. subclavian a.
Sup. vena cava
Asc. aorta
Pul. a.
Rt. auricle
Lt. auricle
Parietal pericardium (reflected)
Lt. ventricle
Rt. ventricle
Diaphragm

Lt. pulmonary vv.
Transverse sinus
Sup. vena cava
Rt. atrium
Inf. vena cava
Rt. pulmonary vv.
Lt. atrium
Pericardial reflection about veins

TRANSVERSE SINUS SEEN
FROM ABOVE AND BEHIND
(NOTE: ARTERIES – ANTERIOR
VEINS – POSTERIOR)

PERICARDIUM

Vagus n.
Ascend. aorta
Pulmonary a.
Transverse sinus
Sup. vena cava
Rt. pulmonary vv.
Oblique sinus
Inf. vena cava
Lt. pulmonary vv.
Parietal layer of serous pericardium
Fibrous pericardium
Diaphragm

TRANSVERSE AND OBLIQUE SINUSES

Pulmonary a.
Transverse sinus
Aorta
Lt. atrium
Rt. atrium
Pericardial cavity
Visceral (epicardium)
Parietal
Serous pericardium
Fibrous pericardium

X-SECTION

Pericardial reflection
Rt. pulmonary a.
Aorta
Transverse sinus
Lt. atrium
Lt. ventricle
Cardiac muscle

Visceral layer (epicardium)
Parietal layer
Serous pericardium
Fibrous pericardium
Pericardial cavity

SAGITTAL SECTION (LT. HEART)
(After Testut)

Pansky

130. INTERIOR OF THE HEART

I. Right atrium: divided into 2 parts by ridge, the *crista terminalis* (creates an external impression between venae cavae, the *sulcus terminalis*)
A. AURICLE: blind pocket
1. Lined by parallel ridges, the *pectinate muscles*
B. SINUS OF VENAE CAVAE (VENARUM): principal cavity
1. Lining is smooth
2. Openings
a. Superior vena cava from above
b. Inferior vena cava from below
i. Valve of inferior vena cava, a crescent-shaped fold along anterior and left margin of cava
c. Coronary sinus opens between orifice of inferior vena cava and atrioventricular orifice
i. Valve, a fold attached along right and inferior borders of opening
d. Small coronary veins (see p. 284)
e. Atrioventricular orifice, oval-shaped opening into right ventricle
C. INTERATRIAL SEPTUM: the posterior wall of atrium
1. Fossa ovalis, oval depression, where septum is thin
2. Limbus of the fossa ovalis: the thick margin of fossa, distinct except below

II. Left atrium: like the right has a smooth-walled portion and a smaller muscular auricle. The 4 pulmonary veins open into the smooth-walled portion

III. Right ventricle occupies most of sternocostal surface from coronary to anterior longitudinal sulci
A. ATRIOVENTRICULAR (TRICUSPID) VALVE (see p. 290)
B. PARTS, separated by ridge, supraventricular crest
1. Conus arteriosus (infundibulum): smooth cephalic part leading into pulmonary trunk
2. Ventricle proper
a. Trabeculae carneae: irregular bundles of muscle projecting on inner surface. Of 3 types:
i. Ridges along wall
ii. Extension across ventricular cavity: septomarginal trabecula (moderator band), an enlarged bundle near apex
iii. Papillary muscles, conical projections into cavity. *Anterior* (largest) partly from anterior and septal walls; *posterior* from numerous parts on posterior wall; and *septal* near septal end of supraventricular crest
b. Chordae tendineae: small tendinous bands from ventricular wall, at apices of papillary muscles, attached to cusps of atrioventricular valves
c. Interventricular septum: set obliquely; the right convex side encroaches on the right ventricle
i. Muscular part: thickest, major part of septum
ii. Membranous part: thin upper part near atrium

IV. Left ventricle: small part of sternocostal and much of diaphragmatic surface
A. LEFT ATRIOVENTRICULAR (MITRAL OR BICUSPID) VALVE (see p. 290)
B. STRUCTURE SIMILAR TO RIGHT except trabeculae carneae more dense. Two large sets of papillary muscles, *anterior* and *posterior,* are attached to their respective walls
C. Aorta with its semilunar valve

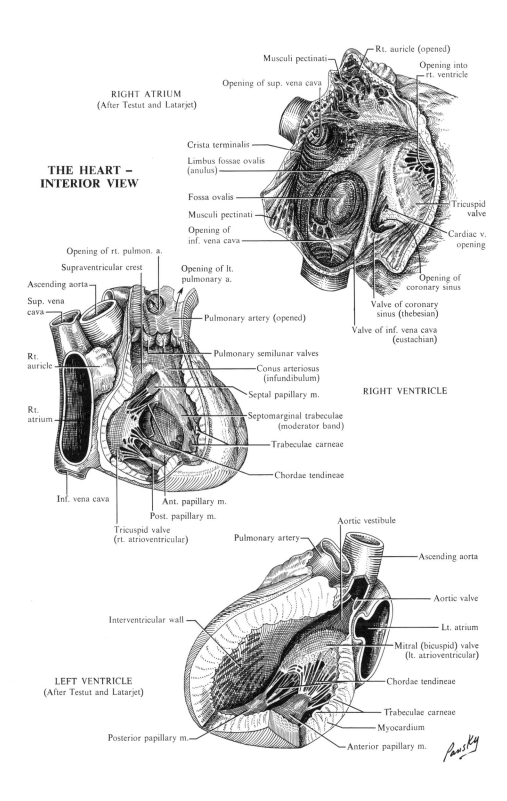

RIGHT ATRIUM
(After Testut and Latarjet)

**THE HEART –
INTERIOR VIEW**

Rt. auricle (opened)

Musculi pectinati

Opening into
rt. ventricle

Opening of sup. vena cava

Crista terminalis

Limbus fossae ovalis
(anulus)

Fossa ovalis

Tricuspid
valve

Musculi pectinati

Cardiac v.
opening

Opening of
inf. vena cava

Opening of
coronary sinus

Valve of coronary
sinus (thebesian)

Valve of inf. vena cava
(eustachian)

Opening of rt. pulmon. a.

Supraventricular crest

Opening of lt.
pulmonary a.

Ascending aorta

Sup. vena
cava

Pulmonary artery (opened)

RIGHT VENTRICLE

Pulmonary semilunar valves

Rt.
auricle

Conus arteriosus
(infundibulum)

Septal papillary m.

Rt.
atrium

Septomarginal trabeculae
(moderator band)

Trabeculae carneae

Chordae tendineae

Inf. vena cava

Ant. papillary m.

Post. papillary m.

Tricuspid valve
(rt. atrioventricular)

Aortic vestibule

Pulmonary artery

Ascending aorta

Aortic valve

Interventricular wall

Lt. atrium

Mitral (bicuspid) valve
(lt. atrioventricular)

Chordae tendineae

LEFT VENTRICLE
(After Testut and Latarjet)

Trabeculae carneae

Myocardium

Posterior papillary m.

Anterior papillary m.

Pansky

131. THE VALVES OF THE HEART

I. **Surface projections of valves** (for valve sounds, see opposite page)
A. PULMONARY: upper border of angle of third costosternal union, on left
B. AORTIC: slightly below and medial to above, opposite articulation of third costosternal joint
C. LEFT ATRIOVENTRICULAR: just to left of midline, opposite fourth costosternal articulation
D. RIGHT ATRIOVENTRICULAR: near sternal margin on right, opposite fourth interspace

II. **Structure**
A. ATRIOVENTRICULAR
1. Right (tricuspid): set in oval aperture 4 cm in longest diameter. Composed of 3 fibrous cusps, thick near attached border, thin near free edges. All cusps are attached to ring (anulus fibrosus) of dense tissue around orifice
   a. Anterior cusp: largest, attached to ventral wall near infundibulum. Receives chordae tendineae from anterior and septal papillary muscles
   b. Posterior cusp: from curved ventricular wall where sternocostal and diaphragmatic surfaces become continuous. Receives chordae from both anterior and posterior papillary muscles
   c. Septal cusp (medial): from septal wall. Receives chordae from posterior and septal papillary muscles
2. Left (bicuspid, mitral): set in ring smaller than right. Composed of 2 cusps
   a. Anterior (aortic): placed anteriorly, attached near aortic orifice. Receives chordae from both papillary muscles of ventricle
   b. Posterior: smaller, attached to posterior wall. Receives chordae from both papillary muscles
B. ARTERIAL
1. Pulmonary semilunar: composed of 3 semilunar cusps (anterior, right, and left) set in a circular orifice at top of the infundibulum
   a. Each cusp attached along a curve that is convex caudally
   b. Sinus: pocket between (behind) cusp and vessel wall at origin of pulmonary artery
   c. Commissure: point where attachments of adjacent cusps come together
   d. Nodule: a thickened area in center of free margin of cusp
   e. Lunula: crescent-shaped, thin part of cusp lying between free margin and a thickened band of fibers in the cusp that arches downward from nodule to commissure
2. Aortic semilunar (right, left, and posterior cusps), same structure as above except nodules thicker, lunulae more distinct, coronary aa. arise from 2 sinuses

III. **Clinical considerations**
A. IT HAS BEEN NOTED that, in general, the free margins of all valves (those that approximate when the valves are closed) are thin. If these become thickened and inflexible or damaged through disease, a close fit will be impossible. Thus, blood can flow back into the chamber from which it has come
B. STENOSIS: a narrowing or constriction of an orifice. Most frequently occurs in pulmonary or aortic opening
C. HEART SOUNDS: 2 audible sounds occur during each heartbeat: lubb-dupp. The first is low-pitched and of long duration caused by closure of both atrioventricular valves and the contraction of the ventricular muscle. The second is short, sharp, and high-pitched and is caused by closure of the semilunar valves

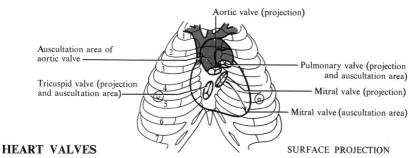

Aortic valve (projection)

Auscultation area of aortic valve

Tricuspid valve (projection and auscultation area)

Pulmonary valve (projection and auscultation area)

Mitral valve (projection)

Mitral valve (auscultation area)

HEART VALVES   SURFACE PROJECTION

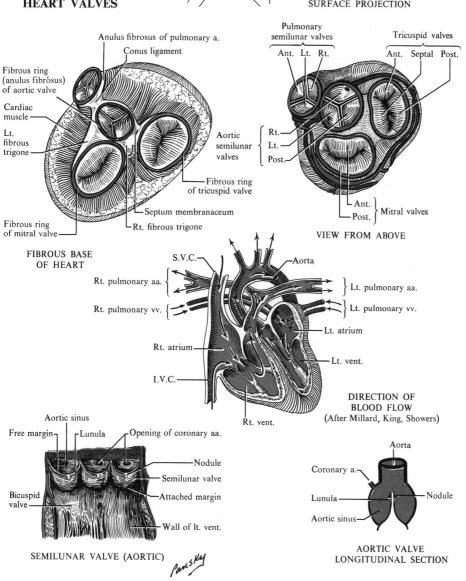

Anulus fibrosus of pulmonary a.

Conus ligament

Fibrous ring (anulus fibrosus) of aortic valve

Cardiac muscle

Lt. fibrous trigone

Aortic semilunar valves — Rt. / Lt. / Post.

Fibrous ring of tricuspid valve

Septum membranaceum

Rt. fibrous trigone

Fibrous ring of mitral valve

FIBROUS BASE OF HEART

Pulmonary semilunar valves — Ant. Lt. Rt.

Tricuspid valves — Ant. Septal Post.

Ant. / Post. } Mitral valves

VIEW FROM ABOVE

S.V.C.

Aorta

Rt. pulmonary aa.

Lt. pulmonary aa.

Rt. pulmonary vv.

Lt. pulmonary vv.

Lt. atrium

Rt. atrium

Lt. vent.

I.V.C.

Rt. vent.

DIRECTION OF BLOOD FLOW
(After Millard, King, Showers)

Aortic sinus

Free margin

Lunula

Opening of coronary aa.

Nodule

Semilunar valve

Bicuspid valve

Attached margin

Wall of lt. vent.

SEMILUNAR VALVE (AORTIC)

Pansky

Aorta

Coronary a.

Nodule

Lunula

Aortic sinus

AORTIC VALVE LONGITUDINAL SECTION

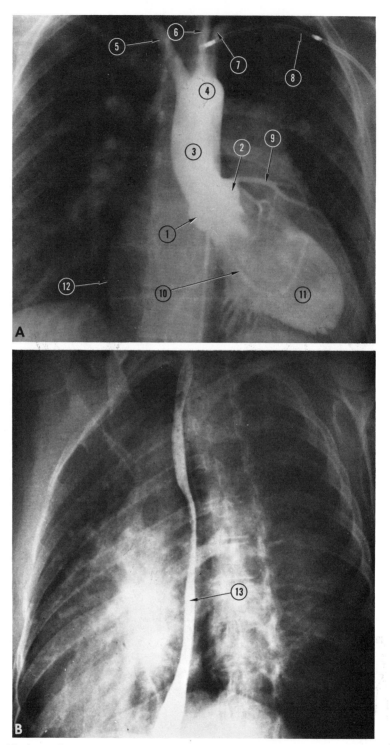

FIGURE 23. **Ascending aorta, coronary arteries, and esophagus. A, Ascending aorta and coronaries; B, esophagus.** *1,* Right aortic sinus; *2,* left aortic sinus; *3,* ascending aorta; *4,* arch of aorta; *5,* brachiocephalic artery; *6,* common carotid artery; *7,* subclavian artery; *8,* catheter; *9,* left coronary artery; *10,* right coronary artery; *11,* cavity of left ventricle; *12,* right margin of heart; *13,* esophagus.

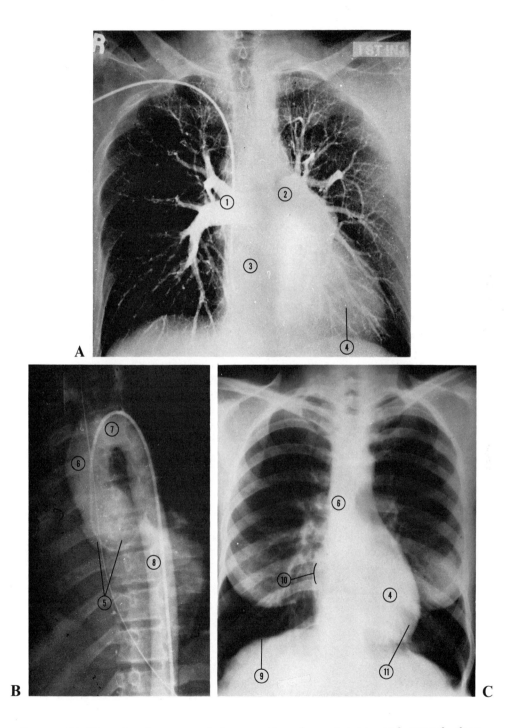

FIGURE 24. **Normal pulmonary arteriogram, thoracic aortogram, and normal chest. A, Normal pulmonary arteriogram, AP view; B, thoracic aortogram; C, normal chest.** *1,* Right pulmonary artery; *2,* left pulmonary artery; *3,* right ventricle; *4,* left ventricle; *5,* aortic valves; *6,* ascending aorta; *7,* aortic arch; *8,* thoracic aorta; *9,* diaphragm; *10,* right border of heart; *11,* apex of heart.

132. THE MEDIASTINUM

I. Definition: a thick partition in the thorax, bounded laterally by the pleurae, anteriorly by the sternum, and posteriorly by the vertebral column

II. Divisions
  A. SUPERIOR: above by plane of first rib, below by horizontal line at level of sternal angle. This passes through the disk between fourth and fifth thoracic vertebrae
  B. INFERIOR: above by superior mediastinum, below by diaphragm
    1. Anterior: above by lower border of superior, below by diaphragm, anteriorly by body of sternum and transverse thoracic muscle, posteriorly by pericardium
    2. Middle: above and below as the anterior. Its posterior and anterior limits are the fibrous pericardium
    3. Posterior: same upper and lower limits as anterior and middle. Lies between the pericardium and the vertebral column. Caudally extends to level of twelfth thoracic vertebra behind diaphragm

III. Superior mediastinum: contents and relations
  A. MOST ANTERIOR, the origins of the sternohyoid and sternothyroid muscles
  B. THYMUS GLAND, usually only fibrous and fatty remnants
  C. VESSELS
    1. Arteries: in lower part, the arch of the aorta; above the arch are the brachiocephalic, left common carotid, and left subclavian arteries
    2. Veins: to the right side and below is the superior vena cava; above and anterior to the arteries are the brachiocephalic veins. Termination of azygos v.
    3. Thoracic duct behind the aortic arch and subclavian artery
  D. VISCERA
    1. Trachea in midline, posterior to major vessels
    2. Esophagus posterior and slightly to left of trachea
  E. NERVES
    1. Both vagus nerves related to the brachiocephalic artery on the right and to the subclavian artery on the left
    2. Cardiac nerves
    3. Left recurrent nerve, near groove between trachea and esophagus
    4. Both phrenic nerves
  F. MOST POSTERIORLY, lower part of longus colli muscle

IV. Anterior mediastinum
  A. NO MAJOR STRUCTURES. Areolar tissue, transverse thoracic muscle, a few lymph nodes; small blood vessels and lymphatics

V. Middle mediastinum
  A. PERICARDIUM AND HEART (see p. 286)
  B. GREAT VESSELS
    1. Superior vena cava
    2. Ascending aorta
    3. Pulmonary trunk with the origins of the right and left pulmonary arteries
  C. NERVES
    1. Phrenic: most lateral, actually along the side of the pericardium and thus on the border of (rather than in) the middle mediastinum

VI. Posterior mediastinum (see p. 302)

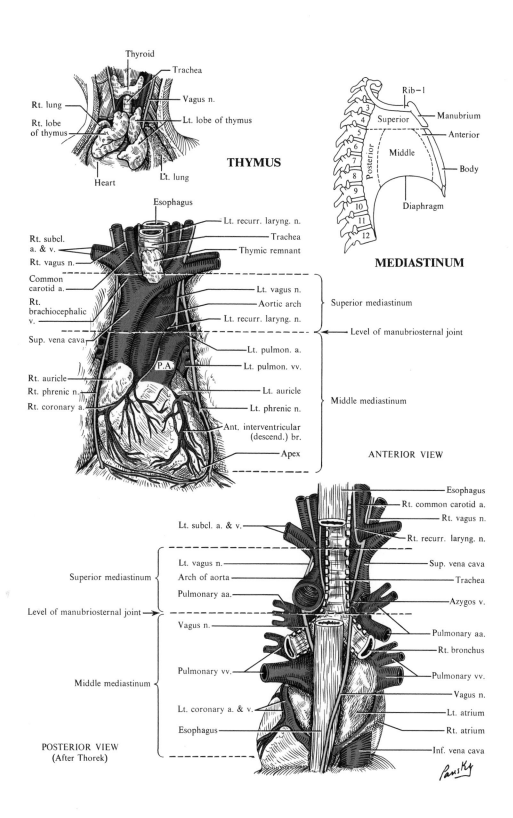

Thyroid

Trachea

Rt. lung

Rt. lobe of thymus

Vagus n.

Lt. lobe of thymus

Heart

Lt. lung

THYMUS

Rib-1

Manubrium

Superior

Anterior

Middle

Body

Posterior

Diaphragm

MEDIASTINUM

Esophagus

Lt. recurr. laryng. n.

Rt. subcl. a. & v.

Trachea

Rt. vagus n.

Thymic remnant

Common carotid a.

Rt. brachiocephalic v.

Lt. vagus n.

Aortic arch

Lt. recurr. laryng. n.

Superior mediastinum

Sup. vena cava

Level of manubriosternal joint

Lt. pulmon. a.

Lt. pulmon. vv.

P.A.

Rt. auricle

Rt. phrenic n.

Rt. coronary a.

Lt. auricle

Lt. phrenic n.

Middle mediastinum

Ant. interventricular (descend.) br.

Apex

ANTERIOR VIEW

Esophagus

Rt. common carotid a.

Rt. vagus n.

Lt. subcl. a. & v.

Rt. recurr. laryng. n.

Lt. vagus n.

Sup. vena cava

Superior mediastinum

Arch of aorta

Trachea

Pulmonary aa.

Azygos v.

Level of manubriosternal joint

Pulmonary aa.

Vagus n.

Rt. bronchus

Pulmonary vv.

Pulmonary vv.

Vagus n.

Middle mediastinum

Lt. coronary a. & v.

Lt. atrium

Esophagus

Rt. atrium

POSTERIOR VIEW
(After Thorek)

Inf. vena cava

Pansky

133. INNERVATION AND CONDUCTION SYSTEM OF HEART

I. **Intrinsic control of cardiac contraction:** cardiac muscle, under proper physiologic conditions has an inherent rhythmicity of contractions. For orderly contractions to occur in proper sequence, a specialized intrinsic system developed

A. SINOATRIAL NODE (SA): a small group of specialized heart muscle cells located near the crista terminalis, between superior vena cava and the right auricle. This initiates a beat above the intrinsic rhythm of the muscle and is known as the pacemaker
   1. Impulses set up in the SA node spread over both atria, causing atrial contraction, and carry the stimulus to the atrioventricular node

B. ATRIOVENTRICULAR NODE (AV): another specialized group of myocardial cells in the septal wall of the right atrium just cephalic to opening of coronary sinus. Because of thickness of the ventricular myocardium and for effectiveness, the ventricular contraction begins at apex. A conduction system from this node is required:
   1. Atrioventricular trunk (bundle of His): a band of specialized myocardium that descends to upper interventricular septum, under septal cusp of tricuspid valve and along lower border of membranous septum, near which it divides
      a. Right branch, under endocardium of septum to innervate the muscle of the septum, anterior papillary muscle, and sternocostal portion of the ventricular wall
      b. Left branch, under endocardium of left side of septum to supply the septum, papillary muscles, and all ventricular wall

II. **Extrinsic control:** for response to the ever-changing physiologic needs of the body, the intrinsic system is under the influence of the autonomic nervous system. The sympathetic division accelerates heart rate and dilates the coronary arteries to make available the increased amounts of oxygen and nutriments. The parasympathetic slows the heart and constricts coronary vessels. The autonomic supply is through 2 plexuses: *superficial,* lying below the aortic arch to right of the *ligamentum arteriosum,* and *deep,* behind arch of aorta over bifurcation of trachea. Both contain fibers and ganglia in which pregang. fibers of the parasympathetics synapse. The sympathetic fibers are postgang. and make no synapse here. Right vagal and sympathetics go to SA node, left vagal and sympathetics to AV node

A. CONNECTIONS AND DISTRIBUTION OF THE CARDIAC PLEXUSES
   1. Superficial: receives superior cervical cardiac branch of left sympathetic and inferior cervical cardiac branch of left vagus. Distributed to left anterior pulmonary plexus, right coronary plexus, and left half of deep cardiac plexus.
   2. Deep: receives superior, middle, and inferior cervical cardiac nerves from right cervical sympathetic; middle and inferior cervical cardiac nerves from left cervical sympathetic; superior and inferior cervical and thoracic branches of right vagus; superior and thoracic branches of left vagus; and visceral rami from upper 4 or 5 thoracic sympathetic ganglia. Distribution:
      a. Right half to right anterior pulmonary plexus, right atrium, and right and left coronary plexuses
      b. Left half to left atrium, left anterior pulmonary and coronary plexuses

III. **Clinical considerations**

A. THE CONDUCTION CURRENT and contraction waves can be amplified and recorded as the electrocardiogram. These readings can give valuable clues concerning the heart and its conduction system. Contraction of the atria causes the P wave; ventricular contraction causes the QRS wave; as some of the ventricular fibers relax, another reflection, the T wave, occurs; from T to P both atria and ventricles are inactive

B. THE SA NODE is supplied by the right or left coronary artery and the right vagus nerve

C. THE AV NODE is supplied by the right coronary artery but by the left vagus nerve

INNERVATION OF HEART

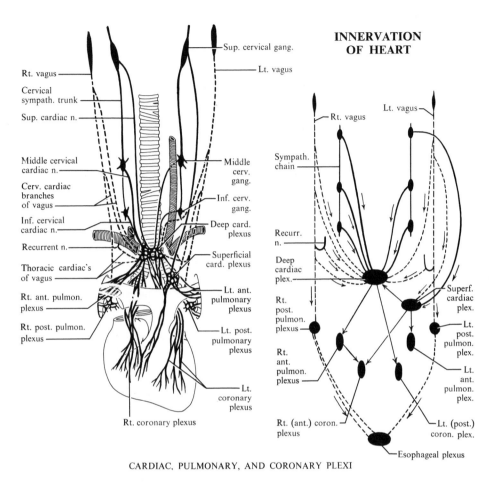

Rt. vagus
Cervical sympath. trunk
Sup. cardiac n.
Middle cervical cardiac n.
Cerv. cardiac branches of vagus
Inf. cervical cardiac n.
Recurrent n.
Thoracic cardiac's of vagus
Rt. ant. pulmon. plexus
Rt. post. pulmon. plexus
Rt. coronary plexus

Sup. cervical gang.
Lt. vagus
Middle cerv. gang.
Inf. cerv. gang.
Deep card. plexus
Superficial card. plexus
Lt. ant. pulmonary plexus
Lt. post. pulmonary plexus
Lt. coronary plexus

Rt. vagus
Lt. vagus
Sympath. chain
Recurr. n.
Deep cardiac plex.
Rt. post. pulmon. plexus
Rt. ant. pulmon. plexus
Rt. (ant.) coron. plexus

Superf. cardiac plex.
Lt. post. pulmon. plex.
Lt. ant. pulmon. plex.
Lt. (post.) coron. plex.
Esophageal plexus

CARDIAC, PULMONARY, AND CORONARY PLEXI

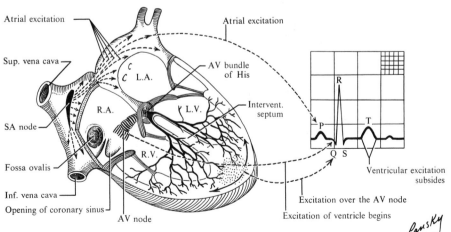

Atrial excitation
Sup. vena cava
SA node
Fossa ovalis
Inf. vena cava
Opening of coronary sinus
AV node

Atrial excitation
AV bundle of His
Intervent. septum
L.A.
R.A.
L.V.
R.V.

P Q R S T
Ventricular excitation subsides
Excitation over the AV node
Excitation of ventricle begins

CORRELATION OF ECG WITH CONDUCTION MECHANISM
(After Millard, King, and Showers)

134. TRACHEA AND BRONCHI

I. Trachea
A. EXTENT: from C6 to upper T5 vertebrae
B. DIMENSIONS: 11 cm long, 2.0–2.5 cm in diameter, larger in males
C. SHAPE: nearly cylindrical, slightly flattened posteriorly
D. STRUCTURE: consisting of hyaline cartilage, muscle, connective tissue, mucous membrane, and glands
   1. Cartilage: 16–20 "rings," incomplete posteriorly, filled in where trachea adjoins esophagus by muscle and connective tissue. Rings arranged in series with fibrous tissue between
      a. Sometimes 2 or more completely or partly fused; sometimes ends are bifid
      b. Last in series, at bifurcation of trachea is atypical, *carina* ending in an imperfect ring for each bronchus
E. RELATIONS
   1. In neck. Anteriorly, from cephalic to caudal: isthmus of thyroid, inferior thyroid veins, sternothyroid and sternohyoid muscles, cervical fascia and anastomotic veins between the 2 anterior jugular veins. Laterally: common carotid (in sheath), lobes of thyroid gland, inferior thyroid arteries, and recurrent nerves. Posteriorly: the esophagus
   2. In thorax. Anteriorly, from superficial to deep: manubrium of sternum, thymus, left brachiocephalic vein, arch of aorta, brachiocephalic and left common carotid arteries, and deep cardiac plexus. Laterally: on right, pleura, right vagus, and brachiocephalic artery; on left, left recurrent nerve, arch of aorta, left common carotid and subclavian arteries
F. SURFACE PROJECTIONS
   1. Bifurcation of trachea behind sternal angle
   2. Upper part of primary bronchus in second interspace at edge of sternum
   3. Arch of aorta crosses trachea behind manubrium, opposite first interspace

II. Primary bronchi
A. RIGHT: wider and shorter than left and makes smaller angle with axis of trachea
   1. Azygos vein arches over it; pulmonary artery at first below, then anterior to it
B. LEFT: longer, narrower, diverges from axis of trachea at greater angle
   1. Passes under arch of aorta and anterior to esophagus and thoracic aorta

III. Clinical considerations
A. FOREIGN OBJECTS are more likely to enter right bronchus because of its larger size and less acute angle
B. THE LEFT BRONCHUS, in spite of its greater length, is more difficult to handle surgically than the right one because of its vascular relations
C. EMERGENCY TRACHEOTOMY may be needed when rima glottidis has closed completely due to spasmodic contraction of laryngeal muscles after severe irritation of the mucosa (foreign bodies, alcohol, etc.) as in "swallowing the wrong way." An opening is made in midline, between lower parts of the muscular triangles. Below 4th tracheal ring, the isthmus of thyroid is avoided. Some venous bleeding from the anterior jugular system or inferior thyroid veins may be encountered

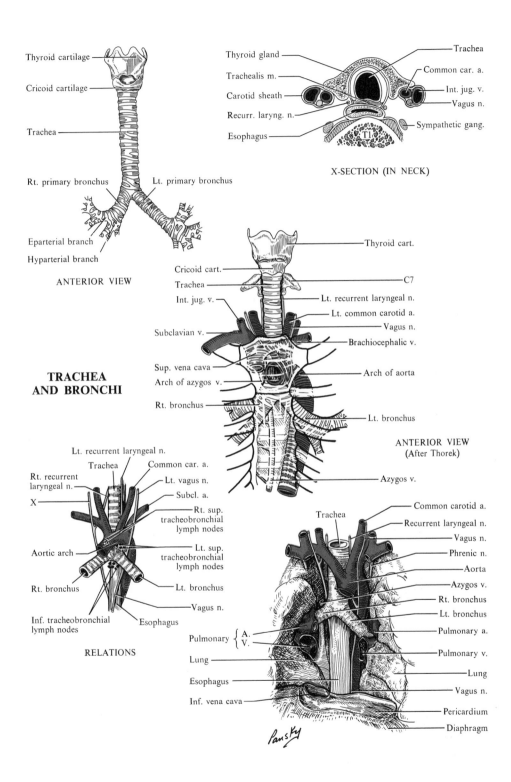

Thyroid cartilage
Cricoid cartilage
Trachea
Rt. primary bronchus
Lt. primary bronchus
Eparterial branch
Hyparterial branch

ANTERIOR VIEW

Thyroid gland
Trachealis m.
Carotid sheath
Recurr. laryng. n.
Esophagus
Trachea
Common car. a.
Int. jug. v.
Vagus n.
Sympathetic gang.
T1

X-SECTION (IN NECK)

**TRACHEA
AND BRONCHI**

Thyroid cart.
Cricoid cart.
Trachea
Int. jug. v.
Subclavian v.
Sup. vena cava
Arch of azygos v.
Rt. bronchus
C7
Lt. recurrent laryngeal n.
Lt. common carotid a.
Vagus n.
Brachiocephalic v.
Arch of aorta
Lt. bronchus
Azygos v.

ANTERIOR VIEW
(After Thorek)

Lt. recurrent laryngeal n.
Trachea
Common car. a.
Rt. recurrent
laryngeal n.
Lt. vagus n.
X
Subcl. a.
Rt. sup.
tracheobronchial
lymph nodes
Lt. sup.
tracheobronchial
lymph nodes
Aortic arch
Rt. bronchus
Lt. bronchus
Vagus n.
Inf. tracheobronchial
lymph nodes
Esophagus

RELATIONS

Trachea
Common carotid a.
Recurrent laryngeal n.
Vagus n.
Phrenic n.
Aorta
Azygos v.
Rt. bronchus
Lt. bronchus
Pulmonary { A.
             V.
Pulmonary a.
Pulmonary v.
Lung
Lung
Esophagus
Vagus n.
Inf. vena cava
Pericardium
Diaphragm

Pansky

135. THE ESOPHAGUS

I. **Extent:** from pharynx at level of cricoid cartilage (sixth cervical vertebra) to stomach at level of tenth thoracic vertebra

II. **Course:** generally vertically downward but with 2 curvatures. Deviates to left, remains to left through root of neck, gradually reaches midline at fifth thoracic vertebrae, again shifts toward left as it moves anteriorly toward esophageal hiatus in diaphragm. Has posteroanterior flexures corresponding to curves of vertebral column

III. **Relations**
  A. CERVICAL PORTION
    1. Anterior: trachea. The recurrent nerves ascend in the groove between the trachea and the esophagus
    2. Posterior: vertebral column and prevertebral musculature
    3. Lateral: thyroid gland and carotid sheath
  B. THORACIC PORTION
    1. Anterior: trachea, left bronchus, pericardium, left vagus, and diaphragm
    2. Posterior: vertebral column, longus colli muscle, right aortic intercostal arteries, thoracic duct, right vagus, and aorta (at caudal end)
    3. Left side: aortic arch, left subclavian artery, thoracic duct, pleura, and descending aorta
    4. Right side: azygos vein, pleura, right vagus, and thoracic duct

IV. **Blood supply**
  A. ARTERIES
    1. Cephalically: inferior thyroid branch of thyrocervical trunk
    2. Small branches of thoracic aorta (4 or 5)
    3. Bronchial arteries
    4. Ascending branch from left gastric artery
    5. Ascending branch from inferior phrenic artery
  B. VEINS: drain into inferior thyroid, azygos, hemiazygos, and gastric veins
    1. Drainage into gastric veins is one link between portal and systemic systems (see p. 348)

V. **Nerve supply**
  A. RECURRENT NERVES supply the striated muscle in upper third of organ
  B. PARASYMPATHETIC
    1. Esophageal branches: from vagi, above root of lung
    2. Esophageal plexus: below root of lung, vagi split into several bundles, which form a network around esophagus. Postganglionic sympathetic fibers also enter plexus
  C. SYMPATHETIC
    1. From upper 4 or 5 thoracic ganglia directly or through branches from cardiac or aortic plexuses
    2. From lower thoracic ganglia, possibly through splanchnic nerves

VI. **Clinical considerations**
  A. THE VEINS of the esophagus include a plexus on its surface and another in the submucosa. Both sets of vessels anastomose below the diaphragm with veins of the stomach (portal system veins). In consequence, they may become greatly dilated when there is portal hypertension
    1. One cause of death from portal hypertension is esophageal hemorrhage from rupture of varicose submucosal veins

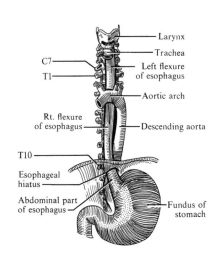

Larynx
Trachea
C7
T1
Left flexure of esophagus
Aortic arch
Rt. flexure of esophagus
Descending aorta
T10
Esophageal hiatus
Abdominal part of esophagus
Fundus of stomach

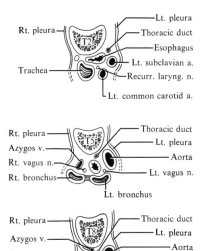

Rt. pleura
Trachea
Lt. pleura
Thoracic duct
Esophagus
Lt. subclavian a.
Recurr. laryng. n.
Lt. common carotid a.

Rt. pleura
Azygos v.
Rt. vagus n.
Rt. bronchus
Thoracic duct
Lt. pleura
Aorta
Lt. vagus n.
Lt. bronchus

Rt. pleura
Azygos v.
Rt. vagus n.
Esophagus
Thoracic duct
Lt. pleura
Aorta
Lt. vagus n.

ESOPHAGUS

Pleura
Rt. crus of diaphragm
Inf. vena cava
Liver
Liver
Lt. pleural cavity
Aorta
Stomach
Esophagus
Peritoneal cavity

X-SECTIONS

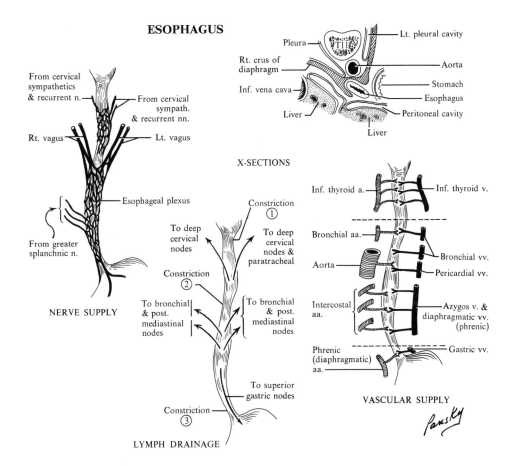

From cervical sympathetics & recurrent n.
From cervical sympath. & recurrent nn.
Rt. vagus
Lt. vagus
Esophageal plexus
From greater splanchnic n.

NERVE SUPPLY

Constriction ①
To deep cervical nodes
To deep cervical nodes & paratracheal
Constriction ②
To bronchial & post. mediastinal nodes
To bronchial & post. mediastinal nodes
To superior gastric nodes
Constriction ③

LYMPH DRAINAGE

Inf. thyroid a.
Inf. thyroid v.
Bronchial aa.
Bronchial vv.
Aorta
Pericardial vv.
Intercostal aa.
Azygos v. & diaphragmatic vv. (phrenic)
Phrenic (diaphragmatic) aa.
Gastric vv.

VASCULAR SUPPLY

Pansky

136. THE POSTERIOR MEDIASTINUM AND THORACIC SYMPATHETICS

I. Vessels

A. THORACIC AORTA (as well as the origins of the intercostal arteries): most posterior structure and situated on the left side

B. AZYGOS VEIN: most posterior structure to the right along with the terminations of the intercostal veins
   1. Hemiazygos vein ascends posteriorly on left and then crosses to join the azygos at the level of T9, passing behind the esophagus, thoracic duct, and aorta

C. THORACIC DUCT lies in front of the bodies of the vertebrae, to the left of midline

II. Viscera

A. ESOPHAGUS descends along the right side of the thoracic aorta and then crosses in front and to the left of the aorta

B. TRACHEA runs in the midline ventral to the esophagus to the level of the upper fifth thoracic vertebra, where it branches into primary bronchi
   1. Right principal bronchus (short and wide): azygos vein arches over it; pulmonary artery at first lies inferior to it and then ventral to it, and the pulmonary vein lies below the artery and bronchus
   2. Left principal bronchus (long and narrow) crosses anterior to esophagus and thoracic aorta, runs at first cephalic, then posterior, and finally caudal to the pulmonary artery. The pulmonary veins are anterior and caudal to the bronchus

C. NERVES
   1. Both vagi
      a. Right runs along trachea behind root of lung through the posterior pulmonary plexus and then to posterior surface of esophagus
      b. Left passes between aorta and left pulmonary artery, then posterior to the root of lung, and finally passes through the posterior pulmonary plexus and down on the anterior surface of esophagus
   2. Splanchnic nerves
      a. Greater arises in cord segments 5–9 and passes through the thoracic sympathetic ganglia and along vertebral border
      b. Lesser arises in cord segments 10–11 and passes through thoracic sympathetic ganglia
      c. Lowest (least, imus) (inconstant) passes through the last thoracic chain ganglion

III. Thoracic portion of sympathetic is not considered to lie in mediastinum

A. COMPONENTS
   1. Series of fusiform ganglia, up to 12 in number, corresponding to each thoracic nerve. Usually, there are fewer due to fusion of 2 or more. In upper thorax they lie against neck of ribs; more caudally they lie at sides of vertebrae
   2. Interganglionic cords interconnect the ganglia longitudinally. These cross the intercostal vessels anteriorly. The cords and ganglia are covered by costal pleura

B. PREGANGLIONIC FIBERS arise in the intermediolateral cell column of entire thoracic cord and leave cord through ventral roots of spinal nerves. They leave the spinal nerves through *white rami communicantes* to join thoracic chain. These may synapse in a ganglion at level of origin; they may pass cephalad or caudad, terminating in other ganglia of chain; or they may ascend to cervical trunk or descend into abdomen. Some merely pass through the thoracic chain and go directly to viscera

C. POSTGANGLIONICS: from the thoracic chain they may go as *gray rami* back to the spinal nerves or reach the cardiac, pulmonary, or esophageal plexuses directly

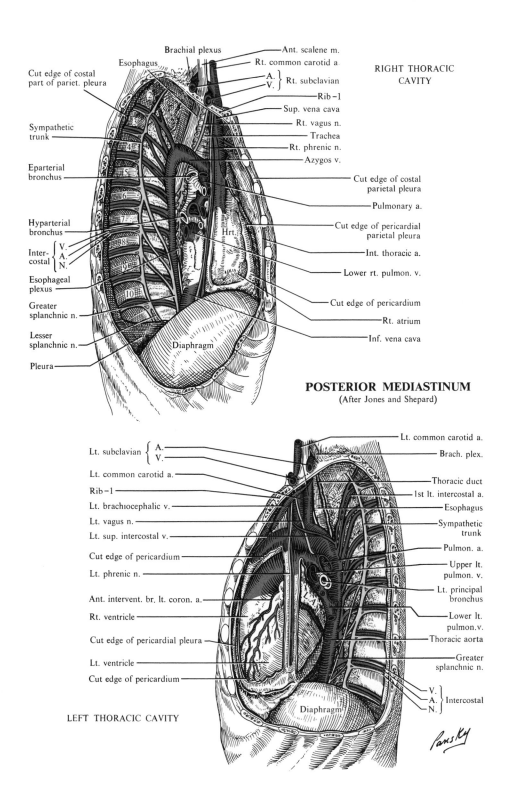

Brachial plexus
Esophagus
Cut edge of costal part of pariet. pleura
Sympathetic trunk
Eparterial bronchus
Hyparterial bronchus
Inter-costal { V. A. N. }
Esophageal plexus
Greater splanchnic n.
Lesser splanchnic n.
Pleura

Ant. scalene m.
Rt. common carotid a.
A. } Rt. subclavian
V.
Rib-1
Sup. vena cava
Rt. vagus n.
Trachea
Rt. phrenic n.
Azygos v.
Cut edge of costal parietal pleura
Pulmonary a.
Cut edge of pericardial parietal pleura
Int. thoracic a.
Lower rt. pulmon. v.
Cut edge of pericardium
Rt. atrium
Inf. vena cava

Hrt.

Diaphragm

RIGHT THORACIC CAVITY

POSTERIOR MEDIASTINUM
(After Jones and Shepard)

Lt. subclavian { A. V. }
Lt. common carotid a.
Rib-1
Lt. brachiocephalic v.
Lt. vagus n.
Lt. sup. intercostal v.
Cut edge of pericardium
Lt. phrenic n.
Ant. intervent. br. lt. coron. a.
Rt. ventricle
Cut edge of pericardial pleura
Lt. ventricle
Cut edge of pericardium

Lt. common carotid a.
Brach. plex.
Thoracic duct
1st lt. intercostal a.
Esophagus
Sympathetic trunk
Pulmon. a.
Upper lt. pulmon. v.
Lt. principal bronchus
Lower lt. pulmon.v.
Thoracic aorta
Greater splanchnic n.
V. A. N. } Intercostal

Diaphragm

LEFT THORACIC CAVITY

Pansky

-303-

137. THE VENOUS SYSTEM OF THE THORACIC WALL

I. Sympathetic trunk (see p. 302)

II. Veins

A. AZYGOS
  1. Origin: in abdomen from the right ascending lumbar vein
  2. Course: passes into thorax through aortic hiatus, ascends along right side of vertebral column; at level of fourth thoracic vertebra it arches over the root of the lung to end in the superior vena cava
  3. Tributaries: subcostal; all right intercostal veins, the cephalic 3 or 4 of which form a common stem—the *right superior (highest) intercostal vein;* esophageal, pericardial, rt. bronchial

B. HEMIAZYGOS
  1. Origin: from the left ascending lumbar vein of abdomen
  2. Course: through crus of diaphragm, ascends along left side of vertebral column to ninth thoracic vertebra, crosses this behind the aorta, esophagus, and thoracic duct to end in azygos vein
  3. Tributaries: lower 4 or 5 left intercostal veins, esophageal and mediastinal veins

C. ACCESSORY HEMIAZYGOS
  1. Origin: from 3 or 4 left intercostals above those drained by hemiazygos vein
  2. Course: descends along left side of vertebral column to end in hemiazygos vein
  3. Tributaries: 2 to 5 intercostal. Sometimes left bronchial vein

D. LEFT SUPERIOR (HIGHEST) INTERCOSTAL
  1. Origin: from the most cephalic intercostal veins on the left side
  2. Course: crosses arch of aorta and ends in the left brachiocephalic vein
  3. Tributaries: 2 to 5 intercostal veins, left bronchial vein

III. Thoracic duct (see p. 361): common trunk for entire body except right side of head, neck, thorax, and right upper extremity

A. ORIGIN: from the cephalic end of the cisterna chyli at the aortic hiatus

B. COURSE: ascends in posterior mediastinum on vertebral bodies, crossing the intercostal arteries and hemiazygos vein. Esophagus and pericardium lie in front. At the level of the fifth thoracic vertebra, it enters the superior mediastinum to pass behind aortic arch and subclavian artery between esophagus and left pleura. In the neck it arches above clavicle, crosses anterior to subclavian artery, vertebral vessels, and thyrocervical trunk. The left common carotid artery, vagus nerve, and internal jugular vein run anterior to it

C. TERMINATION: junction of left subclavian and internal jugular veins

D. TRIBUTARIES: from upper lumbar nodes, from lymph nodes in lower posterior intercostal spaces, nodes in posterior intercostal spaces of left side, posterior mediastinal nodes, left jugular and subclavian trunks

IV. Clinical considerations

A. SINCE THE THORACIC DUCT IS FILLED with colorless or white lymph depending on the fat content, it is not as easily seen as is a blood vessel, and it may be injured inadvertently during surgery in the posterior mediastinum. It may also be ruptured by violence
  1. Rupture or surgical section of the duct and the pleura overlying it allows the duct to empty its lymph into the pleural cavity (*chylothorax*) at the rate of about 60 to 190 ml per hour, resulting in collapse of the lung and pressure on the heart

B. THE HEMIAZYGOS SYSTEM varies markedly in its development from person to person

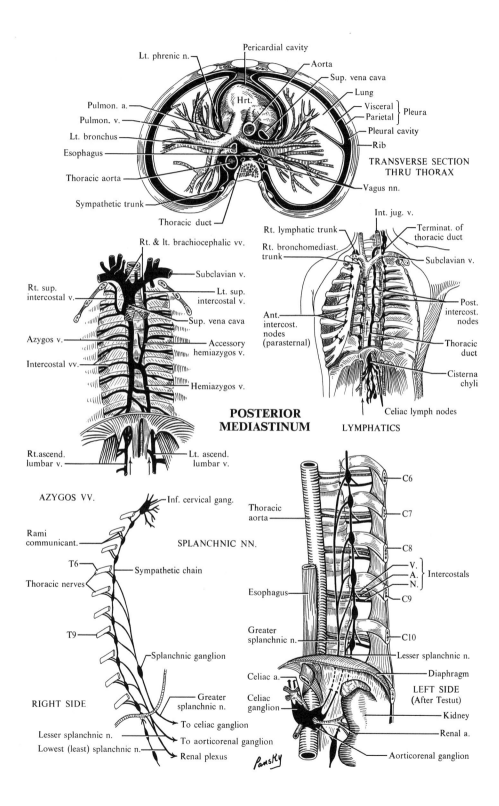

Lt. phrenic n.

Pericardial cavity

Aorta

Sup. vena cava

Lung

Pulmon. a.

Pulmon. v.

Visceral

Parietal ⎱ Pleura

Hrt.

Lt. bronchus

Pleural cavity

Esophagus

Rib

Thoracic aorta

TRANSVERSE SECTION
THRU THORAX

Sympathetic trunk

Vagus nn.

Thoracic duct

Int. jug. v.

Rt. lymphatic trunk

Terminat. of
thoracic duct

Rt. & lt. brachiocephalic vv.

Rt. bronchomediast.
trunk

Subclavian v.

Subclavian v.

Rt. sup.
intercostal v.

Lt. sup.
intercostal v.

Ant.
intercost.
nodes
(parasternal)

Post.
intercost.
nodes

Sup. vena cava

Azygos v.

Accessory
hemiazygos v.

Thoracic
duct

Intercostal vv.

Cisterna
chyli

Hemiazygos v.

POSTERIOR
MEDIASTINUM

Celiac lymph nodes

LYMPHATICS

Rt.ascend.
lumbar v.

Lt. ascend.
lumbar v.

AZYGOS VV.

Inf. cervical gang.

C6

Thoracic
aorta

C7

Rami
communicant.

SPLANCHNIC NN.

C8

T6

Sympathetic chain

V.

A.

N.

⎱ Intercostals

Thoracic nerves

Esophagus

C9

T9

C10

Splanchnic ganglion

Greater
splanchnic n.

Lesser splanchnic n.

Celiac a.

Diaphragm

RIGHT SIDE

Greater
splanchnic n.

Celiac
ganglion

LEFT SIDE
(After Testut)

To celiac ganglion

Kidney

Lesser splanchnic n.

To aorticorenal ganglion

Renal a.

Lowest (least) splanchnic n.

Renal plexus

Aorticorenal ganglion

Pansky

-305-

138. SUMMARY OF RESPIRATORY MOVEMENTS

I. Movements of the thorax in respiration are based on several anatomic peculiarities

A. THE ANTERIOR ENDS AND MIDDLE BODIES of ribs lie at a more caudal level than their posterior ends

B. THE CURVE OF EACH SUCCESSIVE RIB is greater than that of the one above it. Thus, when the ribs are pulled upward, the diameters of the thorax increase, resulting in increased volume and decreased pressure in the thorax

C. RIBS 1 and 2 are less mobile than the others and act as a unit with the manubrium. When tension is placed on these, the entire unit is raised, thus increasing the diameter in the superior portion. Raising and fixing of ribs 1 and 2 make possible greater elevation of the ribs below this level, an important feature in forced inspiration

II. The action of the muscles in respiration*

A. QUIET RESPIRATION

Raise Ribs	*Lower Ribs*
External intercostal muscles (ribs 3–10)	No muscle, passive

B. DEEP RESPIRATION

Raise Ribs	*Lower Ribs*
External intercostal muscles Scalene muscles Sternocleidomastoid muscles Levator costarum muscles Serratus posterior superior muscles	No muscle, passive

C. FORCED RESPIRATION

Raise Ribs	*Lower Ribs*
All the muscles listed above for deep respiration. The levator scapulae, trapezius, and rhomboids raise and fix the scapula so that the pectoral muscles and serratus anterior muscles can raise ribs	Quadratus lumborum, internal intercostals, subcostals, transverse thoracic, and serratus posterior inferior muscles

*For the action of the diaphragm in respiration, see p. 364.

BREATHING MECHANISM

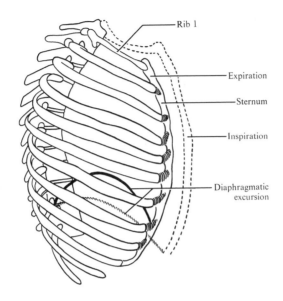

- Rib 1
- Expiration
- Sternum
- Inspiration
- Diaphragmatic excursion

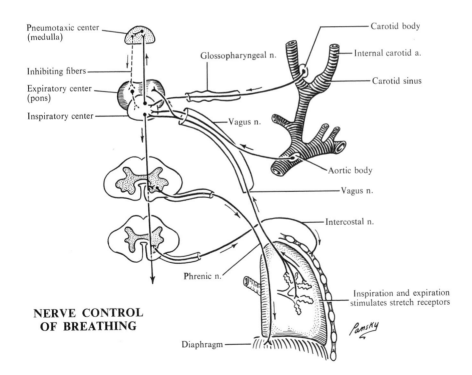

Pneumotaxic center (medulla)

Inhibiting fibers

Expiratory center (pons)

Inspiratory center

Glossopharyngeal n.

Vagus n.

Carotid body

Internal carotid a.

Carotid sinus

Aortic body

Vagus n.

Intercostal n.

Phrenic n.

Inspiration and expiration stimulates stretch receptors

NERVE CONTROL OF BREATHING

Diaphragm

Pansky

Abdomen and Pelvis

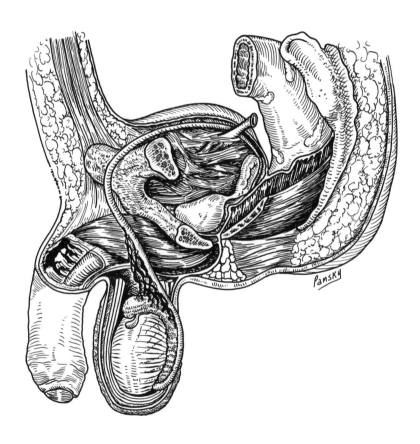

Pansky

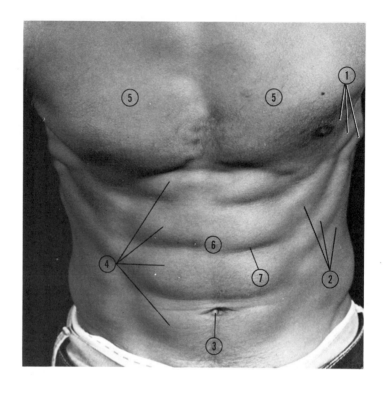

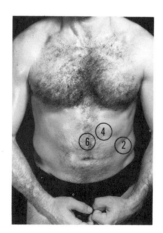 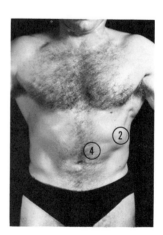

FIGURE 25. **Surface anatomy of abdomen.** *1,* Serratus anterior m.; *2,* external oblique m.; *3,* umbilicus; *4,* rectus abdominis m.; *5,* pectoralis major m.; *6,* linea alba; *7,* tendinous intersections.

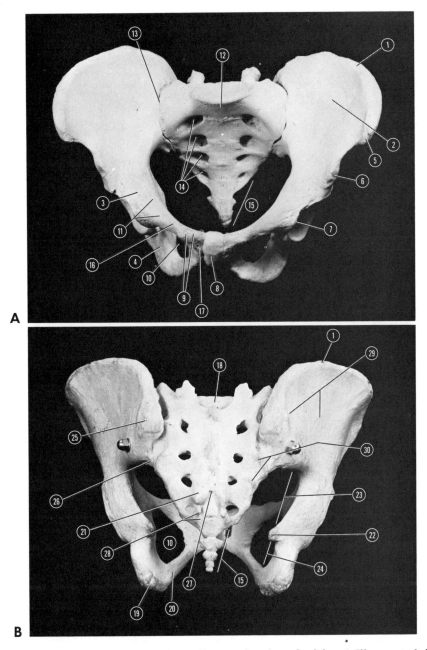

A

B

FIGURE 26. **A, The pelvis from in front; B, posterior view of pelvis.** *1,* Iliac crest; *2,* ilium; *3,* pubis; *4,* ischium; *5,* anterior superior iliac spine; *6,* anterior inferior iliac spine; *7,* acetabulum; *8,* symphysis pubis; *9,* pubic tubercle and crest; *10,* obturator foramen; *11,* iliopubic eminence and arcuate line; *12,* promontory; *13,* sacroiliac joint; *14,* pelvic or anterior sacral foramina; *15,* coccygeal bones; *16,* pecten pubis; *17,* inferior ramus of pubis; *18,* vertebral (sacral) canal; *19,* ischial tuberosity; *20,* ramus of ischium; *21,* articular tubercle; *22,* ischial spine; *23,* greater sciatic notch; *24,* lesser sciatic notch; *25,* posterior superior iliac spine; *26,* posterior inferior iliac spine; *27,* sacral hiatus; *28,* sacral and coccygeal cornua; *29,* gluteal lines; *30,* transverse tubercles of sacrum.

139. TOPOGRAPHY OF THE ABDOMEN

I. **Surface lines** are all vertical on thorax
A. MIDSTERNAL: in midline
B. LATERAL STERNAL: line along the lateral margin of sternum
C. MIDCLAVICULAR: line drawn caudally from point halfway between the middle of the jugular notch and tip of acromion
D. PARASTERNAL: line about halfway between A and C, above

II. **Surface lines:** on abdomen
A. HORIZONTAL
   1. Transpyloric: halfway between the jugular notch and the upper border of the symphysis; crosses tips of the ninth costal cartilages anteriorly and lower first lumbar vertebra posteriorly
   2. Transtubercular: at level of iliac tubercles; crosses the body of the fifth lumbar vertebra posteriorly
B. VERTICAL
   1. Midsagittal: directly in midline of body
   2. Lateral (2): drawn upward from middle of inguinal ligament

III. **Zones created by lines:** on abdomen
A. EPIGASTRIC: above transpyloric, between the 2 lateral lines
B. HYPOCHONDRIAC (right and left): above transpyloric, but lateral to the right and left vertical lines
C. UMBILICAL: between transpyloric, transtubercular, and the 2 vertical lines
D. LATERAL (right and left): between the 2 horizontal lines, but lateral to the vertical lines
E. HYPOGASTRIC: below transtubercular line and between vertical lines
F. ILIAC (right and left): below transtubercular, but lateral to the vertical lines

IV. **Surface projections and locations of certain major viscera**
A. STOMACH: *cardiac orifice* is behind the seventh costal cartilage, 2.5 cm lateral to the left border of sternum; *pylorus* is on the transpyloric line, 1 cm to the right of the midline
B. DUODENUM: *superior part* lies on the transpyloric line to the right of the midline; *duodenojejunal flexure* lies on the transpyloric line, 2.5 cm to the left of the midline
C. ILEOCECAL JUNCTION: just below and medial to the junction of the right vertical and transtubercular lines
D. BASE OF APPENDIX: on the right vertical line at the level of the anterior iliac spine
E. LIVER: upper limit of right lobe at xiphisternal junction, in midline; this line is continued to the right to the 5th costal cartilage in the midclavicular line and then curves to the right and down to the seventh rib at the side of the thorax. The right border continues downward to a point 1 cm below the costal arch. The upper left limit extends from the xiphisternal point to the left 6th costal cartilage, 5 cm from the midline. The lower limiting line runs upward, parallel to, and 1 cm below the thoracic border to the 9th costal cartilage. It then extends obliquely upward toward the left, crossing the midline just above the transpyloric plane to the 8th costal cartilage. From here, a line, slightly convex to the left, is drawn to the upper left limit
F. FUNDUS OF GALLBLADDER: just behind the 9th right costal cartilage

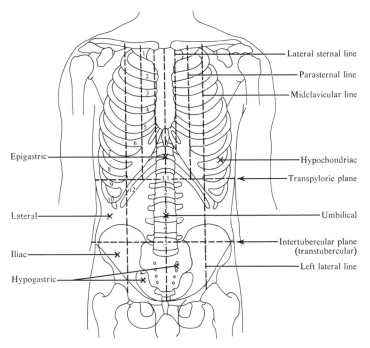

**SURFACE LINES AND REGIONS
OF ABDOMEN AND THORAX**

- Lateral sternal line
- Parasternal line
- Midclavicular line
- Epigastric
- Hypochondriac
- Transpyloric plane
- Lateral
- Umbilical
- Intertubercular plane (transtubercular)
- Iliac
- Left lateral line
- Hypogastric

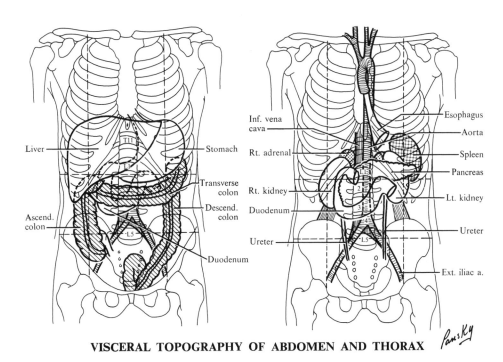

- Liver
- Stomach
- Transverse colon
- Descend. colon
- Ascend. colon
- Duodenum

- Inf. vena cava
- Esophagus
- Aorta
- Rt. adrenal
- Spleen
- Pancreas
- Rt. kidney
- Duodenum
- Lt. kidney
- Ureter
- Ureter
- Ext. iliac a.

VISCERAL TOPOGRAPHY OF ABDOMEN AND THORAX *Pansky*

140. SUPERFICIAL VEINS OF THORAX AND ABDOMEN

I. Thorax has very extensive superficial plexuses of veins, which communicate freely with those on abdominal wall
A. VENTRAL AND LATERAL CHEST WALL are drained by:
1. Lateral thoracic vein receives tributaries from the ventral wall, including the venous plexus of the mammary gland. Terminates in the axillary vein
2. Costoaxillary vein receives tributaries mainly from lateral aspect of chest. Terminates in the lateral thoracic vein
3. Thoracoepigastric vein drains lower lateral wall. Terminates in both the lateral thoracic and superficial epigastric veins
B. MEDIAL CHEST WALL is drained by vessels that penetrate the intercostal spaces to terminate in:
1. The internal thoracic (mammary) veins and their anterior intercostal tributaries
2. Superior epigastric vein

II. Abdomen has extensive venous plexuses, which communicate with other systems, especially with those of thorax, thigh, and genital region
A. INFERIOR LATERAL PARTS are drained by:
1. Superficial circumflex iliac vein, which ends in the great saphenous vein
B. MOST OF THE ABDOMINAL WALL is drained by:
1. Superficial epigastric vein, which also directly communicates with the lateral thoracic vein through the thoracoepigastric vein. It terminates in the great saphenous vein
C. LOWER AND MEDIAL PART OF THE ABDOMEN AND THE EXTERNAL GENITALIA are drained by:
1. Superficial external pudendal vein, which terminates in the great saphenous vein

III. Clinical considerations
A. OCCLUSION OF THE INFERIOR VENA CAVA: venous blood may bypass the block by flowing through the above-mentioned superficial venous plexuses, especially the link that exists between the superficial epigastric through the thoracoepigastric veins leading into the lateral thoracic vein and eventually terminating in the axillary system of veins
B. AT THE UMBILICUS, the superficial epigastric vein connects with veins that are in turn connected with the portal system. In consequence, when either the portal system or the inferior vena cava is obstructed at certain levels, the veins around the umbilicus that connect the two systems may become enlarged and tortuous, forming the *caput medusae*

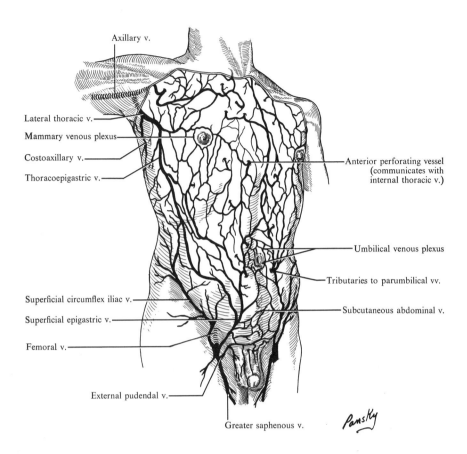

Axillary v.

Lateral thoracic v.

Mammary venous plexus

Costoaxillary v.

Thoracoepigastric v.

Anterior perforating vessel
(communicates with
internal thoracic v.)

Umbilical venous plexus

Tributaries to parumbilical vv.

Superficial circumflex iliac v.

Superficial epigastric v.

Subcutaneous abdominal v.

Femoral v.

External pudendal v.

Greater saphenous v.

Pansky

**SUPERFICIAL VEINS OF
ANTERIOR THORAX AND ABDOMEN**
(After Jones and Shepard)

141. THE MUSCLES OF THE ABDOMEN

I.

Name	Origin	Insertion	Action	Nerve
Extern. abdominal oblique	Lower 8 ribs	Aponeurosis to linea alba from xiphoid to symphysis	Compresses abdomen Flexes and laterally rotates spine Depresses ribs	Intercost. 8–12 Iliohypo-gastric Ilioinguinal
Int. abdominal oblique	Lat. inguinal lig. Ant. iliac crest Lumbar aponeurosis (of thoracolumbar fascia)	Lower border of ribs 9–12 With aponeurosis to linea alba Pubis and pectineal line	Same as above	Same as above
Transversus abdominis	Lat. inguinal lig. Iliac crest Thoracolumbar fascia Cartilages of lower ribs	Through aponeur. to linea alba Pubis and pectineal line	Compresses abdomen Depresses ribs	Intercost. 7–12 and same as above
Rectus abdominis	Crest of pubis Interpubic lig.	Cartilages of ribs 5–7 Xiphoid process	Compresses abdomen Flexes spine	Intercost. 7–12
Quadratus lumborum	Iliolumbar lig. Post. iliac crest	12th rib Tips of transverse processes L1–4	Fixes last 2 ribs Flexes spine laterally	T12, L1
Psoas major	Trans. processes of L1–5 Sides of bodies and fibrocarti-lage T12–L5	Lesser trochanter of femur	Flexes thigh and rotates it laterally Flexes spine and bends spine laterally	L2–3
Iliacus	Upper iliac fossa Iliac crest Ant. sacroiliac lig. Base of sacrum	Tendon of psoas maj. and directly to lesser trochanter	Flexes and laterally rotates thigh	Femoral (L2–3)
Pyramidalis	Pubis and ant. pubic lig.	Linea alba	Tenses linea alba	T12
Cremaster	Middle inguinal lig. Int. oblique m.	Tubercle and crest of pubis	Draws up testis	Genito-femoral

II. Special features

A. ACTION OF THE PSOAS MAJOR and iliacus muscles as rotators is controversial. When the limb is free, the iliopsoas probably acts as a lateral rotator of thigh; when the limb is fixed, it may help to swing the pelvis as the other limb goes forward

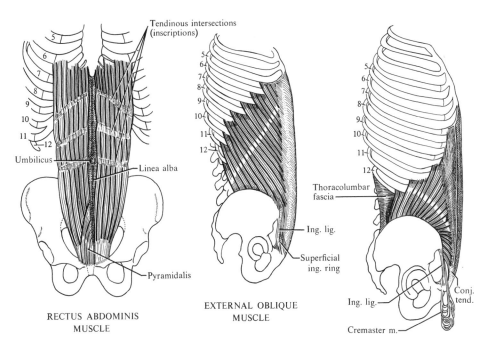

Tendinous intersections
(inscriptions)

5
6
7
8
9
10
11
12

Umbilicus

Linea alba

Pyramidalis

RECTUS ABDOMINIS
MUSCLE

5
6
7
8
9
10
11
12

Ing. lig.

Superficial
ing. ring

EXTERNAL OBLIQUE
MUSCLE

5
6
7
8
9
10
11
12

Thoracolumbar
fascia

Ing. lig.

Conj.
tend.

Cremaster m.

ABDOMINAL MUSCLES

INTERNAL OBLIQUE
MUSCLE

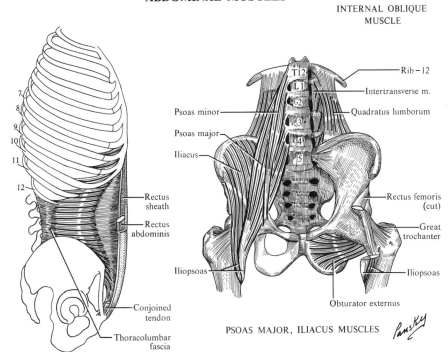

7
8
9
10
11
12

Rectus
sheath

Rectus
abdominis

Conjoined
tendon

Thoracolumbar
fascia

TRANSVERSUS ABDOMINIS MUSCLE

T12
L1
L2
L3
L4
L5

Rib – 12

Intertransverse m.

Quadratus lumborum

Psoas minor

Psoas major

Iliacus

Rectus femoris
(cut)

Great
trochanter

Iliopsoas

Iliopsoas

Obturator externus

PSOAS MAJOR, ILIACUS MUSCLES

Pansky

142. ABDOMINAL APONEUROSIS, RECTUS SHEATH, AND DEEP EPIGASTRIC VESSELS

I. **The aponeurosis of the external abdominal oblique muscle** is a strong membrane of collagenous fibers, which are the major part of the tendon of insertion of that muscle. It covers the entire abdomen and forms part of the anterior wall of the rectus sheath

A. WHERE THE FIBERS FROM BOTH SIDES interlace in the midline, the *linea alba* is formed

B. INGUINAL LIGAMENT is caudal to the border of the aponeurosis and extends from the anterior superior iliac spine to the pubic tubercle. It is the lowest part of the external oblique aponeurosis

C. LACUNAR LIGAMENT: where the medial end of the inguinal ligament is curved under the spermatic cord to attach to the pectineal line beyond the tubercle

D. REFLECTED INGUINAL LIGAMENT: where the fibers pass upward and medially from the bony attachment of the lacunar ligament to reach the linea alba

E. SUPERFICIAL (SUBCUTANEOUS) INGUINAL RING: triangular gap in the aponeurosis, above and lateral to the pubis for the passage of the spermatic cord. Its sides are the *lateral and medial crura*

II. **Rectus sheath**

A. BASIC FORMATION: composed of the aponeurosis of the external and internal oblique and transversus abdominis muscles. At the lateral border of the rectus abdominis, *posterior and anterior layers* are formed, which pass on either surface of the rectus muscle

B. BECAUSE OF THE PECULIARITIES OF ARRANGEMENT, the sheath is best described regionally
   1. Cephalic part (above arcuate line)
      a. Aponeurosis of the external oblique passes anterior
      b. Aponeurosis of internal oblique splits, 1 layer passes anterior, the other posterior
      c. Aponeurosis of transversus runs entirely posterior to rectus
      d. Transversalis fascia runs posterior to rectus
   2. Caudal part (below arcuate line)
      a. Aponeurosis of external oblique passes anterior to rectus
      b. Aponeurosis of internal oblique passes anterior to rectus
      c. Aponeurosis of transversus passes anterior to rectus
      d. Transversalis fascia remains posterior to rectus

C. ARCUATE LINE (SEMICIRCULAR LINE OF DOUGLAS). This is a line on the posterior wall of the rectus sheath below which only the transversalis fascia makes up the posterior layer of the rectus sheath

III. **Epigastric arteries**

A. SUPERIOR descends as 1 of the terminal branches of the internal thoracic artery, passes anterior to the diaphragm, and enters rectus sheath. In the sheath, it is at first behind and then enters the muscle. It anastomoses with the inferior epigastric artery

B. INFERIOR arises from the external iliac artery, above the inguinal ligament, courses upward and medially, pierces transversalis fascia, and enters the posterior rectus sheath. As it ascends, it lies medial to the deep inguinal (internal abdominal) ring

IV. **Special features**

A. THERE HAVE BEEN MANY DIVERSE OPINIONS on the exact anatomy and on the preferred names of structures in the region of insertion of the internal oblique and transversus muscles, particularly on whether a *conjoined tendon or falx inguinalis* (in the sense of a combination of the two mentioned muscles) is ever formed lateral to the lateral border of the rectus muscle

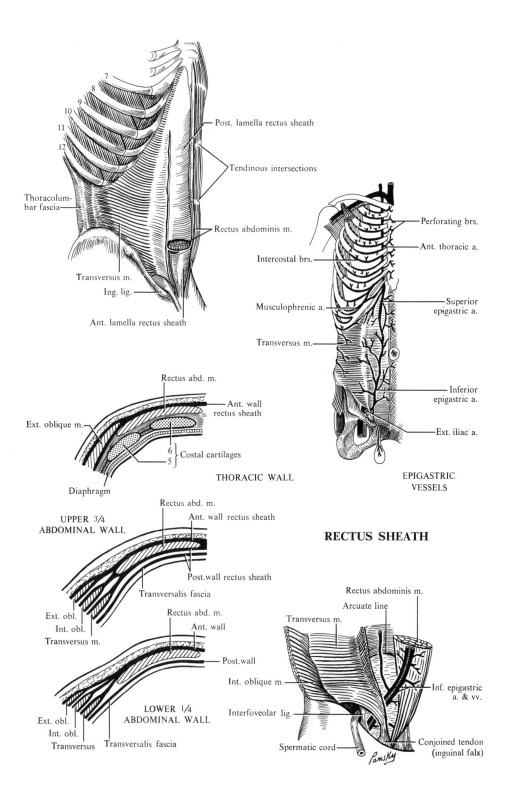

7
8
9
10
11
12

Post. lamella rectus sheath

Tendinous intersections

Thoracolumbar fascia

Rectus abdominis m.

Transversus m.

Ing. lig.

Ant. lamella rectus sheath

Perforating brs.

Ant. thoracic a.

Intercostal brs.

Musculophrenic a.

Transversus m.

Superior epigastric a.

Inferior epigastric a.

Ext. iliac a.

Rectus abd. m.

Ant. wall rectus sheath

Ext. oblique m.

6
5 } Costal cartilages

Diaphragm

THORACIC WALL

EPIGASTRIC VESSELS

UPPER 3/4 ABDOMINAL WALL

Rectus abd. m.

Ant. wall rectus sheath

Post.wall rectus sheath

Transversalis fascia

Ext. obl.
Int. obl.
Transversus m.

Rectus abd. m.

Ant. wall

Post.wall

Int. oblique m.

Ext. obl.
Int. obl.
Transversus   Transversalis fascia

LOWER 1/4 ABDOMINAL WALL

RECTUS SHEATH

Transversus m.

Rectus abdominis m.

Arcuate line

Inf. epigastric a. & vv.

Interfoveolar lig.

Spermatic cord

Conjoined tendon (inguinal falx)

Pansky

143. SPERMATIC CORD, INGUINAL CANAL, AND HERNIA

I. Special features of the lower abdominal wall (see p. 318)

A. CONJOINED TENDON (INGUINAL FALX): fibers of the internal abdominal oblique muscle, which arise from the lateral end of the inguinal ligament, arch over the spermatic cord (or round ligament), and terminate in a tendinous band, common to it and the transversus abdominis. It inserts on the pubis and medial pectineal line dorsal to the lacunar ligament

B. CREMASTER MUSCLE arises from the inguinal ligament as the lowermost fibers of the internal oblique muscle. Sends long loops along the spermatic cord into the scrotum. Inserts on the pubis

C. TRANSVERSALIS FASCIA: the deep fascia that lines the entire abdomen deep to the transversus abdominis muscle
   1. Below arcuate (semicircular) line is thickened and forms the posterior lamina of the rectus sheath

D. INTERFOVEOLAR LIGAMENT: a thickening of the transversalis fascia forming the medial boundary of the internal abdominal ring
   1. Divides fossa above the inguinal ligament into *medial and lateral inguinal fossae* (*fovea*)
   2. In the medial fovea is *Hesselbach's triangle,* bounded medially by the edge of the rectus abdominis, laterally by the inferior epigastric artery, and below by the inguinal ligament

II. Inguinal canal

A. EXTENT: 4 cm, from deep (abdominal) inguinal ring to subcutaneous ring

B. COURSE: oblique, running parallel to and just above the inguinal ligament

C. WALLS
   1. Anterior: aponeurosis of external oblique along entire length; internal oblique over lateral part
   2. Posterior: from medial to lateral—reflected inguinal ligament, inguinal falx, and transversalis fascia
   3. Cephalic: arched fibers of the internal oblique and transversus abdominis muscles
   4. Caudal: inguinal ligament and lacunar ligament

D. CONTENTS: spermatic cord (ductus deferens or round ligament, deferential vessels, testicular artery, pampiniform plexus of veins, lymphatics, and autonomic nerves), ilioinguinal and genital branch of genitofemoral nerve, cremasteric artery and muscle, and internal spermatic fascia

III. Coverings of spermatic cord below subcutaneous ring

A. EXTERNAL SPERMATIC FASCIA: prolongation of the deep abdominal fascia from the external oblique muscle

B. CREMASTERIC (MIDDLE SPERMATIC) FASCIA: derived from the internal oblique muscle and its fascia

C. INTERNAL SPERMATIC FASCIA: from the transversalis fascia at deep ring

IV. Hernia

A. INDIRECT (LATERAL): sac of hernia leaves abdomen in lateral fovea, through deep ring, lateral to the inferior epigastric vessels. Passes entire length of inguinal canal

B. DIRECT (MEDIAL): sac pushes forward from abdomen in inguinal (Hesselbach's) triangle, medial to the inferior epigastric vessels, and only passes through the medial end of the inguinal canal to the subcutaneous ring

C. AN INGUINAL HERNIA typically contains part of a viscus, most commonly a part of the small or large intestine, and an associated peritoneal diverticulum or sac

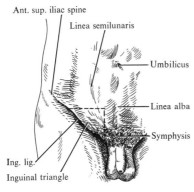

Ant. sup. iliac spine
Linea semilunaris
Umbilicus
Linea alba
Symphysis
Ing. lig.
Inguinal triangle

INGUINAL TOPOGRAPHY

Ant. sup. iliac spine
Aponeurosis ext. obliq. m.
Med. (sup.) crus
Intercrural fibers
Lat. (inf.) crus
Ing. lig. (Poupart's)
Superficial (subcutaneous) ing. ring
Lacuna musculorum
Iliopectineal arch
Reflected inguinal lig.
Iliopectineal eminence
Pecten pubis
Lacunar lig. (Gimbernat's)
Lacuna vasorum

INGUINAL LIG.
SUPERFICIAL INGUINAL RING

LOWER ABDOMINAL WALL

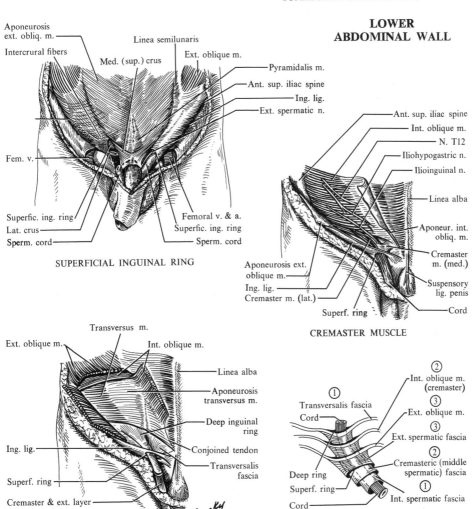

Aponeurosis ext. obliq. m.
Intercrural fibers
Linea semilunaris
Med. (sup.) crus
Ext. oblique m.
Pyramidalis m.
Ant. sup. iliac spine
Ing. lig.
Ext. spermatic n.
Fem. v.
Superfic. ing. ring
Lat. crus
Sperm. cord
Femoral v. & a.
Superfic. ing. ring
Sperm. cord

SUPERFICIAL INGUINAL RING

Ant. sup. iliac spine
Int. oblique m.
N. T12
Iliohypogastric n.
Ilioinguinal n.
Linea alba
Aponeur. int. obliq. m.
Cremaster m. (med.)
Suspensory lig. penis
Cord
Aponeurosis ext. oblique m.
Ing. lig.
Cremaster m. (lat.)
Superf. ring

CREMASTER MUSCLE

Transversus m.
Ext. oblique m.
Int. oblique m.
Linea alba
Aponeurosis transversus m.
Deep inguinal ring
Conjoined tendon
Ing. lig.
Transversalis fascia
Superf. ring
Cremaster & ext. layer

DEEP INGUINAL RING

② Int. oblique m. (cremaster)
③ Ext. oblique m.
① Transversalis fascia
Cord
③ Ext. spermatic fascia
② Cremasteric (middle spermatic) fascia
Deep ring
① Int. spermatic fascia
Superf. ring
Cord

CORD LAYERS

144. PERITONEUM

I. Definition: the serous layer of the abdomen. Has a parietal and a visceral layer

II. Peritoneal cavity: potential space between the layers of the peritoneum

III. Ligaments and mesenteric folds of peritoneum
  A. FALCIFORM LIGAMENT AND LIGAMENTUM TERES OF LIVER (see p. 324)
  B. CORONARY LIGAMENTS OF LIVER
    1. Anterior layer: continuous with right side of falciform ligament, where peritoneum is reflected from the diaphragm to right lobe of liver
    2. Posterior layer: from the back of right lobe to right suprarenal and kidney
  C. LEFT AND RIGHT TRIANGULAR: where ant. and post. coronary ligs. meet
    1. Anterior layer: continuous with left side of falciform; the peritoneum is reflected from the diaphragm to left lobe of liver
    2. Posterior layer: posterior to above, where peritoneum is reflected from the left lobe of liver to diaphragm. The 2 layers join in a sharp fold at the left
  D. HEPATOGASTRIC AND HEPATODUODENAL: this is the *lesser omentum,* extending from the hepatic porta to the stomach and duodenum

IV. Bare area of liver: an area of the liver not covered by peritoneum, chiefly lies between the layers of the coronary ligaments

V. Great omentum: double layer of peritoneum, hangs from greater curvature of stomach, crosses transverse colon, and descends in front of abdominal viscera
  A. GASTROCOLIC LIGAMENT: the omentum between the stomach and transverse colon
  B. PHRENICOLIENAL (LIENORENAL) LIGAMENT: part of the dorsal mesentery from left kidney to spleen
  C. GASTROLIENAL LIGAMENT: part of the dorsal mesentery joining spleen to stomach

VI. Transverse mesocolon: dorsal mesentery of transverse colon

VII. Phrenicocolic ligament: fold of peritoneum from the left colic flexure to diaphragm, helps to support spleen (sustentaculum lienis)

VIII. Mesentery proper: broad, fanlike fold of peritoneum suspending jejunum and ileum from the posterior body wall

IX. Omental bursa (lesser sac): diverticulum from main cavity, behind stomach
  A. BOUNDARIES: Anteriorly, caudate lobe of liver, lesser omentum, stomach, great omentum; posteriorly, great omentum, transverse colon and mesocolon, left suprarenal and left kidney; to the right, opens into the greater sac; to the left are the phrenicocolic ligament, hilum of spleen, and gastrolienal ligament
  B. SUBDIVISIONS: vestibule, between the epiploic foramen and gastropancreatic fold; superior recess, between caudate lobe of liver and diaphragm; lienal recess, between spleen and stomach; inferior recess, includes all the remaining lower part

X. Epiploic foramen: opening between the main cavity and lesser sac
  A. BOUNDARIES: Anteriorly, edge of lesser omentum; posteriorly, inferior vena cava; cephalically, caudate lobe of liver; caudally, duodenum
    1. In free edge of lesser omentum are found: common bile duct to the right with the hepatic artery just to the left of the duct; the portal vein is behind both

XI. Clinical consideration
  A. PERITONITIS: an inflammation of the peritoneum as a result of infection. If patient survives, adhesions may result

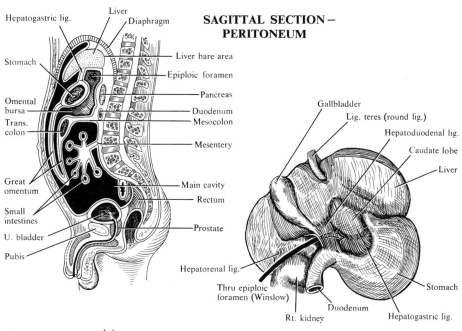

SAGITTAL SECTION –
PERITONEUM

Hepatogastric lig.
Liver
Diaphragm
Stomach
Liver bare area
Epiploic foramen
Pancreas
Omental bursa
Duodenum
Trans. colon
Mesocolon
Mesentery
Great omentum
Main cavity
Rectum
Small intestines
U. bladder
Prostate
Pubis

Gallbladder
Lig. teres (round lig.)
Hepatoduodenal lig.
Caudate lobe
Liver
Stomach
Hepatorenal lig.
Thru epiploic foramen (Winslow)
Duodenum
Rt. kidney
Hepatogastric lig.

LESSER OMENTUM

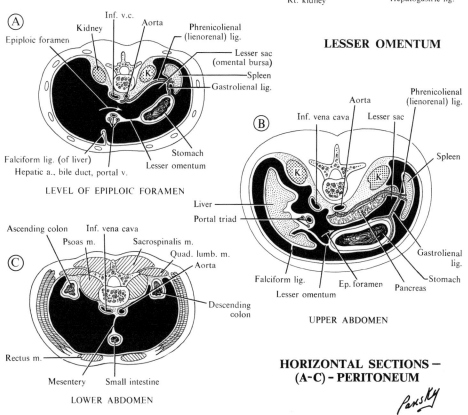

(A)
Inf. v.c.
Kidney
Aorta
Phrenicolienal (lienorenal) lig.
Epiploic foramen
Lesser sac (omental bursa)
Spleen
Gastrolienal lig.
Falciform lig. (of liver)
Hepatic a., bile duct, portal v.
Stomach
Lesser omentum

LEVEL OF EPIPLOIC FORAMEN

(B)
Aorta
Inf. vena cava
Lesser sac
Phrenicolienal (lienorenal) lig.
Spleen
Liver
Portal triad
Gastrolienal lig.
Falciform lig.
Ep. foramen
Stomach
Lesser omentum
Pancreas

UPPER ABDOMEN

(C)
Ascending colon
Inf. vena cava
Psoas m.
Sacrospinalis m.
Quad. lumb. m.
Aorta
Descending colon
Rectus m.
Mesentery
Small intestine

LOWER ABDOMEN

HORIZONTAL SECTIONS –
(A-C) – PERITONEUM

Pansky

145. PERITONEUM OF ABDOMINAL WALL AND PELVIS

I. **Posterior abdominal wall.** With all viscera cut away, the lines of reflection of peritoneum demonstrate the continuity of the peritoneum both on the dorsal wall and also on the upper ventral wall and diaphragm. From above downward: the falciform ligament is continuous with the ventral layer of the coronary and triangular ligaments of liver; the dorsal layer of coronary and triangular ligaments is continuous with the lesser omentum at the abdominal end of the esophagus; the splitting of lesser omentum around the esophagus (and stomach) is reunited and becomes continuous with the dorsal mesentery as the phrenicolienal ligament; there is a fusion of the dorsal mesentery (the great omentum) to the transverse mesocolon; the great omentum passes around the pylorus to become continuous with the free edge of the lesser omentum; the transverse mesocolon dorsally becomes continuous with the peritoneum covering the ascending and descending colons; the peritoneum of the descending colon, farther caudally, merges with the sigmoid mesocolon

II. **Anterior abdominal wall**
   A. ABOVE UMBILICUS: starting at umbilicus and extending cephalically toward liver is the *falciform ligament containing the ligamentum teres*
   B. BELOW UMBILICUS
      1. Median umbilical fold, in midline: a fold caused by the presence of the middle umbilical ligament (urachus), which extends from umbilicus to apex of the bladder
      2. Medial umbilical folds, 1 on either side: due to peritoneum covering lateral umbilical ligaments, which begin at umbilicus, diverge laterally and caudally to join superior vesical branch of internal iliac artery. The ligaments are remnants of the fetal umbilical arteries
      3. Lateral umbilical (epigastric) folds: folds of peritoneum covering inferior epigastric vessels

III. **Fossae or fovea of anterior abdominal wall:** produced by the folds described above
   A. LATERAL AND MEDIAL INGUINAL FOSSAE (FOVEAE): depressions in peritoneum on either side of the lateral umbilical folds
   B. SUPRAVESICAL FOSSAE (FOVEAE): between middle and medial umbilical folds

IV. **Pelvic peritoneum:** extending down from the abdominal walls, this peritoneum covers the pelvic viscera and related structures

V. **Special features**
   A. DEEP INGUINAL RING: with peritoneum intact, a slight depression may be seen in lateral inguinal fossa just lateral to the lateral umbilical fold, above middle of inguinal ligament. Site of indirect (lateral) hernia
   B. INGUINAL (HESSELBACH'S) TRIANGLE: an area bounded laterally by lateral umbilical fold, medially by the border of rectus abdominis, and below by the inguinal ligament. Most frequent site of direct (medial) hernia

VI. **Clinical considerations**
   A. CULDOCENTESIS: aspiration of fluids accumulated in rectouterine pouch
   B. CULDOSCOPY: introduction of an endoscope through the posterior vaginal fornix into the rectouterine pouch for viewing of the pelvic viscera

PERITONEUM

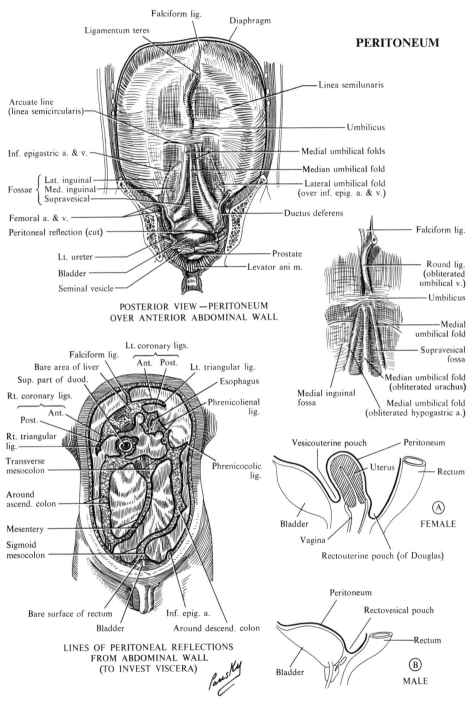

Falciform lig.
Ligamentum teres
Diaphragm

Linea semilunaris

Arcuate line
(linea semicircularis)

Umbilicus

Inf. epigastric a. & v.

Medial umbilical folds

Median umbilical fold

Fossae { Lat. inguinal
Med. inguinal
Supravesical

Lateral umbilical fold
(over inf. epig. a. & v.)

Femoral a. & v.

Ductus deferens

Peritoneal reflection (cut)

Lt. ureter

Prostate

Bladder

Levator ani m.

Seminal vesicle

POSTERIOR VIEW — PERITONEUM
OVER ANTERIOR ABDOMINAL WALL

Falciform lig.

Round lig.
(obliterated
umbilical v.)

Umbilicus

Medial
umbilical fold

Supravesical
fossa

Medial inguinal
fossa

Median umbilical fold
(obliterated urachus)

Medial umbilical fold
(obliterated hypogastric a.)

Lt. coronary ligs.
Falciform lig.
Ant. Post.
Bare area of liver
Lt. triangular lig.
Sup. part of duod.
Esophagus
Rt. coronary ligs.
Phrenicolienal
Ant.
lig.
Post.
Rt. triangular
lig.
Transverse
mesocolon
Phrenicocolic
lig.
Around
ascend. colon
Mesentery
Sigmoid
mesocolon

Bare surface of rectum
Inf. epig. a.
Bladder
Around descend. colon

LINES OF PERITONEAL REFLECTIONS
FROM ABDOMINAL WALL
(TO INVEST VISCERA)

Vesicouterine pouch
Peritoneum
Uterus
Rectum
Bladder
Vagina
Rectouterine pouch (of Douglas)

Ⓐ
FEMALE

Peritoneum
Rectovesical pouch
Rectum
Bladder

Ⓑ
MALE

SAGITTAL VIEW — PELVIC PERITONEUM

146. STOMACH STRUCTURE: PARTS AND RELATIONS

I. Orifices (for surface projections, see p. 312)

A. CARDIAC: between abdominal end of esophagus and stomach and opens toward left. Right side of esophagus is continuous with the lesser curvature and left side with the greater curvature. The latter is indicated by an acute angle—*incisura cardiaca*

B. PYLORIC opens into duodenum toward the right and is indicated by the *duodenopyloric constriction*

II. Curvatures

A. LESSER: continuation of right side of esophagus, is the right or concave border. The *incisura angularis,* a notch in this border, divides the stomach into right and left portions

    1. Hepatogastric ligament (lesser omentum) attaches to it. Between the layers of this ligament run the right and left gastric arteries and the superior gastric (lt. gastric) lymph nodes

B. GREATER: a continuation of the left side of the esophagus, is the left or convex border. The *pyloric vestibule,* a dilatation opposite the angular incisure, is limited on the right by the *sulcus intermedius.* The *pyloric antrum* is that area between the sulcus intermedius and the pylorus

    1. Gastrolienal ligament attached to its left portion

    2. Great omentum attached to its right portion. Between the layers of the omentum are the gastroepiploic vessels and inferior gastric (rt. gastroepiploic) lymph nodes

III. Parts of stomach

A. BODY lies to the left of a vertical line passing through the angular incisure

    1. Fundus: that part of body lying superior to a horizontal line through the cardiac opening

B. PYLORIC PORTION lies to right of a line through angular incisure. Is further subdivided into:

    1. Pyloric vestibule and pyloric antrum, by a line drawn through the sulcus intermedius

IV. Relations of stomach

A. ANTERIOR SURFACE: entire surface is covered with peritoneum. Left half is in contact with diaphragm; right half is in contact with left and quadrate lobes of liver and abdominal wall

B. POSTERIOR SURFACE: entire surface is covered with peritoneum except near the cardiac opening, where the *gastrophrenic ligament* is attached. Is in contact with the diaphragm, spleen, left suprarenal gland, part of left kidney, pancreas, left colic flexure, and upper surface of the transverse mesocolon

V. Special feature

A. MUSCULAR COAT made up of 3 layers of smooth muscle—an inner oblique, chiefly at cardia and spreading to anterior and posterior surfaces; circular, well developed over entire organ; and longitudinal, concentrated along curvatures

VI. Clinical considerations

A. STOMACH ULCER: due to an excess of acid secretion associated with vagal nerve involvement. The bleeding *peptic ulcer* is usually located posterior, and the perforating type is located anterior

B. PARTIAL GASTRECTOMY (STOMACH REMOVAL): most common operation performed on stomach in cases of duodenal ulcer, gastric ulcer, or malignancy

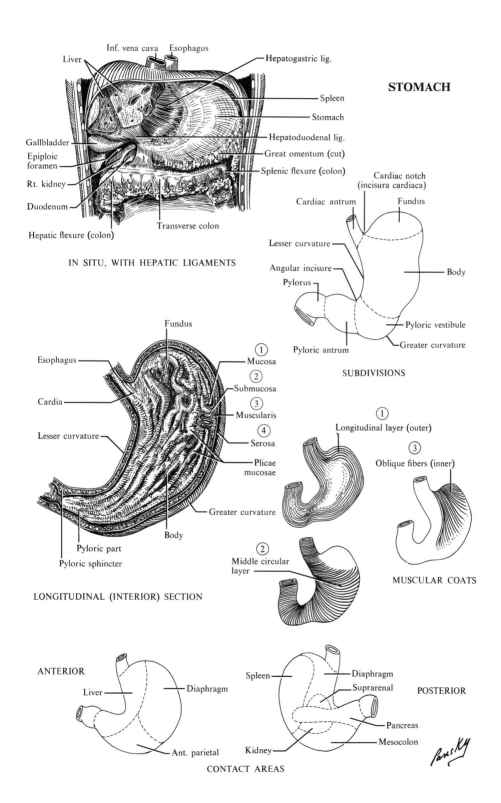

STOMACH

Inf. vena cava
Esophagus
Liver
Hepatogastric lig.
Spleen
Stomach
Hepatoduodenal lig.
Gallbladder
Great omentum (cut)
Epiploic foramen
Splenic flexure (colon)
Rt. kidney
Duodenum
Hepatic flexure (colon)
Transverse colon

IN SITU, WITH HEPATIC LIGAMENTS

Cardiac notch
(incisura cardiaca)
Cardiac antrum
Fundus
Lesser curvature
Angular incisure
Body
Pylorus
Pyloric vestibule
Pyloric antrum
Greater curvature

SUBDIVISIONS

Fundus
Esophagus
① Mucosa
② Submucosa
Cardia
③ Muscularis
④ Serosa
Lesser curvature
Plicae mucosae
Greater curvature
Body
Pyloric part
Pyloric sphincter

LONGITUDINAL (INTERIOR) SECTION

① Longitudinal layer (outer)
③ Oblique fibers (inner)
② Middle circular layer

MUSCULAR COATS

ANTERIOR
Liver
Diaphragm
Ant. parietal

CONTACT AREAS

Spleen
Diaphragm
Suprarenal
POSTERIOR
Pancreas
Kidney
Mesocolon

Pansky

-327-

147. BLOOD SUPPLY, LYMPH DRAINAGE, AND INNERVATION OF THE STOMACH

I. Blood supply
A. ARTERIES: all derived directly or indirectly from the celiac artery
1. Left gastric: directly from celiac, runs upward and to left across dorsal wall of omental bursa to cephalic end of lesser curvature, which it follows
2. Right gastric: branch from common hepatic artery, runs upward along the lesser curvature to anastomose with 1, above
3. Right gastroepiploic: 1 of terminal branches of the gastroduodenal artery from the common hepatic artery, runs toward the left on the greater curvature. Has a large *pyloric branch*
4. Left gastroepiploic: from splenic artery, through gastrolienal ligament, runs from left to right along the greater curvature to meet 3, above
5. Short gastric arteries: 5 to 7 small branches from the splenic to that part of great curvature above the left gastroepiploic artery
B. VEINS: venous drainage directly or indirectly into portal vein
1. Short gastric veins from greater curvature and fundus to lienal vein
2. Left gastroepiploic along greater curvature to lienal vein
3. Right gastroepiploic from right end of greater curvature to superior mesenteric vein
4. Left gastric (coronary) vein runs length of lesser curvature from cardia to portal vein; accompanies left gastric artery
5. Right gastric is a small vein which accompanies the right gastric artery
6. Pyloric along pyloric part of lesser curvature to portal vein

II. Nerve supply
A. PARASYMPATHETIC: preganglionics from posterior vagal trunk (right vagus) posteriorly and anterior vagal trunk (left vagus) anteriorly. These synapse within the walls of the stomach. Thus the postganglionics are very short
B. SYMPATHETIC: preganglionic fibers mainly in the thoracic splanchnic nerves; postganglionics arise in ganglia of celiac plexus

III. Lymphatic drainage
A. VISCERAL NODES
1. Gastric
   a. Superior (left gastric) along left gastric artery
   b. Inferior along right half of greater curvature
2. Hepatic. Subdivided into groups: along hepatic artery, near neck of gallbladder, and in the angle between superior and descending duodenum
3. Pancreaticolienal (splenic) along splenic artery
B. VESSELS OF STOMACH may follow along the left gastric artery to superior gastric nodes, from fundus and body (to left of esophagus) along the left gastroepiploic artery to pancreaticolienal nodes, from right part of greater curvature to inferior gastric and hepatic nodes, and from pyloric region into hepatic and superior gastric nodes

IV. Clinical considerations
A. THE DIRECTION OF LYMPH FLOW and the position of the major lymph nodes are essential in understanding the possible spread of malignancy from stomach
B. THE VAGUS NERVES largely control the secretion of acid by the parietal cells of the stomach. Since excess acid secretion is associated with peptic ulcers (either in the stomach or duodenum), section of the vagus trunks as they enter the abdomen is often carried out to reduce acid production (often in conjunction with resection of the ulcerated area)

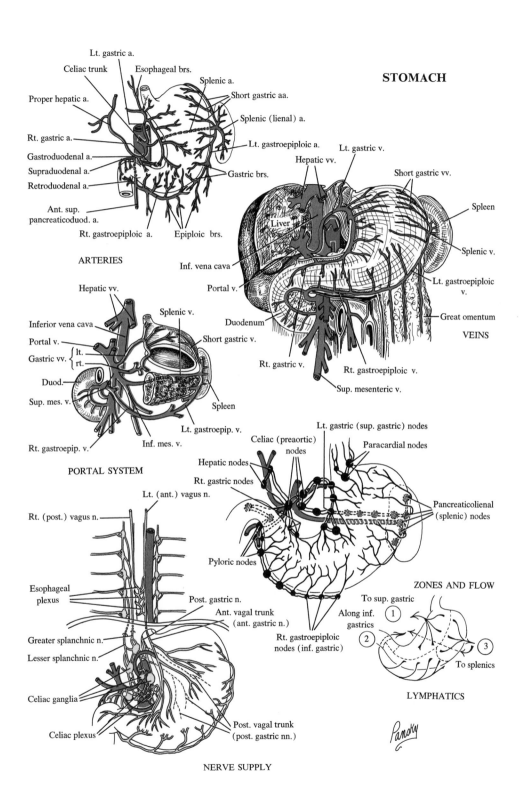

STOMACH

ARTERIES

Lt. gastric a.
Celiac trunk
Esophageal brs.
Splenic a.
Short gastric aa.
Proper hepatic a.
Splenic (lienal) a.
Rt. gastric a.
Lt. gastroepiploic a.
Gastroduodenal a.
Supraduodenal a.
Retroduodenal a.
Gastric brs.
Ant. sup. pancreaticoduod. a.
Rt. gastroepiploic a.
Epiploic brs.

VEINS

Lt. gastric v.
Hepatic vv.
Short gastric vv.
Spleen
Liver
Splenic v.
Inf. vena cava
Lt. gastroepiploic v.
Portal v.
Great omentum
Duodenum
Rt. gastric v.
Rt. gastroepiploic v.
Sup. mesenteric v.

PORTAL SYSTEM

Hepatic vv.
Inferior vena cava
Portal v.
Splenic v.
Gastric vv. { lt. rt. }
Short gastric v.
Duod.
Sup. mes. v.
Spleen
Rt. gastroepip. v.
Lt. gastroepip. v.
Inf. mes. v.

NERVE SUPPLY

Lt. (ant.) vagus n.
Rt. (post.) vagus n.
Esophageal plexus
Post. gastric n.
Ant. vagal trunk (ant. gastric n.)
Greater splanchnic n.
Lesser splanchnic n.
Rt. gastroepiploic nodes (inf. gastric)
Celiac ganglia
Celiac plexus
Post. vagal trunk (post. gastric nn.)

LYMPHATICS

Lt. gastric (sup. gastric) nodes
Celiac (preaortic) nodes
Paracardial nodes
Hepatic nodes
Rt. gastric nodes
Pancreaticolienal (splenic) nodes
Pyloric nodes

ZONES AND FLOW

To sup. gastric
Along inf. gastrics
① ② ③
To splenics

Panshy

-329-

148. DUODENUM

I. Definition: first and shortest part of small intestine

II. Extent: from pylorus to duodenojejunal flexure, 25 cm long

III. Parts and relations
A. SUPERIOR extends from pylorus to the right, under quadrate lobe of liver to neck of gallbladder, where it bends sharply caudad. Nearly completely covered by peritoneum except at neck of gallbladder. Hepatoduodenal ligament attached to upper border of same region. Related above and anteriorly to liver and gallbladder; posteriorly to gastroduodenal artery, common bile duct, and portal vein; below and posteriorly to pancreas
B. DESCENDING from level of neck of gallbladder at first lumbar vertebra, along right side of vertebral column to upper body of L4. Covered over anteriorly by peritoneum except where crossed by transverse mesocolon. Related posteriorly to medial side of right kidney and structures at its hilum (vessels, ureter), inferior vena cava and psoas major muscle; anteriorly to liver, transverse colon, coils of jejunum; medially to head of pancreas and common duct; and laterally to right colic flexure
  1. Common duct and major pancreatic duct pierce wall about 7.0 cm below pylorus. Accessory duct is 2 cm cephalic to this
C. INFERIOR (HORIZONTAL) passes from right to left, with slight cephalic deviation along upper border of L4. Covered anteriorly by peritoneum, except near midline, where crossed by vessels. Related anteriorly to superior mesenteric vessels, which cross it; posteriorly to right crus of diaphragm, inferior vena cava, and aorta; and cephalically to pancreas
D. ASCENDING rises cephalically to left of aorta to upper border of L2, where it turns sharply to join jejunum. Covered anteriorly by peritoneum. Related posteriorly to left psoas major muscle and left renal vessels; and on the right to the uncinate process of pancreas

IV. Structural characteristics
A. SEROSA ONLY PARTLY COMPLETE
B. HIGH FOLDS OF THE CONNECTIVE TISSUE of the submucosa form numerous, nearly circular projections into lumen. These are the *circular folds* (plicae), which are numerous and well developed in duodenum, beginning about 2.5 cm from pylorus
C. DUODENAL GLANDS (OF BRUNNER): compound tubuloalveolar glands of mucous type found in the submucosa
D. VILLI are numerous and large
E. MAJOR DUODENAL PAPILLA with *sphincter of Oddi* around common duct and major duct of pancreas (hepatopancreatic ampulla) within wall of duodenum

V. Clinical considerations
A. MOST OF THE DUODENAL ULCERS OCCUR within 5 cm of the pylorus and more frequently on the anterior wall
B. DUODENECTOMY: because attachment of the duodenum to the posterior body wall structures is secondary, both it and the associated head of the pancreas can be separated from underlying viscera (right kidney) without endangering the blood supply or ducts of the kidney
C. THE FIRST PART of the duodenum has the poorest blood supply, for it is not supplied from the arcades but by small branches from the gastroduodenal artery

DUODENUM

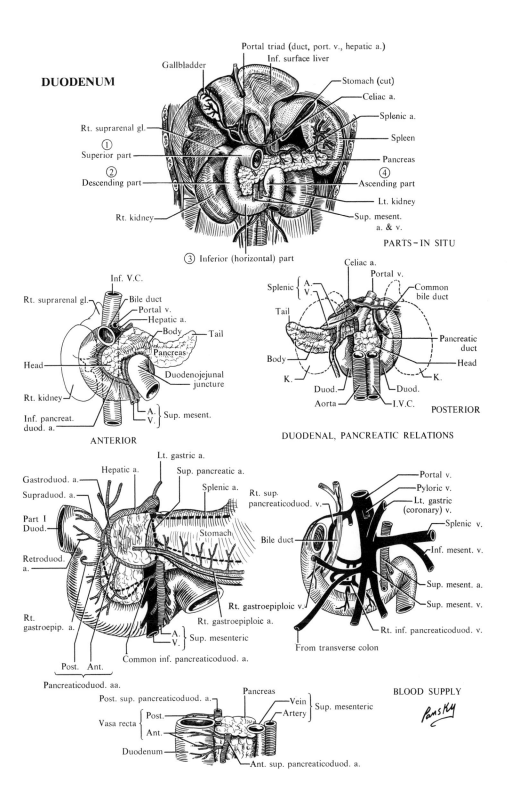

Portal triad (duct, port. v., hepatic a.)
Inf. surface liver
Gallbladder
Stomach (cut)
Celiac a.
Splenic a.
Rt. suprarenal gl.
Spleen
① Superior part
Pancreas
② Descending part
④ Ascending part
Lt. kidney
Rt. kidney
Sup. mesent. a. & v.

PARTS – IN SITU

③ Inferior (horizontal) part

Inf. V.C.
Rt. suprarenal gl.
Bile duct
Portal v.
Hepatic a.
Body
Tail
Head
Pancreas
Duodenojejunal juncture
Rt. kidney
Inf. pancreat. duod. a.
A. V. } Sup. mesent.

ANTERIOR

Celiac a.
Portal v.
Splenic { A. V.
Common bile duct
Tail
Pancreatic duct
Body
Head
K.
K.
Duod.
Duod.
Aorta
I.V.C.

POSTERIOR

DUODENAL, PANCREATIC RELATIONS

Lt. gastric a.
Hepatic a.
Sup. pancreatic a.
Gastroduod. a.
Splenic a.
Supraduod. a.
Rt. sup. pancreaticoduod. v.
Part I Duod.
Stomach
Retroduod. a.
Bile duct
Portal v.
Pyloric v.
Lt. gastric (coronary) v.
Splenic v.
Inf. mesent. v.
Sup. mesent. a.
Sup. mesent. v.
Rt. gastroepip. a.
Rt. gastroepiploic v.
Rt. gastroepiploic a.
A. V. } Sup. mesenteric
Common inf. pancreaticoduod. a.
Rt. inf. pancreaticoduod. v.
From transverse colon
Post. Ant.
Pancreaticoduod. aa.

BLOOD SUPPLY

Pancreas
Post. sup. pancreaticoduod. a.
Vein
Artery } Sup. mesenteric
Vasa recta { Post. Ant.
Duodenum
Ant. sup. pancreaticoduod. a.

Pansky

149. PANCREAS

I. Type of gland
A. ENDOCRINE: islets of Langerhans
B. EXOCRINE: compound tubuloalveolar, serous secretion

II. Parts and their relations
A. HEAD: the broad, right extremity. Lies within the curve of the duodenum
   1. Uncinate process: a prolongation of the left and caudal borders of the head
      a. Superior mesenteric artery with superior mesenteric vein on its right side crosses the uncinate process
   2. Anterior surface: most of the right side separated from the transverse colon by areolar tissue (no peritoneum); lower part of surface below transverse colon is covered by peritoneum; is in contact with coils of the small intestine
   3. Posterior surface: without peritoneum, is in contact with inferior vena cava, common bile duct, renal veins, right crus of diaphragm, and aorta
B. NECK: a constricted portion to the left of the head. Above, it adjoins the pylorus. Behind, it is related to the origin of the portal vein and the gastroduodenal artery
C. BODY
   1. Anterior surface: separated from the stomach by the omental bursa
   2. Posterior surface: nonperitoneal; related to the aorta, splenic vein, left kidney and vessels; left suprarenal; origin of superior mesenteric artery, and crura of diaphragm
   3. Inferior surface: peritoneal; related to duodenojejunal flexure, coils of jejunum, and left colic flexure
   4. Anterior border: layers of transverse mesocolon diverge along this
   5. Superior border: related to celiac artery, with hepatic artery to right and splenic artery to left
D. TAIL: left extremity, extending to surface of spleen, in phrenicolienal ligament

III. Ducts
A. MAJOR (OF WIRSUNG) extends toward right, reaches neck of pancreas, where it turns caudally and dorsally and comes in contact with the common bile duct, forming the hepatopancreatic ampulla (of Vater). These ducts pass obliquely through the wall of the descending duodenum and open through a common orifice into its lumen
B. ACCESSORY (MINOR OF SANTORINI) drains part of head, enters the duodenum above the major

IV. Blood supply
A. ARTERIES
   1. Numerous small branches from the splenic artery
   2. Retroduodenal branch of gastroduodenal artery
   3. Superior pancreaticoduodenal from the gastroduodenal artery
   4. Inferior pancreaticoduodenal artery from the superior mesenteric artery
B. VEINS drain into both the splenic and superior mesenteric veins

V. Lymphatic drainage (see p. 360) follows the course of blood vessels to terminate in the pancreaticolienal, pancreaticoduodenal, and celiac nodes

VI. Nerves: autonomics by way of the splenic division of the celiac plexus

VII. Clinical considerations
A. HYPERTROPHY OF HEAD may cause portal or bile duct obstruction
B. DEGENERATION OF THE ISLETS OF LANGERHANS leads to diabetes mellitus
C. PANCREATITIS is a serious inflammatory condition of the exocrine pancreas

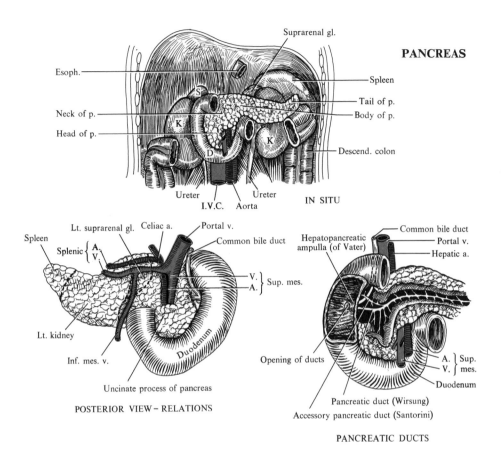

PANCREAS

Suprarenal gl.

Esoph.

Spleen

Tail of p.

Neck of p.

Body of p.

Head of p.

Descend. colon

Ureter   Ureter

I.V.C.   Aorta

IN SITU

Lt. suprarenal gl.   Celiac a.   Portal v.

Spleen

Splenic { A. / V. }

Common bile duct

V. } Sup. mes.
A. }

Lt. kidney

Inf. mes. v.

Duodenum

Uncinate process of pancreas

POSTERIOR VIEW – RELATIONS

Hepatopancreatic ampulla (of Vater)

Common bile duct

Portal v.

Hepatic a.

Opening of ducts

A. } Sup.
V. } mes.

Duodenum

Pancreatic duct (Wirsung)

Accessory pancreatic duct (Santorini)

PANCREATIC DUCTS

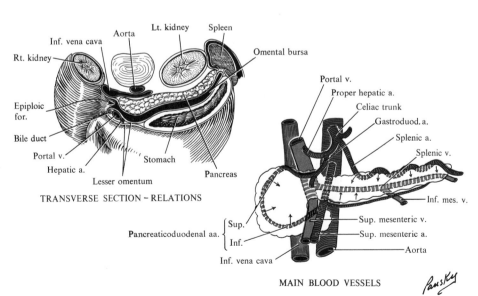

Inf. vena cava   Aorta   Lt. kidney   Spleen

Rt. kidney

Omental bursa

Epiploic for.

Bile duct

Portal v.

Hepatic a.

Lesser omentum

Stomach

Pancreas

TRANSVERSE SECTION – RELATIONS

Portal v.

Proper hepatic a.

Celiac trunk

Gastroduod. a.

Splenic a.

Splenic v.

Inf. mes. v.

Sup. mesenteric v.

Sup. mesenteric a.

Aorta

Pancreaticoduodenal aa. { Sup. / Inf. }

Inf. vena cava

MAIN BLOOD VESSELS

- 333 -

150. DUODENUM-JEJUNUM-ILEUM, DUODENAL FOSSAE, AND SUPERIOR MESENTERIC ARTERY

I. **Duodenojejunal flexure:** point where ascending part of duodenum turns sharply anteriorly and caudally to join jejunum
   A. LOCATION: upper border, second lumbar vertebra, on left side
   B. SUSPENSORY MUSCLE (LIGAMENT OF TREITZ): musculofibrous band, from tissue around celiac artery and right crus of diaphragm to flexure and ascending duodenum, which continues into mesentery. Acts as suspensory ligament

II. **Duodenal recesses (fossae)**
   A. INFERIOR at level of third lumbar vertebra, on left side of ascending duodenum, opens cephalically, and extends down behind ascending duodenum
   B. SUPERIOR ventral to body of second lumbar vertebra on left of ascending duodenum, opens caudally, lies behind peritoneal fold, superior duodenal (*duodenojejunal fold*)
   C. DUODENOJEJUNAL lies below pancreas, between the aorta and the left kidney. The renal vein lies beneath the fossa
   D. PARADUODENAL (rarely found): small pocket that may appear behind ascending branch of left colic artery
   E. RETRODUODENAL (rarely found) lies behind transverse and ascending parts of duodenum, anterior to aorta

III. **Intestinal arteries:** usually 12–15
   A. ORIGIN: superior mesenteric artery
   B. DISTRIBUTION: jejunum and ileum
   C. COURSE: in mesentery, running parallel with each other
   D. BRANCHING: each divides into 2 branches, which unite with branches from adjoining arteries, forming arches with convexities toward the intestine
      1. In upper part of mesentery, from this arch, fairly long, straight arteries arise which go to the gut
      2. As the intestinal tract descends, branches arise from the first set of arches to unite with similar branches from the arch above or below, forming a second set of arches, located nearer the intestine than the first. The straight arteries arising from these are shorter. Still farther caudad, 3, 4, or 5 more arches are added so that in the lower ileum the last arch is close to the gut and the straight arteries are very short

IV. **Meckel's diverticulum:** present in a small percentage of cases
   A. LOCATION: usually about 1 meter above ileocolic valve on the antimesenteric border of ileum. End may be free or attached to abdominal wall
   B. CAUSE: failure of entire omphalovitelline duct of fetus to atrophy

V. **The jejunum—ileum**
   A. SUPPORT: from dorsal body wall by the *mesentery*
   B. JEJUNUM, upper two fifths (2.4 meters); ILEUM, lower three fifths (3.6 meters)
   C. JEJUNUM: greater diameter, thicker wall, and more vascular
   D. ILEUM: distal portion always in pelvis. More fat in mesentery. Aggregated lymph nodules (Peyer's patches) in antimesenteric border

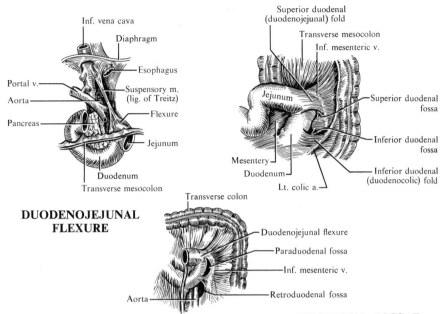

DUODENOJEJUNAL FLEXURE

Inf. vena cava
Diaphragm
Esophagus
Portal v.
Suspensory m. (lig. of Treitz)
Aorta
Flexure
Pancreas
Jejunum
Duodenum
Transverse mesocolon

Superior duodenal (duodenojejunal) fold
Transverse mesocolon
Inf. mesenteric v.
Jejunum
Superior duodenal fossa
Inferior duodenal fossa
Mesentery
Duodenum
Inferior duodenal (duodenocolic) fold
Lt. colic a.

Transverse colon
Duodenojejunal flexure
Paraduodenal fossa
Inf. mesenteric v.
Retroduodenal fossa
Aorta

DUODENAL FOSSAE AND FOLDS

① ABOUT 0.9 METER FROM DUODENOJEJUNAL FLEXURE

② ABOUT 2.1 METERS FROM DUODENOJEJUNAL FLEXURE (Lunette's)

VASCULAR CHANGES FROM JEJUNUM TO ILEUM
(After Thorek)

③ ABOUT 3.0–3.6 METERS FROM FLEXURE (TERTIARY LOOPS)

④ TERMINAL ILEUM (Fat)

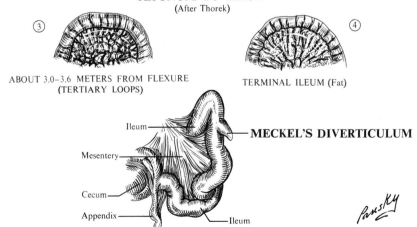

Ileum
Mesentery
Cecum
Appendix
Ileum

MECKEL'S DIVERTICULUM

Pansky

-335-

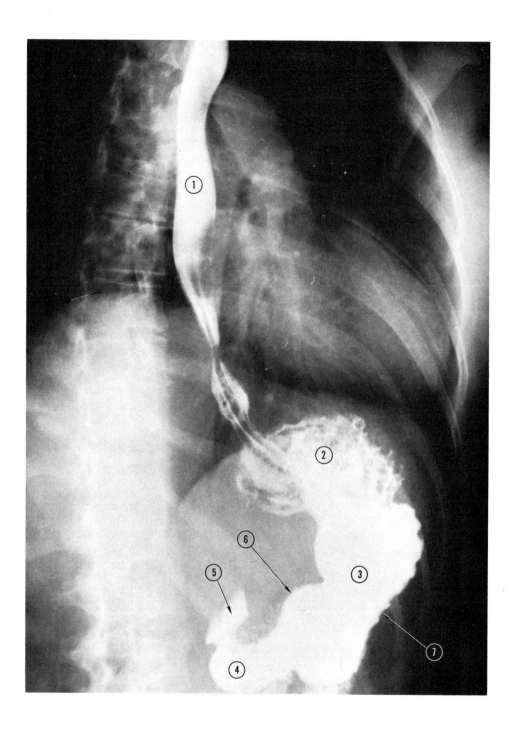

FIGURE 27. **Esophagus and stomach.** *1,* Esophagus; *2,* fundus of stomach; *3,* body of stomach; *4,* pylorus; *5,* duodenal cap; *6,* lesser curvature; *7,* greater curvature.

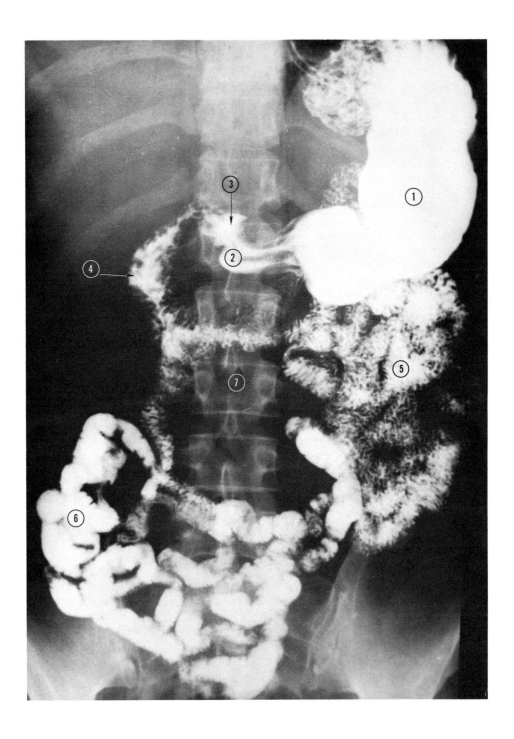

FIGURE 28. **Stomach and small intestine after barium meal.** *1,* Body of stomach; *2,* pylorus; *3,* duodenal cap; *4,* descending duodenum; *5,* jejunum; *6,* ileum; *7,* body of L3.

151. THE LARGE INTESTINE

I. Extent: from terminal ileum to anus, 1.5 meters

II. Parts
A. CECUM: blind pouch that extends caudally below ileocolic valve, located in right iliac fossa above inguinal ligament. Its shape varies. Usually covered by peritoneum
   1. Appendix: long and narrow tube (8.2 cm), which begins at apex of cecum. Position varies. Fixed by *mesoappendix* containing the *appendicular artery*
   2. Ileocecal valve: ileum opens into junction between cecum and ascending colon. Opening is guarded by 2 lips, which project into lumen. The lips merge on either side of opening, forming membranous ridges, the *frenula of the valve*
B. COLON
   1. Begins at iliocecal valve and ascends through right lateral (lumbar) and hypochondriac regions to visceral surface of liver, to right of gallbladder. It then bends sharply to the left as the *right colic* (*hepatic*) *flexure*. Its ventral surface and sides are covered by peritoneum. Dorsally, it is separated from the iliacus, quadratus lumborum, transversus abdominis muscles, and lateral part of right kidney
   2. Transverse: longest and most movable. From right hypochondriac region arches to come across umbilical zone and then upward into left hypochondriac region, where it bends caudally at *left colic flexure* (*splenic*) below spleen. It is invested in peritoneum and is suspended from body wall by *transverse mesocolon*
   3. Descending extends caudally through left hypochondriac and lateral (lumbar) regions along lateral border of left kidney. At caudal end of kidney bends medially and descends in groove between psoas and quadratus lumborum muscles to crest of ileum. Covered anteriorly and on sides with peritoneum, which helps to fix it
   4. Iliac colon: in left iliac fossa, from iliac crest to brim of true pelvis. Anterior to iliacus and psoas muscles. Covered anteriorly and at sides with peritoneum
   5. Sigmoid: begins at pelvic brim, crosses sacrum, and then curves to midline at third sacral segment, where it enters rectum. Usually, completely invested with peritoneum and has a mesocolon. Posteriorly are the left external iliac vessels, left piriformis muscle, and left sacral plexus. Anteriorly, coils of small intestine

III. Special structural characteristics
A. TAENIAE COLI. The longitudinal smooth muscle coat of the colon is incomplete, being collected in 3 bands, the taeniae. These create sacculations (haustra)
   1. Semilunar folds, the crescent-shaped folds between the haustra
B. EPIPLOIC APPENDAGES. Small fat-filled sacs of peritoneum attached along taeniae

IV. Peritoneal recesses (fossae) (retroperitoneal fossae)
A. SUPERIOR ILEOCECAL: fold of peritoneum over the branch of the ileocolic artery
B. INFERIOR ILEOCECAL: behind the ileocecal junction produced by the ileocecal fold (bloodless fold of Treves) from the antimesenteric border of the ileum, crosses the ileocecal junction and joins the mesoappendix
C. RETROCECAL (CECAL): behind the cecum

V. Clinical considerations
A. TAENIAE converge at root of appendix and may be used to locate its root
B. McBURNEY'S POINT: located by drawing a line from the right anterior superior iliac spine to the umbilicus. The midpoint of this line locates root of the appendix
C. COLOSTOMY: an artificial opening between the colon and skin
D. DIVERTICULITIS: chiefly in sigmoid colon; inflammation of abnormal outpocketings

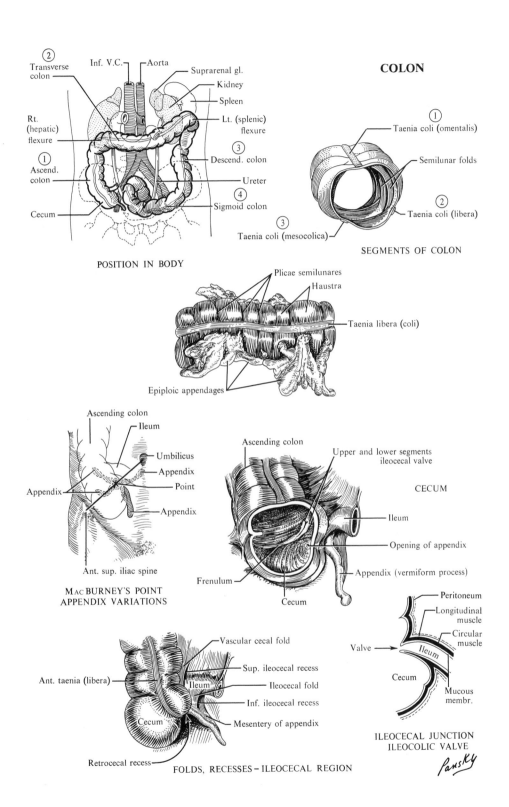

② Transverse colon
Inf. V.C.
Aorta
Suprarenal gl.
Kidney
Spleen
Lt. (splenic) flexure
Rt. (hepatic) flexure
③ Descend. colon
① Ascend. colon
Ureter
④ Sigmoid colon
Cecum

POSITION IN BODY

COLON

① Taenia coli (omentalis)
Semilunar folds
② Taenia coli (libera)
③ Taenia coli (mesocolica)

SEGMENTS OF COLON

Plicae semilunares
Haustra
Taenia libera (coli)
Epiploic appendages

Ascending colon
Ileum
Umbilicus
Appendix
Point
Appendix
Appendix
Ant. sup. iliac spine

MACBURNEY'S POINT
APPENDIX VARIATIONS

Ascending colon
Upper and lower segments ileocecal valve
CECUM
Ileum
Opening of appendix
Appendix (vermiform process)
Frenulum
Cecum

Vascular cecal fold
Sup. ileocecal recess
Ileum
Ileocecal fold
Ant. taenia (libera)
Inf. ileocecal recess
Cecum
Mesentery of appendix
Retrocecal recess

FOLDS, RECESSES – ILEOCECAL REGION

Peritoneum
Longitudinal muscle
Circular muscle
Valve
Ileum
Cecum
Mucous membr.

ILEOCECAL JUNCTION
ILEOCOLIC VALVE

Pansky

152. BLOOD SUPPLY OF THE COLON AND APPENDIX

I. Arteries

A. ILEOCOLIC ARTERY
   1. Origin: lowest branch of superior mesenteric artery
   2. Course: caudally and to the right, into the right iliac fossa
   3. Branches
      a. Superior ascends along ascending colon to join right colic artery
      b. Inferior runs toward ileocolic junction. Branches: *ascending* (*colic*), to ascending colon; *cecal,* anterior and posterior to cecum; *appendicular,* descends posterior to terminal ileum to mesoappendix of appendix; *ileal,* passes to left on ileum to anastomose with last intestinal branch of superior mesenteric artery

B. RIGHT COLIC
   1. Origin: from middle of superior mesenteric artery
   2. Course: to right, behind peritoneum, crossing anterior to internal spermatic or ovarian vessels, right ureter, and psoas major muscle
   3. Branches
      a. Descending: anastomoses with superior branch of ileocolic artery
      b. Ascending: course cephalically to join middle colic artery

C. MIDDLE COLIC
   1. Origin: from superior mesenteric artery, just below pancreas
   2. Course: caudally and anteriorly in transverse mesocolon
   3. Branches: near border of transverse colon
      a. Right: anastomoses with ascending branch of right colic artery
      b. Left: anastomoses with ascending branch of left colic artery

D. LEFT COLIC
   1. Origin: from inferior mesenteric artery
   2. Course: toward the left, behind the peritoneum, in front of left psoas major muscle, and crosses left ureter and internal spermatic vessels
   3. Branches
      a. Ascending ascends in front of left kidney to enter transverse mesocolon; and anastomoses with left branch of middle colic artery
      b. Descending descends to join highest sigmoid artery

E. SIGMOID ARTERIES: 2 or 3
   1. Origin: from inferior mesenteric artery
   2. Course and branches: runs caudally and laterally behind peritoneum, but anterior to psoas major muscle, ureter, and internal spermatic vessels. Enters sigmoid mesocolon and is distributed to that region. *Superior sigmoidal* anastomoses with left colic artery. *Inferior sigmoidal* anastomoses with superior rectal artery

F. SUPERIOR RECTAL
   1. Origin: continuation of inferior mesenteric artery
   2. Course: into pelvis in sigmoid mesocolon and crosses left common iliac vessels
   3. Branches: divides at third sacral segment giving 1 branch to either side of rectum

II. Veins

A. PORTAL VEIN formed by union of splenic and superior mesenteric veins
   1. Splenic vein receives inf. mesenteric and veins from pancreas and stomach
      a. Inferior mesenteric begins in rectum as superior rectal vein and drains sigmoid and descending colon
   2. Superior mesenteric vein receives veins from stomach, pancreas, duodenum, jejunum, ileum, cecum, appendix, ascending and transverse colon

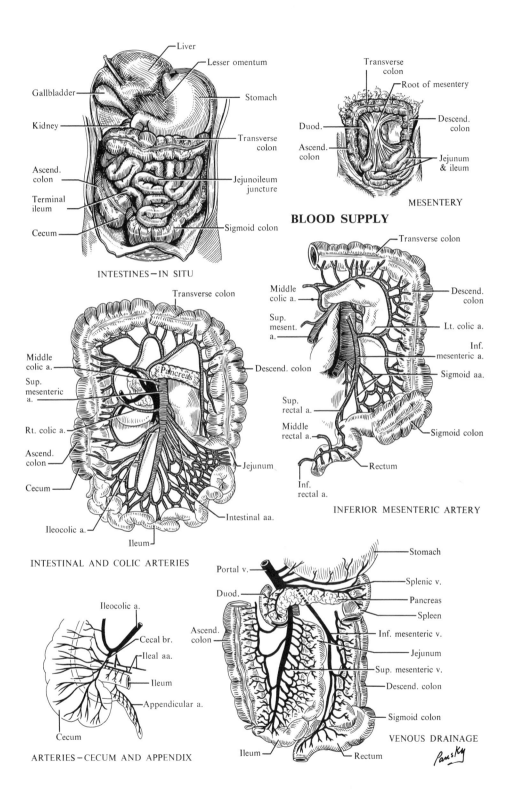

Liver

Lesser omentum

Gallbladder

Stomach

Kidney

Transverse colon

Ascend. colon

Jejunoileum juncture

Terminal ileum

Cecum

Sigmoid colon

INTESTINES — IN SITU

Transverse colon

Root of mesentery

Duod.

Descend. colon

Ascend. colon

Jejunum & ileum

MESENTERY

BLOOD SUPPLY

Transverse colon

Middle colic a.

Sup. mesenteric a.

Pancreas

Descend. colon

Rt. colic a.

Ascend. colon

Cecum

Jejunum

Ileocolic a.

Intestinal aa.

Ileum

INTESTINAL AND COLIC ARTERIES

Transverse colon

Middle colic a.

Sup. mesent. a.

Descend. colon

Lt. colic a.

Inf. mesenteric a.

Sigmoid aa.

Sup. rectal a.

Middle rectal a.

Sigmoid colon

Rectum

Inf. rectal a.

INFERIOR MESENTERIC ARTERY

Ileocolic a.

Cecal br.

Ileal aa.

Ileum

Appendicular a.

Cecum

ARTERIES — CECUM AND APPENDIX

Stomach

Portal v.

Splenic v.

Duod.

Pancreas

Spleen

Ascend. colon

Inf. mesenteric v.

Jejunum

Sup. mesenteric v.

Descend. colon

Sigmoid colon

VENOUS DRAINAGE

Ileum

Rectum

Pansky

153. GALLBLADDER AND DUCTS

I. **Surface projections:** fundus lies at right border of rectus muscle at end of ninth costal cartilage

II. **Location:** in fossa for gallbladder on visceral surface of liver. Cephalically is joined to liver by connective tissue. Caudally is covered by peritoneum

III. **Parts**
A. FUNDUS directed caudally, anteriorly, and to right
B. BODY extends cephalically, posteriorly, and to left
C. NECK in form of S-shaped curve

IV. **Relations**
A. FUNDUS: abdominal wall
B. BODY: cephalically, the liver; caudally, transverse colon, and descending duodenum

V. **Duct system**
A. CYSTIC: about 4 cm long, runs posteriorly, caudally, and toward left from neck of gallbladder and joins hepatic duct to form common bile duct. From neck to common duct contains spiral folds, *spiral folds* (*valves of Reister*)
B. HEPATIC DUCT arises from a major duct of the right and left lobes of liver, which leave through porta hepatis and join to form the main hepatic duct. Runs caudally and to the right in lesser omentum to join the cystic duct, forming common bile duct
C. COMMON DUCT: 7.5 cm in length. Runs caudally in free edge of hepatoduodenal ligament, with hepatic artery to its left and portal vein behind. Passes posterior to superior duodenum, then passes near right border of the head of pancreas. Major duct of pancreas approaches common duct, and the 2 run side by side in an oblique course through wall of duodenum
  1. Walls of terminal ends of 2 ducts are thickened by smooth muscle—*sphincter of ampulla* (*of Oddi*)
    a. Major duodenal papilla: an elevation into lumen of descending duodenum caused by above sphincter
    b. Common duct and duct of pancreas have common opening on major papilla in most cases

VI. **Blood supply and innervation**
A. ARTERIES: cystic from right hepatic artery. Gives a superficial branch, which supplies the free, inferior surface, and a deep branch to cephalic surface. (The origin and number of the cystic artery [ies] are variable)
B. VEINS
  1. Into liver capillaries
  2. Cystic vein opens into the right branch of the portal vein
C. LYMPHATICS (see p. 360) drain into hepatic nodes from bladder and into hepatic and pancreaticoduodenal nodes from the common bile duct
D. NERVES: autonomics by way of celiac plexus

VII. **Clinical consideration**
A. OBSTRUCTION OF THE DUCT SYSTEM as a result of "stone" formation (gallstone) leads to pain and jaundice

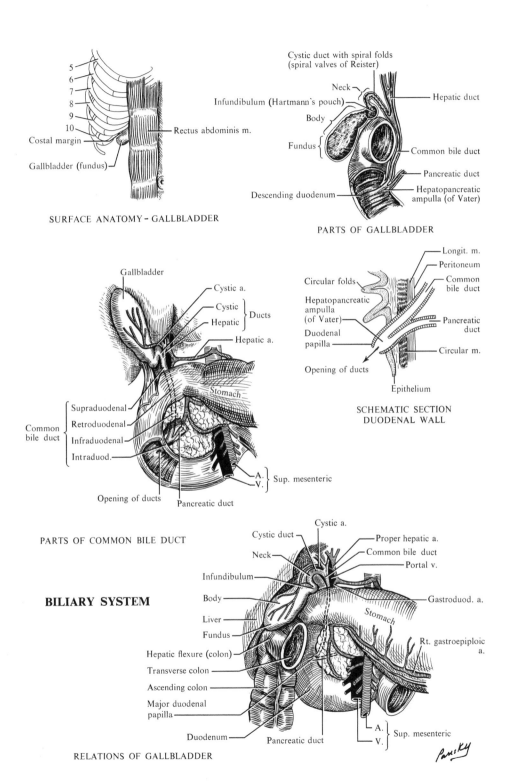

5
6
7
8
9
10
Costal margin
Gallbladder (fundus)

Rectus abdominis m.

SURFACE ANATOMY - GALLBLADDER

Cystic duct with spiral folds
(spiral valves of Reister)
Neck
Infundibulum (Hartmann's pouch)
Body
Fundus {
Descending duodenum

Hepatic duct
Common bile duct
Pancreatic duct
Hepatopancreatic
ampulla (of Vater)

PARTS OF GALLBLADDER

Gallbladder
Cystic a.
Cystic } Ducts
Hepatic
Hepatic a.

Stomach

Supraduodenal
Retroduodenal
Common
bile duct { Infraduodenal
Intraduod.

Opening of ducts
Pancreatic duct

A. } Sup. mesenteric
V.

PARTS OF COMMON BILE DUCT

Circular folds
Hepatopancreatic
ampulla
(of Vater)
Duodenal
papilla
Opening of ducts
Epithelium

Longit. m.
Peritoneum
Common
bile duct
Pancreatic
duct
Circular m.

SCHEMATIC SECTION
DUODENAL WALL

BILIARY SYSTEM

Cystic a.
Cystic duct
Neck
Infundibulum
Body
Liver
Fundus
Hepatic flexure (colon)
Transverse colon
Ascending colon
Major duodenal
papilla
Duodenum

Proper hepatic a.
Common bile duct
Portal v.
Gastroduod. a.

Stomach

Rt. gastroepiploic
a.

A. } Sup. mesenteric
V.

Pancreatic duct

Pansky

RELATIONS OF GALLBLADDER

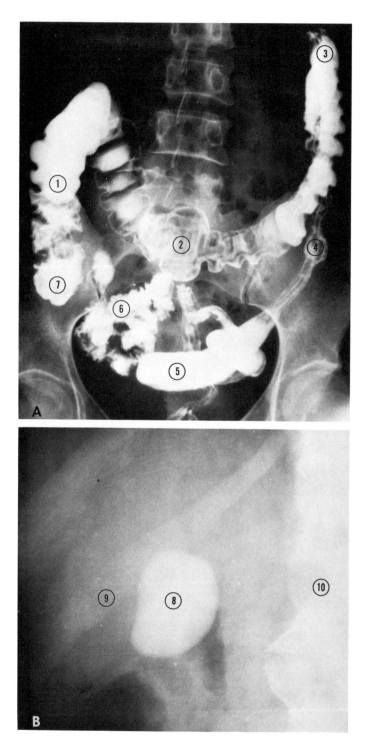

FIGURE 29. **Colon and gallbladder. A, Colon; B, gallbladder.** *1,* Ascending colon; *2,* transverse colon; *3,* splenic flexure; *4,* descending colon; *5,* sigmoid colon; *6,* ileum; *7,* cecum; *8,* gallbladder; *9,* rib; *10,* vertebral column.

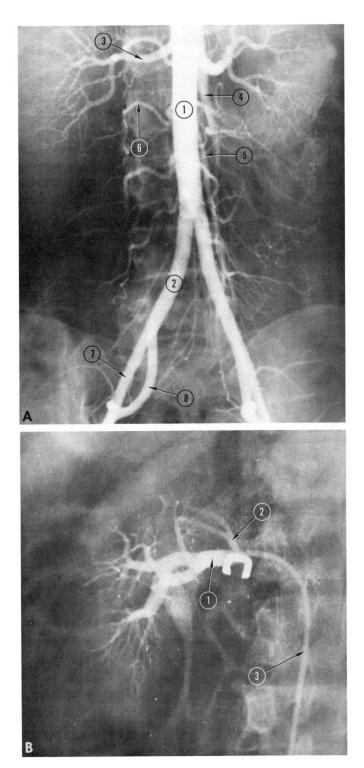

FIGURE 30. **Aortogram and renal arteriogram. A, Aortogram.** *1,* Abdominal aorta; *2,* right common iliac artery; *3,* right renal artery; *4,* superior mesenteric artery; *5,* inferior mesenteric artery; *6,* lumbar artery; *7,* right external iliac artery; *8,* right internal iliac artery. **B, renal arteriogram.** *1,* Renal artery; *2,* suprarenal artery; *3,* catheter.

-345-

154. LIVER

I. Surfaces

A. DIAPHRAGMATIC

    1. Anterior faces diaphragm, which separates it from sixth to tenth ribs and cartilages on right and from seventh and eighth cartilages, on left. Covered by peritoneum except along attachment of falciform ligament

    2. Superior under dome of diaphragm, which separates it from lungs on right and heart on left. Covered by peritoneum except dorsally, at edge of bare area

    3. Posterior (dorsal) fitted against vertebral column and crura of diaphragm. Large area not covered by peritoneum—*bare area*

       a. Concavity for vertebral column

       b. Sulcus (fossa) for inferior vena cava to right of this

       c. Fossa for ductus venosus to left of vena cava

       d. Suprarenal impression to right of vena cava

       e. Esophageal impression to left of fossa for ductus venosus

B. VISCERAL faces posteriorly, caudally, and toward left. Covered by peritoneum except at gallbladder and porta

    1. Porta: fissure in left central part for blood vessels and bile duct

    2. Right portion shows colic, renal, and duodenal impressions

    3. Fossa for gallbladder and fossa for umbilical vein

    4. Left portion: gastric impression, tuber omentale, and caudate process

II. Division into lobes

A. ON VISCERAL SURFACE there is an H-shaped arrangement of fossae and fissures, which delimit the lobes. The left sagittal fossa is made up of the fossa for the umbilical vein (ligamentum teres) anteriorly and the fossa for the ductus venosus posteriorly. The right fossa is divided into an anterior part (fossa for the gallbladder) and a posterior part (fossa for the inferior vena cava) by the *caudate process,* a band of liver tissue joining the caudate and right lobes. The *porta* is the transverse part of the H

B. LOBES

    1. Right: largest and lies to the right of the right sagittal fossa

    2. Quadrate: lies between the gallbladder and the fossa for the umbilical vein. Is ventral to porta

    3. Caudate: lies between vena cava and fossa for the ductus venosus. Is posterior to the porta and is joined to the right lobe by the *caudate process*

    4. Left: to the left of the left sagittal fossa

III. Relations: the various fossae and impressions given above indicate the relations except for the *tuber omentale,* which is directed toward the lesser curvature of the stomach and overlies the lesser omentum

IV. Liver segments: further subdivisions of the liver lobes into smaller segments based on the liver vascular pattern

V. Clinical considerations

A. JAUNDICE, in relation to the liver, is an accumulation of bile pigment in the blood stream frequently as a result of obstruction of the duct system

B. BECAUSE OF ITS GREAT VASCULARITY, the liver is a good site for secondary carcinoma from almost any other body site

C. CIRRHOSIS OF THE LIVER is due to an atrophy of the parenchyma and a hypertrophy of the connective tissue

LIVER

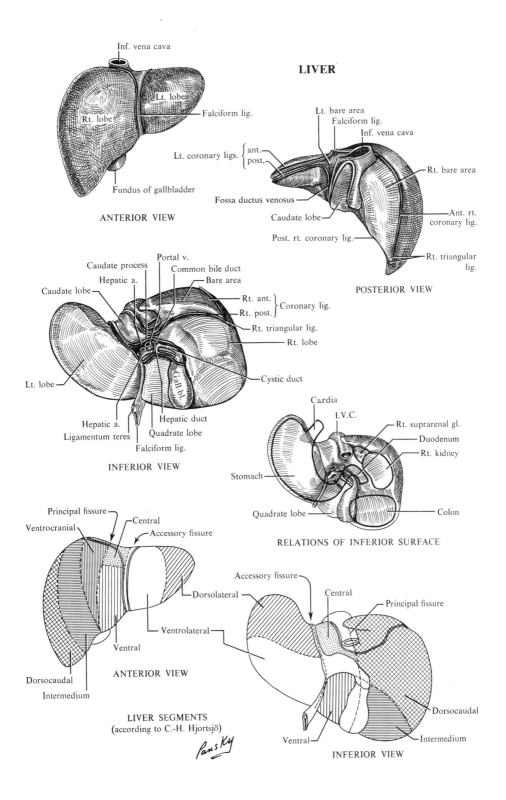

Inf. vena cava

Lt. lobe

Rt. lobe

Falciform lig.

Fundus of gallbladder

ANTERIOR VIEW

Lt. bare area
Falciform lig.
Inf. vena cava
Lt. coronary ligs. { ant.
post.
Rt. bare area
Fossa ductus venosus
Caudate lobe
Ant. rt. coronary lig.
Post. rt. coronary lig.
Rt. triangular lig.

POSTERIOR VIEW

Portal v.
Caudate process
Common bile duct
Hepatic a.
Bare area
Caudate lobe
Rt. ant. } Coronary lig.
Rt. post.
Rt. triangular lig.
Rt. lobe
Lt. lobe
Gall bl.
Cystic duct
Hepatic a.
Hepatic duct
Ligamentum teres
Quadrate lobe
Falciform lig.

INFERIOR VIEW

Cardia
I.V.C.
Rt. suprarenal gl.
Duodenum
Rt. kidney
Stomach
Quadrate lobe
Colon

RELATIONS OF INFERIOR SURFACE

Principal fissure
Central
Ventrocranial
Accessory fissure
Dorsolateral
Ventrolateral
Dorsocaudal
Ventral
Intermedium

ANTERIOR VIEW

LIVER SEGMENTS
(according to C.-H. Hjortsjö)

Accessory fissure
Central
Principal fissure
Dorsocaudal
Ventral
Intermedium

INFERIOR VIEW

155. LIVER LOBULE AND PORTAL CIRCULATION

I. **Liver lobule:** unit of structure
A. SHAPE: polygonal, with scant connective tissue between adjoining lobules
B. COMPOSITION
    1. Hepatic cells: arranged in "cords" or "plates" with bile canaliculi compressed between 2 adjoining cells
    2. Sinusoids: narrow, endothelial-lined channels between liver cords
    3. Central vein: large venous channel running longitudinally, in center of lobule
    4. Portal triad: in connective tissue, usually at one of the angles of the lobule, consisting of a group of 3 structures: bile duct, branch of portal vein, and branch of hepatic artery
C. BILE FLOW: bile is formed in hepatic cells, drains toward periphery of lobule through the canaliculi between the cells, and empties into small duct of triad
D. CIRCULATION: venous blood, carrying materials absorbed from alimentary canal, enters liver through portal vein and passes through branchings to reach the portal triad. From these branches, the portal blood enters sinusoids to reach central vein of lobule. Arterial blood enters through hepatic artery, carrying oxygenated blood through branches in portal triad, from which it enters sinusoids to reach central vein. The central veins are the actual beginnings of the hepatic vein system. Central veins from several lobules enter *sublobular* veins. Sublobular veins unite into increasingly larger trunks, which finally converge to form 3 hepatic veins that enter the vena cava

II. **Collateral portal circulation:** important clinically in case of obstruction of the portal vein. These are areas where portal and systemic systems anastomose
A. GASTRIC VEINS OF PORTAL and esophageal veins of azygos system
B. INFERIOR MESENTERIC OF PORTAL and rectal of internal iliac system
C. BRANCHES ALONG FALCIFORM (PARUMBILICAL) tributaries of the portal vein and superior and inferior epigastric veins
D. INTESTINAL VEINS OF PORTAL (RT. COLIC, ILEOCOLIC, LT. COLIC) with retroperitoneal (lumbar) tributaries of the inferior vena cava

III. **Clinical considerations**
A. PORTAL OBSTRUCTION: in an attempt to get portal blood back to the heart by bypassing the obstruction, the veins in the region of anastomoses may be engorged, creating esophageal varicosities, hemorrhoids, and varicosities on the abdominal wall around umbilicus (caput medusae)
B. PORTACAVAL SHUNT: a procedure for relief of portal vein obstruction. Basically, it consists of either anastomosing the portal vein directly with the inferior vena cava or anastomosing one of its tributaries into the vena cava or into another systemic vein, for example, the splenic vein into the left renal vein

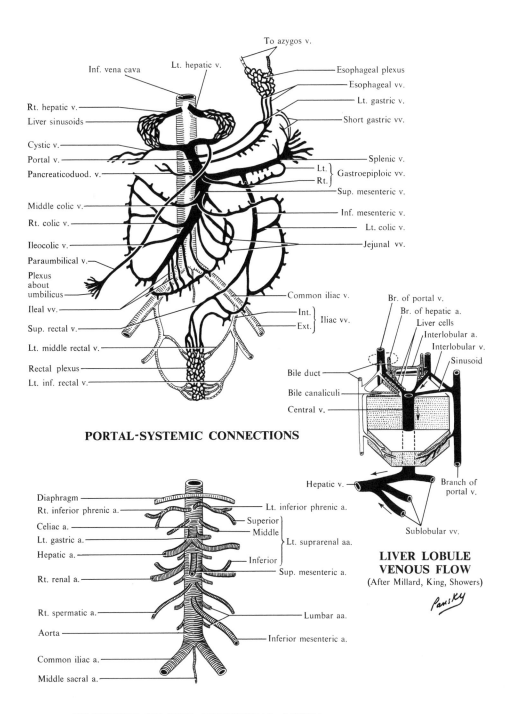

To azygos v.

Inf. vena cava

Lt. hepatic v.

Esophageal plexus

Esophageal vv.

Lt. gastric v.

Rt. hepatic v.

Short gastric vv.

Liver sinusoids

Cystic v.

Splenic v.

Portal v.

Pancreaticoduod. v.

Lt. } Gastroepiploic vv.
Rt. }

Sup. mesenteric v.

Middle colic v.

Inf. mesenteric v.

Rt. colic v.

Lt. colic v.

Ileocolic v.

Jejunal vv.

Paraumbilical v.

Plexus
about
umbilicus

Common iliac v.

Ileal vv.

Int. }
Ext. } Iliac vv.

Sup. rectal v.

Lt. middle rectal v.

Rectal plexus

Lt. inf. rectal v.

PORTAL-SYSTEMIC CONNECTIONS

Br. of portal v.

Br. of hepatic a.

Liver cells

Interlobular a.

Interlobular v.

Sinusoid

Bile duct

Bile canaliculi

Central v.

Hepatic v.

Branch of
portal v.

Sublobular vv.

**LIVER LOBULE
VENOUS FLOW**
(After Millard, King, Showers)

Pansky

Diaphragm

Rt. inferior phrenic a.

Lt. inferior phrenic a.

Celiac a.

Superior
Middle

Lt. gastric a.

Lt. suprarenal aa.

Hepatic a.

Inferior

Rt. renal a.

Sup. mesenteric a.

Rt. spermatic a.

Lumbar aa.

Aorta

Inferior mesenteric a.

Common iliac a.

Middle sacral a.

BRANCHES OF THE ABDOMINAL AORTA

156. SPLEEN

I. Type of structure: a lymphatic organ interposed in the blood stream

II. Surface projection
A. LONG AXIS: in line of tenth rib
B. ENDS: medially 4 cm from dorsal midline, laterally in midaxillary line in ninth interspace
C. BORDERS: cephalic, upper ninth rib; caudal, lower eleventh rib

III. Relations
A. DIAPHRAGMATIC: convex, smooth, facing upward, backward, and to left. Related to diaphragm, which separates it from ninth, tenth, and eleventh ribs, and left lung and pleura
B. VISCERAL
   1. Gastric: concave; faces anteriorly, cephalically, and medially. Contacts posterior side of stomach and tail of pancreas
   2. Renal: flattened; faces medially and caudally. Related to upper anterior surface of left kidney
   3. Colic: small and slightly concave, at anterior extremity, related to left colic flexure
C. POSTERIOR (SUPERIOR) EXTREMITY: directed toward vertebral column
D. ANTERIOR (INFERIOR) EXTREMITY: rests on left colic flexure and phrenicocolic ligament

IV. Support
A. PHRENICOLIENAL (LIENORENAL) LIGAMENT: reflection of peritoneum running from diaphragm and anterior aspect of left kidney to hilum of spleen
   1. Contents: splenic vessels
B. GASTROLIENAL: dorsal mesentery between spleen and stomach
   1. Contents: short gastric and left gastroepiploic vessels
C. PHRENICOCOLIC LIGAMENT: beneath caudal end of spleen

V. Blood supply
A. SPLENIC ARTERY (from celiac artery): large; tortuous course across the posterior wall of the omental bursa and runs through the phrenicolienal ligament to hilum of spleen. Divides into 6 or more branches in hilum
B. SPLENIC VEIN arises from the union of 6 or more veins that emerge from hilum. Vein runs in groove on back of pancreas, below the artery, and ends behind the neck of the pancreas by joining the superior mesenteric vein to form the *portal vein*

VI. Lymphatic drainage into pancreaticolienal nodes (see p. 360)

VII. Nerves: chiefly postganglionics of the sympathetic system from the celiac plexus to the blood vessels, capsule, and trabeculae of the organ

VIII. Functions: storage of red blood cells, which can be forced back into the circulation in a respiratory crisis by contraction of the smooth muscle in the capsule and trabeculae; destruction of worn-out red blood cells; removal of foreign material from the blood stream; and production of mononuclear leukocytes

IX. Clinical consideration
A. A HYPERTROPHIC SPLEEN due to overactivity of its macrophage system can be removed without any *apparent* ill effects

SPLEEN

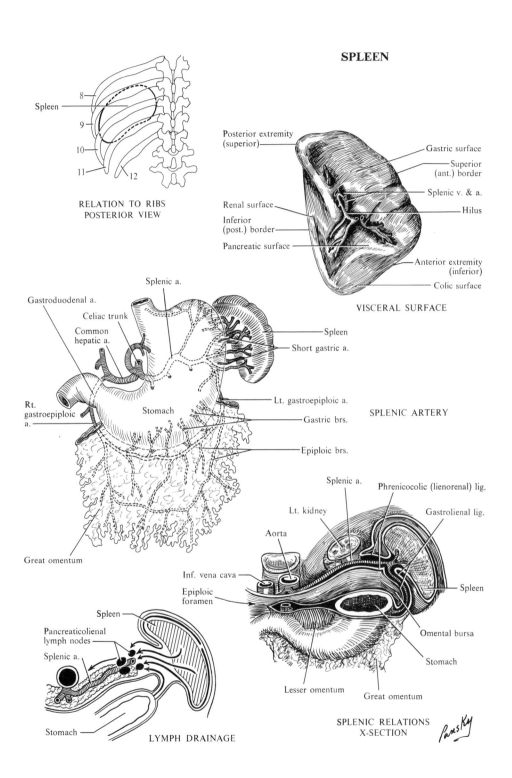

Spleen

8
9
10
11 12

**RELATION TO RIBS
POSTERIOR VIEW**

Posterior extremity
(superior)

Gastric surface

Superior
(ant.) border

Splenic v. & a.

Hilus

Renal surface

Inferior
(post.) border

Pancreatic surface

Anterior extremity
(inferior)

Colic surface

VISCERAL SURFACE

Splenic a.

Gastroduodenal a.

Celiac trunk

Common
hepatic a.

Spleen

Short gastric a.

Lt. gastroepiploic a.

Rt.
gastroepiploic
a.

Stomach

Gastric brs.

SPLENIC ARTERY

Epiploic brs.

Great omentum

Splenic a.

Phrenicocolic (lienorenal) lig.

Lt. kidney

Gastrolienal lig.

Aorta

Spleen

Inf. vena cava

Epiploic
foramen

Omental bursa

Stomach

Spleen

Pancreaticolienal
lymph nodes

Splenic a.

Lesser omentum

Great omentum

Stomach

LYMPH DRAINAGE

**SPLENIC RELATIONS
X-SECTION**

Pansky

-351-

157. THE KIDNEY AND ITS BLOOD VESSELS

I. Nephron: the functional unit of the kidney

A. PARTS: glomerulus, an arterial capillary net; Bowman's capsule surrounding the capillary net; proximal convoluted tubule; Henle's loop; distal convoluted tubule

II. Duct system: nephron joins the duct system, which eventually opens at the apex of the renal pyramid into the minor calyces; the minor calyces unite to form 2 or 3 major calyces, which, in turn, join to form the renal pelvis, the expanded cephalic end of the ureter

III. General structure

A. CAPSULE: tough, fibrous tissue

B. HILUM: medial fissure, which expands into a large central cavity, the *renal sinus,* which contains the proximal part of the renal pelvis, the calyces, branches of renal vessels, nerves, and fat

C. MEDULLA: 8 to 18 pyramidal masses, striated in appearance. Contains collecting ducts and portions of Henle's loop

D. CORTEX lies beneath capsule, overlies bases of medullary pyramids and dips down between them as the *renal columns*
   1. Pars radiata: conical projections from bases of pyramids into cortex. Contains ducts and parts of Henle's loop
   2. Pars convoluta: surrounds radiate part. Contains glomeruli, capsules, and convoluted parts of nephron

IV. Blood vessels

A. ARTERIES. Renal artery branches into *interlobar arteries* between pyramids. At bases of pyramids the interlobar arteries form the *arcuate arteries.* Along their course, the arcuate arteries send branches into the cortex as the *interlobular arteries.* The latter give rise chiefly to *afferent glomerular arteries* and to some *nutrient and perforating capsular arteries.* The capillary net of the glomerulus coalesces to form the *efferent glomerular arteries.* These vessels break up into a true capillary net around the nephrons and also give rise to a few *arteriolae rectae,* which enter the medulla and run toward the pelvis

B. VEINS. Begin in venous plexuses draining the capillary bed around the tubules. These open into *venae rectae,* then into *interlobular veins, arcuate veins, interlobar veins,* and finally the *renal vein*
   1. Stellate veins lie beneath the capsule and drain part of the area supplied by perforating capsular arteries. These veins drain into the interlobular veins

V. Anastomoses between renal and systemic vessels occur in fat around kidney where the perforating capsular arteries join branches from suprarenal, spermatic (or ovarian), superior and inferior mesenteric arteries

VI. Special features

A. WITHIN THE RENAL SINUS, each arterial ramus rebranches, and although the pattern is quite variable, it is said that the distribution is constant enough to allow the division of the kidney into vascular segments that correspond to the prevailing vascular pattern. *Five segmental arteries* and therefore *five renal segments* are described

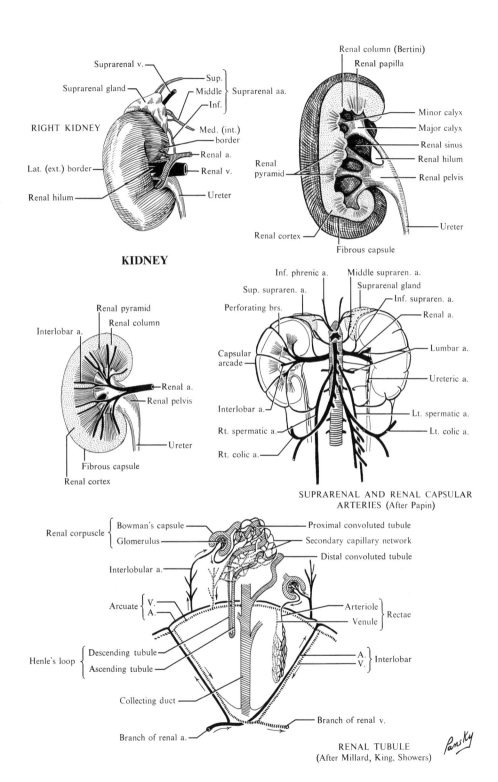

Suprarenal v.

Suprarenal gland

Sup.
Middle } Suprarenal aa.
Inf.

RIGHT KIDNEY

Med. (int.) border

Renal a.

Lat. (ext.) border

Renal v.

Renal hilum

Ureter

KIDNEY

Renal column (Bertini)
Renal papilla

Minor calyx
Major calyx
Renal sinus
Renal hilum
Renal pelvis

Renal pyramid

Ureter

Renal cortex

Fibrous capsule

Renal pyramid
Renal column

Interlobar a.

Renal a.
Renal pelvis

Ureter

Fibrous capsule
Renal cortex

Inf. phrenic a.   Middle supraren. a.
Sup. supraren. a.   Suprarenal gland
Perforating brs.   Inf. supraren. a.
Renal a.

Capsular arcade

Lumbar a.

Ureteric a.

Interlobar a.

Rt. spermatic a.

Lt. spermatic a.
Lt. colic a.

Rt. colic a.

SUPRARENAL AND RENAL CAPSULAR
ARTERIES (After Papin)

Renal corpuscle { Bowman's capsule
Glomerulus

Proximal convoluted tubule
Secondary capillary network
Distal convoluted tubule

Interlobular a.

Arcuate { V.
A.

Arteriole
Venule } Rectae

Henle's loop { Descending tubule
Ascending tubule

A.
V. } Interlobar

Collecting duct

Branch of renal v.

Branch of renal a.

RENAL TUBULE
(After Millard, King, Showers)

Pansky

158. RELATIONS AND FIXATION OF THE KIDNEY

I. Relations
A. RIGHT
1. Anterior: convex, faces ventrally and to the right. Related to *suprarenal gland, visceral surface of the liver, colic flexure,* and the *small intestine*
2. Posterior: more flattened. Embedded in fat with no peritoneum. Cephalic pole lies on twelfth rib; below this are the diaphragm, lumbocostal arches, psoas and quadratus lumborum muscles, and tendon of transversus abdominis muscle. Crossed by upper lumbar arteries, T12, iliohypogastric, and ilioinguinal nerves
3. Lateral border: convex without important relations
4. Medial border: *hilum* for renal vessels and ureter near center. Above the hilum, it is in contact with the suprarenal, and below the hilum, with the ureter
B. LEFT
1. Anterior: convex; faces anteriorly and to the left. Related to *suprarenal, spleen, body of pancreas* (splenic vessels), *stomach, left colic flexure, small intestine*
2. Posterior: less convex. Embedded in fat with no peritoneum. Superior pole rests on eleventh rib, the twelfth rib more caudally. In other respects, similar to right
3. Left border: similar to right
4. Medial: similar to right

II. Relations of structures at hilum: renal vein most anterior, artery intermediate, and ureter most posterior. Branches of arteries and veins may pass posterior to the ureter

III. Renal vessels
A. VEINS terminate in the vena cava
1. Left: longer than right, crosses anterior side of aorta just below the superior mesenteric artery and opens into the inferior vena cava above the right vein
    a. Tributaries: left inferior phrenic, left internal spermatic, and left suprarenal
2. Right: short, lies in front of renal artery. No extrarenal tributaries
B. ARTERIES arise from the aorta at right angles, at level of disk between L1 and L2
1. Left: slightly above level of right renal artery. Lies posterior to renal vein, body of pancreas, and splenic vein. Inferior mesenteric vein crosses it anteriorly
2. Right: longer than left. Passes behind inferior vena cava and right renal vein with head of pancreas and descending duodenum overlying the veins

IV. Renal fascia
A. FORMATION: from subserous extraperitoneal fascia; splits near the lateral border of the kidney
1. Anterior layer over anterior surface and continues over renal vessels and aorta to join similar layer of the other side
2. Posterior layer continues beneath the kidney, but passes posterior to the aorta to the other side, but is adherent to the deep fascia
B. CONNECTIONS TO KIDNEY: fibrous strands of renal fascia connect to capsule
C. ADIPOSE CAPSULE (PERIRENAL FAT): fatty tissue lying between the fascia and surface of the kidney
1. Pararenal fat: fat behind renal fascia

V. Support: renal fascia and vessels, adipose capsule, and pararenal fat

VI. Clinical considerations
A. CONGENITAL ANOMALIES
1. Unascended kidney: kidney may remain near original site of development and might be found near pelvic brim with common iliac vessels
2. Horseshoe kidney: the caudal poles of the two kidneys are joined

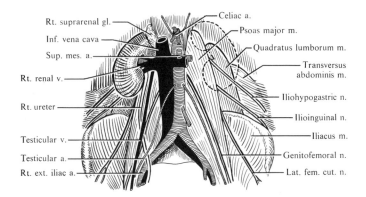

Rt. suprarenal gl. — Celiac a.
Inf. vena cava — Psoas major m.
Sup. mes. a. — Quadratus lumborum m.
Rt. renal v. — Transversus abdominis m.
Rt. ureter — Iliohypogastric n.
— Ilioinguinal n.
Testicular v. — Iliacus m.
Testicular a. — Genitofemoral n.
Rt. ext. iliac a. — Lat. fem. cut. n.

KIDNEY RELATIONS

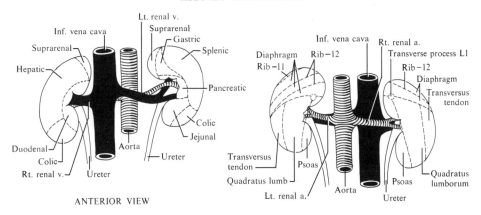

Inf. vena cava — Lt. renal v.
Suprarenal — Suprarenal
Hepatic — Gastric
— Splenic
— Pancreatic
Duodenal — Colic
Colic — Jejunal
Rt. renal v. — Colic
Ureter — Aorta — Ureter

ANTERIOR VIEW

Inf. vena cava — Rt. renal a.
Diaphragm — Transverse process L1
Rib-11 — Rib-12
Rib-12 — Diaphragm
— Transversus tendon
Transversus tendon — Psoas
Quadratus lumb — Quadratus lumborum
Lt. renal a. — Aorta — Ureter — Psoas

POSTERIOR VIEW

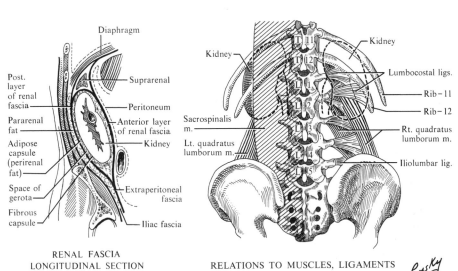

Diaphragm
Post. layer of renal fascia — Suprarenal
Pararenal fat — Peritoneum
Adipose capsule (perirenal fat) — Anterior layer of renal fascia
— Kidney
Space of gerota — Extraperitoneal fascia
Fibrous capsule — Iliac fascia

Kidney — Kidney
Kidney — Lumbocostal ligs.
Sacrospinalis m. — Rib-11
Lt. quadratus lumborum m. — Rib-12
— Rt. quadratus lumborum m.
— Iliolumbar lig.

RENAL FASCIA LONGITUDINAL SECTION

RELATIONS TO MUSCLES, LIGAMENTS

Pansky

-355-

159. URETERS

I. Origin: renal pelvis at the level of the spine of L1

II. Parts and relations

A. ABDOMINAL: both lie beneath the peritoneum, embedded in subserous tissue on the medial part of psoas major muscle; both are crossed by the spermatic (ovarian) vessels. This part ends as it enters the true pelvis crossing the bifurcation of the common iliac vessels
   1. Right: near origin, covered by descending duodenum, to the right of the inferior vena cava. Crossed by right colic and ileocolic vessels, the mesentery, and terminal ileum
   2. Left: crossed by left colic vessels and sigmoid mesocolon

B. PELVIC
   1. Male: runs caudally on the lateral pelvic wall along the anterior border of the greater sciatic notch. Is anterior to internal iliac artery and medial to obturator, inferior vesical, and middle rectal arteries. At level of lower part of sciatic notch, it turns medially to reach the lateral angle of the bladder. Here it lies anterior to the seminal vesicle. The vas crosses over it as it approaches the bladder
   2. Female (see p. 380): forms the posterior boundary of the ovarian fossa. Runs medially and anteriorly on lateral aspect of cervix and upper vagina to fundus of the bladder. In part of its course is accompanied by uterine artery, which then crosses over the ureter

C. INTRAMURAL (see p. 366) runs obliquely through the bladder wall for a distance of 2 cm to open at lateral angles of trigone of bladder

III. Constrictions: areas of diminished diameter

A. URETEROPELVIC JUNCTION
B. AT CROSSING OF ILIAC VESSELS
C. AT JUNCTION WITH BLADDER

IV. Vessels and nerves

A. ARTERIES: ureteric branches of renal, internal spermatic, superior and inferior vesical arteries
B. VEINS: follow correspondingly named arteries and terminate in correspondingly named veins
C. LYMPHATICS pass to the lumbar and internal iliac nodes
D. NERVES: spermatic, renal, and hypogastric plexuses

V. Clinical considerations

A. KIDNEY STONES may descend in the ureter and become lodged, particularly in the areas of ureteric constriction, and result in a great deal of pain and urinary retention, which can damage the kidney structure
B. CONGENITAL ANOMALIES: ureters may be double over part or all of extent
C. SURGICAL APPROACH TO BOTH KIDNEY AND ADRENAL: from back and side, by an incision below and parallel to the 12th rib. If necessary, it can be extended to the front of the abdomen, paralleling the inguinal ligament. In this way, the entire procedure can remain retroperitoneal; it avoids cutting nerves since the incision is parallel to their course; and the kidney structures can be separated from overlying structures such as duodenum and pancreas, since kidney belongs to the dorsal body wall, whereas other organs have become only secondarily adherent
D. OBSTRUCTION of the ureter at any level leads to dilatation of the parts above, including the renal pelvis and calices, resulting in *hydronephrosis*

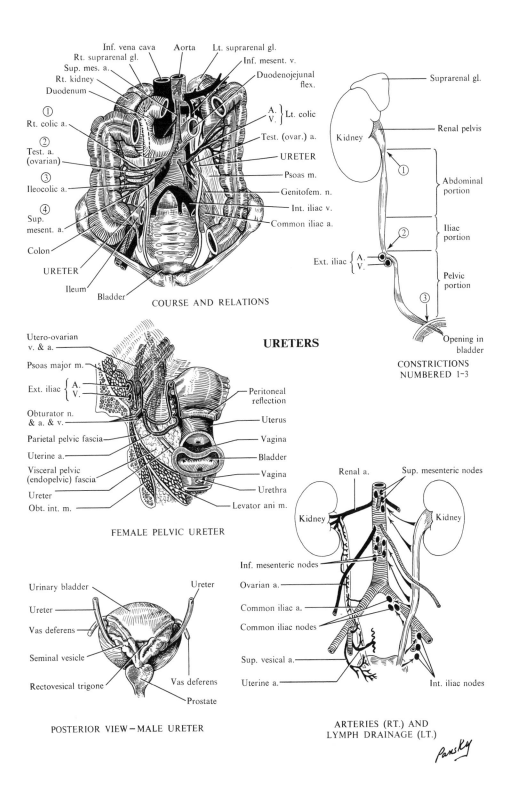

COURSE AND RELATIONS

Inf. vena cava
Rt. suprarenal gl.
Sup. mes. a.
Rt. kidney
Duodenum
Aorta
Lt. suprarenal gl.
Inf. mesent. v.
Duodenojejunal flex.
A. }
V. } Lt. colic
Test. (ovar.) a.
URETER
Psoas m.
Genitofem. n.
Int. iliac v.
Common iliac a.

① Rt. colic a.
② Test. a. (ovarian)
③ Ileocolic a.
④ Sup. mesent. a.
Colon
URETER
Ileum
Bladder

URETERS

Suprarenal gl.
Kidney
Renal pelvis
① Abdominal portion
Ext. iliac { A. V. }
② Iliac portion
Pelvic portion
③ Opening in bladder

CONSTRICTIONS NUMBERED 1–3

Utero-ovarian v. & a.
Psoas major m.
Ext. iliac { A. V. }
Obturator n. & a. & v.
Parietal pelvic fascia
Uterine a.
Visceral pelvic (endopelvic) fascia
Ureter
Obt. int. m.
Peritoneal reflection
Uterus
Vagina
Bladder
Vagina
Urethra
Levator ani m.

FEMALE PELVIC URETER

Urinary bladder
Ureter
Vas deferens
Seminal vesicle
Rectovesical trigone
Ureter
Vas deferens
Prostate

POSTERIOR VIEW – MALE URETER

Renal a.
Sup. mesenteric nodes
Kidney
Kidney
Inf. mesenteric nodes
Ovarian a.
Common iliac a.
Common iliac nodes
Sup. vesical a.
Uterine a.
Int. iliac nodes

ARTERIES (RT.) AND LYMPH DRAINAGE (LT.)

Pansky

160. SUPRARENAL GLAND

I. Position, size, shape, and relations
A. LOCATION: at cranial pole of kidney, at L1, within renal fascia
B. SIZE: length and width, 3.0–5.0 cm; thickness, 0.4–0.6 cm; weight, 3.5–5.0 g, heavier in male than female
C. RELATIONS AND SHAPE
  1. Right: pyramidal, with hilum below apex near anterior border
    a. Anterior: medially, inferior vena cava without peritoneum; laterally liver, without peritoneum above, with peritoneum below
    b. Posterior: diaphragm above; cranial pole and anterior surface of right kidney below
  2. Left: semilunar in shape; hilum near caudal end of anterior surface. Slightly larger than right
    a. Anterior: peritoneum of omental bursa above; pancreas and splenic vein, without peritoneum, below
    b. Posterior: medially, left crus of diaphragm; laterally, anterior surface of left kidney

II. Blood supply
A. ARTERIES
  1. Superior suprarenal artery from inferior phrenic artery (from aorta)
  2. Middle suprarenal artery, directly from aorta
  3. Inferior suprarenal artery from renal artery (from aorta)
B. VEINS
  1. Suprarenal vein receives blood from all parts of the gland and leaves through the hilum
    a. Right: enters inferior vena cava
    b. Left: enters the left renal vein

III. Parts
A. THICK CORTEX derived from the mesoderm
B. MEDULLA derived from the embryonic neural crest ectoderm

IV. Secretions
A. CORTEX: cortisol, corticosterone, aldosterone, 11-dehydroepiandrosterone, progesterone, estradiol, and estrone
B. MEDULLA: epinephrine and norepinephrine

V. General functions
A. CORTEX: gluconeogenesis; enhances water diuresis, probably by increasing glomerular filtration; maintains electrolyte balance, thus maintaining blood volume and pressure; fat deposition; has effect on lymphocytes (lympholytic); anti-inflammatory and anti-allergic. Normally has little effect on sex
B. MEDULLA: essential to the "fight-or-flight" mechanism; raises blood pressure; increases heart rate; dilates bronchi and breaks down glycogen, thus elevating blood sugar

VI. Clinical considerations (chief pathology involves cortex only)
A. HYPERACTIVITY: Cushing's disease; adrenogenital syndrome; primary or secondary aldosteronism
B. HYPOFUNCTION: acute adrenal insufficiency (Waterhouse-Friderichsen syndrome); chronic adrenal insufficiency (Addison's disease)

SUPRARENAL GLANDS

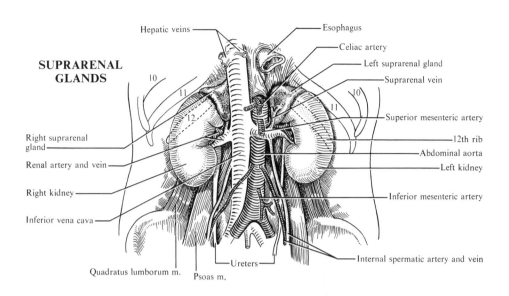

Hepatic veins
Esophagus
Celiac artery
Left suprarenal gland
Suprarenal vein
10
11
12
10
11
Superior mesenteric artery
Right suprarenal gland
12th rib
Abdominal aorta
Renal artery and vein
Left kidney
Right kidney
Inferior mesenteric artery
Inferior vena cava
Internal spermatic artery and vein
Quadratus lumborum m.
Psoas m.
Ureters

SUPRARENAL VESSELS

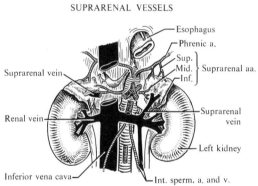

Esophagus
Phrenic a.
Sup.
Mid. } Suprarenal aa.
Inf.
Suprarenal vein
Suprarenal vein
Renal vein
Left kidney
Inferior vena cava
Int. sperm. a. and v.

SUPRARENAL ARTERIES

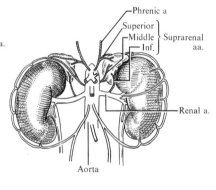

Phrenic a
Superior
Middle } Suprarenal aa.
Inf.
Renal a.
Aorta

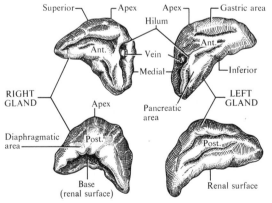

Superior
Apex
Apex
Gastric area
Hilum
Ant.
Ant.
Vein
Medial
Inferior
RIGHT GLAND
Apex
LEFT GLAND
Pancreatic area
Diaphragmatic area
Post.
Post.
Base (renal surface)
Renal surface

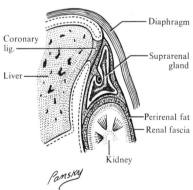

Diaphragm
Coronary lig.
Suprarenal gland
Liver
Perirenal fat
Renal fascia
Kidney

Pansky

161. LYMPHATICS OF THE ABDOMEN

I. Lymph nodes
A. PARIETAL
    1. Epigastric: along inferior epigastric vessels
    2. Lumbar
        a. Right lateral aortic: anterior to inferior vena cava at level of renal vessels, posterior to inferior vena cava, on the origins of the right psoas muscle and right crus of diaphragm. Afferents: from common iliac nodes; ovary, testis, uterus, kidney, suprarenal, and abdominal muscles. Efferents: chiefly form the right lumbar trunk, but some pass to pre- and retroaortic nodes or thoracic duct
        b. Left lateral aortic: on left side of aorta, on origin of left psoas muscle and left crus of diaphragm. Afferents and efferents similar to above but for left side
        c. Preaortic: in front of aorta around origin of 3 major arterial branches and named accordingly: celiac, superior mesenteric, and inferior mesenteric. Afferents: few from lateral aortics, mainly from viscera supplied by the related arteries. Efferents: few to retroaortics, mainly to intestinal trunk
        d. Retroaortic: on bodies of third and fourth lumbar vertebrae behind aorta. Afferents: from lateral and preaortic nodes. Efferents to cisterna chyli

II. Lymphatic drainage of viscera
A. STOMACH: see p. 328
B. LIVER
    1. Convex surface: to posterior mediastinal nodes, superior gastric nodes, celiac group of preaortic nodes, and hepatic nodes
    2. Visceral surface: hepatic and posterior mediastinal nodes
C. GALLBLADDER: to hepatic and pancreaticoduodenal nodes; to hepatic nodes from common duct
D. DUODENUM: to pancreaticoduodenal nodes and thence to hepatic and preaortic (superior mesenteric) nodes
E. JEJUNUM AND ILEUM: vessels are the lacteals to mesenteric nodes in mesentery and then to superior mesenteric group of preaortic nodes
F. COLON
    1. Ascending and transverse: through right colic and middle colic nodes to mesenteric nodes to superior mesenteric group of preaortics
    2. Descending and sigmoid; left colic nodes along left colic and sigmoid arteries to inferior mesenteric group of preaortics
G. PANCREAS: to pancreaticolienal nodes (thence to celiac group of preaortics), to pancreaticoduodenal nodes, and to superior mesenteric group of preaortic nodes

III. Cisterna chyli
A. LOCATION: in front of second lumbar vertebra, behind and to right of aorta beside right crus of diaphragm
B. FORMATION: right and left lumbar trunks and intestinal trunk
C. TERMINATION: narrows down and passes through aortic hiatus of diaphragm to become thoracic duct

IV. Clinical considerations
A. REGIONAL LYMPH NODES: a surgeon may judge the extent of metastases from a malignancy by examining nodes draining the area; for example, from the sigmoid colon, first check for nodes in sigmoid mesocolon and then examine the inferior mesenteric group found at the origin of the inferior mesenteric artery

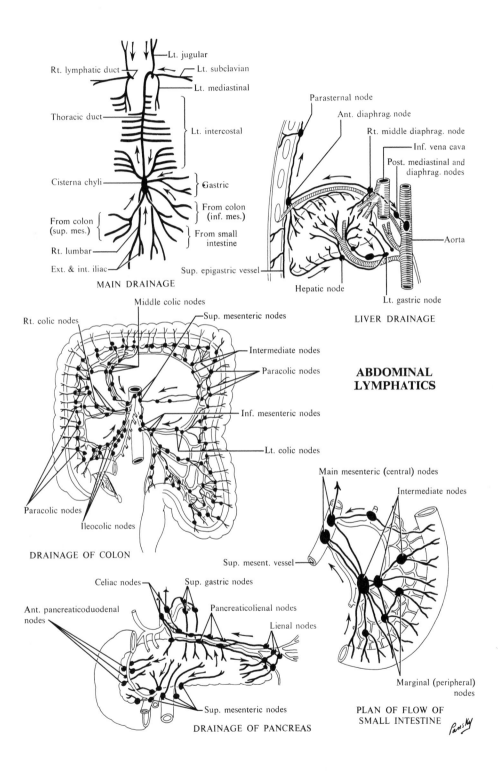

MAIN DRAINAGE

Lt. jugular
Rt. lymphatic duct
Lt. subclavian
Lt. mediastinal
Thoracic duct
Lt. intercostal
Cisterna chyli
Gastric
From colon (inf. mes.)
From colon (sup. mes.)
From small intestine
Rt. lumbar
Ext. & int. iliac

LIVER DRAINAGE

Parasternal node
Ant. diaphrag. node
Rt. middle diaphrag. node
Inf. vena cava
Post. mediastinal and diaphrag. nodes
Aorta
Sup. epigastric vessel
Hepatic node
Lt. gastric node

DRAINAGE OF COLON

Middle colic nodes
Rt. colic nodes
Sup. mesenteric nodes
Intermediate nodes
Paracolic nodes
Inf. mesenteric nodes
Lt. colic nodes
Paracolic nodes
Ileocolic nodes

ABDOMINAL LYMPHATICS

PLAN OF FLOW OF SMALL INTESTINE

Main mesenteric (central) nodes
Intermediate nodes
Sup. mesent. vessel
Marginal (peripheral) nodes

DRAINAGE OF PANCREAS

Celiac nodes
Sup. gastric nodes
Ant. pancreaticoduodenal nodes
Pancreaticolienal nodes
Lienal nodes
Sup. mesenteric nodes

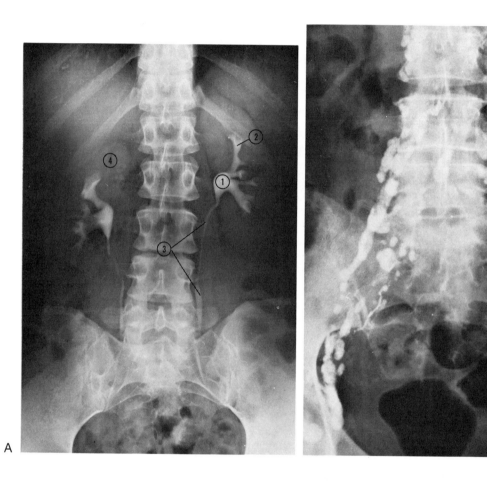

A

B

FIGURE 31. **Intravenous pyelogram and lymphangiogram. A, Intravenous pyelogram; B, lymphangiogram, showing iliac nodes on the right side.** *1,* Renal pelvis; *2,* major calyx; *3,* ureter; *4,* kidney.

-362-

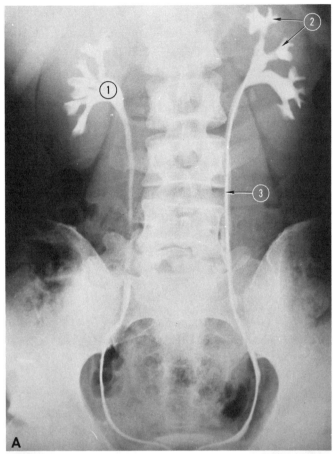

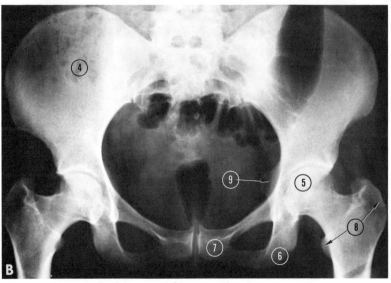

FIGURE 32. **Retrograde pyelogram and normal pelvis. A, Retrograde pyelogram; B, normal pelvis, adult.** *1,* Pelvis of kidney; *2,* major and minor calyces; *3,* ureter; *4,* ilium; *5,* head of femur; *6,* ischium; *7,* pubis; *8,* greater and lesser trochanters of femur; *9,* spine of ischium.

162. THE DIAPHRAGM

I. Location, configuration, and composition
A. SERVES AS A SEPTUM between the thoracic and abdominal cavities
B. Dome-shaped, with concavity facing caudally
C. COMPOSED OF SKELETAL MUSCLE and dense collagenous connective tissue

II. Origin of muscular fibers
A. STERNAL PART: 2 muscular bands from the dorsal side of the xiphoid process
B. COSTAL PART: from costal cartilages and bone of ribs 7–12
C. LUMBAR PART: from the lumbocostal arches and crura
   1. Lumbocostal ligaments (arches)
      a. Medial arcuate ligament: tendinous arch crossing the psoas muscle. It is attached medially to the body of first (and second) lumbar vertebrae and laterally to the anterior transverse process of first (and second) lumbar vertebrae
      b. Lateral arcuate ligament: tendinous arch crossing the quadratus lumborum muscle. It is attached medially to the anterior transverse process of the first lumbar vertebra and laterally to the tip of rib 12
   2. Crura
      a. Right: larger than left and arises from the bodies and disks of lumbar vertebrae 1–3. The most medial fibers cross in front of the aorta to the left side
      b. Left: arises from the bodies and disks of lumbar vertebrae 1 and 2. Some of its medial fibers cross to other side

III. Insertion: central tendon, where muscular fibers converge and become tendinous near the center of diaphragm

IV. Innervation and action
A. INNERVATION: phrenic nerve of cervical plexus, C4 (also C3 and C5)
B. ACTION: contraction of the muscle causes descent of the central tendon. This decreases pressure and increases the volume of the thoracic cavity resulting in air being "pushed" into the lungs

V. Orifices in the diaphragm, with the major structures passing through
A. ESOPHAGEAL HIATUS: at the level of the tenth thoracic vertebra. Transmits esophagus and right and left vagus nerves
B. AORTIC HIATUS: at the level of the twelfth thoracic vertebra, just to the left of the midline. Transmits aorta, azygos vein, and thoracic duct. (Some anatomists consider this hiatus as being behind diaphragm)
C. VENA CAVAL FORAMEN (HIATUS): about at the level of the disk between the eighth and ninth thoracic vertebrae, to the right of the midline. Transmits inferior vena cava and small branches of right phrenic nerve
D. MINOR OPENINGS
   1. Right crus: 2, for the right greater and lesser splanchnic nerves
   2. Left crus: 3, for the left greater and lesser splanchnic nerves and the hemiazygos vein
   3. Anteriorly: between sternal and costal parts of the diaphragm for the passage of the superior epigastric artery

VI. Clinical consideration
A. HIATUS HERNIA: an opening, usually on the left side, of the diaphragm near the esophageal hiatus permitting the abdominal viscera to ascend into the thorax

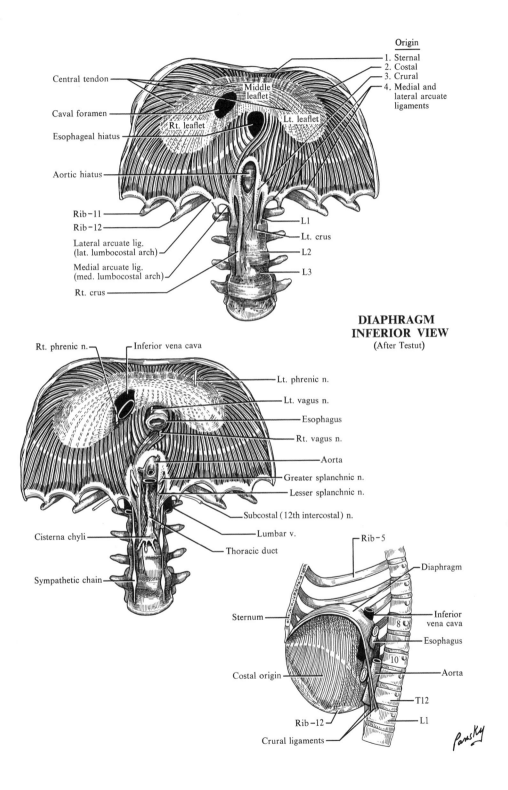

Origin
1. Sternal
2. Costal
3. Crural
4. Medial and lateral arcuate ligaments

Central tendon
Caval foramen
Esophageal hiatus
Middle leaflet
Rt. leaflet
Lt. leaflet
Aortic hiatus

Rib-11
Rib-12
Lateral arcuate lig. (lat. lumbocostal arch)
Medial arcuate lig. (med. lumbocostal arch)
Rt. crus

L1
Lt. crus
L2
L3

**DIAPHRAGM
INFERIOR VIEW**
(After Testut)

Rt. phrenic n.
Inferior vena cava
Lt. phrenic n.
Lt. vagus n.
Esophagus
Rt. vagus n.
Aorta
Greater splanchnic n.
Lesser splanchnic n.
Subcostal (12th intercostal) n.
Lumbar v.
Thoracic duct

Cisterna chyli
Sympathetic chain

Rib-5
Diaphragm
Sternum
Inferior vena cava
Esophagus
Aorta
Costal origin
T12
L1
Rib-12
Crural ligaments

Pansky

-365-

163. BLADDER AND MALE URETHRA

I. Bladder, male: surfaces and relations
A. FUNDUS (posterior): triangular, directed caudally and posteriorly. Separated from rectum by rectovesical septum, seminal vesicles, and vas deferens
B. APEX directed toward pubic symphysis
    1. Median umbilical ligament (urachus) continued up abdominal wall from the apex to the umbilicus
C. SUPERIOR SURFACE bounded laterally by lateral borders that delimit it from inferior surface, bounded posteriorly by a line connecting the ureters. Is covered by peritoneum and is related to the sigmoid colon and coils of ileum
D. INFEROLATERAL SURFACE: directed caudally and is without peritoneum; separated from pubis by prevesical cleft (space of Retzius)
E. NECK: triangular, in contact with base of prostate, contains urethral orifice

II. Bladder, female: surfaces and relations
A. FUNDUS separated above from anterior surface of uterus by vesicouterine pouch; below and behind it is related to cervix and upper vaginal wall
B. SUPERIOR SURFACE: uterus rests on this when bladder is empty
C. INFERIOR SURFACE rests on pelvic and urogenital diaphragms

III. Fixation (see p. 369): held in place by ligaments attached to its inferior surface
A. TRUE LIGAMENTS
    1. Pubovesicals: between pubis and bladder, directly in the female; in the male are attached to the prostate as the medial and lateral puboprostatic ligaments
    2. Rectovesical: from bladder to sides of rectum and sacrum
    3. Median umbilical: from apex of bladder to abdominal wall
B. FALSE LIGAMENTS: a group of peritoneal folds from the bladder to the abdominal or pelvic walls: 1 median, 2 medial, 2 lateral, and 2 sacrogenital (posterior false ligaments)

IV. Special features of interior
A. TRIGONE: a smooth triangular area above the urethral orifice. The posterolateral angles are formed by the ureteric orifices; the base is formed by the interureteric ridge, between the orifices; and the anterior angle is at the internal urethral orifice

V. Vessels and nerves (see p. 368)
A. ARTERIES: superior and inferior vesical, middle rectal
B. VEINS: vesical plexus to vesical veins to internal iliac veins, communications with prostatic plexus
C. LYMPHATICS: to external, internal, sacral, and median common iliac nodes
D. NERVES: via inferior hypogastric and vesical plexuses

VI. Distended bladder, male: surface and relations
A. FUNDUS: little change
B. SUMMIT directed anteriorly and cephalically above attachment of median umbilical ligament to apex, resulting in a peritoneal pouch between the anterior body wall and summit
C. POSTEROSUPERIOR SURFACE directed cephalically and posteriorly, covered by peritoneum, separated from rectum by rectovesical pouch
D. ANTEROINFERIOR SURFACE: nonperitoneal, below, related to pubic bones; above, with anterior abdominal wall
E. LATERAL SURFACES: below, nonperitoneal; related to lateral walls of pelvis

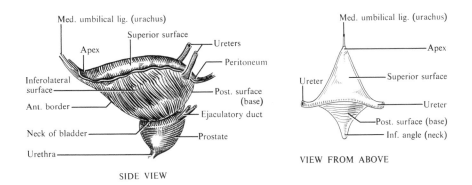

Med. umbilical lig. (urachus)
Superior surface
Apex
Ureters
Peritoneum
Inferolateral surface
Post. surface (base)
Ant. border
Ejaculatory duct
Neck of bladder
Prostate
Urethra

SIDE VIEW

Med. umbilical lig. (urachus)
Apex
Ureter
Superior surface
Ureter
Post. surface (base)
Inf. angle (neck)

VIEW FROM ABOVE

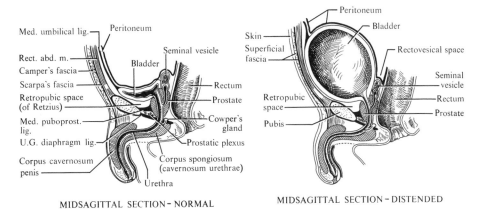

Med. umbilical lig.
Peritoneum
Seminal vesicle
Rect. abd. m.
Bladder
Camper's fascia
Scarpa's fascia
Rectum
Retropubic space (of Retzius)
Prostate
Med. puboprost. lig.
Cowper's gland
U.G. diaphragm lig.
Prostatic plexus
Corpus cavernosum penis
Corpus spongiosum (cavernosum urethrae)
Urethra

MIDSAGITTAL SECTION – NORMAL

Peritoneum
Bladder
Skin
Superficial fascia
Rectovesical space
Seminal vesicle
Retropubic space
Rectum
Pubis
Prostate

MIDSAGITTAL SECTION – DISTENDED

BLADDER

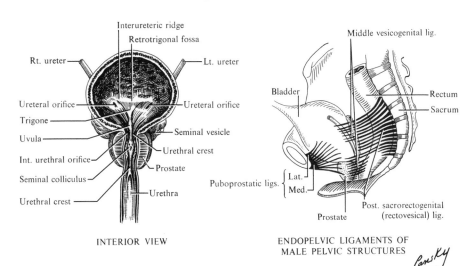

Interureteric ridge
Retrotrigonal fossa
Rt. ureter
Lt. ureter
Ureteral orifice
Ureteral orifice
Trigone
Uvula
Seminal vesicle
Int. urethral orifice
Urethral crest
Seminal colliculus
Prostate
Urethral crest
Urethra

INTERIOR VIEW

Middle vesicogenital lig.
Bladder
Rectum
Sacrum
Puboprostatic ligs. { Lat. Med.
Post. sacrorectogenital (rectovesical) lig.
Prostate

ENDOPELVIC LIGAMENTS OF
MALE PELVIC STRUCTURES

Pansky

164. LYMPHATICS, NERVES, AND FASCIA OF BLADDER AND PROSTATE

I. Lymphatics

A. BLADDER. Arise from the submucous plexus and form 3 groups of vessels
  1. From superior and inferolateral surfaces to external iliac nodes
  2. From fundus to external and internal iliac nodes
  3. From neck, with vessels of prostate, to sacral and medial common iliac nodes
B. PROSTATE, with those of 3, above, and from seminal vesicle to sacral, internal iliac, and common iliac nodes

II. Nerves

A. AUTONOMIC
  1. Sympathetic: preganglionics from lower thoracic and upper lumbar levels. Postganglionics from superior hypogastric plexus, which is a caudal continuation of the aortic and inferior mesenteric plexuses via the hypogastric nerves into the inferior hypogastric plexus. Some postganglionics are derived from the sacral trunk
  2. Parasympathetic: preganglionics run from S2, S3, and S4 as white rami communicantes, through the vesical division of pelvic plexus, with ganglia located on or in wall of bladder
B. SOMATIC MOTOR to external sphincter via branches from pudendal nerve arising from S2, S3, and S4
C. AFFERENT (visceral) travel with both sympathetic and parasympathetic fibers. Those associated with emptying the bladder travel with the parasympathetics

III. Micturition (emptying of bladder)

A. AFFERENT LIMB begins in stretch receptors in muscular wall and travels as IIC, above
B. EFFERENT LIMB: mainly the result of excitation brought to involuntary muscle through the pelvic splanchnics. This not only compresses the bladder but opens its internal sphincter. There is a simultaneous relaxation of the external sphincter

IV. Urinary retention. Mainly through sympathetic excitation to internal sphincter and through voluntary control via the pudendal nerve to the external sphincter

V. Endopelvic fascia (subserous). This is the fascia located between the peritoneum (above) and the fascia of the pelvic wall and floor. It invests the pelvic viscera and their vascular pedicles. Condensations of this fascia acquire special terminology

A. HYPOGASTRIC SHEATH (stalk): around hypogastric vessels
B. SUPERIOR AND INFERIOR ALAR OR VESICAL WINGS: around superior and inferior vesical vessels
C. MACKENRODT'S LIGAMENT (cardinal ligament): in the female, around uterine vessels
D. MIDDLE RECTAL LIGAMENT: around middle rectal vessels
E. SACROGENITAL LIGAMENT: around the nerve plexuses from the sacral region to viscera
F. RECTOVESICAL (MALE) AND RECTOVAGINAL (FEMALE) FASCIA: between respective organs
G. PUBOPROSTATIC LIGAMENTS: between pubis and the bladder and prostate
H. UTEROVESICAL AND VESICOVAGINAL LIGAMENTS: between uterus, vagina, and bladder
I. PRESACRAL LIGAMENT: around the superior rectal vessels, hypogastric (presacral) nerves, and sympathetic trunk

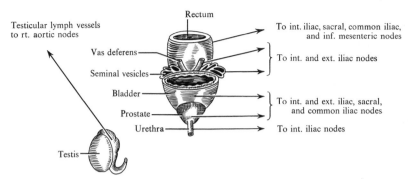

Testicular lymph vessels to rt. aortic nodes

Rectum

Vas deferens

Seminal vesicles

Bladder

Prostate

Urethra

Testis

To int. iliac, sacral, common iliac, and inf. mesenteric nodes

To int. and ext. iliac nodes

To int. and ext. iliac, sacral, and common iliac nodes

To int. iliac nodes

LYMPH DRAINAGE OF MALE PELVIS

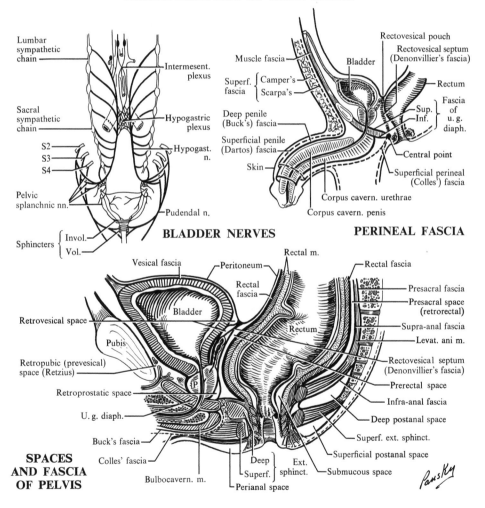

Lumbar sympathetic chain

Intermesent. plexus

Sacral sympathetic chain

Hypogastric plexus

S2
S3
S4

Hypogast. n.

Pelvic splanchnic nn.

Pudendal n.

Sphincters { Invol. / Vol. }

BLADDER NERVES

Muscle fascia

Superf. fascia { Camper's / Scarpa's }

Deep penile (Buck's) fascia

Superficial penile (Dartos) fascia

Skin

Bladder

Rectovesical pouch

Rectovesical septum (Denonvillier's fascia)

Rectum

Fascia of u. g. diaph.

Sup.
Inf.

Central point

Superficial perineal (Colles') fascia

Corpus cavern. urethrae

Corpus cavern. penis

PERINEAL FASCIA

Vesical fascia

Peritoneum

Rectal m.

Rectal fascia

Rectal fascia

Bladder

Rectum

Presacral fascia

Presacral space (retrorectal)

Supra-anal fascia

Levat. ani m.

Retrovesical space

Pubis

Retropubic (prevesical) space (Retzius)

Retroprostatic space

U. g. diaph.

Buck's fascia

Colles' fascia

Bulbocavern. m.

Deep / Superf. } Ext. sphinct.

Perianal space

Rectovesical septum (Denonvillier's fascia)

Prerectal space

Infra-anal fascia

Deep postanal space

Superf. ext. sphinct.

Superficial postanal space

Submucous space

SPACES AND FASCIA OF PELVIS

Pansky

165. VAS DEFERENS, SEMINAL VESICLES, AND PROSTATE

I. Ductus (vas) deferens: the excretory duct of the testis

A. COURSE: ascends along posterior border of the testis; in spermatic cord, passes through the inguinal canal to the deep inguinal ring; bends around inferior epigastric artery; curves posteriorly and caudally; crosses anterior to the external iliac vessels and enters pelvis; descends on medial side of lateral umbilical ligament and obturator vessels and nerve; crosses to medial side of ureter to run between fundus of bladder and seminal vesicles; to base of prostate, where its terminal part widens into an *ampulla.* The ampulla is joined by the duct of the seminal vesicle to form the *ejaculatory duct,* which runs through the prostate to open into the urethra

II. Seminal vesicle: bilateral lobulated sacs consisting of irregular pouches

A. RELATIONS: anterior surface against fundus of bladder; posterior surface separated from rectum by rectovesical fascia; superiorly, related to vas deferens and ureters; inferiorly, joins vas deferens at the posterior surface of the prostate

III. Prostate: a cone-shaped glandular body, the size of a chestnut, containing much connective tissue and smooth muscle

A. PARTS AND RELATIONS

1. Base faces cephalad against neck of bladder
2. Apex directed caudad against urogenital diaphragm
3. Posterior surface separated from rectum by rectovesical septum
4. Anterior surface separated from symphysis pubis by a venous plexus and fat. Urethra opens through this surface just above apex
5. Inferolateral surfaces separated from levator ani muscles by venous plexus
6. Lobes

   a. Anterior: small, nonglandular area in front of urethra

   b. Posterior: posterior to urethra and ejaculatory ducts and behind middle lobe

   c. Two lateral (right and left) occupy almost entire base of gland, lateral and anterior to urethra

   d. Median (middle): glandular, posterior to urethra but anterior to ejaculatory ducts; here are found the subtrigonal and cervical glands (Albarran's)

IV. Special features of the prostatic urethra

A. URETHRAL CREST: a longitudinal ridge on the posterior wall

B. PROSTATIC SINUS: depression on sides of crest into which the prostatic ducts open

C. SEMINAL COLLICULUS: summit of the urethral crest on which open the ejaculatory ducts and a median blind-ending sac, the *prostatic utricle*

V. Fixation: by puboprostatic ligaments, urogenital diaphragm, and levator ani mm.

VI. Vessels and nerves

A. ARTERIES: middle rectal and inferior vesical

B. VEINS: prostatic plexus to internal iliac vein

C. LYMPHATICS terminate in the internal iliac and sacral nodes

D. NERVES: prostatic plexus from inferior hypogastric plexus

VII. Clinical considerations

A. ANTERIOR LOBE: adenomas rare; no urethral encroachment

B. POSTERIOR LOBE: adenomas rare; lobe encountered in digital examination

C. LATERAL LOBE: hypertrophy causes urinary obstruction

D. MEDIAN LOBE: important clinically; enlargement of mucous glands leads to obstruction; adenomas frequent, encroaching into urethra, blocking internal orifice

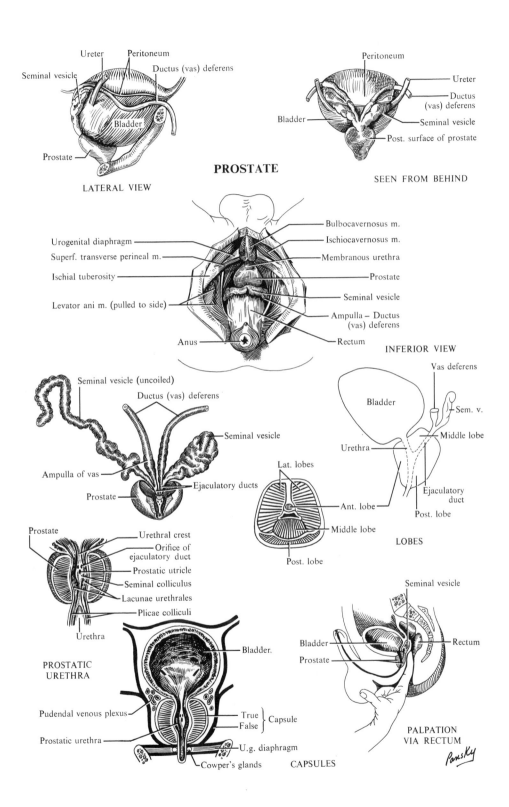

PROSTATE

LATERAL VIEW

Ureter
Peritoneum
Seminal vesicle
Ductus (vas) deferens
Bladder
Prostate

SEEN FROM BEHIND

Peritoneum
Ureter
Ductus (vas) deferens
Bladder
Seminal vesicle
Post. surface of prostate

Urogenital diaphragm
Superf. transverse perineal m.
Ischial tuberosity
Levator ani m. (pulled to side)
Anus

Bulbocavernosus m.
Ischiocavernosus m.
Membranous urethra
Prostate
Seminal vesicle
Ampulla – Ductus (vas) deferens
Rectum

INFERIOR VIEW

Seminal vesicle (uncoiled)
Ductus (vas) deferens
Seminal vesicle
Ampulla of vas
Prostate
Ejaculatory ducts

Vas deferens
Bladder
Sem. v.
Urethra
Middle lobe
Ejaculatory duct
Post. lobe

Lat. lobes
Ant. lobe
Middle lobe
Post. lobe

LOBES

Prostate
Urethral crest
Orifice of ejaculatory duct
Prostatic utricle
Seminal colliculus
Lacunae urethrales
Plicae colliculi
Urethra

PROSTATIC URETHRA

Pudendal venous plexus
Prostatic urethra

Bladder.
True
False
Capsule
U.g. diaphragm
Cowper's glands

CAPSULES

Seminal vesicle
Bladder
Prostate
Rectum

PALPATION VIA RECTUM

Pansky

-371-

166. RECTUM AND ANAL CANAL

I. Rectum

A. COURSE AND EXTENT: from 3rd sacral segment to slightly below tip of coccyx. Here bends abruptly posteriorly to anal canal. Total length, 12 cm

B. CURVATURES
   1. Two posteroanterior: upper convex posteriorly; lower convex anteriorly
   2. Two lateral: to right, at junction of third and fourth sacral segments; to left, at sacrococcygeal articulation

C. DIAMETER
   1. Cephalic end similar to colon; caudal end dilated to form *rectal ampulla*

D. SPECIAL STRUCTURAL CHARACTERISTICS
   1. Transverse rectal folds (Houston's valves): permanent, 3 transverse folds, which project into lumen
      a. Upper, from right side, near cephalic end
      b. Middle, 3 cm below first, extends inward from left
      c. Lower, largest, opposite bladder and extends posteriorly from anterior wall
   2. Unlike colon, the rectum has a complete outer longitudinal muscle coat

E. PERITONEAL RELATIONSHIP. Upper two thirds has some peritoneum: most cephalic portion has peritoneum anteriorly and laterally, lower down it has peritoneum only anteriorly. The lower third has no peritoneum

F. RELATIONS
   1. Posteriorly: superior rectal vessels, left piriformis muscle, left sacral plexus of nerves, and the fascia covering the sacrum, coccyx, and levator ani muscle
   2. Anteriorly
      a. Male: separated by coils of intestine in rectovesical fossa from fundus of bladder, and by rectovesical septum from the triangular area of the fundus of bladder, seminal vesicles, ductus deferens, and prostate
      b. Female: separated by intestinal coils in rectouterine fossa from the uterus and by rectovaginal septum from the posterior wall of the vagina

II. Anal canal

A. COURSE AND EXTENT: origin at level of apex of prostate, directed posteriorly and caudally. Length, 2.5–4.0 cm

B. SURROUNDED BY INTERNAL AND EXTERNAL ANAL SPHINCTERS. Supported by levator ani muscle

C. SPECIAL FEATURES
   1. Anal columns: vertical folds due to dilated veins of rectal plexus
   2. Anal sinuses: furrows between columns
   3. Anal valves: folds joining columns at caudal end of anal sinuses

D. RELATIONS
   1. Posteriorly: anococcygeal body (musculofibrous tissue)
   2. Anteriorly
      a. Male: separated by central tendon of perineum from membranous and bulbar urethra
      b. Female: central tendon of perineum separates it from vagina

III. Anal sphincters

A. INTERNAL: a thickening of the intrinsic smooth muscle coat of the intestinal wall

B. EXTERNAL: composed of striated, voluntary muscle
   1. Subcutaneous: portion immediately around anal orifice
   2. Superficial part (main portion) arises from the anococcygeal raphé, splits to encircle canal, and inserts in central tendon of perineum
   3. Deep part (a true sphincter) encircles the anus, the fibers of the 2 sides decussating ventral and dorsal to anus

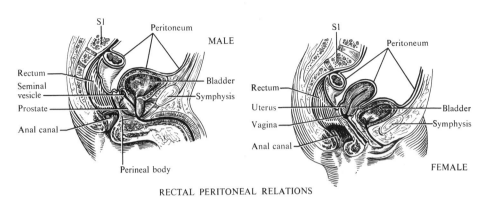

MALE

S1

Peritoneum

Rectum
Seminal vesicle
Prostate
Anal canal

Bladder
Symphysis

Perineal body

FEMALE

S1

Peritoneum

Rectum
Uterus
Vagina
Anal canal

Bladder
Symphysis

RECTAL PERITONEAL RELATIONS

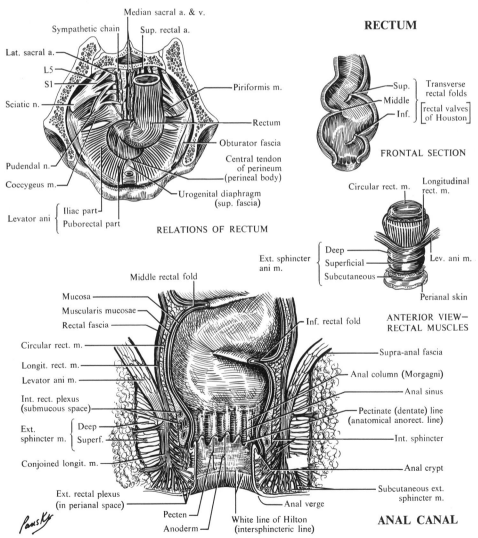

RECTUM

Median sacral a. & v.

Sympathetic chain
Sup. rectal a.

Lat. sacral a.
L5
S1
Sciatic n.

Pudendal n.
Coccygeus m.

Levator ani { Iliac part
Puborectal part

Piriformis m.

Rectum

Obturator fascia

Central tendon
of perineum
(perineal body)

Urogenital diaphragm
(sup. fascia)

RELATIONS OF RECTUM

Sup.
Middle
Inf.

Transverse
rectal folds

[rectal valves
of Houston]

FRONTAL SECTION

Circular rect. m.

Longitudinal
rect. m.

Ext. sphincter
ani m. { Deep
Superficial
Subcutaneous

Lev. ani m.

Perianal skin

ANTERIOR VIEW—
RECTAL MUSCLES

Middle rectal fold

Mucosa
Muscularis mucosae
Rectal fascia

Circular rect. m.

Longit. rect. m.

Levator ani m.

Int. rect. plexus
(submucous space)

Ext.
sphincter m. { Deep
Superf.

Conjoined longit. m.

Ext. rectal plexus
(in perianal space)

Pecten
Anoderm

Inf. rectal fold

Supra-anal fascia

Anal column (Morgagni)

Anal sinus

Pectinate (dentate) line
(anatomical anorect. line)

Int. sphincter

Anal crypt

Subcutaneous ext.
sphincter m.

Anal verge

White line of Hilton
(intersphincteric line)

ANAL CANAL

167. BLOOD SUPPLY AND LYMPH DRAINAGE OF RECTUM

I. Blood supply
A. ARTERIES
1. Superior rectal arises from the inferior mesenteric artery (see p. 340). Two branches descend on either side of rectum. Above anus, each gives rise to several small branches, which pass straight caudally, regularly spaced, to the level of the internal sphincter, at which point they form loops around the caudal rectum and anastomose with other vessels
2. Middle rectal arises either directly from the internal iliac artery or by a common stem with the inferior vesical artery, which approaches rectum from side and joins loop at caudal end of rectum and upper anal canal
3. Inferior rectal arises from internal pudendal artery, pierces wall of the pudendal (Alcock's) canal, and gives 2 or 3 branches, which pass medially through the ischiorectal fossa to muscle and skin around anus
B. VEINS
1. Rectal plexus: network of vessels around anal canal. Usually, at cephalic border of canal, some veins are dilated or sacculated. This plexus drains:
   a. Chiefly through 6 ascending vessels, which lie between muscularis and mucosa for about 12.0 cm. At this level they unite to form the superior rectal vein, a tributary of the inferior mesenteric vein
   b. Middle rectal: from plexus, with tributaries from bladder, prostate, and seminal vesicle, swings laterally on pelvic surface of levator ani to internal iliac vein
   c. Inferior rectal: from lower plexus into internal pudendal vein and then into internal iliac vein

II. Lymphatic drainage
A. FROM ANUS: with lymphatics of superficial perineum and scrotum into superficial inguinal nodes
B. FROM ANAL CANAL: accompany middle rectal artery and end in internal iliac nodes around internal iliac vessels; from these to common iliac nodes and then to lateral aortic group
C. FROM RECTUM: through pararectal nodes which lie on rectal muscles and sigmoid mesocolon to the inferior mesenteric group of preaortic nodes

III. Nerves (see p. 386): derived from the inferior mesenteric and hypogastric plexuses

IV. Clinical considerations
A. THE ANAL CANAL presents 4 landmarks. The *anocutaneous line* marks the lower end of the gastrointestinal tract. *Hilton's white line* marks the interval between the external and internal anal sphincters. The *pectin* is the mucocutaneous junction, internal hemorrhoids developing above and external hemorrhoids below this line; this line is also a lymphatic dividing line between the flow of lymph upward into the pelvis, primarily to the internal iliac nodes, or downward to the subinguinal nodes. The *anorectal line* is the line above the anal crypts and sinuses, marking the beginning of the anal canal
B. THE LINE around the anal canal that can be traced by following the anal valves and the bases of the anal columns is usually referred to by clinicians as the *pectinate, dentate,* or *mucocutaneous line.* It is an important landmark
1. The change between columnar or cuboidal epithelium in the upper part of the canal and the stratified epithelium in the lower part occurs at or close to this level and is important because carcinomas from the two types of epithelium differ
2. The line is only about 0.25 to 1.0 cm below a divide in the nerve supply, with the afferent innervation above it through fibers of the pelvic plexus (visceral type) and that below of somatic nerve fibers in the pudendal nerve

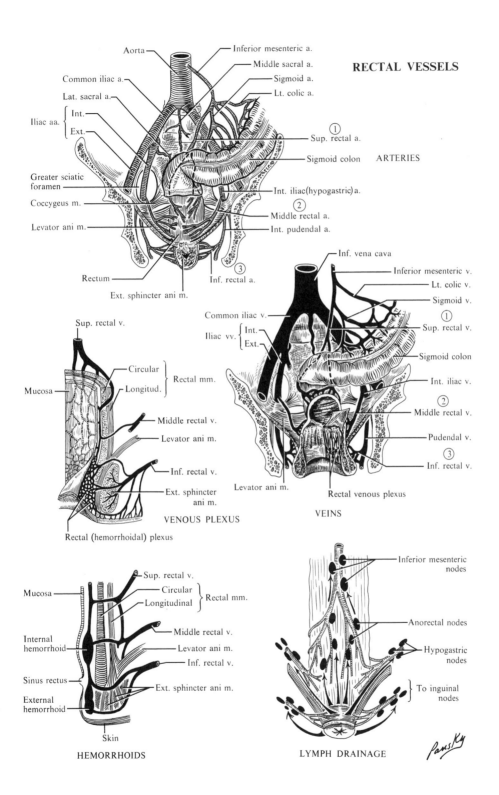

RECTAL VESSELS

Aorta
Inferior mesenteric a.
Common iliac a.
Middle sacral a.
Lat. sacral a.
Sigmoid a.
Lt. colic a.
Iliac aa. { Int.
Ext.
Sup. rectal a. ①
Sigmoid colon   ARTERIES
Greater sciatic foramen
Int. iliac(hypogastric) a.
Coccygeus m.
Middle rectal a. ②
Levator ani m.
Int. pudendal a.
Rectum
Inf. rectal a. ③
Ext. sphincter ani m.

Inf. vena cava
Inferior mesenteric v.
Lt. colic v.
Sigmoid v.
Common iliac v.
Iliac vv. { Int.
Ext.
Sup. rectal v. ①
Sigmoid colon
Int. iliac v.
Middle rectal v. ②
Pudendal v.
Inf. rectal v. ③
Levator ani m.
Rectal venous plexus

VEINS

Sup. rectal v.
Circular
Longitud. } Rectal mm.
Mucosa
Middle rectal v.
Levator ani m.
Inf. rectal v.
Ext. sphincter ani m.
VENOUS PLEXUS
Rectal (hemorrhoidal) plexus

Mucosa
Sup. rectal v.
Circular
Longitudinal } Rectal mm.
Middle rectal v.
Internal hemorrhoid
Levator ani m.
Inf. rectal v.
Sinus rectus
Ext. sphincter ani m.
External hemorrhoid
Skin
HEMORRHOIDS

Inferior mesenteric nodes
Anorectal nodes
Hypogastric nodes
To inguinal nodes
LYMPH DRAINAGE

Pansky

168. THE PELVIC DIAPHRAGM

I. Composition: the levator ani and coccygeus muscles with their superior and inferior fasciae

II. Location: forms a sling across pelvic cavity

III. Function: it is the pelvic floor that holds in and supports the viscera

IV. Muscles

A. LEVATOR ANI, innervated by S4 (sometimes S3 or S5) through the pudendal plexus, consists of the following parts:
1. Pubococcygeus arises from pubis and its superior ramus to insert into anococcygeal raphé and coccyx
    a. Puborectalis: most medial portion of the pubococcygeus originating at pubis and the superior layer of the urogenital diaphragm to pass along the side of the rectum to meet its counterpart behind that organ, forming a sling. Some of its fibers continue to the tip of the coccyx
2. Iliococcygeus arises from the arcus tendineus and the ischial spine. Inserts on last segments of coccyx and anococcygeal raphé.

B. COCCYGEUS innervated by S4 and S5. Arises from the spine of the ischium and sacrospinous ligament. Inserts into the sides of lower sacrum and coccyx

V. Parietal pelvic fascia (fascia lining walls of pelvis): is continuous with transversalis fascia over brim of the pelvis and is attached to bones along rim of true pelvis. Has 2 primary regional divisions

A. PIRIFORM covers piriformis muscle and leaves pelvis with that muscle through greater sciatic foramen. Sacral nerves are behind this layer; internal iliac vessels are in front of it

B. OBTURATOR: divisible into intrapelvic and extrapelvic portions by the arcus tendineus of pelvic fascia (a thickening of fascia from ischial spine to pubic bone near obturator membrane). Ensheaths obturator vessels and nerve. At arcus, splits into 3 layers
1. Supra-anal covers inner (superior) surface of levator ani and coccygeus muscles
2. Infra-anal covers inferior surface of levator ani and coccygeus muscles
3. Extrapelvic continuation of obturator covers obturator muscle below arcus and lines the lateral wall of ischiorectal fossa
    a. Pudendal (Alcock's) canal: a split in the fascia to invest the pudendal nerve and internal pudendal vessels

VI. Visceral pelvic (subserous endopelvic) fascia (see p. 368) invests pelvic viscera and their vessels

VII. Ischiorectal fossa: a wedge-shaped space at either side of anal canal. Its base is directed caudally, and its apex is cephalic, along the arcus tendineus. It contains fat, connective tissue, rectal vessels, and nerves. It is bounded medially by the infra-anal fascia over the levator ani and external anal sphincter muscles, laterally by extrapelvic obturator fascia over the obturator internus muscle and the ischial tuberosities, posteriorly by the gluteus maximus muscle and sacrotuberous ligament, anteriorly by the posterior edge of the urogenital diaphragm, inferiorly by fascia and skin

A. ANTERIOR RECESS: a continuation of fossa anteriorly, between urogenital diaphragm below and the levator and obturator internus muscles above and lateral

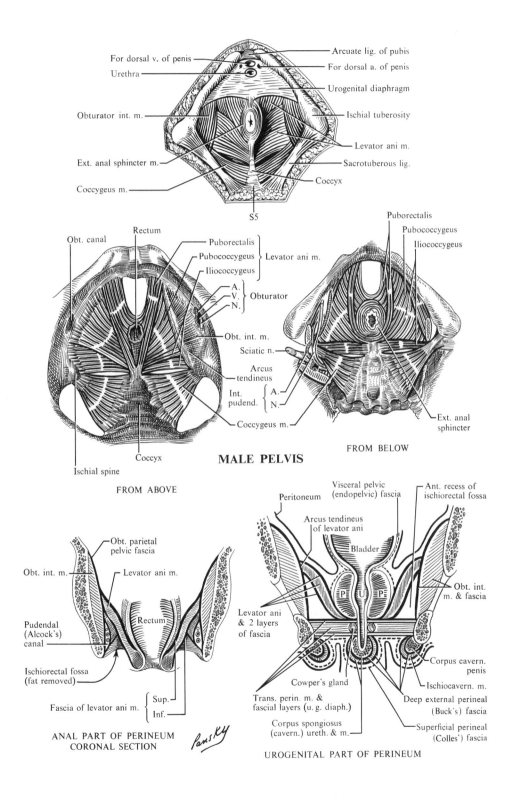

For dorsal v. of penis

Urethra

Obturator int. m.

Ext. anal sphincter m.

Coccygeus m.

Arcuate lig. of pubis

For dorsal a. of penis

Urogenital diaphragm

Ischial tuberosity

Levator ani m.

Sacrotuberous lig.

Coccyx

S5

Obt. canal

Rectum

Puborectalis

Pubococcygeus } Levator ani m.

Iliococcygeus

A.
V. } Obturator
N.

Obt. int. m.

Sciatic n.

Arcus tendineus

Int. pudend. { A.
N.

Coccygeus m.

Ischial spine

Coccyx

FROM ABOVE

Puborectalis

Pubococcygeus

Iliococcygeus

Ext. anal sphincter

FROM BELOW

MALE PELVIS

Obt. parietal pelvic fascia

Obt. int. m.

Levator ani m.

Rectum

Pudendal (Alcock's) canal

Ischiorectal fossa (fat removed)

Fascia of levator ani m. { Sup.
Inf.

ANAL PART OF PERINEUM
CORONAL SECTION

Peritoneum

Visceral pelvic (endopelvic) fascia

Ant. recess of ischiorectal fossa

Arcus tendineus of levator ani

Bladder

P U P

Obt. int. m. & fascia

Levator ani & 2 layers of fascia

Cowper's gland

Trans. perin. m. & fascial layers (u. g. diaph.)

Corpus spongiosus (cavern.) ureth. & m.

Corpus cavern. penis

Ischiocavern. m.

Deep external perineal (Buck's) fascia

Superficial perineal (Colles') fascia

UROGENITAL PART OF PERINEUM

Pansky

169. RELATIONS OF THE UTERUS, TUBES, AND OVARY— PART I

I. Ovary: the germinal and endocrine gland of the female

A. LOCATION AND RELATIONS: in the ovarian fossa on the lateral wall of the pelvis, behind the broad ligament, with the external iliac vessels above, the ureter posteroinferiorly, and covered by the fimbria of the uterine tube medially

B. SURFACES AND BORDERS: lateral and medial surfaces, upper (tubal) extremity, lower (uterine) extremity, anterior (mesovarian) border, posterior (free) border

C. FIXATION: *suspensory ligament,* a peritoneal fold running from the upper extremity to the iliac vessels; *proper ligament of the ovary,* from the lower extremity through the mesometrium (broad ligament) to the lateral angle of the uterus; *mesovarium,* a mesentery joining the anterior border to the posterior side of the broad ligament

II. Uterine tube: duct to carry ova from ovaries to uterus

A. LOCATION AND RELATIONS: lies in the free, cephalic border of the broad ligament—*the mesosalpinx*

B. PARTS: *infundibulum,* with abdominal ostium surrounded by *fimbria; ampulla,* the middle, wide part, which curves over ovary; *isthmus,* the constricted medial part, which enters the uterus; and *interstitial* or *intrauterine* part

III. Uterus: organ adapted for the development of the fertilized ovum

A. PARTS

1. Body
   a. Vesical (anterior) surface: lies on the superior surface of the bladder, covered with peritoneum, which is reflected onto bladder forming the vesicouterine pouch
   b. Intestinal (posterior) surface: related to the sigmoid colon and coils of small intestine, covered with peritoneum
   c. Fundus: directed anteriorly and cephalically, related to the coils of the small intestine
   d. Lateral margins: mesometrium (broad), round, with ovarian ligaments attached here, receives uterine tubes

2. Cervix: the constricted part of the uterus demarcated from the body by the *isthmus,* which indicates the position of the internal os. The axis is a curve with the concavity directed anteriorly. Divided into 2 parts by the vagina
   a. Supravaginal: separated from the bladder anteriorly by the parametrium (fatty, fibrous tissue, which is continuous with the tissue between the layers of the broad ligament); posteriorly it is covered with peritoneum and is separated from the rectum by coils of small intestine
   b. Vaginal: protrudes into the vagina
      i. Ostium (external os): the opening of cervix into vagina
      ii. Fornices (anterior, posterior, and lateral): the groove that lies between the walls of the vagina and cervix. The posterior fornix is deepest

IV. Clinical considerations

A. PROLAPSE: when supports of the uterus become stretched and very lax, the cervix may descend, for varying degrees, into the vagina or even out into the vestibule

B. THE RELATIONS of the body of the uterus are markedly changed by pregnancy, progressive enlargement bringing the uterus high into the abdominal cavity

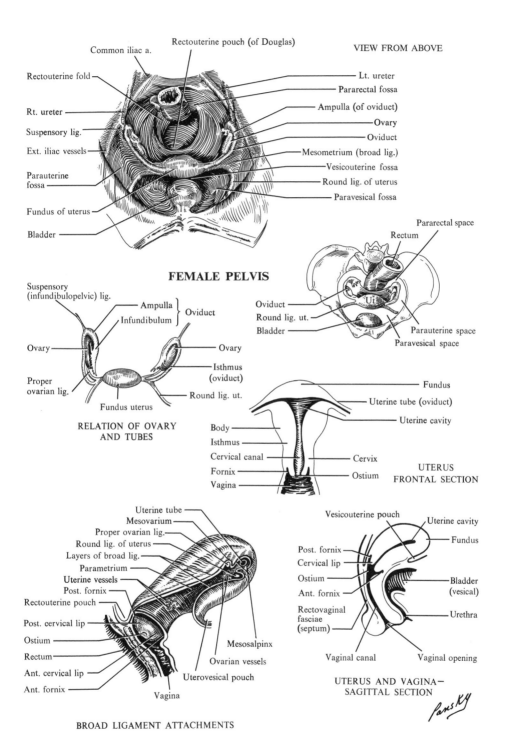

VIEW FROM ABOVE

Rectouterine pouch (of Douglas)

Common iliac a.

Rectouterine fold

Rt. ureter

Suspensory lig.

Ext. iliac vessels

Parauterine fossa

Fundus of uterus

Bladder

Lt. ureter

Pararectal fossa

Ampulla (of oviduct)

Ovary

Oviduct

Mesometrium (broad lig.)

Vesicouterine fossa

Round lig. of uterus

Paravesical fossa

FEMALE PELVIS

Pararectal space

Rectum

Suspensory (infundibulopelvic) lig.

Ampulla

Infundibulum

Oviduct

Ovary

Proper ovarian lig.

Fundus uterus

Ovary

Isthmus (oviduct)

Round lig. ut.

RELATION OF OVARY AND TUBES

Oviduct

Round lig. ut.

Bladder

Ut.

Parauterine space

Paravesical space

Fundus

Uterine tube (oviduct)

Uterine cavity

Body

Isthmus

Cervical canal

Fornix

Vagina

Cervix

Ostium

UTERUS FRONTAL SECTION

Uterine tube

Mesovarium

Proper ovarian lig.

Round lig. of uterus

Layers of broad lig.

Parametrium

Uterine vessels

Post. fornix

Rectouterine pouch

Post. cervical lip

Ostium

Rectum

Ant. cervical lip

Ant. fornix

Vagina

Mesosalpinx

Ovarian vessels

Uterovesical pouch

BROAD LIGAMENT ATTACHMENTS

Vesicouterine pouch

Uterine cavity

Fundus

Post. fornix

Cervical lip

Ostium

Ant. fornix

Rectovaginal fasciae (septum)

Vaginal canal

Bladder (vesical)

Urethra

Vaginal opening

UTERUS AND VAGINA— SAGITTAL SECTION

ParisKY

170. RELATIONS OF THE UTERUS, TUBES, AND OVARY—PART II

B. FIXATION AND SUPPORT
1. Mesometrium (broad ligaments): folds of peritoneum extending across the pelvis containing uterus, parametrium, vessels, nerves, uterine tubes, ureters, round ligaments, paroophoron, and epoophoron
2. Round ligaments: fibrous cords attached to superior lateral borders of uterus, pass over external iliac vessels and inguinal ligaments to leave the abdomen through the deep inguinal rings. They traverse the inguinal canal and are anchored in the labia majora
3. Cardinal ligaments: subserous tissue (endopelvic fascia) around vagina and cervix, which extends laterally across the pelvis within the lowest part of the broad ligaments to attach to the deep fascia covering the levator ani muscles. Vaginal and uterine vessels lie in it
4. Uterosacral ligaments: subserous tissue (endopelvic tissue) attached to the cervix, running posteriorly to join the deep fascia over the sacrum
5. Levator ani muscles and fascia support it from below
6. Peritoneal ligaments (false). These are similar to the false ligaments of the bladder and are reflections of the peritoneum from uterus to other organs or are folds covering deeper fibrous bands: vesicouterine (anterior ligament) to bladder, rectovaginal (posterior ligament) to rectum, sacrogenital covering uterosacral ligaments

V. **Vagina:** that part of the birth canal extending from the uterus to the vestibule
A. INCLINATION: cephalically and posteriorly, forms an angle of about 90° with uterus. The posterior wall is 2.5–3.0 cm longer than the anterior
B. RELATIONS: upper end embraces cervix; anteriorly are the fundus of bladder and urethra; posteriorly lie the rectouterine peritoneal pouch, rectovaginal fascia, and perineal body. The pelvic ureters lie close to the lateral fornices and, as these reach the bladder, lie anterior to the anterior fornix

VI. **Vessels, nerves, and lymphatics** (see p. 382)

VII. **Clinical considerations**
A. THE UTERUS, normally anteflexed and anteverted, may assume a variety of abnormal positions, for example, retroversion, retroflexion, or prolapse (a falling or sinking into the vagina)
B. IN HYSTERECTOMY (removal of uterus), the relations of the ureters to the cervix, vagina, and uterine vessels must be recalled
C. THE RELATIONS OF THE POSTERIOR FORNIX to the bottom of the rectouterine pouch makes possible easier palpation of the pelvic viscera, drainage of abdominal fluids, and inspection of pelvic viscera without abdominal incisions
D. HYSTERECTOMY IS EITHER PARTIAL, in which at least a part of the cervix is left, or complete (panhysterectomy)
E. SALPINGITIS is an inflammation of the uterine tube(s)
F. SALPINGECTOMY is a removal of the uterine tube(s)
G. TUBAL PREGNANCY is a situation in which the fertilized ovum implants in the uterine tube
H. OVARIAN CYST FORMATION (single or multiple) may occur as a result of lack of ovulation of the follicle and its continued growth
I. A BULGING OF THE BLADDER into the anterior vaginal wall is known as *cystocele,* and a bulging into the posterior vaginal wall is known as *rectocele.* Each involves laceration of the intervening connective tissue and adjacent vaginal wall

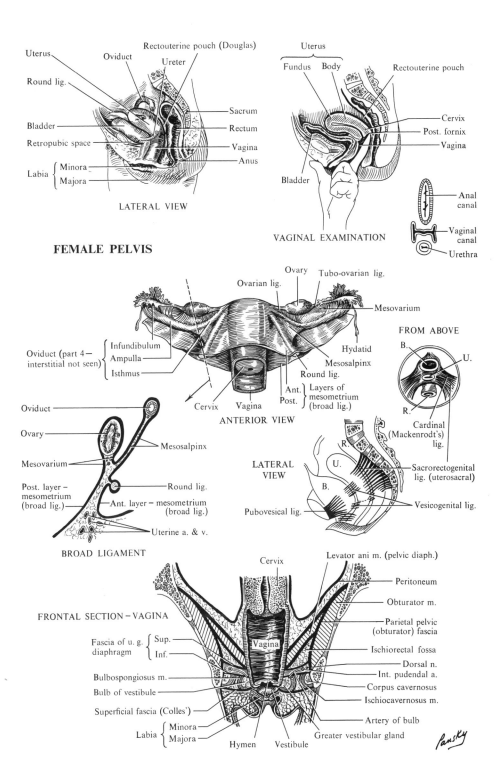

FEMALE PELVIS

LATERAL VIEW

Uterus
Oviduct
Rectouterine pouch (Douglas)
Ureter
Round lig.
Sacrum
Rectum
Bladder
Retropubic space
Vagina
Anus
Labia { Minora, Majora }

VAGINAL EXAMINATION

Uterus
Fundus   Body
Rectouterine pouch
Cervix
Post. fornix
Vagina
Bladder

Anal canal
Vaginal canal
Urethra

ANTERIOR VIEW

Ovary   Tubo-ovarian lig.
Ovarian lig.
Mesovarium
Oviduct (part 4 — interstitial not seen) { Infundibulum, Ampulla, Isthmus }
Hydatid
Mesosalpinx
Round lig.
Cervix   Vagina
Ant. } Layers of
Post. } mesometrium (broad lig.)

FROM ABOVE

B.
U.
R.
Cardinal (Mackenrodt's) lig.

BROAD LIGAMENT

Oviduct
Ovary
Mesovarium
Mesosalpinx
Post. layer — mesometrium (broad lig.)
Ant. layer — mesometrium (broad lig.)
Round lig.
Uterine a. & v.

LATERAL VIEW

R.
U.
B.
Pubovesical lig.
Sacrorectogenital lig. (uterosacral)
Vesicogenital lig.

FRONTAL SECTION — VAGINA

Cervix
Levator ani m. (pelvic diaph.)
Peritoneum
Obturator m.
Parietal pelvic (obturator) fascia
Ischiorectal fossa
Dorsal n.
Int. pudendal a.
Corpus cavernosus
Ischiocavernosus m.
Artery of bulb
Greater vestibular gland
Fascia of u. g. diaphragm { Sup., Inf. }
Vagina
Bulbospongiosus m.
Bulb of vestibule
Superficial fascia (Colles')
Labia { Minora, Majora }
Hymen   Vestibule

Pansky

171. BLOOD SUPPLY, INNERVATION, AND LYMPH DRAINAGE OF FEMALE GENITAL SYSTEM

I. Blood supply
A. OVARY
  1. Artery: ovarian artery from aorta descends in the suspensory ligament of the ovary to broad ligament, sending branches to ovary and uterine tube. Anastomoses with uterine artery
  2. Vein: pampiniform plexus to ovarian vein, which travels with the ovarian artery to terminate on the right in the vena cava, on the left in the renal vein
B. UTERINE TUBE
  1. Artery: similar to that of the ovary
  2. Vein: similar to that of the ovary, with some flow to uterine plexus
C. UTERUS
  1. Arteries: ovarian; uterine from anterior division of internal iliac crosses ureter to reach side of uterus through broad ligament, where it ascends to level of uterine tube. Supplies cervix, upper vagina, body of uterus, uterine tube, and round ligament
  2. Veins: from uterine plexus on the sides and superior angles of uterus. Communicates with vaginal plexus but is drained chiefly by uterine veins, which end in internal iliac veins
D. VAGINA
  1. Artery: vaginal artery from internal iliac (comparable to inferior vesical in the male). Sends branches to uterus and joins branches from uterine artery to form the azygos artery of the vagina
  2. Vein: from vaginal plexus, which has communications with vesical, rectal, and uterine plexuses, through vaginal vein to internal iliac vein

II. Lymphatic drainage
A. OVARY: vessels following ovarian artery to enter lateral and preaortic nodes
B. UTERINE TUBE: follows ovarian and uterine drainage
C. UTERUS
  1. Cervix: to external, internal, and common iliac nodes
  2. Body and fundus: mostly follow ovarian drainage to lateral and preaortic nodes, some to external iliac and superficial inguinal nodes (along round lig. to labia)
D. VAGINA: upper, middle, and lower portions drain to external, internal, and common iliac nodes; vulvar drainage ends in superficial inguinal nodes

III. Lymph nodes concerned in drainage of female genital system
A. COMMON ILIAC: on sides of common iliac artery at bifurcation of aorta. Most afferents from internal and external iliac nodes; efferents to lateral aortic nodes
B. EXTERNAL ILIAC: along external iliac artery. Afferents mainly from inguinal nodes but also from glans clitoridis and urethra, bladder, cervix, and upper vagina. Efferents to lateral aortic nodes
C. INTERNAL ILIAC: along corresponding artery. Afferents follow all branches of internal iliac artery from all pelvic viscera. Efferents to common iliac nodes

IV. Nerves
A. OVARY: receives fibers from hypogastric and ovarian plexuses
B. UTERINE TUBE: fibers from hypogastric and ovarian plexuses
C. UTERUS: uterovaginal portion of hypogastric plexus
D. VAGINA: uterovaginal portion of hypogastric plexus
E. MOST OF THE PAIN FIBERS FROM THE VAGINA travel with the sacral parasympathetic fibers, as do those from the cervix uteri, and enter the cord via nerves S2, S3, and S4
F. THE LOWER 2.5 CM OF THE VAGINA receives its innervation from the pudendal nerves (also S2, S3, and S4)

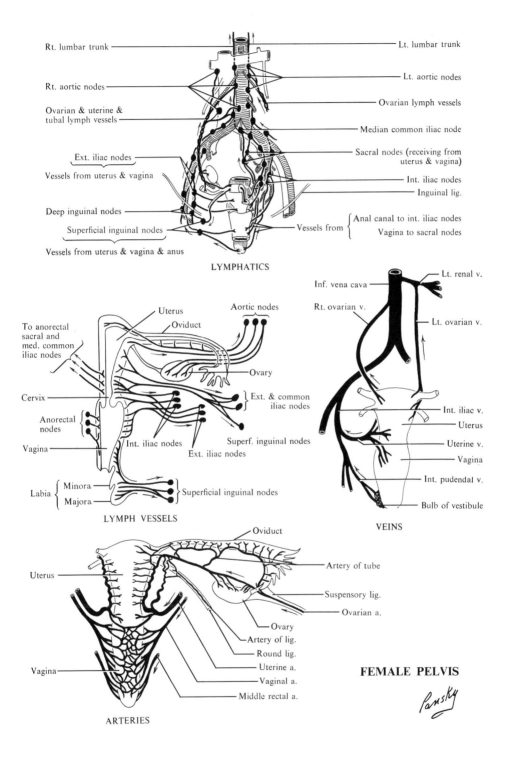

LYMPHATICS

Rt. lumbar trunk

Rt. aortic nodes

Ovarian & uterine & tubal lymph vessels

Ext. iliac nodes

Vessels from uterus & vagina

Deep inguinal nodes

Superficial inguinal nodes

Vessels from uterus & vagina & anus

Lt. lumbar trunk

Lt. aortic nodes

Ovarian lymph vessels

Median common iliac node

Sacral nodes (receiving from uterus & vagina)

Int. iliac nodes

Inguinal lig.

Vessels from { Anal canal to int. iliac nodes / Vagina to sacral nodes

LYMPH VESSELS

Uterus

Oviduct

Aortic nodes

To anorectal sacral and med. common iliac nodes

Cervix

Anorectal nodes

Vagina

Int. iliac nodes

Ext. iliac nodes

Ovary

Ext. & common iliac nodes

Superf. inguinal nodes

Labia { Minora / Majora } Superficial inguinal nodes

VEINS

Inf. vena cava

Rt. ovarian v.

Lt. renal v.

Lt. ovarian v.

Int. iliac v.

Uterus

Uterine v.

Vagina

Int. pudendal v.

Bulb of vestibule

ARTERIES

Uterus

Vagina

Oviduct

Artery of tube

Suspensory lig.

Ovarian a.

Ovary

Artery of lig.

Round lig.

Uterine a.

Vaginal a.

Middle rectal a.

FEMALE PELVIS

Pansky

-383-

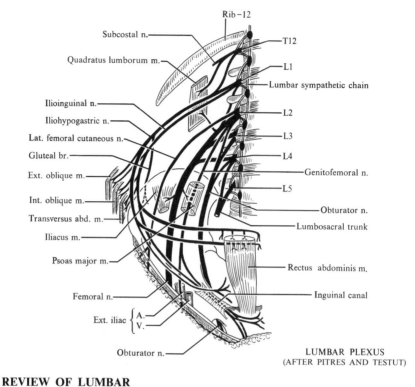

LUMBAR PLEXUS
(AFTER PITRES AND TESTUT)

REVIEW OF LUMBAR AND SACRAL PLEXUSES

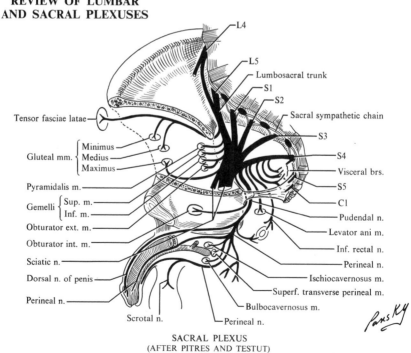

SACRAL PLEXUS
(AFTER PITRES AND TESTUT)

FIGURE 33. **Lumbosacral plexus.**

– 384 –

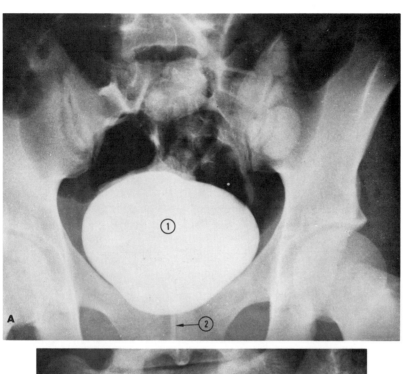

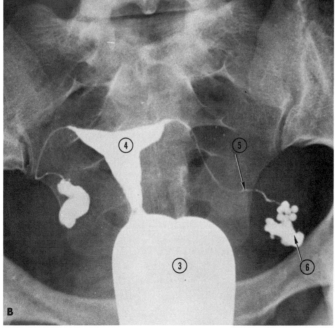

FIGURE 34. **Cystogram and female genital tract.** **A, Cystogram; B, female genital tract after barium sulfate injection.** *1,* Bladder; *2,* catheter in urethra; *3,* vagina; *4,* uterus; *5,* uterine tube; *6,* infundibulum of tube.

172. ABDOMINAL AND PELVIC AUTONOMICS

I. Lumbar part of sympathetic trunk continuous with the thoracic trunk under medial lumbocostal arch; lies anterior to bodies of lumbar vertebrae; on right, partly behind inferior vena cava; on left, behind aorta

II. Lumbar sympathetic ganglia: number 2–6 with a variety of fusions
A. Roots
    1. White rami, as far as L2 to each of upper 3 ganglia
B. BRANCHES
    1. Gray rami to each of the lumbar nerves
    2. Lumbar splanchnics to aortic network from upper 3 lumbar nerves
    3. Visceral branches

III. Pelvic sympathetic trunk is a continuation of the lumbar trunk, lies on sacrum medial to sacral foramina, with communications across midline
A. GANGLIA: number 4 or 5, with frequent fusions; lowest ganglion, the coccygeal, of each side usually fuses in midline as *ganglion impar*
B. BRANCHES: gray rami to all sacral and coccygeal nerves, visceral branches to hypogastric and pelvic plexuses

IV. Celiac plexus: a network of fibers and ganglia surrounding the aorta in the region of origin of the celiac artery and its branches at level of L1
A. GANGLIA: 2 large groups of interconnected ganglia on either side of the aorta. The preganglionics of the thoracic splanchnic nerves synapse here. The vagal fibers merely pass through
B. ASSOCIATED GANGLIA: aorticorenal at the origin of the renal artery, and the superior mesenteric at the origin of the superior mesenteric artery
C. ROOTS OF PLEXUS (see pp. 302, 305): preganglionics of the great splanchnic nerve go to the celiac ganglion, and those of the lesser splanchnic nerve to the aorticorenal ganglion. Postganglionics are distributed with the celiac plexus along the major branches of the abdominal aorta

V. Intermesenteric (aorta) plexus surrounds aorta between the mesenteric vessels
A. INFERIOR MESENTERIC GANGLION: around origin of the inferior mesenteric artery
B. ROOTS: celiac, from celiac plexus; lumbar splanchnics from L1, L2
C. BRANCHES: to kidney, testis or ovary, ureter, colon, and rectum

VI. Superior hypogastric plexus: caudal prolongation of the intermesenteric plexus in region of aortic bifurcation
A. ROOTS: last 3 lumbar splanchnics and intermesenteric branches
B. BRANCHES: hypogastric nerves, which cross pelvic brim and spread out as a plexus in the pelvis, the inferior hypogastric or pelvic plexus, for the pelvic viscera

VII. Clinical considerations
A. LUMBAR SYMPATHETIC GANGLIONECTOMY: a trunk resection to reduce peripheral arterial deficiency resulting from arteriosclerosis, thromboangiitis obliterans, or femoral embolus
B. RESECTION OF SUPERIOR HYPOGASTRIC PLEXUS to restore normal bladder function in cases of retention resulting from lesions of the spinal cord

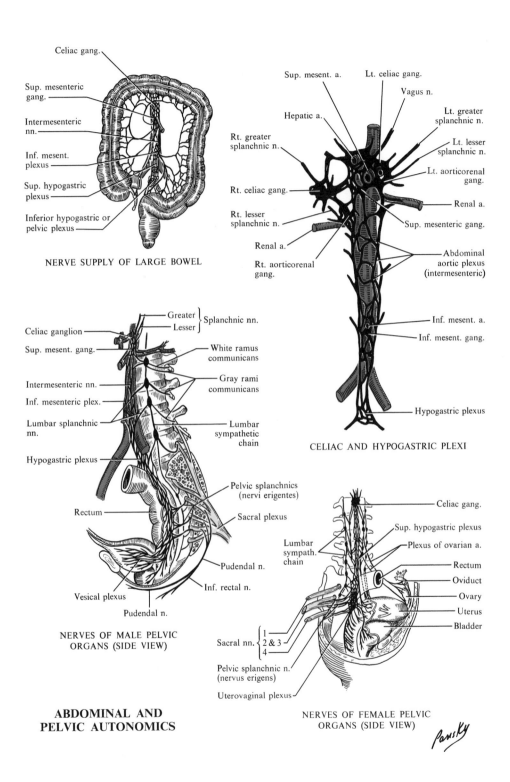

Celiac gang.

Sup. mesenteric gang.

Intermesenteric nn.

Inf. mesent. plexus

Sup. hypogastric plexus

Inferior hypogastric or pelvic plexus

NERVE SUPPLY OF LARGE BOWEL

Sup. mesent. a.   Lt. celiac gang.

Vagus n.

Hepatic a.

Rt. greater splanchnic n.

Rt. celiac gang.

Rt. lesser splanchnic n.

Renal a.

Rt. aorticorenal gang.

Lt. greater splanchnic n.

Lt. lesser splanchnic n.

Lt. aorticorenal gang.

Renal a.

Sup. mesenteric gang.

Abdominal aortic plexus (intermesenteric)

Inf. mesent. a.

Inf. mesent. gang.

Hypogastric plexus

CELIAC AND HYPOGASTRIC PLEXI

Greater } Splanchnic nn.
Lesser

Celiac ganglion

Sup. mesent. gang.

White ramus communicans

Gray rami communicans

Intermesenteric nn.

Inf. mesenteric plex.

Lumbar splanchnic nn.

Lumbar sympathetic chain

Hypogastric plexus

Pelvic splanchnics (nervi erigentes)

Rectum

Sacral plexus

Pudendal n.

Inf. rectal n.

Vesical plexus

Pudendal n.

NERVES OF MALE PELVIC ORGANS (SIDE VIEW)

Lumbar sympath. chain

Sacral nn. { 1
2 & 3
4

Pelvic splanchnic n. (nervus erigens)

Uterovaginal plexus

Celiac gang.

Sup. hypogastric plexus

Plexus of ovarian a.

Rectum

Oviduct

Ovary

Uterus

Bladder

NERVES OF FEMALE PELVIC ORGANS (SIDE VIEW)

ABDOMINAL AND PELVIC AUTONOMICS

- 387 -

ABDOMINOPELVIC NERVES

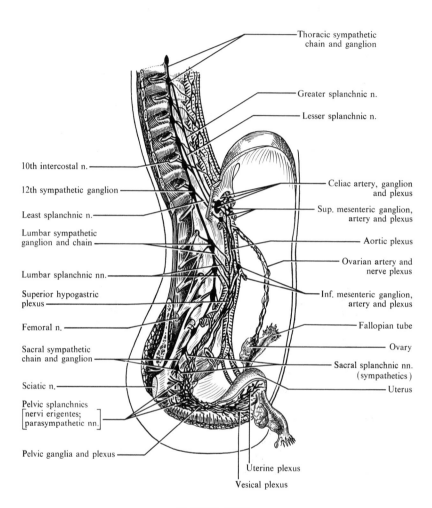

Thoracic sympathetic chain and ganglion

Greater splanchnic n.

Lesser splanchnic n.

10th intercostal n.

12th sympathetic ganglion

Least splanchnic n.

Lumbar sympathetic ganglion and chain

Lumbar splanchnic nn.

Superior hypogastric plexus

Femoral n.

Sacral sympathetic chain and ganglion

Sciatic n.

Pelvic splanchnics [nervi erigentes; parasympathetic nn.]

Pelvic ganglia and plexus

Celiac artery, ganglion and plexus

Sup. mesenteric ganglion, artery and plexus

Aortic plexus

Ovarian artery and nerve plexus

Inf. mesenteric ganglion, artery and plexus

Fallopian tube

Ovary

Sacral splanchnic nn. (sympathetics)

Uterus

Uterine plexus

Vesical plexus

SAGITTAL VIEW

FIGURE 35. **Abdominal pelvic nerves.**

-388-

NERVES OF UTERUS AND PERINEUM

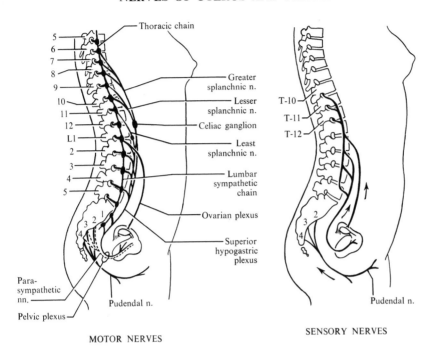

MOTOR NERVES

SENSORY NERVES

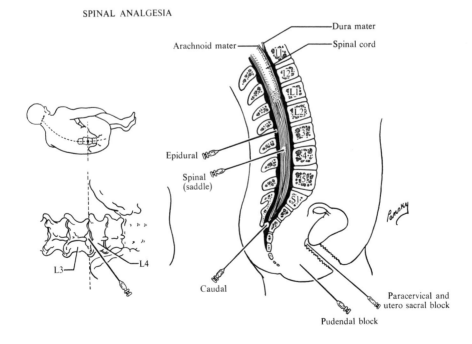

SPINAL ANALGESIA

FIGURE 36. **Nerves of uterus and perineum.**

173. PELVIC VESSELS; PERINEAL VESSELS AND NERVES

I. **Arteries of pelvis.** Originate from internal iliac artery which descends from the region of the lumbosacral articulation to the greater sciatic notch, and divides into:

A. ANTERIOR BRANCH, from which arise:
1. Superior vesical to upper bladder and vas deferens (*artery to vas deferens*)
2. Middle vesical to fundus of bladder and seminal vesicle
3. Inferior vesical to fundus of bladder, prostate, and seminal vesicles
4. Middle rectal to rectum
5. Obturator leaves pelvis to medial thigh via obturator canal
6. Internal pudendal: 1 of terminal branches, leaves pelvis through greater sciatic foramen, below piriformis muscle
7. Inferior gluteal: larger of terminal branches, curves posteriorly between first and second or second and third sacral nerves, then runs between piriformis and coccygeus muscles, through greater sciatic foramen into gluteal region
8. Uterine (see p. 382) to uterus and cervix
9. Vaginal (see p. 382) to cervix and vagina

B. POSTERIOR BRANCHES
1. Iliolumbar ascends posterior to obturator nerve and ext. iliac vessels to medial border of psoas m.; divides into a lumbar branch to psoas and quadratus lumborum mm. and to spinal cord; and an iliac branch to iliac, gluteal, abdominal mm.
2. Lateral sacral
   a. Superior: large, runs medially and enters first or second sacral foramen
   b. Inferior: crosses to sacral foramina; descends; sends branches through foramina
3. Superior gluteal runs posteriorly between lumbosacral trunk and first sacral nerve, and leaves pelvis through greater sciatic foramen above the piriformis muscle

II. **Veins of pelvis.** Correspond to the named arteries. Originate in venous plexuses

III. **Vessels of the female perineum** (for male, see p. 398)

A. ARTERIES: Internal pudendal artery. Crosses spine of ischium and enters ischiorectal fossa through lesser sciatic foramen. Lies in pudendal (Alcock's) canal. Gives *inferior rectal artery,* which crosses ischiorectal fossa to lower rectum and anal canal; *perineal artery,* which enters superficial perineal compartment to supply ischiocavernosus and bulbospongiosus and superficial transverse perineal muscles to end as *posterior labial arteries.* From Alcock's canal, the internal pudendal artery enters the deep perineal compartment and gives branches to the bulb, urethra, and greater vestibular glands as well as the *deep* and *dorsal arteries* to the *clitoris*

B. VEINS: originate in deep veins of clitoris, receive tributaries from bulb and run through the perineum as the internal pudendal veins, to end in the int. iliac veins

IV. **Nerves of the female perineum**

A. PERINEAL BRANCH OF POSTERIOR FEMORAL CUTANEOUS to labia
B. PUDENDAL. From S2–4, leaves pelvis through greater sciatic foramen, passes over spine of ischium, and enters ischiorectal fossa through lesser sciatic foramen
1. Inferior rectal nerve: crosses ischiorectal fossa to external sphincter and skin
2. Perineal nerve enters deep compartment and divides into:
   a. Superficial branches pierce inferior layer of U.G. diaphragm to labium majus
   b. Deep branches mainly to muscles of perineum including sphincter urethrae. Nerve to bulb comes from the branch to the bulbospongiosus muscle and also supplies the mucous membrane of urethra
3. Dorsal nerve of clitoris: terminal branch, which runs with dorsal artery

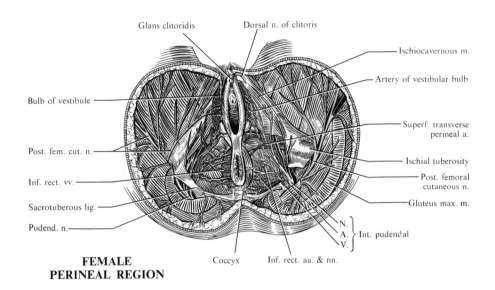

Glans clitoridis — Dorsal n. of clitoris

Ischiocavernous m.

Artery of vestibular bulb

Bulb of vestibule

Superf. transverse perineal a.

Post. fem. cut. n.

Ischial tuberosity

Inf. rect. vv.

Post. femoral cutaneous n.

Sacrotuberous lig.

Gluteus max. m.

Pudend. n.

N.
A. } Int. pudendal
V.

**FEMALE
PERINEAL REGION**

Coccyx   Inf. rect. aa. & nn.

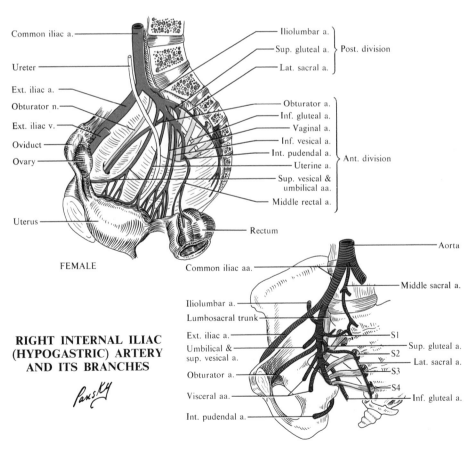

Common iliac a.

Iliolumbar a.
Sup. gluteal a. } Post. division
Lat. sacral a.

Ureter

Ext. iliac a.

Obturator n.

Obturator a.
Inf. gluteal a.
Vaginal a.

Ext. iliac v.

Inf. vesical a.
Int. pudendal a. } Ant. division
Uterine a.

Oviduct

Ovary

Sup. vesical & umbilical aa.

Middle rectal a.

Uterus

Rectum

FEMALE

Aorta

Common iliac aa.

Middle sacral a.

Iliolumbar a.
Lumbosacral trunk

**RIGHT INTERNAL ILIAC
(HYPOGASTRIC) ARTERY
AND ITS BRANCHES**

Ext. iliac a.
Umbilical & sup. vesical a.

S1
Sup. gluteal a.
S2
Lat. sacral a.
S3

Obturator a.

Visceral aa.

S4
Inf. gluteal a.

Int. pudendal a.

Pansky

174. THE FEMALE PERINEUM

I. **Boundaries:** superficial and deep fascia, superficial and deep compartments, similar to male (see p. 394)

II. **Central tendon of perineum** (perineal body, central point, in the female "the perineum"): a mass of fibrous and muscular tissue in midline between anus and vagina (anus and bulb in the male), may be 4 cm in diameter (2 cm in the male). In this center, fasciae of perineum fuse; muscles also arise, interdigitate, or insert here

III. **Muscles:** except for the differences listed below, are similar to the male (see p. 394)
   A. SUPERFICIAL TRANSVERSE PERINEAL similar to, but smaller than, male
   B. BULBOSPONGIOSUS (SPHINCTER VAGINAE) surrounds orifice of vagina and covers vestibular bulb. Arises in central tendon and inserts on corpora cavernosa of clitoris. It narrows vaginal orifice and maintains erection of clitoris
   C. DEEP TRANSVERSE PERINEAL inserts into sides of vagina

IV. **External genitalia**
   A. MONS PUBIS: eminence overlying symphysis pubis, composed mainly of fat, covered with hair
   B. LABIA MAJORA: longitudinal folds of skin that extend caudally and posteriorly from mons. Skin of outer surface is pigmented and set with hair. Inner layer is smooth with large sebaceous glands. Between layers are loose connective tissue, fat, blood vessels, nerves, and glands
      1. Pudendal cleft (rima): between labia
      2. Anterior labial commissure: where labia join anteriorly
      3. Posterior labial commissure: mainly skin connecting the posterior ends of labia
   C. LABIA MINORA: 2 small folds between labia majora, surrounding vaginal orifice
      1. Frenulum of labia (fourchette): where these folds join posteriorly
      2. Prepuce: most anterior part of labia, which have split to pass above clitoris
      3. Frenulum of clitoris: parts of labia that pass below clitoris
   D. CLITORIS: located beneath anterior labial commissure and partly hidden by prepuce and frenulum of labia minora. Has 2 corpora and a glans but no urethra
   E. VESTIBULE: area posterior to clitoris and between labia minora. Has several openings: external urethral; vaginal, behind the urethra; ducts of major vestibular glands
      1. Hymen: membrane of variable size and form partly blocking the vaginal orifice in the virgin
   F. BULB OF VESTIBULE: 2 masses of erectile tissue on either side of vaginal orifice
   G. GREATER VESTIBULAR GLANDS lie beneath posterior ends of bulb. Each has duct that opens into the vestibule in a groove between hymen and labia minora

V. **Clinical considerations**
   A. VAGINAL HYSTERECTOMY: for repair of procedentia and prolapse associated with cystocele and rectocele
   B. POSTERIOR COLPOTOMY: in treatment of pelvic abscess
   C. RECTO- AND VESICOVAGINAL FISTULAE (abnormal connections between bladder and rectum and vagina): from injuries during childbirth or from faulty development
   D. CYSTS OF THE MAJOR VESTIBULAR GLANDS (Bartholin) can occur
   E. IMPERFORATE HYMEN: a failure of an opening to occur in the hymen, which may not be recognized until puberty when menstrual fluid dilates the genital canal creating pressure (hematocolpos)
   F. EPISIOTOMY is an incision in the vulva which is made to direct the tear in the perineum to the side during labor

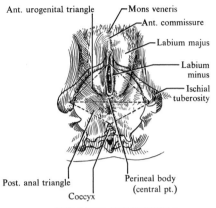

Ant. urogenital triangle — Mons veneris
— Ant. commissure
— Labium majus
— Labium minus
— Ischial tuberosity
Post. anal triangle
Coccyx
Perineal body (central pt.)

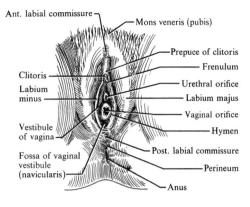

Ant. labial commissure —
— Mons veneris (pubis)
Clitoris — — Prepuce of clitoris
— Frenulum
Labium minus — — Urethral orifice
— Labium majus
— Vaginal orifice
Vestibule of vagina — — Hymen
— Post. labial commissure
Fossa of vaginal vestibule (navicularis) — — Perineum
— Anus

FEMALE EXTERNAL GENITALIA

FEMALE PERINEUM
SUPERFICIAL TO DEEP
(A-E)

A

Labium majus — — Labium minus
— Ischiocav. m.
— Bulbospon. m.
— Ischial tuberosity
Ischial tuberosity
Colles' fascia (deep superfic.) —
Deep   Superf.
Layers of superficial fascia

B

— Body of clitoris
— Glans of clitoris
U. — — Bulbospongiosus m.
V. — — Ischiocavernosus m.
— Ischial tuberosity
Labium minus — — Inf. fascia of u. g. diaphragm
Central tendon of perineum (perineal body) — Superficial transverse perineal m.
— Ext. sphincter m.

C

— Clitoris
— Corporus cavernosum clitoridis
Bulb of vestibule — — Inf. fascia of u. g. diaphragm
Great vestibular gland (Bartholin) —

D
Deep transverse perineal m. (sphincter urethrae m. not seen)
Urethra —
Vagina —
Inf.   Sup.
Fascia layers of urogenital (u. g.) diaphragm

E
Urethra —
Vagina —
Deep transverse perineal m. (cut) —
Levator ani m.

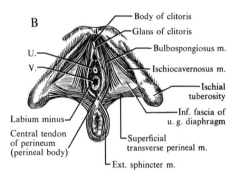
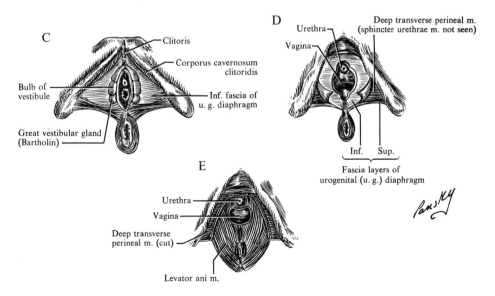

Pansky

175. THE MALE PERINEUM

I. **Boundaries:** ventrally, pubic arch and arcuate ligaments; dorsally, tip of the coccyx; laterally, inferior rami of pubis and ischium and sacrotuberous ligaments

II. **Divisions:** dorsal, anal region; ventral, urogenital region

III. **Superficial fascia:** 2 layered as on abdomen
   A. SUPERFICIAL LAYER. In U.G. region continuous with Camper's fascia, superficial fascia of thigh, and superficial anal fascia; medially continuous with the dartos of scrotum. In anal region it is fatty and fills ischiorectal fossa
   B. DEEP LAYER (COLLES' FASCIA) continues with Scarpa's fascia of abdomen, attached to ischiopubic rami, curves around superficial transverse perineal muscle to fuse with deep fascia and central tendon. In anal region adherent to superficial layer

IV. **Muscles:** all innervated by the perineal branch of pudendal nerve

Name	Origin	Insertion	Action
A. IN SUPERFICIAL PERINEAL POUCH			
Superfic. trans. perineal	Ischial tuberos.	Central tendon	Fixes central tendon
Bulbospongiosus	Central. tend., med. raphé	Side and dorsum of penis	Compresses urethra, helps in erection
Ischiocavernosus	Ischial tuberos.	Crura of penis	Maintains erection
B. IN DEEP PERINEAL COMPARTMENT			
Deep trans. perineal	Ramus of ischium	Joins muscle of other side	Compresses memb. urethra
Sphinct. ureth.	Ramus of ischium	Same as above	Same as above

V. **Deep fascia:** obturator and infra-anal
   A. OBTURATOR (see p. 376); over obturator m. on lat. wall of ischiorectal fossa
   B. INFRA-ANAL covers inferior surface of levator ani and coccygeus muscles, joins obturator along arcus tendineus. At border of urogenital region splits into 3 sheets
     1. Deep (external) perineal attached to ischiopubic rami and urogenital diaphragm. Is continuous with deep abdominal fascia, fascia lata, deep penile (Buck's) fascia. Covers superficial perineal muscles
     2. Inferior (superficial) layer of urogenital diaphragm. Attached to ischiopubic rami, leaving a slight space between the ventral edge of this layer and the arcuate ligament to permit the passage of the dorsal vein of penis
        a. Pierced by urethra, ducts of bulbourethral glands, arteries to bulb, deep arteries to penis, dorsal arteries and nerves of penis
     3. Superior (deep) layer of urogenital diaphragm. Similar to above, with deeper attachments. Prostate rests on it and capsule attached to it. Pierced by urethra

VI. **Perineal compartments**
   A. SUPERFICIAL: space between deep external perineal fascia (Gallaudet's fascia) and the inferior layer of the urogenital diaphragm. It contains crura and bulb of penis, ischio- and bulbospongiosus muscles, superficial transverse perineal muscle, deep artery of penis, vessels and nerves of all the muscles, major vestibular glands of female
   B. DEEP: space between the inferior and superior fascia of the urogenital diaphragm. Contains deep transverse perineal and sphincter urethrae muscles, bulbourethral glands, internal pudendal vessels, nerves, and their branches
   C. SUPERFICIAL PERINEAL FASCIAL CLEFT: between deep external perineal fascia (Gallaudet's) and deep layer of superficial fascia of perineum (Colles')

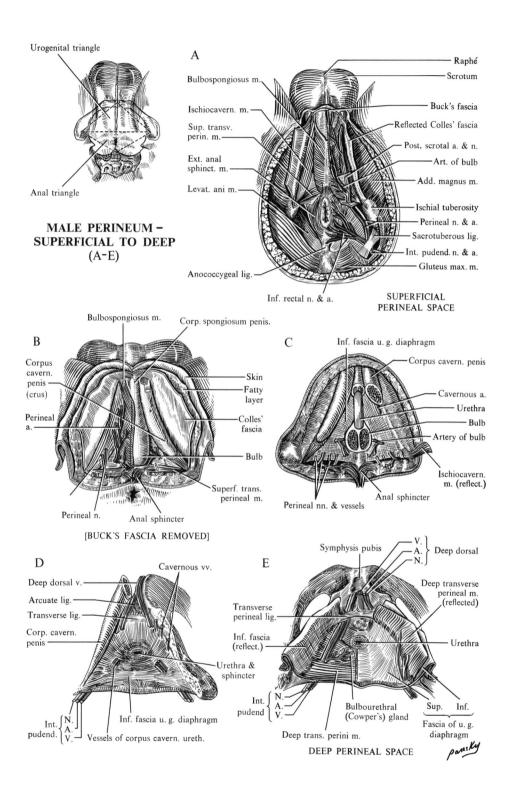

Urogenital triangle

Anal triangle

MALE PERINEUM –
SUPERFICIAL TO DEEP
(A-E)

A

Bulbospongiosus m.
Ischiocavern. m.
Sup. transv. perin. m.
Ext. anal sphinct. m.
Levat. ani m.

Anococcygeal lig.

Inf. rectal n. & a.

Raphé
Scrotum
Buck's fascia
Reflected Colles' fascia
Post. scrotal a. & n.
Art. of bulb
Add. magnus m.
Ischial tuberosity
Perineal n. & a.
Sacrotuberous lig.
Int. pudend. n. & a.
Gluteus max. m.

SUPERFICIAL
PERINEAL SPACE

B

Bulbospongiosus m.
Corp. spongiosum penis.

Corpus cavern. penis (crus)

Perineal a.

Perineal n.   Anal sphincter

Skin
Fatty layer
Colles' fascia
Bulb
Superf. trans. perineal m.

[BUCK'S FASCIA REMOVED]

C

Inf. fascia u. g. diaphragm

Corpus cavern. penis
Cavernous a.
Urethra
Bulb
Artery of bulb
Ischiocavern. m. (reflect.)

Perineal nn. & vessels   Anal sphincter

D

Cavernous vv.

Deep dorsal v.
Arcuate lig.
Transverse lig.
Corp. cavern. penis

Int. pudend. { N. A. V. }

Inf. fascia u. g. diaphragm

Vessels of corpus cavern. ureth.

Urethra & sphincter

E

Symphysis pubis

V.
A.   Deep dorsal
N.

Transverse perineal lig.
Inf. fascia (reflect.)

Deep transverse perineal m. (reflected)

Urethra

Int. pudend { N. A. V. }

Deep trans. perini m.

Bulbourethral (Cowper's) gland

Sup.   Inf.
Fascia of u. g. diaphragm

DEEP PERINEAL SPACE

pansky

176. SCROTUM AND TESTIS

I. Scrotum: cutaneous sac to contain testis. Divided by a median raphé, which is continuous on penis and along midline to anus. Composed of 2 layers
A. SKIN: thin and pigmented, with hair and sebaceous glands
B. DARTOS: superficial fascia containing scattered smooth muscle. Is closely bound to skin but only loosely attached to deeper layers

II. Coverings of testis
A. EXTERNAL SPERMATIC FASCIA (intercrural, intercolumnar). At subcutaneous inguinal ring is continuous with deep abdominal fascia of external oblique muscle which is attached to crura of ring
B. CREMASTERIC LAYER (middle spermatic fascia). This is composed of cremasteric muscle fibers interconnected by fascia. It is derived from the internal abdominal oblique muscle and its fascia
C. INTERNAL SPERMATIC FASCIA is derived from and continuous with the transversalis fascia at the deep inguinal ring
D. TUNICA VAGINALIS. Serous membrane derived from the peritoneum of the abdomen. Has 2 layers
  1. Visceral covers testis and epididymis. Posteriorly is reflected onto scrotal lining
  2. Parietal covers more area than visceral layer, for it lines scrotum and extends a short distance into the spermatic cord
  3. Cavity of tunica vaginalis: space between the visceral and parietal layers

III. Testis
A. STRUCTURE
  1. Capsule, the tunica albuginea: a dense, fibrous membrane
  2. Mediastinum testis: reflection of capsule along posterior border into the interior of gland, forming a partial septum
  3. Trabeculae (septa): partial partitions radiating from the front and sides of mediastinum. These divide testis into numerous conical lobules
  4. Lobules: each contains several highly convoluted seminiferous tubules, which give rise to sperm
  5. Tubuli recti: at apex of lobule, next to mediastinum, seminiferous tubules straighten out, carry sperm to rete testis
  6. Rete testis: network of irregular spaces that transmit sperm across mediastinum to efferent ducts (ductules)
  7. Efferent ductules: lead from rete, penetrate tunica albuginea to the head of the epididymis, where each of the 12–15 ductules becomes highly convoluted
  8. Duct of epididymis: the efferent ducts open into this. In the head and body of the epididymis this duct is convoluted. In the tail of the epididymis it becomes straighter where it leads into the vas deferens

IV. Epididymis. An important storehouse for sperm. In general, all parts are covered by the visceral layer of serous membrane, which helps to bind this structure to the testis. As noted above, it contains efferent ductules and the duct of the epididymis

V. Clinical considerations
A. HYDROCELE: an accumulation of fluid in the cavity of tunica vaginalis
B. ORCHITIS: an inflammation of the glandular structure of the testis, may occur as a complication of mumps and can lead to sterility
C. VARICOCELE is a condition of enlargement (varicosity) of the veins of the spermatic cord

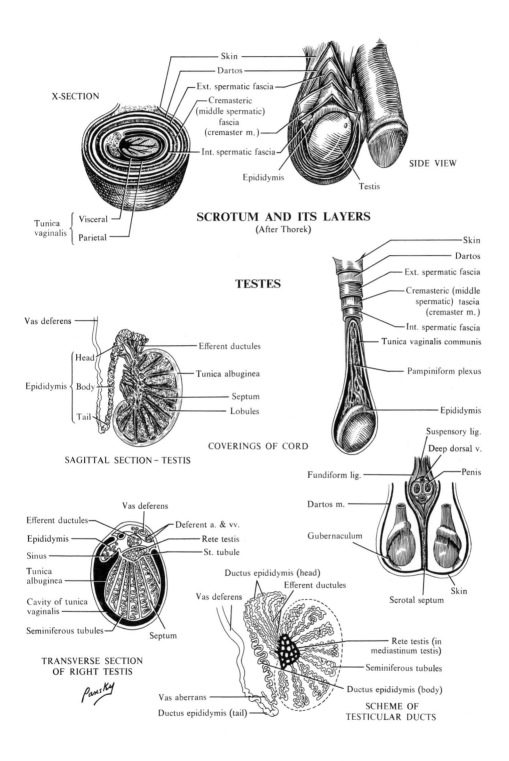

X-SECTION

Skin
Dartos
Ext. spermatic fascia
Cremasteric
(middle spermatic)
fascia
(cremaster m.)
Int. spermatic fascia

SIDE VIEW

Epididymis

Testis

Tunica
vaginalis { Visceral
Parietal

SCROTUM AND ITS LAYERS
(After Thorek)

TESTES

Vas deferens

Efferent ductules

Head

Body

Tail

Epididymis

Tunica albuginea

Septum

Lobules

SAGITTAL SECTION – TESTIS

COVERINGS OF CORD

Skin
Dartos
Ext. spermatic fascia
Cremasteric (middle
spermatic) fascia
(cremaster m.)
Int. spermatic fascia
Tunica vaginalis communis
Pampiniform plexus
Epididymis

Suspensory lig.
Deep dorsal v.
Penis
Fundiform lig.
Dartos m.
Gubernaculum
Skin
Scrotal septum

Vas deferens
Efferent ductules
Epididymis
Sinus
Tunica
albuginea
Cavity of tunica
vaginalis
Seminiferous tubules

Deferent a. & vv.
Rete testis
St. tubule

Septum

TRANSVERSE SECTION
OF RIGHT TESTIS

Pansky

Ductus epididymis (head)
Efferent ductules
Vas deferens

Rete testis (in
mediastinum testis)
Seminiferous tubules

Ductus epididymis (body)

Vas aberrans
Ductus epididymis (tail)

SCHEME OF
TESTICULAR DUCTS

177. THE PENIS

I. Composition: 2 dorsal cylindrical masses, the corpora cavernosa penis, and a single ventral mass, the corpus spongiosum (corpus cavernosum urethrae)

A. CORPORA CAVERNOSA PENIS: anterior three fourths composed of erectile tissue contained in a strong fibrous capsule. Proximally, the capsules of the adjacent sides of the corpora form a complete *septum penis;* distally, the septum is incomplete. The posterior one fourth of each corpus diverges as the *crura of the penis.* The latter consists of tendonlike tissue; serves as attachments of the penis to the ischiopubic rami

B. CORPUS SPONGIOSUM: erectile tissue containing penile urethra. 2 expansions.
  1. Glans: conical tip that forms a cap over the other corpora. The *corona glandis* is the widest, proximal portion
  2. Bulbus penis. The proximal enlargement is entered by the urethra

II. Ligaments

A. FUNDIFORM: sling of Scarpa's fascia extending to the sides and dorsum of the penis

B. SUSPENSORY: deep fascia from linea alba, symphysis pubis, and arcuate ligament

III. Skin: thin, hairless, and dark. Forms fold over glans called the prepuce (foreskin)

IV. Fascia

A. SUPERFICIAL continuous with Scarpa's fascia and the dartos of the scrotum

B. DEEP (FASCIA PENIS, BUCK'S) starts from deep (external) perineal fascia and fascia investing muscles of penis, which extend distally as a single sheet covering the corpora to the glans

V. Vessels and nerves (see also p. 395)

A. ARTERIES: from the internal pudendal artery in the deep perineal compartment
  1. Dorsal arteries pierce urogenital diaphragm, pass through suspensory ligament to lie on either side of the deep dorsal vein beneath deep (Buck's) fascia
  2. Deep arteries pierce urogenital diaphragm and enter crura to run in center of corpora cavernosa penis
  3. Artery of bulb pierces urogenital diaphragm to bulb and corpus spongiosum
  4. Urethral a. pierces U.G. diaphragm to glans in corpus spongiosum penis

B. VEINS: all begin in cavernous spaces of erectile tissue
  1. Superficial dorsal vein drains prepuce and skin, runs in subcutaneous tissue, and ends in superficial external pudendal vein (of saphenous)
  2. Deep vein drains erectile tissue, runs under deep fascia between dorsal arteries, passes through suspensory lig. and into pelvis between arcuate lig. and urogenital diaphragm and drains into int. pudendal v. and then to int. iliac v.

C. LYMPHATICS: superficial and deep drain to superficial inguinal nodes, glans drains to deep inguinal and external iliac nodes

D. NERVES: anterior scrotal of ilioinguinal, perineal and dorsal nerves from pudendal nerve. Autonomics from hypogastric and pelvic plexuses

VI. Erection caused by contraction of ischio- and bulbospongiosus muscles, which compress veins; at the same time, elevated blood pressure and dilatation of arteries increase flow of blood into the cavernous spaces of the erectile tissue

VII. Ejaculation consists of:

A. EMISSION: reflex wave of muscle contraction in ductus deferens with contraction of seminal vesicle and prostate. Under sympathetic control

B. EJACULATION PROPER: rhythmic contraction of bulbocavernosus (and ischiocavernosus) muscles. Ejaculation and erection are under parasympathetic control

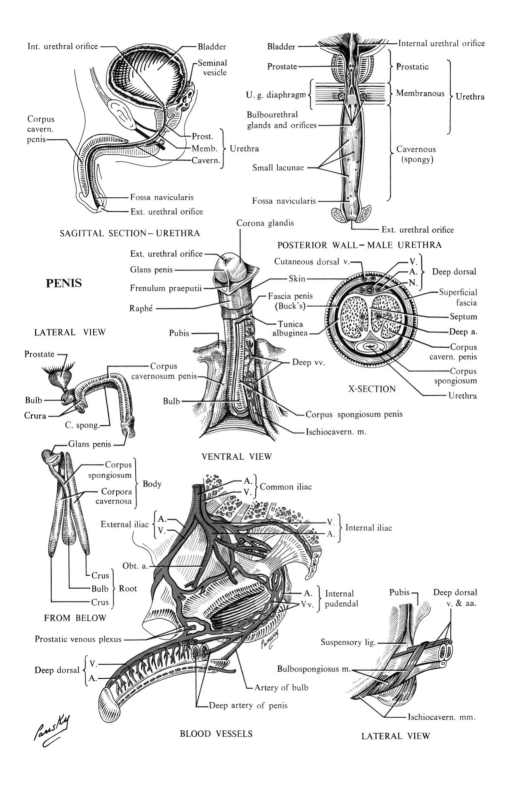

PENIS

Int. urethral orifice — Bladder
Seminal vesicle
Corpus cavern. penis
Prost.
Memb. } Urethra
Cavern.
Fossa navicularis
Ext. urethral orifice

SAGITTAL SECTION – URETHRA

Bladder — Internal urethral orifice
Prostate
Prostatic
U. g. diaphragm
Membranous } Urethra
Bulbourethral glands and orifices
Cavernous (spongy)
Small lacunae
Fossa navicularis
Ext. urethral orifice

POSTERIOR WALL – MALE URETHRA

Corona glandis
Ext. urethral orifice
Glans penis
Frenulum praeputii
Raphé

LATERAL VIEW

Cutaneous dorsal v.
Skin
Fascia penis (Buck's)
Tunica albuginea
Deep vv.

V.
A. } Deep dorsal
N.
Superficial fascia
Septum
Deep a.
Corpus cavern. penis
Corpus spongiosum
Urethra

X-SECTION

Prostate
Bulb
Crura
C. spong.

Pubis
Corpus cavernosum penis
Bulb
Corpus spongiosum penis
Ischiocavern. m.

VENTRAL VIEW

Glans penis
Corpus spongiosum
Corpora cavernosa
} Body

Crus
Bulb } Root
Crus

FROM BELOW

A.
V. } Common iliac

External iliac { A.
V.

V.
A. } Internal iliac

Obt. a.

A.
V v. } Internal pudendal

Prostatic venous plexus

Deep dorsal { V.
A.

Artery of bulb
Deep artery of penis

BLOOD VESSELS

Pubis
Deep dorsal v. & aa.
Suspensory lig.
Bulbospongiosus m.
Ischiocavern. mm.

LATERAL VIEW

UNIT SIX
Lower
Extremity

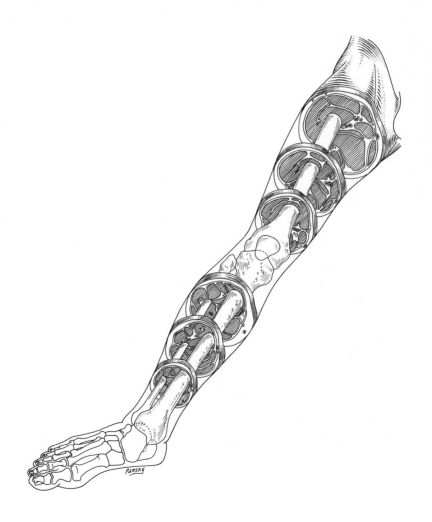

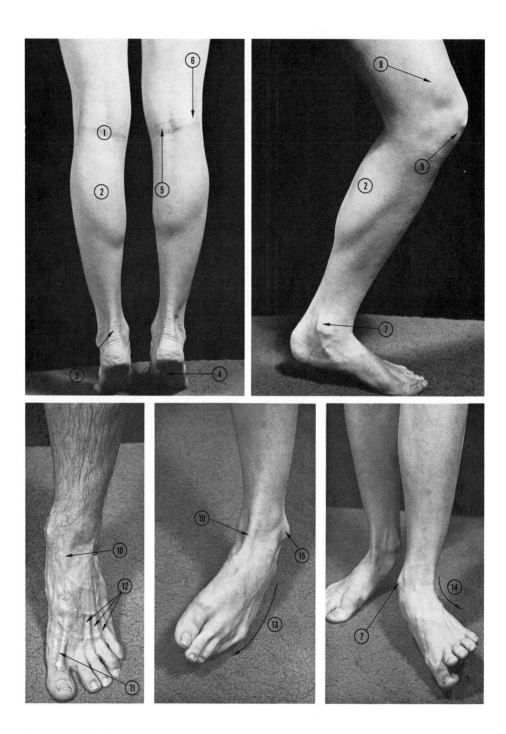

FIGURE 37. **Surface anatomy of lower extremity.** *1,* Popliteal fossa; *2,* gastrocnemius m.; *3,* tendo calcaneus (Achilles); *4,* plantar surface of foot; *5,* semitendinosus m.; *6,* biceps femoris m.; *7,* medial malleolus; *8,* vastus medialis m.; *9,* patella; *10,* tibialis anterior m.; *11,* extensor hallucis longus tendon; *12,* extensor digitorum longus tendons; *13,* inversion; *14,* eversion; *15,* lateral malleolus.

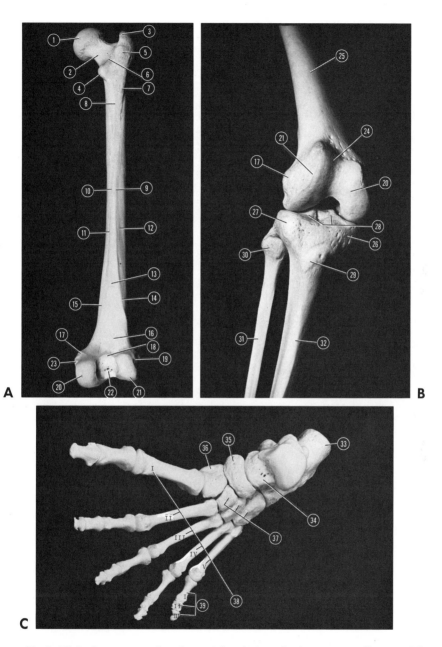

FIGURE 38. **A, Right femur, posterior view.** *1,* head; *2,* neck; *3,* greater trochanter; *4,* lesser trochanter; *5,* quadrate tubercle; *6,* intertrochanteric line; *7,* gluteal tuberosity, *8,* pectineal line; *9* and *10,* lateral and medial lip of linea aspera; *11,* medial surface; *12,* lateral surface; *13,* posterior surface; *14* and *15,* lateral and medial supracondylar lines; *16,* popliteal surface; *17,* adductor tubercle; *18,* intercondylar line; *19,* lateral epicondyle; *20,* medial condyle; *21,* lateral condyle; *22,* intercondylar fossa; 23, medial epicondyle. **B. Right femur, tibia, fibula, anterior view.** *24,* Patellar surface; *25,* anterior surface of shaft; *26,* medial condyle; *27,* lateral condyle; *28,* intercondylar tubercles; *29,* tuberosity; *30,* head of fibula; *31,* shaft of fibula; *32,* shaft of tibia. **C, Left foot.** *33,* Calcaneus; *34,* talus; *35,* navicular; *36,* cuboid; *37,* lateral, intermediate, and medial cuneiforms; *38,* metatarsals; *39,* phalanges of little toe.

-403-

178. CUTANEOUS NERVES AND DERMATOMES OF THE LOWER EXTREMITY

I. Source of cutaneous nerves: lumbar plexus, sacral plexus, posterior primary divisions of lumbar and sacral nerves

II. Origin and distribution

Name	Spinal Component	Distribution
A. FROM LUMBAR PLEXUS (L1–L4)		
1. Iliohypogastric	L1	
Lat. branch		Lat. thigh, ant. gluteal region
Ant. branch		Abdomen
2. Ilioinguinal	L1	Upper med. thigh, ext. genitalia
3. Genitofemoral	L1, 2	
Genital br.		Upper med. thigh, scrotum
Femoral br.		Upper ant. thigh
4. Lat. fem. cutaneous	L2, 3	Lat. ant. thigh, post. lat. thigh
5. Obturator	L2, 3, 4	Med. thigh, just above knee
6. Femoral		
Ant. cutaneous	L2, 3	Ant. and anteromed. thigh
Saphenous	L3, 4	Med. side knee, leg, ankle
B. FROM SACRAL PLEXUS (L4–S3)		
1. Post. fem. cutaneous	S1, 2, 3	Lower buttock, post. thigh, middle post. leg
2. Sural	S1, 2	Post. leg, lat. side and dorsum of foot
3. Lat. sural cutan.	S1, 2	Lateral side of leg
4. Medial calcaneal	S1, 2	Heel
5. Medial plantar	L4, 5	Sole of foot, adjacent sides first 4 toes
6. Lateral plantar	S1, 2	Lat. sole of foot, sides of toes 4 and 5
7. Deep peroneal (fibular)	L4, 5	Adjacent sides of dorsum, toes 1 and 2
8. Superfic. peroneal (fibular)	L4, 5, S1	Ant. lower leg and ankle, dorsum of foot, adjacent sides toes 1–4
C. FROM POSTERIOR PRIMARY DIVISIONS OF LUMBAR AND SACRAL NERVES		
1. Superior clunial	L1, 2, 3	Upper med. buttock
2. Middle clunial	S1, 2, 3	Med. aspect of buttock

III. Dermatomes

A. DEFINITION: these represent *specific skin areas* supplied by *specific spinal nerves* regardless of the named cutaneous nerves that are distributed to the same general territory

B. ARRANGEMENT: the dermatomes of the lower extremity are related to the development of the limb. Their spiral course in the lower extremity is an expression of the rotation of the limb as an adaptation to the erect position

C. CLINICAL SIGNIFICANCE: the skin is easily tested for various sensory modalities (for example, touch or temperature). If certain dermatomes are found insensitive, this can be related to definite spinal cord levels, and thus a lesion in the central nervous system can be more readily located

IV. Clinical considerations

A. SPINAL-VERTEBRAL SEGMENTS: a patient with numbness in middle of dorsum of foot and anterolateral leg may have a lesion in the 5th lumbar cord segment, which lies at the level of 11th thoracic vertebra

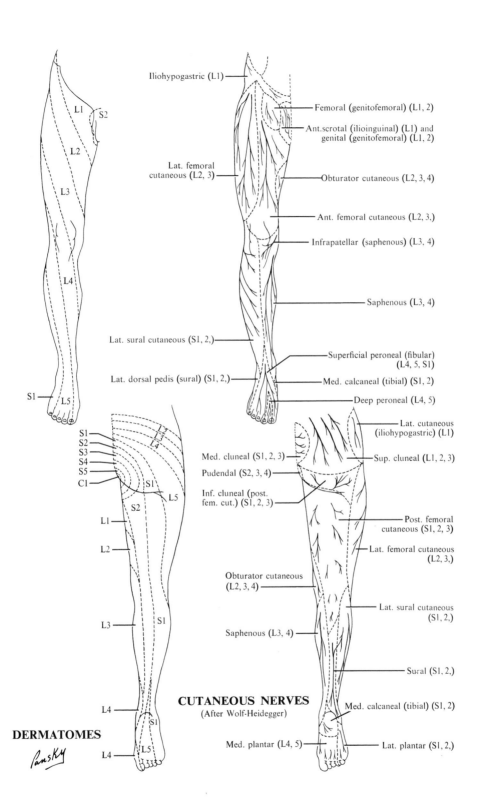

Iliohypogastric (L1)

Femoral (genitofemoral) (L1, 2)

Ant.scrotal (ilioinguinal) (L1) and
genital (genitofemoral) (L1, 2)

Lat. femoral
cutaneous (L2, 3)

Obturator cutaneous (L2, 3, 4)

Ant. femoral cutaneous (L2, 3,)

Infrapatellar (saphenous) (L3, 4)

Saphenous (L3, 4)

Lat. sural cutaneous (S1, 2,)

Superficial peroneal (fibular)
(L4, 5, S1)

Lat. dorsal pedis (sural) (S1, 2,)

Med. calcaneal (tibial) (S1, 2)

Deep peroneal (L4, 5)

Lat. cutaneous
(iliohypogastric) (L1)

Med. cluneal (S1, 2, 3)

Sup. cluneal (L1, 2, 3)

Pudendal (S2, 3, 4)

Inf. cluneal (post.
fem. cut.) (S1, 2, 3)

Post. femoral
cutaneous (S1, 2, 3)

Lat. femoral cutaneous
(L2, 3,)

Obturator cutaneous
(L2, 3, 4)

Lat. sural cutaneous
(S1, 2,)

Saphenous (L3, 4)

Sural (S1, 2,)

CUTANEOUS NERVES
(After Wolf-Heidegger)

Med. calcaneal (tibial) (S1, 2)

DERMATOMES

Med. plantar (L4, 5)

Lat. plantar (S1, 2,)

-405-

179. SUPERFICIAL VEINS OF THE LOWER EXTREMITY

I. General: lie between the layers of the superficial fascia. Valves are more numerous in the veins of the lower extremity than they are in the upper

II. Specific veins

A. GREAT SAPHENOUS VEIN: longest vein in body
1. Origin: medial side of dorsal venous arch of foot
2. Course: ascends in front of medial malleolus, along medial side of leg, behind the medial condyles of tibia and femur, along medial side of thigh to the saphenous hiatus, where it pierces deep fascia
3. Termination: in femoral vein about 3.75 cm below inguinal ligament
4. Tributaries
   a. External pudendal from genital area
   b. Superficial circumflex iliac from upper lateral thigh
   c. Superficial epigastric from abdomen, has communication with lateral thoracic vein by way of *thoracoepigastric vein*
   d. Has variable number of other tributaries frequently unnamed from foot, leg, and thigh

B. SMALL SAPHENOUS VEIN
1. Origin: lateral side of dorsal venous arch of foot
2. Course: from behind lateral malleolus along lateral side of tendo calcaneus (Achilles), crosses the latter to middle of back of leg, runs straight upward to pierce the deep fascia in lower popliteal space
3. Termination: in popliteal vein between heads of gastrocnemius muscle
4. Tributaries: especially from lateral foot and leg, send communication upward to join great saphenous vein

III. Clinical considerations

A. VARICOSE VEINS: abnormal dilation and loss of valvular competence in superficial veins of lower extremity, especially the great saphenous system
1. Prevalent in lower limb because of great weight of the long column of blood (heart to foot). Since venous blood flow depends on muscular movement, long-standing or increased pressure on the more caudal part of the system (as in pregnancy) causes venous stasis. This causes walls to dilate in regions of the numerous valves
2. To relieve this condition, superficial veins can be ligated, stripped, etc., without interference with venous return because of the numerous communications with the deep veins

B. VENIPUNCTURE: the great saphenous vein is most frequently used, especially its lower end. A cut-down is made about 1.25 cm anterior to the medial malleolus. When necessary, the cephalic portion may be used, although here it may be embedded in fat. The site used is often described as 2 fingerbreadths medial and 1 fingerbreadth below the pubic tubercle. Another method of finding the vein is given as 1 fingerbreadth below the inguinal ligament just medial to the point where the pulsating femoral artery may be palpated

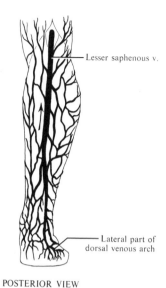

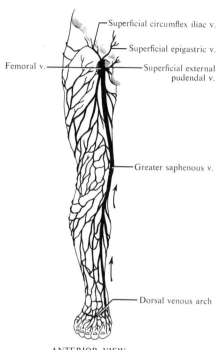

Superficial circumflex iliac v.

Superficial epigastric v.

Femoral v.

Superficial external pudendal v.

Greater saphenous v.

Dorsal venous arch

ANTERIOR VIEW

Lesser saphenous v.

Lateral part of dorsal venous arch

POSTERIOR VIEW

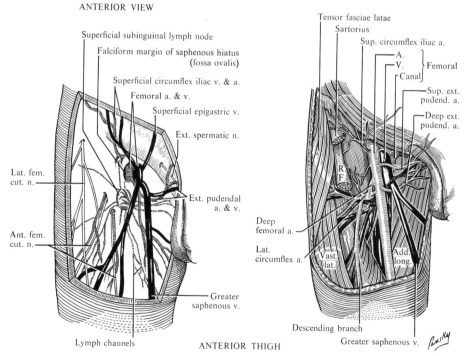

Superficial subinguinal lymph node

Falciform margin of saphenous hiatus (fossa ovalis)

Superficial circumflex iliac v. & a.

Femoral a. & v.

Superficial epigastric v.

Ext. spermatic n.

Lat. fem. cut. n.

Ext. pudendal a. & v.

Ant. fem. cut. n.

Greater saphenous v.

Lymph channels

ANTERIOR THIGH

Tensor fasciae latae

Sartorius

Sup. circumflex iliac a.

A.

V. Femoral

Canal

Sup. ext. pudend. a.

Deep ext. pudend. a.

R. F.

Deep femoral a.

Lat. circumflex a.

Vast. lat.

Add. long.

Descending branch

Greater saphenous v.

-407-

180. LYMPHATIC DRAINAGE OF THE LOWER EXTREMITY

I. Superficial vessels lie in superficial fascia

A. MEDIAL GROUP follow great saphenous vein
1. Origin and course: medial side of dorsum of foot, pass on both sides of medial malleolus, run behind medial femoral condyle
2. Termination: superficial subinguinal nodes

B. LATERAL GROUP
1. Origin and course: lateral side of foot, ascend leg on both posterior and anterior sides; those on anterior side join medial group (see above)
2. Termination: for anterior vessels, see A2, above; the posterior group end in the popliteal nodes

II. Deep vessels accompany deep blood vessels; thus, these are anterior and posterior tibial and peroneal sets, 2 or 3 lymphatics with each artery. They terminate in the popliteal nodes

III. Lymph nodes: 3 sets

A. ANTERIOR TIBIAL (inconstant): when present, receive vessels along anterior tibial artery; their efferents go to popliteal nodes

B. POPLITEAL NODES: usually 6 or 7 scattered in fat of popliteal fossa
1. Afferents: vessels running with small saphenous vein, vessels from knee joint running with genicular arteries, and vessels running with anterior and posterior tibial arteries
2. Efferents: some follow great saphenous vein to superficial subinguinal nodes; most follow femoral vessels to deep inguinal nodes

C. INGUINAL: vary from 12 to over 20, located in proximal part of femoral trigone. Divided into 2 sets by a transverse line drawn across thigh at the level of the end of the great saphenous vein
1. Superficial inguinal lie above the line
   a. Afferents from penis, scrotum, perineum, buttock, and abdominal wall
   b. Efferents to external iliac nodes
2. Subinguinal: below line. Consist of 2 groups
   a. Superficial, on either side of great saphenous vein
      i. Afferents follow superficial vessels of lower limb, especially those with great saphenous vein; a few from external genitalia and buttock
      ii. Efferents to deep subinguinal and external iliac nodes
   b. Deep, under deep fascia on medial side of femoral vein; 1–3 in number. One of these may lie in femoral canal (node of Cloquet)
      i. Afferents: deep lymphatics of lower extremity; from superficial subinguinal nodes
      ii. Efferents: to external iliac nodes

IV. Clinical considerations

A. DRAINAGE of the external genitalia, perineum, buttock, and lower anal canal into the inguinal nodes should be noted

B. THE UTERUS adjoining the attachment of the round ligament sends its lymphatics along the ligament to drain into the inguinal nodes

C. AS THEY COURSE UPWARD in the thigh, the superficial lymphatics from the lateral side run anteriorly and then medially and those from the medial side run anteriorly and then lateral, to converge toward the path of the great saphenous vein. This creates a sort of "lymphatic divide" on the back of the thigh

D. INJECTION OF RADIOPAQUE MATERIAL into the lymphatics of the lower limb can demonstrate the lymphatics as well as lymph nodes

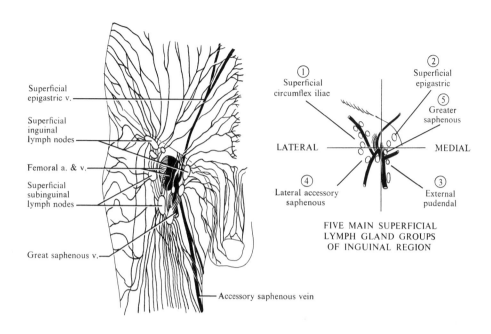

Superficial
epigastric v.

Superficial
inguinal
lymph nodes

Femoral a. & v.

Superficial
subinguinal
lymph nodes

Great saphenous v.

Accessory saphenous vein

① Superficial
circumflex iliac

② Superficial
epigastric

⑤ Greater
saphenous

LATERAL   MEDIAL

④ Lateral accessory
saphenous

③ External
pudendal

FIVE MAIN SUPERFICIAL
LYMPH GLAND GROUPS
OF INGUINAL REGION

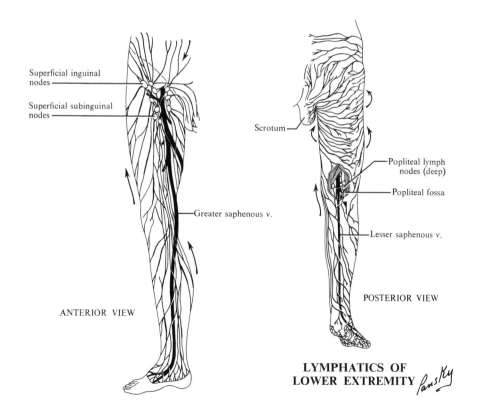

Superficial inguinal
nodes

Superficial subinguinal
nodes

Greater saphenous v.

ANTERIOR VIEW

Scrotum

Popliteal lymph
nodes (deep)

Popliteal fossa

Lesser saphenous v.

POSTERIOR VIEW

LYMPHATICS OF
LOWER EXTREMITY Pansky

181. BONE OF THE HIP

I. Hip bone
A. PARTS: ilium, ischium, and pubis
    1. Ilium: ala showing an arched crest with a tubercle; anterior, posterior, and inferior gluteal lines on the lateral surface; anterior superior and inferior spines projecting ventrally; posterior superior and inferior spines projecting dorsally; and the iliac fossa, arcuate line, auricular surface, and tuberosity, on its medial side
    2. Ischium: shows a spine, tuberosity, and rami
    3. Pubis: shows rami and crest, tubercle, pectin, iliopubic eminence, and symphysis
B. SPECIAL FEATURES
    1. Acetabulum formed superiorly by ilium, posteroinferiorly by ischium, and antero-inferiorly by pubis
    2. Obturator foramen and groove
    3. Greater and lesser sciatic notches

II. Clinical features
A. FRACTURES: most commonly due to crushing forces, occurring at thinnest places—either ala of ilium or ischiopubic rami. Usually little displacement
B. BURSITIS: inflammation, especially of bursa, over ischial tuberosity, from sitting for long hours on hard seat, "tailor's or weaver's bottom"
C. OSSIFICATION: from 8 centers

Location	When Appears	When Closes
Lower ilium	8th–9th fetal week	18th year
Superior ramus, ischium	3rd fetal month	18th year
Superior ramus, pubis	4th–5th fetal month	18th year
Acetabulum (1 or more)	12th year	18th year
Iliac crest	Puberty	20th–25th year
Tuberosity of ischium	Puberty	20th–25th year
Symphysis pubis	Puberty	20th–25th year

III. Special features
A. THE PELVIS serves two purposes: it transmits the weight of the body to the lower limbs, and it also forms the lower part of the abdominal cavity (pelvic cavity) and houses part of the abdominopelvic viscera
B. PELVIC INLET: divides the pelvis into an "open" anterosuperior region and a confined posteroinferior tunnel, the *true pelvis*. The inlet extends from the sacral promontory on to each ala of the sacrum, across a sacroiliac joint and along each iliopectineal line on to the superior aspect of each pubic bone beside the centrally located symphysis
C. PELVIC OUTLET: from the tip of the 5th piece of the sacrum (not the tip of coccyx), then on each side the sacrotuberous ligament (between sacrum and tuberosity of ischium), to the ischial tuberosity, then the ischiopubic ramus, and finally, anteriorly, the anteroinferior aspect of the symphysis pubis (a diamond-shaped area)
D. THE PELVIC CAVITY lies between the inlet and outlet

HIP BONE

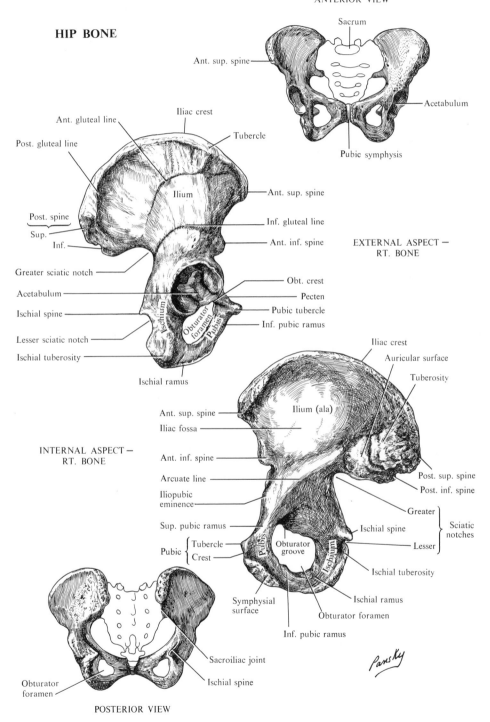

ANTERIOR VIEW

Sacrum

Ant. sup. spine

Iliac crest

Tubercle

Acetabulum

Ant. gluteal line

Post. gluteal line

Ilium

Ant. sup. spine

Pubic symphysis

Inf. gluteal line

Ant. inf. spine

EXTERNAL ASPECT —
RT. BONE

Post. spine

Sup.

Inf.

Greater sciatic notch

Acetabulum

Ischial spine

Lesser sciatic notch

Ischial tuberosity

Obt. crest

Pecten

Pubic tubercle

Inf. pubic ramus

Ischium

Obturator foramen

Pubis

Ischial ramus

Iliac crest

Auricular surface

Tuberosity

Ant. sup. spine

Ilium (ala)

Iliac fossa

Ant. inf. spine

Post. sup. spine

Post. inf. spine

INTERNAL ASPECT —
RT. BONE

Arcuate line

Iliopubic
eminence

Greater

Ischial spine

Sciatic
notches

Lesser

Sup. pubic ramus

Pubic { Tubercle
Crest

Pubis

Obturator
groove

Ischium

Ischial tuberosity

Symphysial
surface

Ischial ramus

Obturator foramen

Inf. pubic ramus

Pansky

Obturator
foramen

Sacroiliac joint

Ischial spine

POSTERIOR VIEW

-411-

182. THE PELVIC GIRDLE

I. Definition
A. BONES THROUGH WHICH LOWER EXTREMITY IS ATTACHED TO THE TRUNK

II. Bones (see p. 410)

III. Articulations
A. SACROILIAC, between auricular surface of sacrum and ilium
  1. Type: amphiarthrodial (syndesmosis)
  2. Ligaments
     a. Anterior from anterolateral sacrum to auricular and preauricular sulcus of ilium
     b. Posterior from first 3 transverse tubercles of sacrum to tuberosity and posterior superior spine of ilium
     c. Interosseous: short fibers between tuberosities of sacrum and ilium
B. SACROCOCCYGEAL, between apex of sacrum and base of coccyx
  1. Type: amphiarthrodial (syndesmosis)
  2. Ligaments: anterior, posterior, lateral, interarticular, and fibrocartilage disk
C. INTERPUBIC, between the pubic bones of opposite sides
  1. Type: amphiarthrodial (symphysis)
  2. Ligaments
     a. Superior pubic connects bones superiorly
     b. Arcuate connects bones inferiorly
     c. Interpubic: fibrocartilage disk
D. LIGAMENTS between sacrum and ischium
  1. Sacrotuberous from posterior inferior iliac spine, fourth and fifth transverse tubercles and sides of sacrum, and side of coccyx to the inner ischial tuberosity
  2. Sacrospinous from side of sacrum and coccyx to spine of ischium

IV. Special features
A. OBTURATOR MEMBRANE: interlacing fibers, attached to bony margins of foramen, which the membrane nearly fills. Obturator internus and externus muscles arise from inner and outer surfaces
B. GREATER SCIATIC FORAMEN
  1. Boundaries: in front and above, ilium and rim of great sciatic notch; behind, sacrotuberous ligament; below, sacrospinous ligament
  2. Structures passing through or lying in it: piriformis muscle, superior gluteal vessels and nerve, inferior gluteal vessels and nerve, internal pudendal vessels and nerve, sciatic nerve, posterior femoral cutaneous nerve, nerves to obturator internus and quadratus femoris muscles
C. LESSER SCIATIC FORAMEN
  1. Boundaries: in front, tuberosity of ischium; above, spine of ischium and sacrospinous ligament; behind, sacrotuberous ligament
  2. Structures passing through: tendon of obturator internus muscle, nerve to obturator internus muscle, internal pudendal vessels and nerve

V. Clinical considerations
A. SACROILIAC STRAIN is probably quite rare, although formerly often diagnosed, due to the great strength of the supporting ligaments

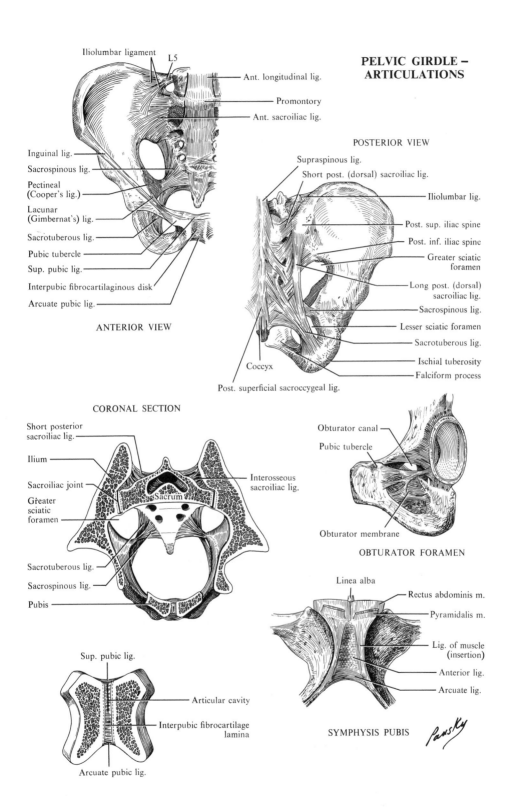

Iliolumbar ligament

L5

Ant. longitudinal lig.

Promontory

Ant. sacroiliac lig.

**PELVIC GIRDLE –
ARTICULATIONS**

POSTERIOR VIEW

Inguinal lig.

Sacrospinous lig.

Pectineal
(Cooper's lig.)

Lacunar
(Gimbernat's) lig.

Sacrotuberous lig.

Pubic tubercle

Sup. pubic lig.

Interpubic fibrocartilaginous disk

Arcuate pubic lig.

ANTERIOR VIEW

Supraspinous lig.

Short post. (dorsal) sacroiliac lig.

Iliolumbar lig.

Post. sup. iliac spine

Post. inf. iliac spine

Greater sciatic
foramen

Long post. (dorsal)
sacroiliac lig.

Sacrospinous lig.

Lesser sciatic foramen

Sacrotuberous lig.

Ischial tuberosity

Falciform process

Coccyx

Post. superficial sacrococcygeal lig.

CORONAL SECTION

Short posterior
sacroiliac lig.

Ilium

Sacroiliac joint

Greater
sciatic
foramen

Sacrum

Interosseous
sacroiliac lig.

Sacrotuberous lig.

Sacrospinous lig.

Pubis

Obturator canal

Pubic tubercle

Obturator membrane

OBTURATOR FORAMEN

Linea alba

Rectus abdominis m.

Pyramidalis m.

Lig. of muscle
(insertion)

Anterior lig.

Arcuate lig.

Sup. pubic lig.

Articular cavity

Interpubic fibrocartilage
lamina

Arcuate pubic lig.

SYMPHYSIS PUBIS

Pansky

183. BONE OF THE THIGH

I. Femur
A. PARTS: proximal extremity, shaft, distal extremity
1. Proximal: head with fovea; neck, constricted portion between head and intertrochanteric crest and line; lesser trochanter, intertrochanteric crest with quadrate tubercle, intertrochanteric line, gluteal tuberosity
2. Shaft: linea aspera with a medial and lateral lip; pectineal line, an upward continuation of medial lip; continuation of lateral lip to gluteal tuberosity; inferior continuation of lips of linea aspera as supracondylar lines; nutrient foramen
3. Distal: popliteal plane (planum popliteum), flat space between supracondylar lines; medial and lateral condyles and epicondyles; intercondylar fossa; adductor tubercle
B. SPECIAL FEATURES
1. Longest, largest, heaviest bone of body
2. Angle between shaft and neck: 120°–125°, smaller in female, greater in children
3. In erect posture, shaft runs obliquely, the medial distal ends of femora in contact
4. Anatomical neck; epiphyseal line between head and neck
5. Surgical neck: junction of shaft and proximal extremity, just below lesser trochanter
C. OSSIFICATION, FROM 5 CENTERS

Location	When Appears	When Closes
Body	7th fetal week	
Distal end	9th fetal week	20th year
Head	1st year	After puberty, last to close
Greater trochanter	4th year	After puberty, 2nd to close
Lesser trochanter	13th–14th year	After puberty, 1st to close

II. Clinical features
A. FRACTURE
1. Subtrochanteric, a break just below lesser trochanter. Due to power of iliopsoas, proximal fragment of bone is flexed, adducted, and laterally rotated. The distal part will be shortened by pull of hamstrings, rectus femoris, adductors, and sartorius muscles; abducted by gluteal muscles
2. Fractures in the lower third. Proximal part is fairly stable. Distal part will override (shorten) by pull of the hamstrings, rectus femoris, sartorius, and adductor magnus muscles and will be pulled backward by attachments of the gastrocnemius and plantaris muscles. This could lead to damage of the popliteal artery, which lies close to the bone in that position
B. THE ANGLE between the neck and body of the femur is called the *angle of inclination.* Marked decrease in this angle, that is, a more transverse direction of the neck, can result from weight-bearing on a bone that is not capable of standing it. Such an abnormal decrease is known as *coxa vara* (bent hip). It is usually due to rickets, but the same effect can result from disease of the bone or a fracture of the neck that is improperly set
C. THE BLOOD SUPPLY OF THE FEMUR consists of multiple arteries entering at each end and one or two *nutrient arteries* entering the body. The latter are derived from upper perforating branches of the profunda femoris (deep femoral)

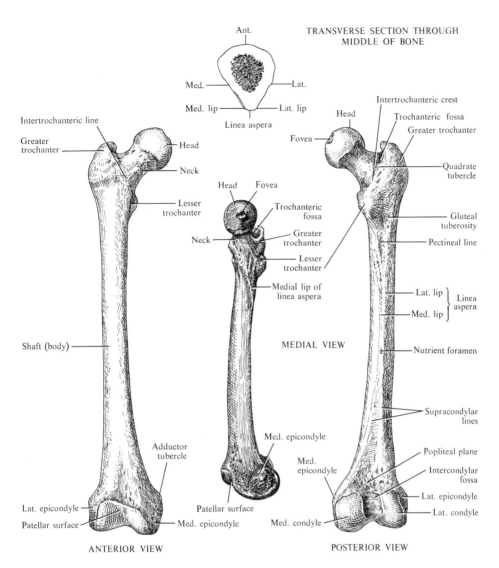

TRANSVERSE SECTION THROUGH
MIDDLE OF BONE

Ant.

Med. — — Lat.

Med. lip — — Lat. lip

Linea aspera

Intertrochanteric line

Greater trochanter

Head

Neck

Lesser trochanter

Shaft (body)

Adductor tubercle

Lat. epicondyle

Patellar surface

Med. epicondyle

ANTERIOR VIEW

Head   Fovea

Trochanteric fossa

Greater trochanter

Neck

Lesser trochanter

Medial lip of linea aspera

MEDIAL VIEW

Med. epicondyle

Patellar surface

Med. epicondyle

Intertrochanteric crest

Trochanteric fossa

Greater trochanter

Head

Fovea

Quadrate tubercle

Gluteal tuberosity

Pectineal line

Lat. lip
Med. lip
} Linea aspera

Nutrient foramen

Supracondylar lines

Popliteal plane

Intercondylar fossa

Lat. epicondyle

Lat. condyle

Med. epicondyle

Med. condyle

POSTERIOR VIEW

RIGHT FEMUR

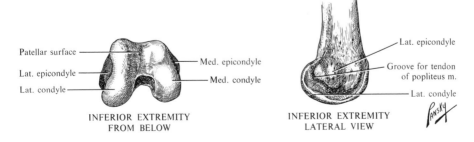

Patellar surface

Lat. epicondyle

Lat. condyle

Med. epicondyle

Med. condyle

INFERIOR EXTREMITY
FROM BELOW

Lat. epicondyle

Groove for tendon of popliteus m.

Lat. condyle

INFERIOR EXTREMITY
LATERAL VIEW

184. MUSCLES OF THE ANTERIOR THIGH

I.

Name	Origin	Insertion	Action	Nerve
Sartorius	Anterior superior iliac spine	Upper medial tibia	Flexes and rotates thigh laterally Flexes leg, rotates it medially	Femoral
Quadriceps				
Rectus femoris	Ant. inf. iliac spine Rim of acetabulum	Base of patella	Flexes thigh Extends leg	Femoral
Vastus lateralis	Intertrochanteric line Great trochanter Glut. tuber. Lat. linea aspera	Lateral patella	Extends leg	Femoral
Vastus medius	Intertrochanteric line Med. linea aspera Supracondylar line	Medial patella	Extends leg	Femoral
Vastus intermedius	Upper ant. shaft of femur	Base of patella	Extends leg	Femoral
Iliopsoas	Out of field, see p. 316	Lesser trochanter	Flexes thigh Lat. rotates and adducts thigh	Nn. to iliopsoas (L2, 3, 4)

II. Special features

A. ILIOPSOAS flexes thigh. When extremity is free, it rotates thigh laterally. When extremity is fixed, as when foot is on ground, it rotates pelvis

III. Clinical considerations (see p. 414)

A. THE MEMBERS OF THE ANTERIOR GROUP of muscles differ considerably in their actions. In general, the iliopsoas and pectineus act only at the hip joint; the sartorius and rectus femoris act both at the hip and knee joints (but with opposite actions at the knee joint); and the vastus muscles act only at the knee joint

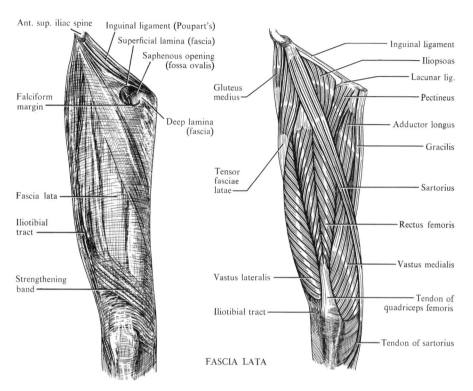

Top left figure (FASCIA LATA):

Ant. sup. iliac spine
Inguinal ligament (Poupart's)
Superficial lamina (fascia)
Saphenous opening (fossa ovalis)
Falciform margin
Deep lamina (fascia)
Fascia lata
Iliotibial tract
Strengthening band

FASCIA LATA

Top right figure:

Inguinal ligament
Iliopsoas
Lacunar lig.
Pectineus
Adductor longus
Gracilis
Sartorius
Rectus femoris
Vastus medialis
Tendon of quadriceps femoris
Tendon of sartorius
Gluteus medius
Tensor fasciae latae
Vastus lateralis
Iliotibial tract

ANTERIOR THIGH MUSCLES

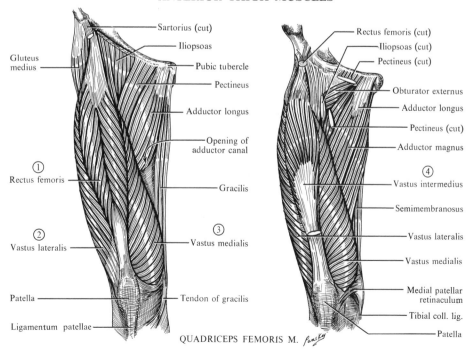

Bottom left figure:

Sartorius (cut)
Iliopsoas
Pubic tubercle
Pectineus
Adductor longus
Opening of adductor canal
Gracilis
Vastus medialis ③
Tendon of gracilis
Gluteus medius
① Rectus femoris
② Vastus lateralis
Patella
Ligamentum patellae

Bottom right figure:

Rectus femoris (cut)
Iliopsoas (cut)
Pectineus (cut)
Obturator externus
Adductor longus
Pectineus (cut)
Adductor magnus
④ Vastus intermedius
Semimembranosus
Vastus lateralis
Vastus medialis
Medial patellar retinaculum
Tibial coll. lig.
Patella

QUADRICEPS FEMORIS M. *Pansky*

185. Fascia, Femoral Triangle, and Femoral Sheath

I. Fascia of the anterior and anteromedial thigh
A. SUPERFICIAL: continuous with that of the abdomen, gluteal region, and leg. Consists of 2 layers, with vessels and nerves between them. The more superficial layer is fatty. The deep layer is adherent to the deep fascia below the inguinal ligament along the medial thigh and at the edge of the saphenous hiatus (fossa ovalis)
   1. Cribriform fascia: the deep layer of the superficial fascia over the saphenous hiatus (fossa ovalis)
B. DEEP FASCIA (fascia lata) is attached above to the os coxa and inguinal ligament, below to the tibia, and is continuous with the crural fascia
C. INTERMUSCULAR SEPTA
   1. Medial divides the extensors from the adductors, extends inward to the bone between the quadriceps and adductor longus and brevis muscles
   2. Lateral extends inward from the fascia to the linea aspera and separates the vastus lateralis from the biceps femoris muscle
D. SAPHENOUS HIATUS (FOSSA OVALIS): gap in the fascia lata, where it is split into a superficial layer that curves laterally and downward from the pubic tubercle and inguinal ligament along the saphenous vein, and then under the vein to fuse to the deep layer medial to the vein. The deep layer is part of the iliopectineal fascia
   1. Structures piercing the cribriform fascia to enter or leave the fossa: great saphenous vein, superficial circumflex iliac vessels, superficial epigastric vessels, and external pudendal vessels

II. Femoral triangle
A. BOUNDED above by the inguinal ligament, laterally by the sartorius muscle, and medially by the adductor longus muscle
B. FLOOR, from lateral to medial: iliacus, psoas major, and pectineus muscles

III. Adductor (Hunter's) canal: a musculofibrous canal in the middle of the thigh, beginning at the apex of the femoral triangle and ending in the tendinous hiatus of the adductor magnus muscle
A. BOUNDED: anterolaterally, vastus medialis; behind, adductor longus and magnus muscles; and covered (roofed) anteromedially by fibrous tissue under the sartorius muscle (vastoadductor fascia)

IV. Interval behind inguinal ligament
A. ILIOPECTINEAL ARCH: A septum of iliopectineal fascia extends between the iliopubic eminence to the inguinal ligament, dividing the area into 2 compartments
   1. Muscular contains iliacus and psoas major muscles, femoral nerve, and lateral femoral cutaneous nerve
   2. Vascular contains femoral sheath and its contents (see below)

V. Femoral sheath: a prolongation of the transversalis fascia from the abdomen, fusing to the iliac fascia dorsally. Funnel-shaped, narrow end fuses with the intrinsic fascia of the vessels contained about 5 cm below the inguinal ligament. Two vertical septa divide the sheath into 3 compartments: lateral, for artery and femoral branch of genitofemoral nerve; intermediate, for vein; and medial, for femoral canal, containing a few small lymphatics and a lymph node (node of Cloquet)
A. THE FEMORAL RING is the cephalic end of the femoral canal

VI. Clinical considerations
A. THE FEMORAL CANAL is the most common site for femoral herniations

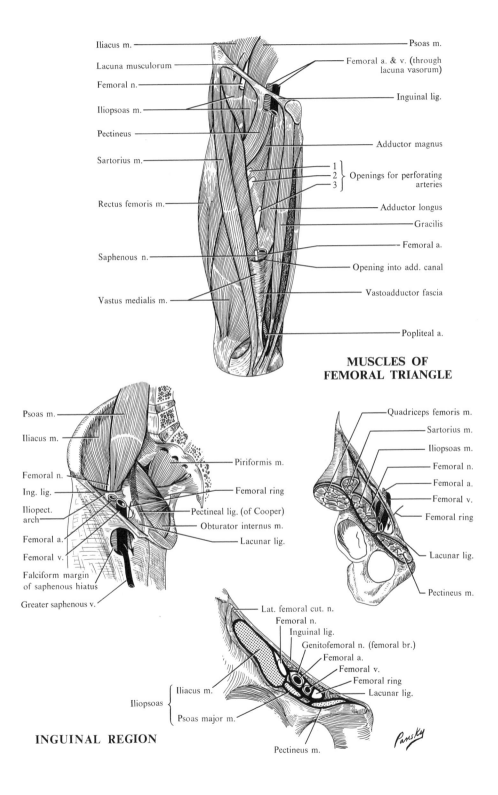

Iliacus m.

Lacuna musculorum

Femoral n.

Iliopsoas m.

Pectineus

Sartorius m.

Rectus femoris m.

Saphenous n.

Vastus medialis m.

Psoas m.

Femoral a. & v. (through lacuna vasorum)

Inguinal lig.

Adductor magnus

1
2 } Openings for perforating
3   arteries

Adductor longus

Gracilis

Femoral a.

Opening into add. canal

Vastoadductor fascia

Popliteal a.

**MUSCLES OF
FEMORAL TRIANGLE**

Psoas m.

Iliacus m.

Femoral n.

Ing. lig.

Iliopect. arch

Femoral a.

Femoral v.

Falciform margin of saphenous hiatus

Greater saphenous v.

Piriformis m.

Femoral ring

Pectineal lig. (of Cooper)

Obturator internus m.

Lacunar lig.

Quadriceps femoris m.

Sartorius m.

Iliopsoas m.

Femoral n.

Femoral a.

Femoral v.

Femoral ring

Lacunar lig.

Pectineus m.

Lat. femoral cut. n.

Femoral n.

Inguinal lig.

Genitofemoral n. (femoral br.)

Femoral a.

Femoral v.

Femoral ring

Lacunar lig.

Iliacus m.

Psoas major m.

Iliopsoas

INGUINAL REGION

Pansky

Pectineus m.

186. MUSCLES OF THE MEDIAL THIGH

I.

Name	Origin	Insertion	Action	Nerve
Pectineus	Pectineal line Pubis, between iliopubic eminence and tubercle	Pectineal line of femur	Flexes thigh Adducts and lat. rotates thigh	Femoral
Adductor longus	Front of pubis	Med. linea aspera of femur	Adducts, flexes, and lat. rotates thigh	Obturator
Gracilis	Lower symphysis Pubic arch	Upper med. tibia, below condyle	Adducts thigh	Obturator
Adductor brevis	Inferior pubic ramus	Pectineal line and upper med. linea aspera of femur	Adducts and flexes thigh	Obturator
Adductor magnus	Inf. ischiopubic rami Outer inf. ischial tuberosity	Med. gluteal tuberosity Med. linea aspera Supracondylar line and adductor tubercle	Adducts thigh Upper part flexes thigh Lower part extends and rotates thigh laterally	Obturator

II. **Special features** (see p. 433)
   A. AT THE INSERTION OF THE ADDUCTOR MAGNUS MUSCLE along and close to the linea aspera are a series of small tendinous arches attached to the bone. The upper 4 are small apertures transmitting the perforating branches of the deep femoral artery, which supply the hamstring muscles at the back of the thigh as well as a nutrient vessel to the femur. The fifth and lowest of the openings is by far the largest, the *tendinous hiatus* (*adductor hiatus*), which represents the caudal end of the adductor canal through which the femoral vessels enter the popliteal space

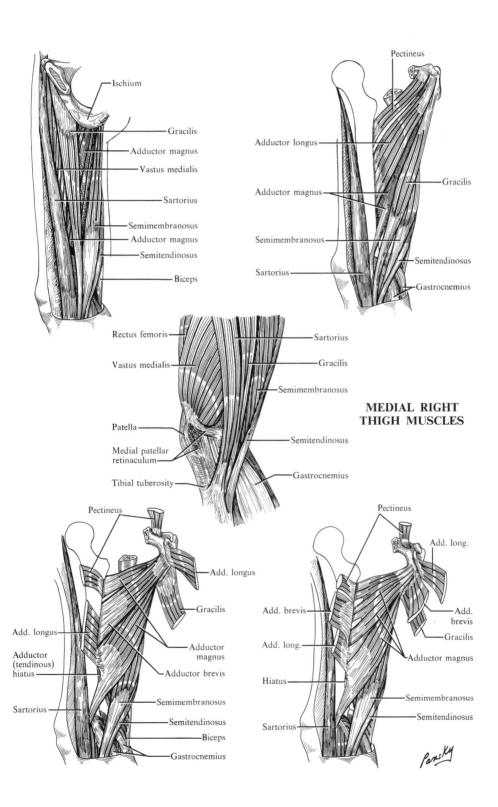

Ischium

Gracilis

Adductor magnus

Vastus medialis

Sartorius

Semimembranosus

Adductor magnus

Semitendinosus

Biceps

Pectineus

Adductor longus

Adductor magnus

Gracilis

Semimembranosus

Semitendinosus

Sartorius

Gastrocnemius

Rectus femoris

Vastus medialis

Patella

Medial patellar retinaculum

Tibial tuberosity

Sartorius

Gracilis

Semimembranosus

Semitendinosus

Gastrocnemius

**MEDIAL RIGHT
THIGH MUSCLES**

Pectineus

Add. longus

Gracilis

Add. longus

Adductor
(tendinous)
hiatus

Sartorius

Adductor
magnus

Adductor brevis

Semimembranosus

Semitendinosus

Biceps

Gastrocnemius

Pectineus

Add. long.

Add. brevis

Add. long.

Hiatus

Sartorius

Add.
brevis

Gracilis

Adductor magnus

Semimembranosus

Semitendinosus

Pansky

187. FEMORAL AND OBTURATOR VESSELS AND NERVES

I. **Femoral artery** (see p. 419)

> A. RELATIONS
>
> *Anterior*
> Fascia lata, nerve to vastus medialis muscle,
> saphenous nerve, sartorius muscle, roof of
> adductor canal
>
> | *Lateral* | **Femoral** | *Medial* |
> |---|---|---|
> | Vastus intermedius muscle | **Artery** | Adductor longus and sartorius muscles, femoral vein |
>
> *Posterior*
> Psoas major, pectineus, adductor longus
> and adductor magnus muscles, deep
> femoral and femoral veins

B. BRANCHES: superficial epigastric, superficial circumflex iliac, external pudendal (with scrotal and inguinal branches), deep femoral and descending genicular arteries

II. **Deep femoral artery** arises from the back of the femoral artery. At first it lies lateral to, then behind, the artery as far as the medial side of the femur. It then runs behind the adductor longus muscle to end as the fourth perforating artery. Branches:

A. MEDIAL CIRCUMFLEX FEMORAL winds around the medial side of the femur between the pectineus and psoas major muscles, then between the obturator externus and adductor brevis muscles to the back of the thigh, where it anastomoses with the inferior gluteal, lateral circumflex femoral, and first perforating arteries

B. LATERAL CIRCUMFLEX FEMORAL arises from the lateral side, behind the sartorius and rectus femoris muscles. Runs in front of the femur to anastomose with the superior gluteal, deep circumflex iliac, and superior lateral genicular arteries

C. THREE PERFORATING ARTERIES pierce adductor magnus muscle to reach the back of thigh, forming anastomotic loops

D. THE TERMINATION OF THE ARTERY is the fourth perforating artery

III. **Femoral nerve.** Distribution:

A. MUSCULAR to pectineus, sartorius, and quadriceps muscles

B. CUTANEOUS: Anterior branches to anterior and medial thigh
   1. Saphenous runs with the femoral artery, crossing the artery from lateral to medial in the adductor canal. It becomes superficial at the medial side of the knee to supply the skin on the medial side of the leg

C. NERVES TO HIP JOINT AND KNEE JOINT

IV. **Femoral vein** begins at the tendinous hiatus of the adductor magnus muscle. Receives muscular tributaries, the deep femoral vein, and the great saphenous vein

V. **Obturator vessels and nerve**

A. ARTERY is distributed to obturator externus, pectineus, adductors, and gracilis muscles, hip joint, and anastomoses with the inf. gluteal and med. circumflex arteries

B. NERVE: Generally, if a nerve crosses or is in vicinity of a joint, it supplies the joint
   1. Anterior branch: lies deep to the pectineus and adductor longus muscles and superficial to the adductor brevis muscle. It supplies the adductor longus, brevis, and gracilis muscles, cutaneous areas on the medial thigh, and hip joint
   2. Posterior branch: as it enters thigh, lies posterior to the adductor brevis muscle but superficial to the adductor magnus muscle. It supplies the obturator externus, adductor magnus, and brevis muscles and also the knee joint

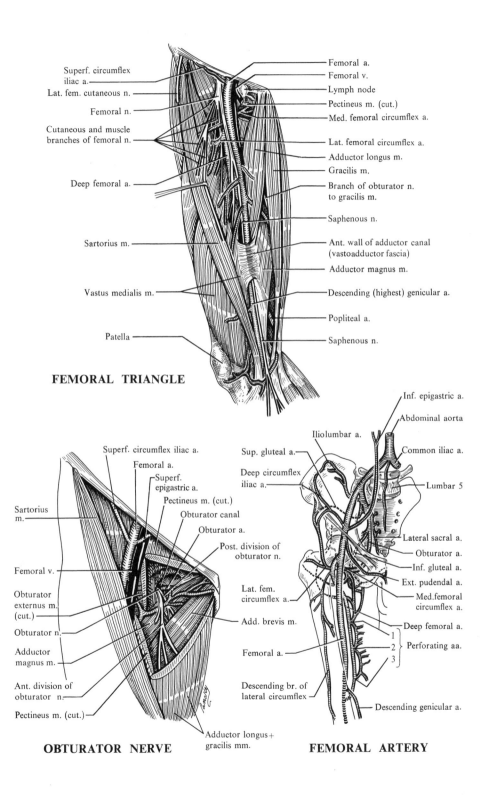

Superf. circumflex iliac a.

Lat. fem. cutaneous n.

Femoral n.

Cutaneous and muscle branches of femoral n.

Deep femoral a.

Sartorius m.

Vastus medialis m.

Patella

Femoral a.

Femoral v.

Lymph node

Pectineus m. (cut.)

Med. femoral circumflex a.

Lat. femoral circumflex a.

Adductor longus m.

Gracilis m.

Branch of obturator n. to gracilis m.

Saphenous n.

Ant. wall of adductor canal (vastoadductor fascia)

Adductor magnus m.

Descending (highest) genicular a.

Popliteal a.

Saphenous n.

FEMORAL TRIANGLE

Superf. circumflex iliac a.

Femoral a.

Superf. epigastric a.

Pectineus m. (cut.)

Obturator canal

Obturator a.

Post. division of obturator n.

Sartorius m.

Femoral v.

Obturator externus m. (cut.)

Obturator n.

Adductor magnus m.

Ant. division of obturator n.

Pectineus m. (cut.)

Lat. fem. circumflex a.

Add. brevis m.

Femoral a.

Descending br. of lateral circumflex

Adductor longus + gracilis mm.

OBTURATOR NERVE

Inf. epigastric a.

Abdominal aorta

Iliolumbar a.

Sup. gluteal a.

Deep circumflex iliac a.

Common iliac a.

Lumbar 5

Lateral sacral a.

Obturator a.

Inf. gluteal a.

Ext. pudendal a.

Med. femoral circumflex a.

Deep femoral a.

1
2 Perforating aa.
3

Descending genicular a.

FEMORAL ARTERY

-423-

188. MUSCLES OF THE GLUTEAL REGION

I.

Name	Origin	Insertion	Action	Nerve
Gluteus maximus	Ilium behind post. glut. line Sacrotuberous lig.	Iliotibial band Glut. tuberosity of femur	Extends and lat. rotates thigh Braces knee	Inf. glut.
Gluteus medius	Outer ilium and crest between post. and inf. glut. lines	Ridge, lat. side great. trochanter	Abducts thigh Ant. part rotates medially Post. part rotates laterally	Sup. glut.
Gluteus minimus	Outer ilium, between ant. and inf. glut. lines Sciatic notch	Ant. border, great. trochanter	Abducts thigh, rotates medially Weakly flexes	Sup. glut.
Tensor fasciae latae	Ant. iliac crest Notch between ant. sup. and ant. inf. iliac spines	Iliotibial band	Flexes, abducts, and rotates thigh med.	Sup. glut.
Piriformis	Ant. sacrum Great. sciatic notch	Upper border great. trochanter	Abducts and rotates thigh laterally	N. to piriformis
Obturator internus	Ischiopubic rami Inner obt. memb.	Med. surface, great. trochanter	Rotates laterally and abducts thigh	N. to obturator int.
Gemellus superior	Outer surface, ischial spine	Tendon obturator int. m.	See obt. int.	N. to obturator int.
Gemellus inferior	Ischial tuberosity	Tendon obturator int. m.	See obt. int.	N. to quadratus femoris
Obturator externus	Ischiopubic rami Outer obt. memb.	Trochanteric fossa	Rotates thigh lat.	Obturator
Quadratus femoris	Ischial tuberosity	Quadrate line	Rotates thigh lat.	N. to quadratus femoris

II. Special features
A. PIRIFORMIS arises inside pelvis and divides the greater sciatic foramen into superior and inferior portions
B. OBTURATOR INTERNUS MUSCLE arises inside pelvis
C. GEMELLI insert into tendon of obturator internus muscle

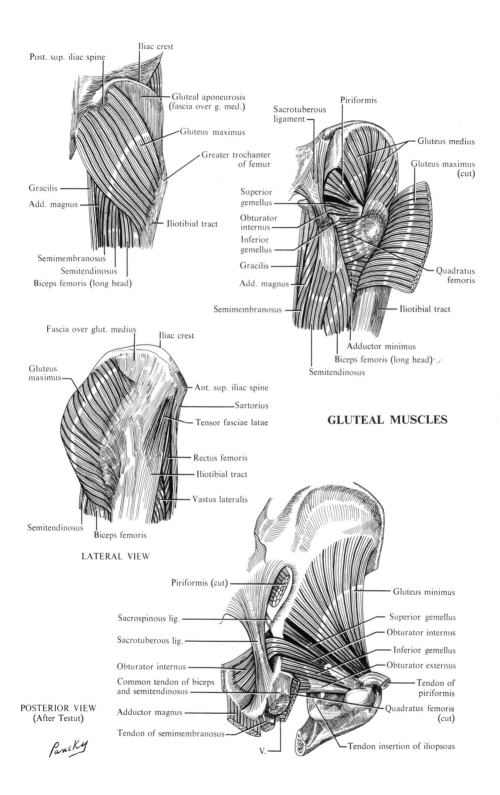

Post. sup. iliac spine

Iliac crest

Gluteal aponeurosis
(fascia over g. med.)

Gluteus maximus

Greater trochanter
of femur

Gracilis

Add. magnus

Iliotibial tract

Semimembranosus

Semitendinosus

Biceps femoris (long head)

Piriformis

Sacrotuberous
ligament

Gluteus medius

Gluteus maximus
(cut)

Superior
gemellus

Obturator
internus

Inferior
gemellus

Gracilis

Add. magnus

Quadratus
femoris

Semimembranosus

Iliotibial tract

Adductor minimus

Biceps femoris (long head)

Semitendinosus

Fascia over glut. medius

Iliac crest

Gluteus
maximus

Ant. sup. iliac spine

Sartorius

Tensor fasciae latae

GLUTEAL MUSCLES

Rectus femoris

Iliotibial tract

Vastus lateralis

Semitendinosus

Biceps femoris

LATERAL VIEW

Piriformis (cut)

Gluteus minimus

Sacrospinous lig.

Superior gemellus

Sacrotuberous lig.

Obturator internus

Obturator internus

Inferior gemellus

Common tendon of biceps
and semitendinosus

Obturator externus

Tendon of
piriformis

Adductor magnus

Quadratus femoris
(cut)

POSTERIOR VIEW
(After Testut)

Tendon of semimembranosus

V.

Tendon insertion of iliopsoas

Pansky

189. VESSELS AND NERVES OF THE GLUTEAL REGION

I. **Superficial nerves** (see p. 404)

II. **Arteries**
A. SUPERIOR GLUTEAL: largest branch of internal iliac artery; leaves pelvis through greater sciatic foramen, above piriformis muscle. Branches:
   1. Superficial branch enters gluteus maximus muscle and anastomoses with inferior gluteal artery
   2. Deep branch runs forward under gluteus medius muscle; superior division continues along upper border of the gluteus minimus muscle to anterior superior iliac spine to anastomose with the deep circumflex iliac and lateral femoral circumflex arteries. The inferior division passes downward over the surface of the gluteus minimus muscle to the greater trochanter and anastomoses with the lateral femoral circumflex artery
B. INFERIOR GLUTEAL: from internal iliac artery, leaves pelvis through greater sciatic foramen below piriformis muscle. Branches:
   1. Ischiadic (sciatic), which runs with the sciatic nerve
   2. Anastomotic: forms the crucial anastomosis by uniting with the lateral and medial circumflex femoral and the first perforating artery from the profunda
C. INTERNAL PUDENDAL: from internal iliac artery, leaves pelvis through greater sciatic foramen, below piriformis muscle, crosses spine of ischium and enters ischiorectal fossa through the lesser sciatic foramen

III. **Nerves**
A. SCIATIC: largest nerve of body, from sacral plexus, leaves pelvis through greater sciatic foramen below the piriformis muscle and is accompanied by the posterior femoral cutaneous nerve, inferior gluteal artery, pudendal vessels, and nerves. No gluteal branches except a branch to hip joint
B. PUDENDAL NERVE follows pudendal vessels
C. POSTERIOR FEMORAL CUTANEOUS NERVE enters region below piriformis muscle with inferior gluteal artery and gives inf. cluneal branches to skin at lower border of gluteus maximus muscle
D. SUPERIOR GLUTEAL NERVE runs with superior gluteal vessels to supply gluteus medius, gluteus minimus, and tensor fasciae latae muscles
E. INFERIOR GLUTEAL NERVE follows inferior gluteal vessels to gluteus maximus muscle
F. OTHER MUSCULAR BRANCHES OF PLEXUS
   1. Nerve to quadratus femoris and inferior gemellus muscles leaves pelvis below piriformis muscle
   2. Nerve to obturator internus and superior gemellus muscles leaves pelvis below piriformis muscle and enters ischiorectal fossa through lesser sciatic foramen
   3. Nerve to piriformis muscle does not leave pelvis

IV. **Clinical considerations**
A. SCIATICA is a neuritis (inflammation of the nerve) of the sciatic nerve characterized by intense pain at back of thigh
B. LESIONS OF THE SCIATIC NERVE lead to paralysis of all of the muscles below the knee. Sensation on the lateral side of the leg and both surfaces of the foot is also lost
C. THE SCIATIC NERVE leaves the buttock by passing just lateral to the ischial tuberosity, and thereafter runs downward in the posterior midline, behind the adductor magnus but deep to the other muscles arising from the tuberosity

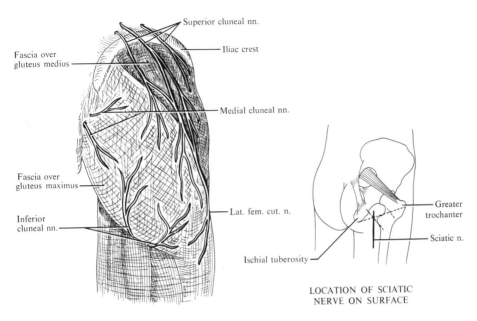

Superior cluneal nn.

Iliac crest

Fascia over
gluteus medius

Medial cluneal nn.

Fascia over
gluteus maximus

Inferior
cluneal nn.

Lat. fem. cut. n.

Greater
trochanter

Sciatic n.

Ischial tuberosity

LOCATION OF SCIATIC
NERVE ON SURFACE

SUPERFICIAL NERVES – GLUTEAL REGION

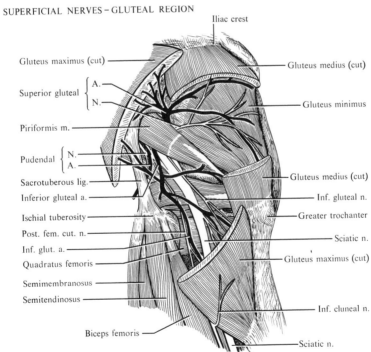

Iliac crest

Gluteus maximus (cut)

Gluteus medius (cut)

Superior gluteal { A.
N.

Gluteus minimus

Piriformis m.

Pudendal { N.
A.

Gluteus medius (cut)

Sacrotuberous lig.

Inferior gluteal a.

Inf. gluteal n.

Ischial tuberosity

Greater trochanter

Post. fem. cut. n.

Sciatic n.

Inf. glut. a.

Gluteus maximus (cut)

Quadratus femoris

Semimembranosus

Semitendinosus

Inf. cluneal n.

Biceps femoris

Sciatic n.

DEEP GLUTEAL STRUCTURES

Pansky

190. MUSCLES OF THE POSTERIOR THIGH

I.

Name	Origin	Insertion	Action	Nerve
Biceps femoris				
Long head	Inner medial ischial tuberosity Sacrotuberous lig.	Head of fibula Lat. condyle of tibia	Extends thigh Flexes leg	Tibial
Short head	Lat. linea aspera Lat. intermuscular septum	Lat. condyle of tibia	Flexes leg	Comm. peroneal
Semitendinosus	Lower medial ischial tuberosity	Medial body of tibia nearly to anterior crest	Extends thigh Flexes leg	Tibial
Semimembranosus	Upper outer ischial tuberosity	Post. medial aspect of tibial condyle	Extends thigh Flexes leg	Tibial

II. Special features
A. POPLITEAL FOSSA (space): a diamond-shaped space behind the knee
 1. Boundaries: laterally and above by biceps femoris muscle, medially and above by the semimembranosus and semitendinosus muscles, laterally and below by the lateral head of the gastrocnemius muscle, and medially and below by the medial head of the gastrocnemius muscle
 2. The floor, from above downward, is formed by popliteal surface of the femur, oblique popliteal ligament, and the popliteus muscle
 3. The roof: covered by fascia lata, superficial fascia, and skin
 4. Contents: popliteal vessels, tibial and common peroneal (fibular) nerves, termination of the small saphenous vein, lower end of the posterior femoral cutaneous nerve, articular branch of the obturator nerve, and a few small lymph nodes and fat

III. Clinical considerations
A. CHARLEY HORSE: this may be expressed as pain or muscle stiffness after a bruise or excessive athletic activity and frequently occurs in the hamstring muscles (see table above)
B. IN THEIR ACTIONS, all the muscles arising from the ischial tuberosity assist the gluteus maximus to extend the hip joint, and since they normally maintain more tone than does the gluteus maximus, patients with paralyzed posterior hamstrings tend to fall forward
C. THE SHORT HEAD OF THE BICEPS can act only at the knee. It is a more efficient flexor at the knee than is the long head, for the long head apparently relaxes during the last half of flexion, while the short head continues to contract

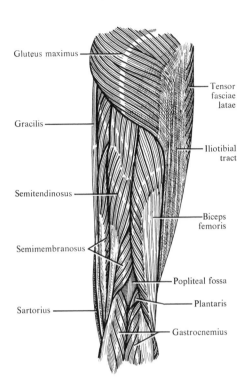

Gluteus maximus

Gracilis

Semitendinosus

Semimembranosus

Sartorius

Tensor fasciae latae

Iliotibial tract

Biceps femoris

Popliteal fossa

Plantaris

Gastrocnemius

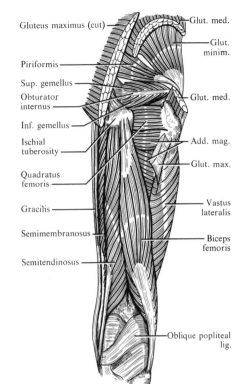

Gluteus maximus (cut)

Piriformis

Sup. gemellus

Obturator internus

Inf. gemellus

Ischial tuberosity

Quadratus femoris

Gracilis

Semimembranosus

Semitendinosus

Glut. med.

Glut. minim.

Glut. med.

Add. mag.

Glut. max.

Vastus lateralis

Biceps femoris

Oblique popliteal lig.

POSTERIOR THIGH MUSCLES

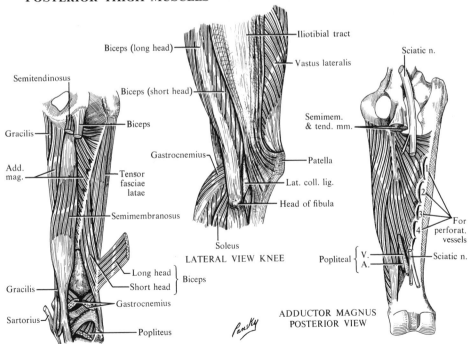

Semitendinosus

Gracilis

Add. mag.

Gracilis

Sartorius

Biceps (long head)

Biceps (short head)

Biceps

Tensor fasciae latae

Semimembranosus

Long head
Short head
} Biceps

Gastrocnemius

Popliteus

Iliotibial tract

Vastus lateralis

Gastrocnemius

Patella

Lat. coll. lig.

Head of fibula

Soleus

LATERAL VIEW KNEE

Sciatic n.

Semimem. & tend. mm.

1
2
3
4
} For perforat. vessels

Popliteal { V.
A.

Sciatic n.

ADDUCTOR MAGNUS POSTERIOR VIEW

Pansky

191. SACRAL PLEXUS AND DERIVATIVES

I. **Formation:** lumbosacral trunk (L4 and L5) plus anterior primary divisions of sacral nerves 1–3 inclusive

  A. LOCATION: posterolateral pelvis, between internal iliac vessels and piriformis muscle

  B. MANNER OF FORMATION: nerves that enter plexus, except S3, have posterior and anterior branches that converge toward greater sciatic foramen

  C. BRANCHES

Anterior Division	*Posterior Division*
N. to quadratus femoris and inf. gemellus mm. (L4, L5, S1)	N. to piriformis m. (S1, S2)
N. to obturator internus and sup. gemellus mm. (L5, S1, S2)	Superior gluteal (L4, L5, S1)
Post. fem. cutaneous (S2, S3)	Inferior gluteal (L5, S1, S2)
Sciatic	Post. fem. cutaneous (S1, S2)
Tibial (L4, L5, S1, S2, S3)	Perforating cutaneous (S2, S3)
	Sciatic
Pudendal (S2, S3, S4)	Common peroneal (fibular) (L4, L5, S1, S2)

  D. DISTRIBUTION

    1. Nerves to individually named muscles need no further comment

    2. Posterior femoral cutaneous receives posterior divisions from S1 and S2 and anterior divisions from S2 and S3. Leaves pelvis with inferior gluteal artery to lower part of gluteus maximus muscle (see p. 427)

    3. Perforating cutaneous nerve pierces sacrotuberous ligament, supplies lower and medial buttock

    4. Superior gluteal follows superior gluteal artery, supplies gluteus medius, gluteus minimus, and tensor fasciae latae muscles

    5. Inferior gluteal follows inferior gluteal vessels, supplies gluteus maximus muscle

    6. Sciatic, actually 2 nerves, tibial and common peroneal (fibular) within the same sheath

      a. Relations: in front, lies on posterior ischium, obturator internus, gemelli, and quadratus femoris muscles; behind lies the gluteus maximus muscle; farther down it runs on the adductor magnus muscle and is crossed by the long head of the biceps muscle

      b. Splits just above popliteal space into its 2 parts (see p. 437)

        i. Tibial supplies all hamstrings except short head of biceps, and then enters leg (see p. 449)

        ii. Common peroneal (fibular) supplies short head of biceps and enters leg (see p. 449)

    7. Pudendal leaves pelvis through greater sciatic foramen, crosses ischial spine, and enters ischiorectal fossa (see p. 395)

II. **Clinical considerations** (see p. 426)

  A. LESIONS OF THE SUPERIOR GLUTEAL NERVE cause a drastic sag of the pelvis toward the side of the "swing phase" in walking

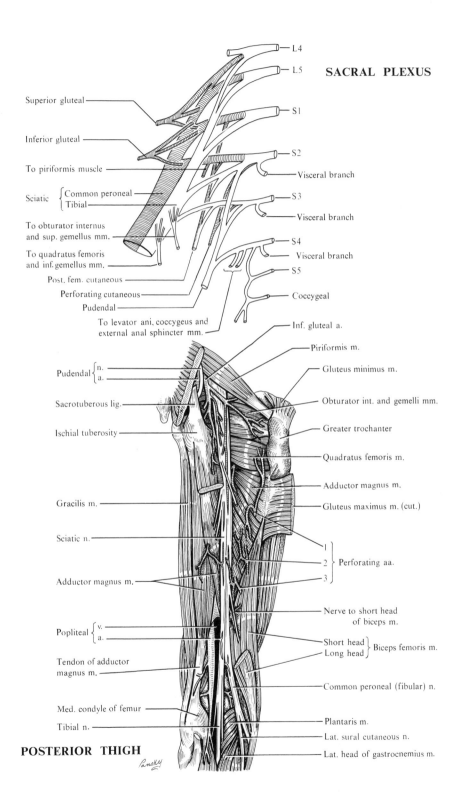

SACRAL PLEXUS

L4
L5
S1
Superior gluteal
Inferior gluteal
To piriformis muscle
S2
Visceral branch
Sciatic { Common peroneal
Tibial
S3
Visceral branch
To obturator internus
and sup. gemellus mm.
To quadratus femoris
and inf. gemellus mm.
S4
Visceral branch
Post. fem. cutaneous
Perforating cutaneous
S5
Pudendal
Coccygeal
To levator ani, coccygeus and
external anal sphincter mm.

Inf. gluteal a.
Piriformis m.
Pudendal { n.
a.
Gluteus minimus m.
Sacrotuberous lig.
Obturator int. and gemelli mm.
Ischial tuberosity
Greater trochanter
Quadratus femoris m.
Adductor magnus m.
Gracilis m.
Gluteus maximus m. (cut.)
Sciatic n.
1
2 } Perforating aa.
3
Adductor magnus m.
Nerve to short head
of biceps m.
Popliteal { v.
a.
Short head } Biceps femoris m.
Long head
Tendon of adductor
magnus m.
Common peroneal (fibular) n.
Med. condyle of femur
Plantaris m.
Tibial n.
Lat. sural cutaneous n.

POSTERIOR THIGH
Lat. head of gastrocnemius m.

Pansky

192. THE HIP JOINT

I. Type: enarthrosis (ball and socket)

II. Bones: head of femur and acetabulum of hip bone

III. Movements: flexion, extension, abduction, adduction, lateral (external) rotation, and medial (internal) rotation

IV. Ligaments
   A. ACETABULAR LABRUM (glenoid labrum) deepens acetabulum, helps to hold head of femur
   B. TRANSVERSE ACETABULAR completes rim of acetabulum
   C. ARTICULAR CAPSULE extends from rim of acetabulum to the intertrochanteric crest and line
      1. Zona orbicularis: band of circularly arranged fibers in capsule
   D. ILIOFEMORAL (Y-shaped ligament of Bigelow) from anterior inferior iliac spine to intertrochanteric line, prevents overextension, abduction, and lateral rotation
   E. ISCHIOFEMORAL from ischium behind the acetabulum to blend with the capsule, checks medial rotation
   F. PUBOFEMORAL from superior pubic ramus, joins iliofemoral ligament, checks abduction
   G. LIGAMENTUM CAPITIS FEMORIS (TERES) (round ligament) from acetabular notch and transverse ligament to fovea of femur; little function as ligament but guides artery to head of femur

V. Synovial membrane: lines articular capsule. Covers labrum, ligamentum teres, neck of femur, from the attachment of capsule, below, to the articular cartilage of head of femur

VI. Muscles acting on hip joint

Flexion	Extension	Abduction	Adduction	Med. Rotate	Lat. Rotate
Iliopsoas	Glut. max.	Glut. med.	Add. mag.	Glut. med.	Piriformis
Sartorius	Semitend.	Glut. min.	Add. long.	Glut. min.	Obturator int.
Pectineus	Semimemb.	Tensor f.l.	Add. brev.	Tensor f.l.	Gemelli
Rectus fem.	Biceps fem.	Piriformis	Gracilis	Add. mag.	Obturator ext.
Add. long.	Add. mag.	Sartorius	Pectineus	(post. part)	Quadrat. fem.
Add. brev.	(post. part)		Obturator ext.		Glut. max.
Add. mag.			Quadrat. fem.		Adductors (all)
(ant. part)					
Tensor f.l.					

VII. Clinical considerations
   A. DISLOCATIONS: since circulation to the head of femur over the ligamentum teres may be disrupted, reduction within 12 to 24 hours is mandatory
      1. Congenital: due to failure of acetabulum to deepen. More common in women
      2. Traumatic: usually caused by a blow on the knee when the thigh is flexed, tearing capsule
         a. The head of the femur is usually dislocated posteriorly, with a tearing of the posterior part of the capsule, and frequently fracture of the acetabulum
         b. In anterior dislocation, much rarer than posterior, the head of the femur passes around the medial edge of the iliofemoral ligament and lodges against the body of the pubic bone or obturator foramen

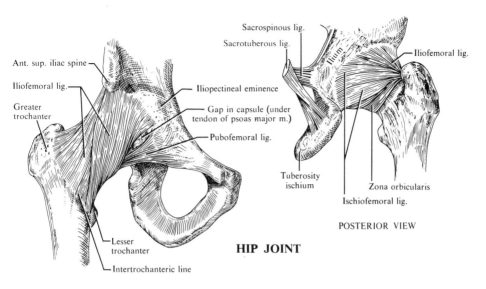

Sacrospinous lig.

Sacrotuberous lig.

Iliofemoral lig.

Ant. sup. iliac spine

Iliofemoral lig.

Greater trochanter

Iliopectineal eminence

Gap in capsule (under tendon of psoas major m.)

Pubofemoral lig.

Ilium

Tuberosity ischium

Zona orbicularis

Ischiofemoral lig.

POSTERIOR VIEW

Lesser trochanter

HIP JOINT

Intertrochanteric line

ANTERIOR VIEW

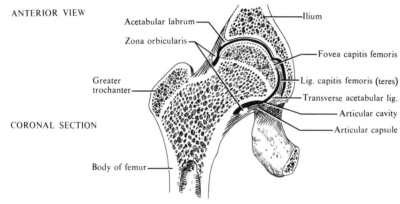

Acetabular labrum

Ilium

Zona orbicularis

Fovea capitis femoris

Greater trochanter

Lig. capitis femoris (teres)

Transverse acetabular lig.

Articular cavity

CORONAL SECTION

Articular capsule

Body of femur

Rectus femoris tendon

MEDIAL VIEW

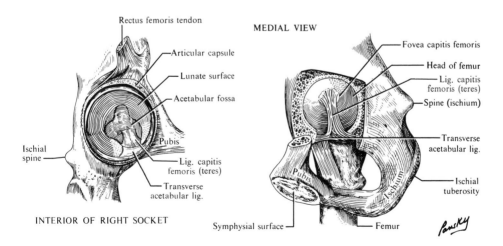

Articular capsule

Fovea capitis femoris

Lunate surface

Head of femur

Acetabular fossa

Lig. capitis femoris (teres)

Spine (ischium)

Ischial spine

Pubis

Transverse acetabular lig.

Lig. capitis femoris (teres)

Ischial tuberosity

Transverse acetabular lig.

Pubis

Ischium

INTERIOR OF RIGHT SOCKET

Symphysial surface

Femur

Pansky

193. BONES OF THE LEG

I. Patella (knee cap) (see p. 441); a sesamoid bone in tendon of quadriceps femoris muscle
A. FLAT AND TRIANGULAR: superior border convex upward, apex downward. Anterior surface convex and rough. Posterior or articular surface is smooth and divided by a ridge into a larger lateral and smaller medial facet

II. Tibia (shin bone): second longest bone of body
A. PARTS: proximal (superior) extremity, shaft (body), and distal (inferior) extremity
B. PROXIMAL EXTREMITY shows medial and lateral condyles, anterior and posterior inter-condylar areas (fossae), an intercondylar eminence, a tuberosity, and an articular facet for head of fibula
C. SHAFT shows an anterior border (crest); broad, convex, smooth medial (subcutaneous) surface; and a lateral (interosseous) border
D. LOWER EXTREMITY shows a medial malleolus; smooth, concave inferior articular surface for talus; fibular notch; posterior groove for flexor hallucis longus muscle; and a medial malleolar sulcus for the tibialis posterior and flexor digitorum longus muscles

III. Fibula (calf bone): most slender of long bones
A. PARTS: body and 2 extremities
B. PROXIMAL EXTREMITY shows a head, with superior articular surfaces, and an apex (styloid process)
C. SHAFT is described as having 3 borders and 3 surfaces
D. DISTAL EXTREMITY shows a lateral malleolus, with a smooth articular and inferior extremity for the talus and a roughened medial area for the tibia

IV. Ossification

Location	When Appears	When Closes
A. TIBIA, from 3 centers		
Body	7th fetal week	——
Prox. end	At time of birth	20th year
Dist. end	2nd year	18th year
B. FIBULA, from 3 centers		
Body	8th fetal week	——
Prox. end	4th year	25th year
Dist. end	2nd year	20th year
C. PATELLA, from 1 center	2nd or 3rd year	At puberty

V. Clinical considerations
A. FRACTURE OF PATELLA can be due to muscle pull. Fragments are usually drawn apart by the muscle pull, and thus closed reduction is often difficult due to tendons being caught in the break
B. FRACTURES OF FIBULAR NECK may damage common peroneal nerve
C. IN CASE OF FRACTURE of either fibula or tibia individually, the unbroken bone helps to splint the other with little displacement. If tibia is broken, look for break in fibula at another level. Fracture of both bones is referred to as Pott's fracture
D. LATERAL MALLEOLUS is often snapped off in overinversion of foot (ankle turn). Over-eversion may break medial malleolus, and if force is strong enough, lateral may also break

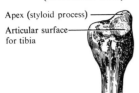

Apex (styloid process)
Articular surface for tibia

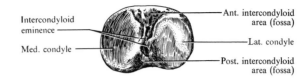

Intercondyloid eminence
Med. condyle

Ant. intercondyloid area (fossa)
Lat. condyle
Post. intercondyloid area (fossa)

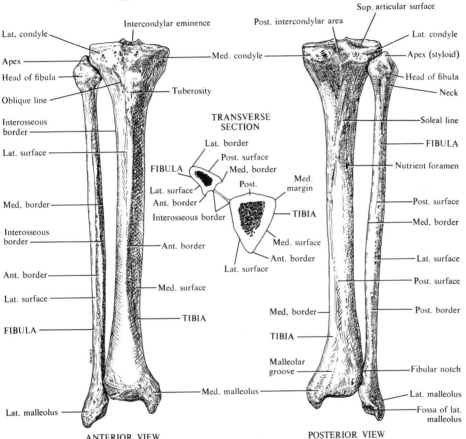

Lat. condyle
Apex
Head of fibula
Oblique line
Interosseous border
Lat. surface
Med. border
Interosseous border
Ant. border
Lat. surface
FIBULA
Lat. malleolus

Intercondylar eminence
Med. condyle
Tuberosity

Post. intercondylar area
Sup. articular surface
Lat. condyle
Apex (styloid)
Head of fibula
Neck
Soleal line
FIBULA
Nutrient foramen
Post. surface
Med. border
Lat. surface
Post. surface
Post. border
Fibular notch
Lat. malleolus
Fossa of lat. malleolus

Med. condyle
Ant. border
Med. surface
TIBIA
Med. malleolus

TRANSVERSE SECTION

Lat. border
Post. surface
Med. border
FIBULA
Lat. surface
Ant. border
Interosseous border
Post.
Med. margin
Interosseous border
TIBIA
Med. surface
Ant. border
Lat. surface

Med. border
TIBIA
Malleolar groove

ANTERIOR VIEW

POSTERIOR VIEW

RIGHT TIBIA AND FIBULA

For talus
Lat. malleolus

For post. talofibular ligament

POSTERIOR
Med. malleolus
Inf. articular surface
ANTERIOR

INFERIOR EXTREMITY—FIBULA

INFERIOR EXTREMITY—TIBIA
(FROM BELOW)

194. POPLITEAL REGION AND ANASTOMOSIS AROUND THE KNEE

I. Popliteal fossa (see p. 428)

II. Popliteal artery: a continuation of the femoral artery as it passes through tendinous hiatus in adductor magnus muscle

> A. RELATIONS
>
> *Anterior*
> Femur, oblique popliteal ligament, popliteus muscle
>
> | *Lateral* | | *Medial* |
> |---|---|---|
> | Biceps and lateral head gastrocnemius muscles; lateral condyle of femur, popliteal vein, tibial nerve | **Popliteal artery** | Semimembranosus, medial head gastrocnemius and plantaris muscles, medial condyle femur, tibial nerve, popliteal vein |
>
> *Posterior*
> Semimembranosus, gastrocnemius and plantaris muscles, popliteal vein, tibial nerve

B. BRANCHES: medial and lateral superior genicular, middle genicular, medial and lateral inferior genicular

III. Tibial nerve
A. RELATIONS: enters fossa from beneath biceps muscle, lateral to popliteal vessels, and crosses superficial to vessels to reach medial side
B. BRANCHES follow the sup., inf. medial, and middle genicular aa. to joint

IV. Common peroneal (fibular) nerve
A. RELATIONS: along lateral side of fossa at border of biceps muscle
B. BRANCHES follow the superior and inferior lateral genicular arteries to joint

V. Popliteal vein formed by anterior and posterior tibial veins at lower border of popliteus muscle. First medial and then cross superficial to the popliteal artery to the lateral side; becomes femoral vein as it passes through tendinous hiatus

VI. Anastomosis (in superficial and deep plexuses around patella) in case of ligation of:

Above Ligation	*Below Ligation*
A. FEMORAL ARTERY, between origins of deep femoral and descending genicular arteries	
1. Desc. br. of lat. circumflex of profunda to:	Lat. sup. and inf. genic. of poplit. Ant. tibial recur. of ant. tibial
	Through deep plexus to med. sup. and inf. genic. brs. of popliteal
B. FEMORAL, at tendinous hiatus	
1. Same as 1, above	
2. Descending (highest) genicular of femoral	
a. Musculoarticular brs. to:	Lat. and med. sup. genic. of poplit.
b. Through saphenous br. to:	Med. inf. genic. of poplit.
C. POPLITEAL, between origins of superior and inferior genicular arteries	
1. Lat. superior genicular to:	Lat. inf. genic. of poplit. Fibular and recurrent tibial brs. of ant. tibial a.
2. Med. superior genicular to:	Med. inf. genicular a.

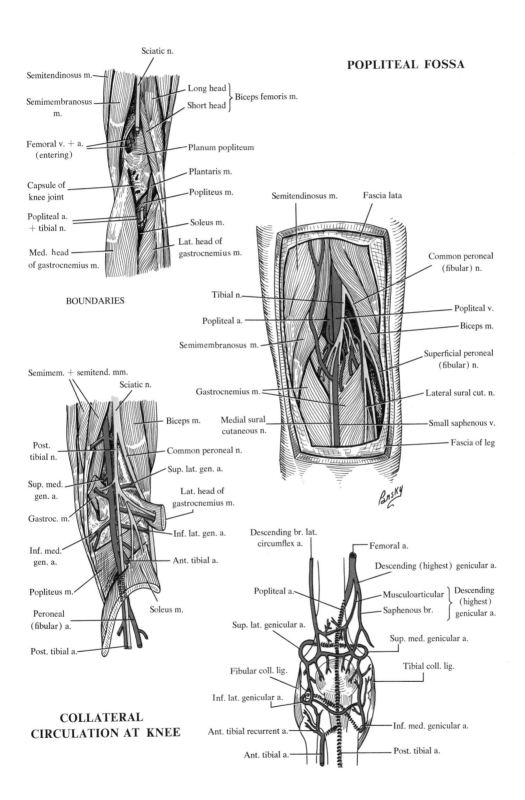

POPLITEAL FOSSA

Sciatic n.

Semitendinosus m.

Semimembranosus m.

Long head
Short head } Biceps femoris m.

Femoral v. + a. (entering)

Planum popliteum

Plantaris m.

Capsule of knee joint

Popliteus m.

Popliteal a. + tibial n.

Soleus m.

Med. head of gastrocnemius m.

Lat. head of gastrocnemius m.

BOUNDARIES

Semitendinosus m.   Fascia lata

Tibial n.

Popliteal a.

Semimembranosus m.

Common peroneal (fibular) n.

Popliteal v.

Biceps m.

Superficial peroneal (fibular) n.

Gastrocnemius m.

Medial sural cutaneous n.

Lateral sural cut. n.

Small saphenous v.

Fascia of leg

Bnsky

Semimem. + semitend. mm.

Sciatic n.

Biceps m.

Post. tibial n.

Sup. med. gen. a.

Gastroc. m.

Inf. med. gen. a.

Popliteus m.

Peroneal (fibular) a.

Post. tibial a.

Common peroneal n.

Sup. lat. gen. a.

Lat. head of gastrocnemius m.

Inf. lat. gen. a.

Ant. tibial a.

Soleus m.

Descending br. lat. circumflex a.

Femoral a.

Descending (highest) genicular a.

Popliteal a.

Musculoarticular
Saphenous br. } Descending (highest) genicular a.

Sup. lat. genicular a.

Sup. med. genicular a.

Fibular coll. lig.

Tibial coll. lig.

Inf. lat. genicular a.

Inf. med. genicular a.

COLLATERAL CIRCULATION AT KNEE

Ant. tibial recurrent a.

Ant. tibial a.

Post. tibial a.

I. Type: ginglymus (hinge) between femur and tibia, arthrodial between femur and patella

II. Divisions: 3 joints in 1: 2 between medial and lateral condyles of femur and tibia; 1 between patella and femur

III. Movements: femur-tibia permits flexion and extension with slight rotation when leg is flexed; femur-patella, sliding up and down

IV. Ligaments

A. ARTICULAR CAPSULE: from femur to tibia, strengthened by fibers from fascia lata, iliotibial tract, and tendons of the vasti, hamstrings, and sartorius muscles

B. LIGAMENTUM PATELLAE: from apex of patella to tuberosity of tibia. Helps hold patella in place; serves as part of tendon of quadriceps muscle

C. OBLIQUE POPLITEAL: from lateral femur, over condyles to posterior head of tibia. Checks extension

D. ARCUATE POPLITEAL: from lateral condyle of femur to styloid process of fibula. May check medial rotation of leg

E. TIBIAL COLLATERAL (medial ligament): from medial side of medial femoral condyle to medial condyle and body of tibia. Prevents lateral bending; checks extension, hyperflexion, and lateral rotation

F. FIBULAR COLLATERAL (lateral ligament): from back of lateral femoral condyle to lateral side of head and styloid process of fibula; checks hyperextension; is relaxed in flexion

G. CORONARY: from capsule to periphery of menisci and tibia. Helps hold menisci in place

H. ANTERIOR CRUCIATE: from medial back of lateral femoral condyle to front of tibial intercondylar eminence. Checks extension, lateral rotation, and anterior slipping of tibia on femur

I. POSTERIOR CRUCIATE: from front and lateral side of medial femoral condyle to posterior intercondylar fossa and posterior end of lateral meniscus. Checks extension, lateral rotation, and posterior slipping of tibia on femur

J. MEDIAL MENISCUS: crescent-shaped (oval); attached to tibia in front of anterior cruciate ligament and in posterior intercondylar fossa. Deepens medial tibial condyle

K. LATERAL MENISCUS: nearly circular; attached to tibia in front of anterior cruciate ligament, blending with latter; posteriorly attached behind intercondylar eminence in front of medial meniscus. Through the anterior meniscofemoral and posterior menisco-femoral (ligament of Wrisberg) ligaments, it is attached to medial femoral condyle. Deepens lateral tibial condyle

L. TRANSVERSE: interconnects anterior parts of 2 menisci

V. Clinical considerations

A. LESIONS OF THE MENISCI: displacements and tears are the most frequent type of damage, the medial being involved 5 to 15 times as often as the lateral. This is true not only because it is longer and less securely attached, but because the abnormal forces which cause the injury are most frequently applied to the lateral side of the knee which violently separate the medial tibial and femoral condyles. The meniscus may be torn or loosened from attachment to the tibial collateral ligament. In either case, all or part becomes jammed between the articular surfaces of the condyles, "locking" the joint

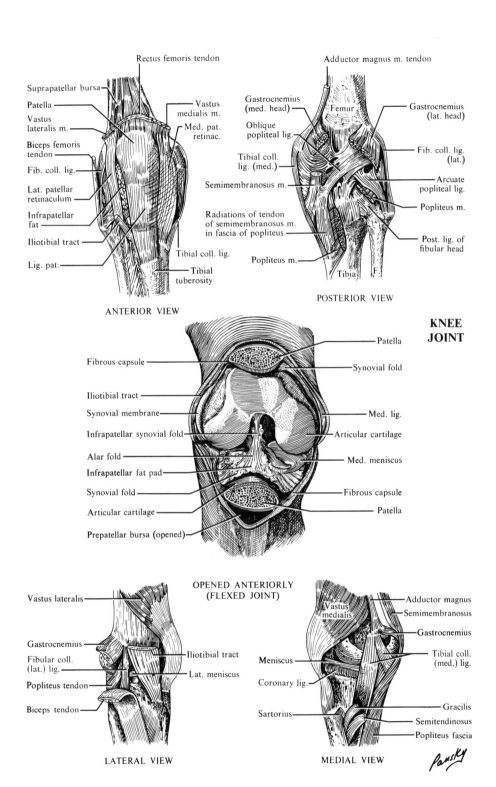

Rectus femoris tendon

Suprapatellar bursa
Patella
Vastus lateralis m.
Biceps femoris tendon
Fib. coll. lig.
Lat. patellar retinaculum
Infrapatellar fat
Iliotibial tract
Lig. pat.

Vastus medialis m.
Med. pat. retinac.

Tibial coll. lig.
Tibial tuberosity

ANTERIOR VIEW

Adductor magnus m. tendon

Gastrocnemius (med. head)
Femur
Oblique popliteal lig.
Tibial coll. lig. (med.)
Semimembranosus m.
Radiations of tendon of semimembranosus m. in fascia of popliteus
Popliteus m.
Tibia

Gastrocnemius (lat. head)
Fib. coll. lig. (lat.)
Arcuate popliteal lig.
Popliteus m.
Post. lig. of fibular head

POSTERIOR VIEW

KNEE JOINT

Fibrous capsule
Iliotibial tract
Synovial membrane
Infrapatellar synovial fold
Alar fold
Infrapatellar fat pad
Synovial fold
Articular cartilage
Prepatellar bursa (opened)

Patella
Synovial fold
Med. lig.
Articular cartilage
Med. meniscus
Fibrous capsule
Patella

OPENED ANTERIORLY
(FLEXED JOINT)

Vastus lateralis
Gastrocnemius
Fibular coll. (lat.) lig.
Popliteus tendon
Biceps tendon

Iliotibial tract
Lat. meniscus

LATERAL VIEW

Vastus medialis
Meniscus
Coronary lig.
Sartorius

Adductor magnus
Semimembranosus
Gastrocnemius
Tibial coll. (med.) lig.
Gracilis
Semitendinosus
Popliteus fascia

MEDIAL VIEW

196. KNEE JOINT—PART II

VI. **Synovial membrane:** largest and most extensive in body
- A. EXTENDS ABOVE PATELLA and laterally and medially beneath vasti muscles
- B. EXTENDS DOWNWARD under patellar ligament; separated from latter by the infrapatellar pad (fat)
- C. SENDS FOLDS, *alar,* into joint cavity, which converge to form the *patellar fold*
- D. LINES CAPSULE, extends over menisci to free border, then under these to tibia
- E. BEHIND LATERAL MENISCUS forms blind sac between latter and tendon of popliteus muscle
- F. REFLECTED IN FRONT OF CRUCIATE LIGAMENTS

VII. **Muscles acting on the joint**

Flexion	Extension	Medial Rotation	Lateral Rotation
Semimembranosus	Quadriceps femoris	Popliteus	Biceps femoris
Semitendinosus	Tensor fasciae latae	Semimembranosus	
Biceps femoris		Semitendinosus	
Sartorius		Sartorius	
Gracilis		Gracilis	
Popliteus			
Gastrocnemius			
Plantaris			

VIII. **Clinical considerations**
- A. TEARS OF THE ANTERIOR CRUCIATE LIGAMENT are fairly frequently associated with tears of other ligaments, particularly the tibial collateral
  1. All the ligaments of the knee contribute to its stability, but the tibial collateral is especially important, for the major part of this strong ligament is tense in all positions of the joint and is thus a very valuable stabilizer
- B. THE CAPSULE OF THE KNEE JOINT is supplied by twigs from all the vessels that enter into anastomosis around the joint. In addition, the middle genicular artery penetrates the capsule posteriorly and is distributed especially to the tissue of the intercondylar region.
- C. THE NERVES TO THE KNEE JOINT are typically derived from the femoral, the obturator, and both parts of the sciatic nerve. Many follow the arteries but some run directly to the capsule

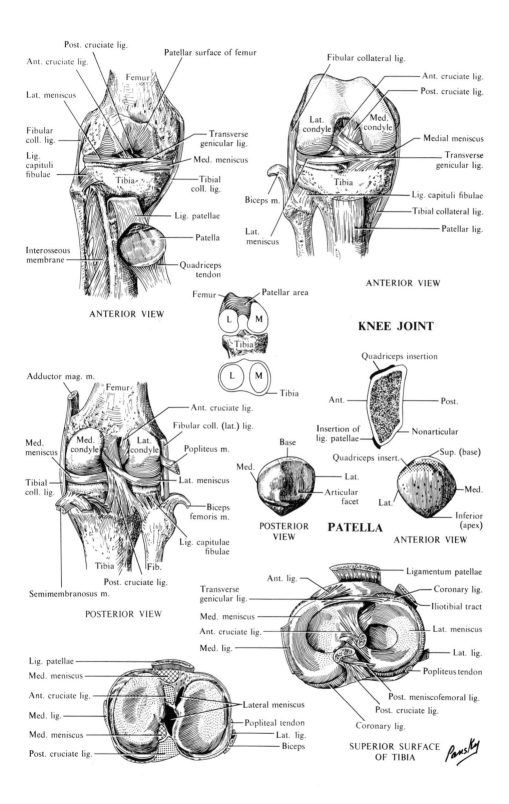

Post. cruciate lig.

Ant. cruciate lig.

Femur

Patellar surface of femur

Lat. meniscus

Fibular coll. lig.

Lig. capituli fibulae

Tibia

Interosseous membrane

Transverse genicular lig.

Med. meniscus

Tibial coll. lig.

Lig. patellae

Patella

Quadriceps tendon

ANTERIOR VIEW

Fibular collateral lig.

Ant. cruciate lig.

Post. cruciate lig.

Lat. condyle

Med. condyle

Medial meniscus

Transverse genicular lig.

Tibia

Biceps m.

Lat. meniscus

Lig. capituli fibulae

Tibial collateral lig.

Patellar lig.

ANTERIOR VIEW

Femur

Patellar area

L   M

Tibia

L   M

Tibia

KNEE JOINT

Adductor mag. m.

Femur

Med. meniscus

Med. condyle

Lat. condyle

Ant. cruciate lig.

Fibular coll. (lat.) lig.

Popliteus m.

Lat. meniscus

Biceps femoris m.

Lig. capitulae fibulae

Tibial coll. lig.

Tibia   Fib.

Post. cruciate lig.

Semimembranosus m.

POSTERIOR VIEW

Quadriceps insertion

Ant.

Insertion of lig. patellae

Post.

Nonarticular

Base

Med.

Lat.

Articular facet

POSTERIOR VIEW

Quadriceps insert.

Sup. (base)

Med.

Lat.

Inferior (apex)

PATELLA

ANTERIOR VIEW

Lig. patellae

Med. meniscus

Ant. cruciate lig.

Med. lig.

Med. meniscus

Post. cruciate lig.

Lateral meniscus

Popliteal tendon

Lat. lig.

Biceps

Ant. lig.

Transverse genicular lig.

Med. meniscus

Ant. cruciate lig.

Med. lig.

Ligamentum patellae

Coronary lig.

Iliotibial tract

Lat. meniscus

Lat. lig.

Popliteus tendon

Post. meniscofemoral lig.

Post. cruciate lig.

Coronary lig.

SUPERIOR SURFACE OF TIBIA

Pansky

-441-

197. KNEE JOINT—PART III, TIBIOFIBULAR ARTICULATION

I. **Bursae of knee:** usually named according to immediately adjacent structures (as tendons, etc.)

A. IN FRONT: 4. (1) Between front of patella and skin, (2) between front of femur and quadriceps femoris muscle (generally communicates with joint cavity), (3) between patellar ligament and upper tibia, (4) between tibial tuberosity and skin

B. LATERALLY: 3. (1) Between fibular collateral ligament and biceps muscle, (2) between fibular collateral ligament and popliteus muscle, (3) between popliteus muscle and lateral condyle of femur (generally communicates with joint cavity)

C. MEDIALLY: 3. (1) Between tibial collateral ligament and tendons of semitendinosus, sartorius, and gracilis muscles, (2) between tibial collateral ligament and tendon of semimembranosus muscle, (3) between tendon of semimembranosus muscle and tibia

D. POSTERIOR: 2. (1) Between lateral head of gastrocnemius muscle and capsule (sometimes communicates with joint), (2) between medial head of gastrocnemius muscle and capsule, extending under tendon of semimembranosus muscle (generally communicates with joint cavity)

II. **Tibiofibular joint** (proximal)

A. TYPE: diarthrodial

B. BONES: head of fibula and fibular facet, below lateral condyle of tibia

C. MOVEMENTS: gliding up and down

D. LIGAMENTS
   1. Articular capsule
   2. Anterior capitular
   3. Posterior capitular

III. **Interosseous membrane** (intermediate)

A. TYPE: synarthrosis

B. FUNCTION: holds shafts together, strengthens fibula, for muscle attachments

C. LIGAMENT: the interosseous membrane. Fibers inclined downward and lateralward from lateral border of tibia to anteromedial border of fibula

IV. **Tibiofibular joint** (distal)

A. TYPE: syndesmosis

B. MOVEMENTS: slight up and down

C. LIGAMENTS
   1. Anterior tibiofibular (lateral malleolar)
   2. Posterior tibiofibular (lateral malleolar)
   3. Interosseous
   4. Inferior transverse

V. **Clinical considerations**

A. BURSITIS: bursa most frequently involved:
   1. Prepatellar: most common; becomes irritated by repeated or prolonged kneeling, thus "housemaid's knee" or "nun's knee"
   2. Gastrocnemius-semimembranosus (popliteal): may be caused by abnormal stress or strain; when distended called "Baker's cyst"
   3. Between semimembranosus tendon and tibial collateral ligament and the tibia: when swollen and painful may be mistaken for lesion of medial meniscus

B. GENU VALGUM: this is "knock knee" with medial condyle and shaft of femur protruding too far medially

C. GENU VARUM: this is "bow leg" with outward curving of the lower limb

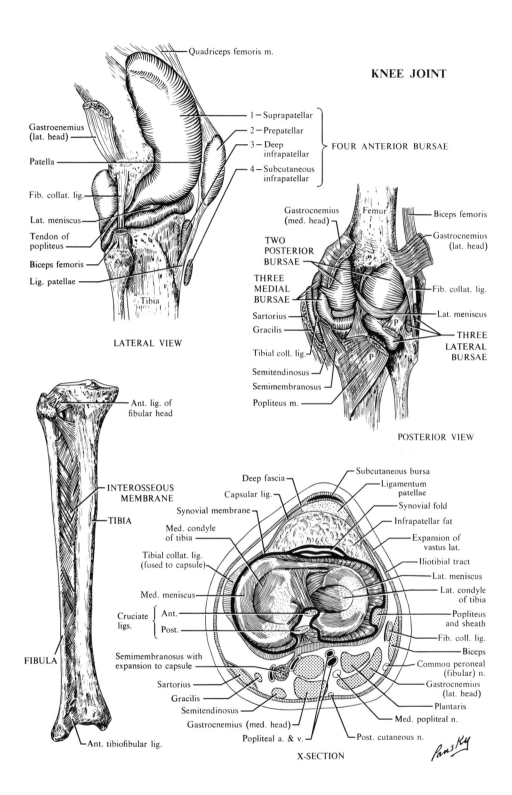

Quadriceps femoris m.

KNEE JOINT

Gastroenemius (lat. head)

Patella

Fib. collat. lig.

Lat. meniscus

Tendon of popliteus

Biceps femoris

Lig. patellae

Tibia

1 – Suprapatellar
2 – Prepatellar
3 – Deep infrapatellar
4 – Subcutaneous infrapatellar

FOUR ANTERIOR BURSAE

LATERAL VIEW

Gastrocnemius (med. head)

Femur

Biceps femoris

Gastrocnemius (lat. head)

TWO POSTERIOR BURSAE

THREE MEDIAL BURSAE

Sartorius

Gracilis

Tibial coll. lig.

Semitendinosus

Semimembranosus

Popliteus m.

Fib. collat. lig.

Lat. meniscus

THREE LATERAL BURSAE

P

P

POSTERIOR VIEW

Ant. lig. of fibular head

INTEROSSEOUS MEMBRANE

TIBIA

FIBULA

Ant. tibiofibular lig.

Deep fascia

Capsular lig.

Synovial membrane

Med. condyle of tibia

Tibial collat. lig. (fused to capsule)

Med. meniscus

Cruciate ligs. { Ant. / Post.

Semimembranosus with expansion to capsule

Sartorius

Gracilis

Semitendinosus

Gastrocnemius (med. head)

Popliteal a. & v.

Subcutaneous bursa

Ligamentum patellae

Synovial fold

Infrapatellar fat

Expansion of vastus lat.

Iliotibial tract

Lat. meniscus

Lat. condyle of tibia

Popliteus and sheath

Fib. coll. lig.

Biceps

Common peroneal (fibular) n.

Gastrocnemius (lat. head)

Plantaris

Med. popliteal n.

Post. cutaneous n.

X-SECTION

Pansky

-443-

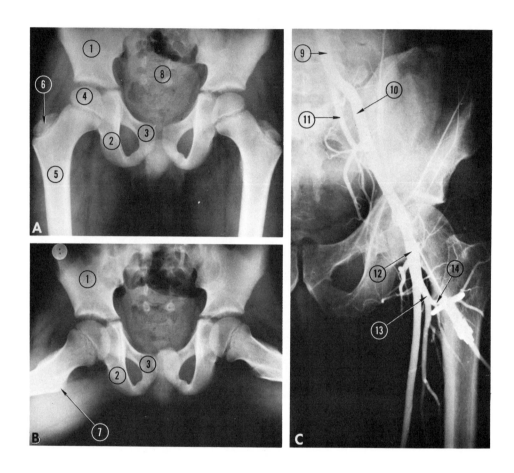

FIGURE 39. **Immature pelvis and arteriograms around hip. A, Immature pelvis, normal position; B, Immature pelvis, thighs abducted; C, arteriogram at hip.** *1,* Ilium; *2,* ischium; *3,* pubis; *4,* head of femur; *5,* shaft of femur; *6,* greater trochanter of femur; *7,* lesser trochanter of femur; *8,* sacrum; *9,* common iliac artery; *10,* external iliac artery; *11,* internal iliac artery; *12,* femoral artery; *13,* deep femoral artery; *14,* lateral circumflex artery.

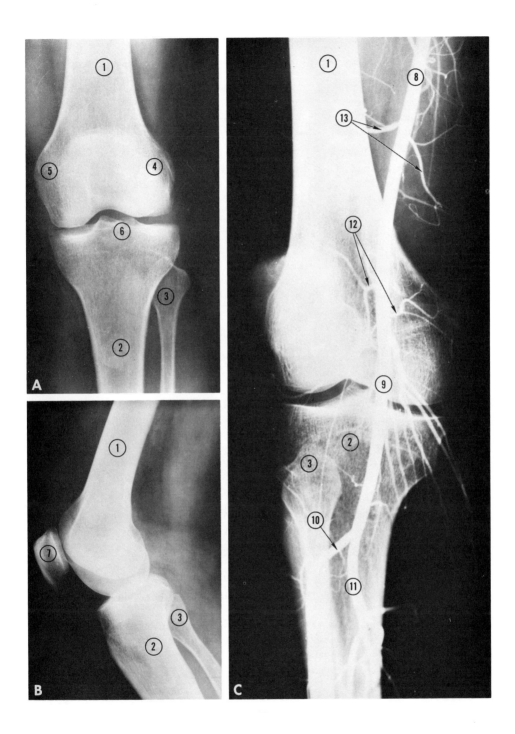

FIGURE 40. **Knee joint and arteriogram at knee. A, Knee joint, anterior view; B, knee joint, lateral view; C, arteriogram around knee.** *1,* Shaft of femur; *2,* tibia; *3,* head of fibula; *4,* lateral condyle; *5,* medial condyle of femur; *6,* intercondylar eminence; *7,* patella; *8,* femoral artery; *9,* popliteal artery; *10,* peroneal artery; *11,* posterior tibial artery; *12,* superior genicular arteries; *13,* muscular arteries.

x

-445-

198. MUSCLES OF THE ANTEROLATERAL LEG

Name	Origin	Insertion	Action	Nerve
I. Anterior				
Tibialis anterior	Lower lateral tibial condyle Lateral tibia Interosseous membrane	Medial side, first cuneiform Base, first metatarsal	Dorsiflexes foot Inverts foot	Deep peroneal (fibular)
Extensor digitorum longus	Lateral tibial condyle Anterior crest fibula Interosseous membrane	Bases of second and terminal phalanges of lateral 4 toes	Extends toes Dorsiflexes foot Everts foot	Deep peroneal (fibular)
Peroneus tertius	Distal fibula Interosseous membrane	Base fifth metatarsal	Dorsiflexes foot Everts foot	Deep peroneal (fibular)
Extensor hallucis longus	Middle anterior fibula Interosseous membrane	Base, distal phalanx of big toe	Extends big toe Everts foot	Deep peroneal (fibular)
II. Lateral				
Peroneus (fibularis) longus	Lateral tibial condyle Head of fibula Anterior capitular ligament Middle lateral fibula	Inferior, first cuneiform Lateral first metatarsal	Plantar flexes foot Everts foot Supports arch	Superficial peroneal (fibular)
Peroneus (fibularis) brevis	Middle lateral fibula	Dorsal surface of tuberosity, fifth metatarsal	Everts foot Plantar flexes foot	Superficial peroneal (fibular)

III. Special features

A. SUPERIOR AND INFERIOR EXTENSOR RETINACULA, flexor retinaculum, and peroneal retinacula (see p. 460)

B. COURSE OF TENDON OF PERONEUS LONGUS: passes behind lateral malleolus with tendon of peroneus brevis muscle but lies posterior to this, under peroneal retinacula. Along lateral side of calcaneus it passes over lateral side of cuboid bone and then in a groove on plantar surface of this bone. It is held in the groove by the *long plantar ligament*. It then crosses foot obliquely to first cuneiform and metatarsal bones

C. THE FOUR ANTERIOR MUSCLES of the leg lie in a fascial compartment between the anterior intermuscular septum and the tibia. All 4 are innervated by the deep peroneal nerve. All have one action in common, for they all dorsiflex the foot, although with varying strength

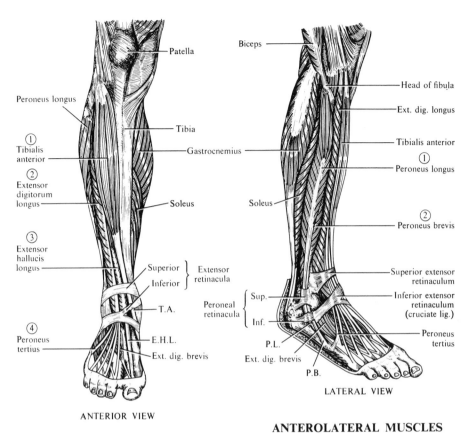

Patella

Peroneus longus

① Tibialis anterior

② Extensor digitorum longus

③ Extensor hallucis longus

④ Peroneus tertius

Tibia

Gastrocnemius

Soleus

Superior
Inferior } Extensor retinacula

T.A.

E.H.L.

Ext. dig. brevis

ANTERIOR VIEW

Biceps

Head of fibula

Ext. dig. longus

Tibialis anterior

① Peroneus longus

② Peroneus brevis

Soleus

Superior extensor retinaculum

Inferior extensor retinaculum (cruciate lig.)

Peroneus tertius

Peroneal retinacula { Sup.
Inf.

P.L.

Ext. dig. brevis

P.B.

LATERAL VIEW

ANTEROLATERAL MUSCLES

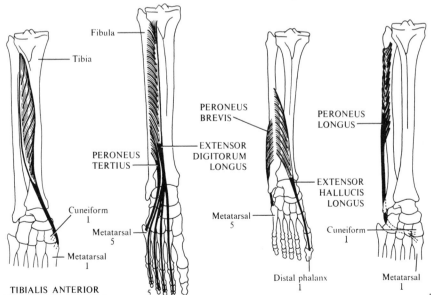

Fibula

Tibia

Tibia

Cuneiform 1

Metatarsal 5

Metatarsal 1

TIBIALIS ANTERIOR
(After Cates and Basmajian)

PERONEUS TERTIUS

PERONEUS BREVIS

EXTENSOR DIGITORUM LONGUS

Metatarsal 5

5 4 3 2

Distal phalanges

PERONEUS LONGUS

EXTENSOR HALLUCIS LONGUS

Cuneiform 1

Distal phalanx 1

Metatarsal 1

Pansky

199. COMPARTMENTS, DEEP VESSELS, AND NERVES OF THE LEG

I. Intermuscular septa: projections of fascia to anterior and lateral crests of fibula

A. ANTERIOR: between peronei and extensor digitorum longus muscles

B. POSTERIOR: between peronei and soleus muscles

II. Compartments. These septa plus tibia, interosseous membrane, and fibula divide the leg into anterior, lateral, and posterior compartments

A. DEEP TRANSVERSE FASCIA crosses posterior compartment, separating the soleus muscle from the other, deeper muscles

III. Anterior tibial artery begins at lower border of popliteus muscle. It passes between 2 heads of tibialis posterior muscle and above interosseous membrane to the front of the leg, where it is joined by the deep peroneal (anterior tibial) nerve

A. RELATIONS

	Anterior Skin, superficial and deep fascia; tibialis anterior, extensor digitorum longus, and extensor hallucis longus muscles; deep peroneal nerve	
Lateral Extensor digitorum longus and extensor hallucis longus muscles; deep peroneal nerve	**Anterior tibial artery**	*Medial* Deep peroneal nerve and extensor hallucis longus muscle
	Posterior Interosseous membrane, tibia, ankle joint	

B. BRANCHES: anterior and posterior tibial recurrent; fibular; anterior medial and anterior lateral malleolar; and muscular

IV. Posterior tibial artery is a direct continuation of the popliteal artery

A. RELATIONS

	Anterior Tibialis posterior and flexor digitorum longus muscles, tibia, ankle joint	
Lateral Tibial nerve	**Posterior tibial artery**	*Medial* Tibial nerve
	Posterior Skin and fascia, gastrocnemius and soleus muscles, tibial nerve and deep transverse fascia	

B. BRANCHES: Peroneal (fibular), posterior medial malleolar, communicating, medial calcaneal, muscular, and nutrient (tibial)

V. Peroneal artery arises from the posterior tibial artery, approaches fibula and lies in a fibrous canal between tibialis posterior and flexor hallucis longus muscles

A. RELATIONS

	Anterior Tibialis posterior muscle, interosseous membrane	
Lateral Flexor hallucis longus, fibula	**Peroneal (fibular) artery**	*Medial* Flexor hallucis longus muscle
	Posterior Soleus, flexor hallucis longus muscle, deep transverse fascia	

B. BRANCHES: Perforating, communicating, lateral calcaneal, muscular, and nutrient

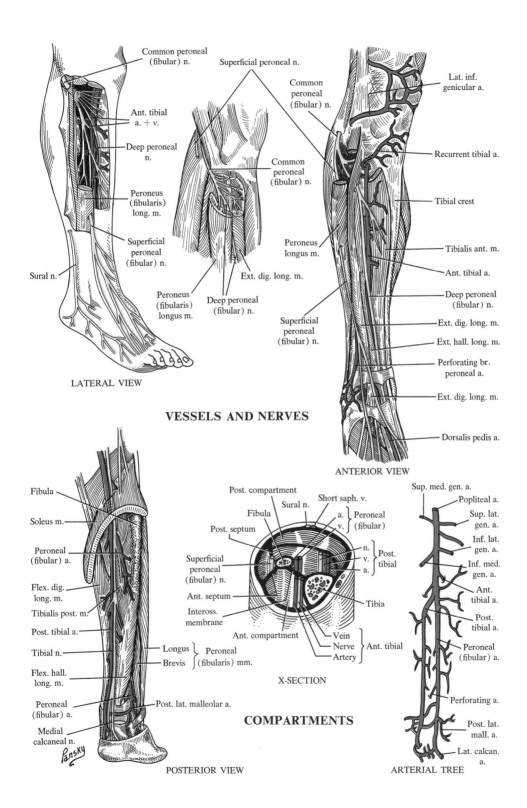

Common peroneal
(fibular) n.

Ant. tibial
a. + v.

Deep peroneal
n.

Peroneus
(fibularis)
long. m.

Superficial
peroneal
(fibular) n.

Sural n.

Peroneus
(fibularis)
longus m.

LATERAL VIEW

Superficial peroneal n.

Common
peroneal
(fibular) n.

Common
peroneal
(fibular) n.

Ext. dig. long. m.

Deep peroneal
(fibular) n.

Peroneus
longus m.

Superficial
peroneal
(fibular) n.

VESSELS AND NERVES

Lat. inf.
genicular a.

Recurrent tibial a.

Tibial crest

Tibialis ant. m.

Ant. tibial a.

Deep peroneal
(fibular) n.

Ext. dig. long. m.

Ext. hall. long. m.

Perforating br.
peroneal a.

Ext. dig. long. m.

Dorsalis pedis a.

ANTERIOR VIEW

Fibula

Soleus m.

Peroneal
(fibular) a.

Flex. dig.
long. m.

Tibialis post. m.

Post. tibial a.

Tibial n.

Flex. hall.
long. m.

Peroneal
(fibular) a.

Medial
calcaneal n.

Pansky

POSTERIOR VIEW

Longus
Brevis
Peroneal
(fibularis) mm.

Post. lat. malleolar a.

Post. compartment

Fibula

Post. septum

Superficial
peroneal
(fibular) n.

Ant. septum

Inteross.
membrane

Ant. compartment

Sural n.

Short saph. v.

a.
v.
Peroneal
(fibular)

n.
v.
a.
Post.
tibial

Tibia

Vein
Nerve
Artery
Ant. tibial

X-SECTION

COMPARTMENTS

Sup. med. gen. a.

Popliteal a.

Sup. lat.
gen. a.

Inf. lat.
gen. a.

Inf. med.
gen. a.

Ant.
tibial a.

Post.
tibial a.

Peroneal
(fibular) a.

Perforating a.

Post. lat.
mall. a.

Lat. calcan.
a.

ARTERIAL TREE

–449–

200. MUSCLES OF THE POSTERIOR LEG

Name	Origin	Insertion	Action	Nerve
I. Superficial group				
Gastro-cnemius				
Medial head	Upper posterior medial condyle and femur above this	Through tendo-calcan. into mid-post. calcan.	Flexes leg Plantar flexes foot	Tibial
Lateral head	Upper posterior lateral condyle and femur above this	See above	See above	Tibial
Soleus	Head and upper fibula Soleal line and upper med. tibia	See above	Plantar flexes foot	Tibial
Plantaris	Lower lateral supracondylar line	Posterior calcaneus	Flexes leg Plantar flexes foot	Tibial
II. Deep group				
Popliteus	Lateral condyle of femur Popliteal lig.	Posterior tibia above soleal line	Flexes leg Rotates leg medially	Tibial
Flexor hallucis longus	Inferior $\frac{2}{3}$ posterior fibula Interosseous membrane	Base, distal phalanx, big toe	Flexes distal phalanx Plantar flexes and inverts foot	Tibial
Flexor digitorum longus	Posterior tibia below soleal line	Base, last phalanx, lateral 4 toes	Flexes toes Plantar flexes and inverts foot	Tibial
Tibialis posterior	Posterior surface, interosseous membrane Posterior shaft, tibia Upper shaft, fibula	Tuberosity navicular, Sustentac. tali 3 cuneiforms, Cuboid Bases metatarsals 2, 3, 4	Plantar flexes, adducts, and inverts foot Supports arch	Tibial

III. Special features

A. NOTE THE LONG, SLENDER PLANTARIS TENDON running on the medial border of the tendocalcaneus. May rupture

B. THE FLEXOR HALLUCIS LONGUS TENDON runs in a groove on the posterior inferior tibia, posterior talus, and under sustentaculum tali

C. THE FLEXOR DIGITORUM LONGUS TENDON runs in a groove behind medial malleolus with the tibialis posterior muscle, obliquely forward and lateralward superficial to the deltoid lig. into the sole of foot, where it crosses below flexor hallucis longus

D. THE TIBIALIS POSTERIOR TENDON runs in front of the flexor digitorum longus in a groove with this muscle as they lie behind the medial malleolus, under the flexor retinaculum, but superficial to deltoid lig. It passes under calcaneonavicular lig.

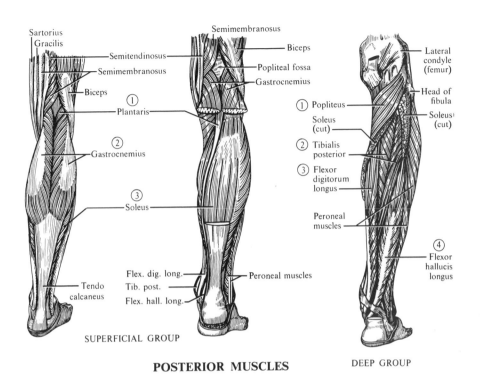

Sartorius
Gracilis
Semitendinosus
Semimembranosus
Biceps
Plantaris ①
Gastrocnemius ②
Soleus ③
Tendo
calcaneus

Semimembranosus
Biceps
Popliteal fossa
Gastrocnemius

Flex. dig. long.
Tib. post.
Flex. hall. long.
Peroneal muscles

SUPERFICIAL GROUP

Lateral
condyle
(femur)
Head of
fibula
Popliteus ①
Soleus
(cut)
Soleus
(cut)
Tibialis ②
posterior
Flexor ③
digitorum
longus
Peroneal
muscles
④
Flexor
hallucis
longus

DEEP GROUP

POSTERIOR MUSCLES

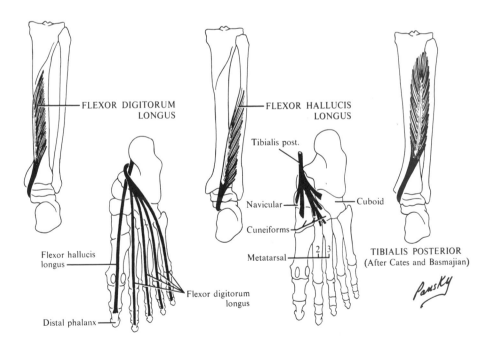

FLEXOR DIGITORUM
LONGUS

Flexor hallucis
longus

Flexor digitorum
longus

Distal phalanx

FLEXOR HALLUCIS
LONGUS

Tibialis post.

Navicular
Cuboid

Cuneiforms

Metatarsal   2   3

TIBIALIS POSTERIOR
(After Cates and Basmajian)

Pansky

201. BONES OF ANKLE AND FOOT

I. Tarsals: talus, calcaneus, navicular, cuboid, 3 cuneiforms
- A. TALUS (astragalus, ankle bone): 3 parts
    1. Body: behind, rests on the anterior part of calcaneus; above, lies under the tibia and is gripped by both malleoli; posteriorly has a medial and lateral tubercle with intervening groove for flexor hallucis longus muscle tendon; inferiorly has deep groove—*sulcus tali*
    2. Head: in front articulates with navicular and calcaneus
    3. Neck: constriction between 1 and 2, above
- B. CALCANEUS (heel bone). Has the following features:
    1. Sustentaculum tali: medial shelf to hold head of talus
    2. Groove for flexor hallucis longus muscle tendon runs below sustentaculum
    3. Posterior third ("heel") projects behind ankle joint
    4. Tuberosity: part in contact with ground; plantar surface has medial and lateral tubercles (processes)
    5. Peroneal trochlea on lateral aspect
    6. Groove corresponding to sulcus tali with which it forms *sinus tarsi*
- C. NAVICULAR (boat-shaped)
    1. Tuberosity on medial side
- D. CUBOID
    1. Groove on plantar surface for peroneus longus muscle tendon
- E. CUNEIFORMS (wedge-shaped), 3: (1) medial, (2) intermediate, (3) lateral

II. Metatarsals: 5, numbered from medial to lateral
- A. EACH CONSISTS OF:
    1. Head (distal end): articulates with proximal phalanx
    2. Body (midportion)
    3. Base (proximal end): articulates with tarsals and bases of other metatarsals
- B. FIRST: shortest, stoutest; 2 sesamoid bones at plantar surface of distal end
- C. FIFTH has a tuberosity projecting posteriorly

III. Phalanges: 14 in number
- A. THREE FOR TOES 2–5: (1) proximal, (2) middle, (3) distal
- B. TWO FOR LARGE TOE (hallux)
- C. EACH PHALANX consists of base (proximal end), body, and head (distal end)

IV. Ossification
- A. TARSALS: all from 1 center except calcaneus, which has a second for heel
    1. Centers appear: calcaneus sixth, talus seventh, cuboid ninth fetal months; third cuneiform, first year; first cuneiform, third year; second cuneiform and navicular, fourth year; heel, tenth year. Latter joins body at puberty
- B. METATARSALS: from 2 centers—body and head for 2–5; body and base for first
    1. Centers appear: body, ninth fetal week; base of first in third year; heads of 2–5, 5–8 years. All join body between 18–20 years
- C. PHALANGES: from 2 centers—body and base
    1. Centers appear: body, tenth fetal week; base, 4–10 years. Join body, 18 years

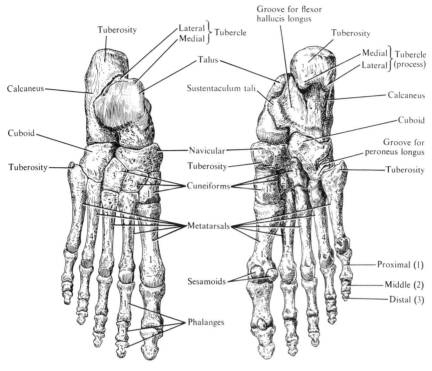

DORSAL VIEW

PLANTAR VIEW

BONES OF ANKLE AND FOOT

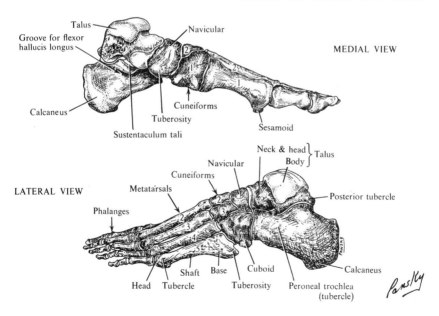

MEDIAL VIEW

LATERAL VIEW

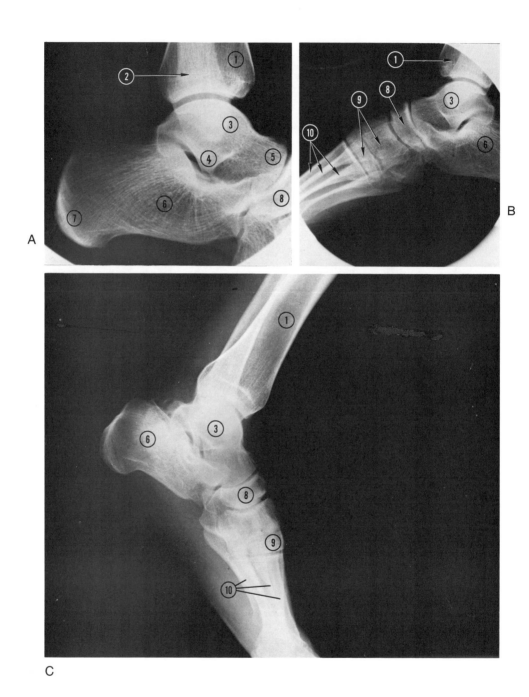

FIGURE 41. **Ankle joint. A, Left ankle joint; B, right ankle joint and foot, medial view; C, left ankle joint and foot.** *1,* Tibia; *2,* fibula; *3,* talus; *4,* sustentaculum tali; *5,* head of talus; *6,* calcaneus; *7,* tuberosity of calcaneus; *8,* navicular; *9,* cuneiforms; *10,* metatarsals.

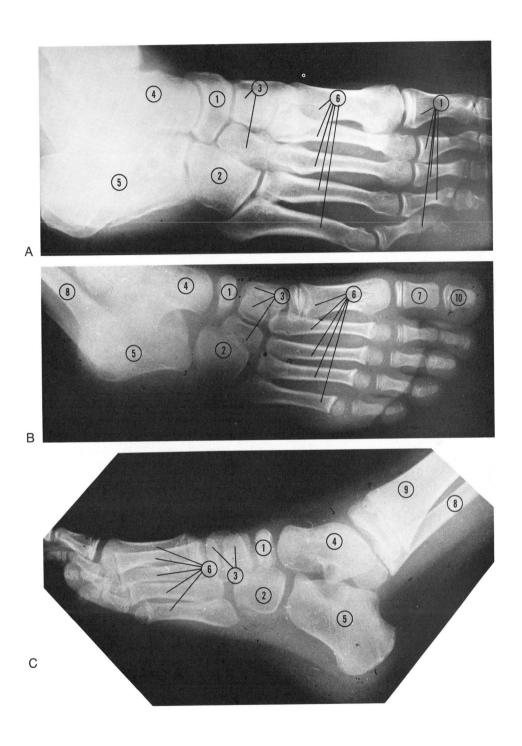

FIGURE 42. Feet. A, Plantar view of left foot; B, plantar view of left foot; C, medial view of right foot. *1*, Navicular; *2*, cuboid; *3*, cuneiforms; *4*, talus; *5*, calcaneus; *6*, metatarsals; *7*, proximal phalanges; *8*, fibula; *9*, tibia; *10*, distal phalanx (toe).

202. FLEXOR RETINACULUM, TENDON SHEATHS, AND VESSELS OF MEDIAL SIDE OF ANKLE

I. **Flexor retinaculum** holds the long flexor tendons at the medial side of ankle

A. EXTENDS FROM MARGINS OF MEDIAL MALLEOLUS TO CALCANEUS; passes over tendon of tibialis posterior muscle; attaches to bone behind this; and then forms superficial and deep layers:

1. Superficial extends to tuberosity of calcaneus
2. Deep passes over flexor digitorum longus and flexor hallucis longus muscles and is attached to bone on either side of these, forming separate osseofibrous canals

B. THIS RETINACULUM converts the bony grooves into 4 osseofibrous canals for the following structures numbered from medial to lateral: tendon of tibialis posterior muscle; tendon of flexor digitorum longus muscle; posterior tibial artery and vein and tibial nerve; and the tendon of the flexor hallucis longus muscle

II. **Tendon sheaths**

A. BEHIND THE MEDIAL MALLEOLUS, the tendon of the tibialis posterior muscle lies in front of, but in the same groove with the flexor digitorum longus muscle. Both have separate sheaths: that for the tibialis posterior starts 5 cm above malleolus and extends to tuberosity of navicular; that for flexor digitorum longus begins just above tip of malleolus and ends opposite first cuneiform. The tendon of the flexor hallucis longus muscle lies dorsal to that of the flexor digitorum longus muscle and crosses it from lateral to medial. Its sheath starts at the tip of the malleolus and extends to base of first metatarsal

III. **Anastomoses about the ankle**

Medial Malleolar Net	Lateral Malleolar Net
Ant. med. malleolar from ant. tibial a.	Ant. lat. malleolar from ant. tibial a.
Med. tarsal from dorsalis pedis a.	Lat. tarsal from dorsalis pedis a.
Post. med. malleolar from post. tibial a.	Perf. peroneal from peroneal (fibular) a.
Med. calcaneal from post. tibial a.	Lat. calcaneal from peroneal (fibular) a.
	Arcuate from dorsalis pedis a.

IV. **The deep transverse fascial septum** separates the deep and superficial groups of the posterior crural muscles. At the ankle it is continuous with the flexor retinaculum

V. **Relations of vessels**

A. DORSALIS PEDIS ARTERY is a continuation of the anterior tibial artery in front of the ankle, has deep peroneal nerve on its lateral side and tendon of flexor hallucis longus muscle on its medial side as they pass under the extensor retinacula

B. THE TIBIAL NERVE lies lateral to the posterior tibial artery as they pass in the third compartment beneath the flexor retinaculum

C. THE PERONEAL VESSELS lie behind the inferior tibiofibular articulation

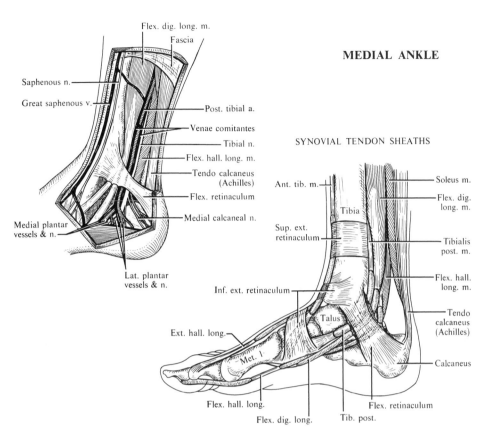

Flex. dig. long. m.
Fascia
Saphenous n.
Great saphenous v.
Post. tibial a.
Venae comitantes
Tibial n.
Flex. hall. long. m.
Tendo calcaneus (Achilles)
Flex. retinaculum
Medial calcaneal n.
Medial plantar vessels & n.
Lat. plantar vessels & n.

MEDIAL ANKLE

SYNOVIAL TENDON SHEATHS

Ant. tib. m.
Tibia
Sup. ext. retinaculum
Inf. ext. retinaculum
Talus
Ext. hall. long.
Met. I
Soleus m.
Flex. dig. long. m.
Tibialis post. m.
Flex. hall. long. m.
Tendo calcaneus (Achilles)
Calcaneus
Flex. hall. long.
Flex. dig. long.
Tib. post.
Flex. retinaculum

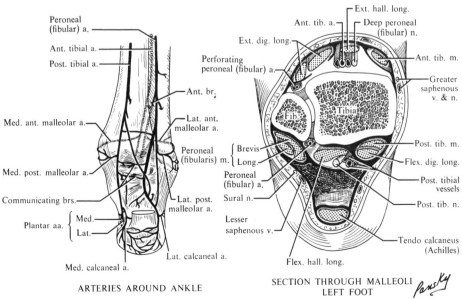

Peroneal (fibular) a.
Ant. tibial a.
Post. tibial a.
Ant. br.
Lat. ant. malleolar a.
Med. ant. malleolar a.
Peroneal (fibularis) m.
Med. post. malleolar a.
Communicating brs.
Lat. post. malleolar a.
Plantar aa. { Med. Lat. }
Lat. calcaneal a.
Med. calcaneal a.

ARTERIES AROUND ANKLE

Ext. hall. long.
Ant. tib. a.
Deep peroneal (fibular) n.
Ext. dig. long.
Perforating peroneal (fibular) a.
Ant. tib. m.
Greater saphenous v. & n.
Fib.
Tibia
Peroneal (fibular) m. { Brevis Long. }
Post. tib. m.
Flex. dig. long.
Peroneal (fibular) a.
Post. tibial vessels
Sural n.
Post. tib. n.
Lesser saphenous v.
Tendo calcaneus (Achilles)
Flex. hall. long.

SECTION THROUGH MALLEOLI LEFT FOOT

Pansky

203. MUSCLES OF DORSUM OF FOOT

I.

Name	Origin	Insertion	Action	Nerve
Extensor hallucis longus	Middle anterior fibula Interosseous membrane	Base, distal phalanx of big toe	Extends big toe Everts foot	Deep peroneal (fibular)
Extensor digitorum longus	Lateral tibial condyle Anterior crest fibula Interosseous membrane	Bases of second and terminal phalanges of lateral 4 toes	Extends toes Dorsiflexes foot Everts foot	Deep peroneal (fibular)
Peroneus (fibularis) tertius	Distal fibula Inteross. memb.	Base fifth metatarsal	Dorsiflexes foot Everts foot	Deep peroneal (fibular)
Tibialis anterior	Lower lateral tibial condyle Lateral tibia Inteross. memb.	Medial side, first cuneiform Base, first metatarsal	Dorsiflexes foot Inverts foot	Deep peroneal (fibular)
Extensor digitorum brevis	Distal lateral and superior surface, calcaneus	Lateral side, long extensor tendons; slips to base of 1st phalanx of medial 4 toes	Extends medial 4 toes	Deep peroneal (fibular)
Dorsal interossei (4)	Adjacent sides and bases of metatarsals 1–4	1st and 2nd on either side, base of 1st phalanx of 2nd toe; 3 and 4 on lateral side, proximal phalanx toes 3 and 4	Abduct toes Flex proximal phalanx Extend distal 2 phalanges	Deep branch of the lateral plantar

II. Special features

A. RETINACULA AROUND ANKLE (see p. 460)

B. SYNOVIAL TENDON SHEATH OF ANTERIOR MUSCLES

    1. Tibialis anterior: from above superior extensor retinaculum, under inferior extensor retinaculum to level of talonavicular joint

    2. Extensor hallucis longus: from just above superior limb of inferior extensor retinaculum, beneath both limbs of retinaculum to level of first tarsometatarsal joint

    3. Tendon of extensor digitorum longus: envelops this and peroneus tertius muscle from above the inferior extensor retinaculum to middle of cuboid bone

C. INSERTIONS OF EXTENSOR TENDONS

    1. Extensor hallucis has medial and lateral expansions covering metatarsophalangeal joint and a prolongation of tendon to base of first phalanx

    2. Extensor digitorum longus: tendons to toes 2, 3, 4 are joined by extensor brevis; each receives fibers from lumbricales and interossei, spreads out on dorsum of first phalanx of each toe, splits into 3 slips, the intermediate to base of second phalanx and the 2 collateral to base of third phalanx

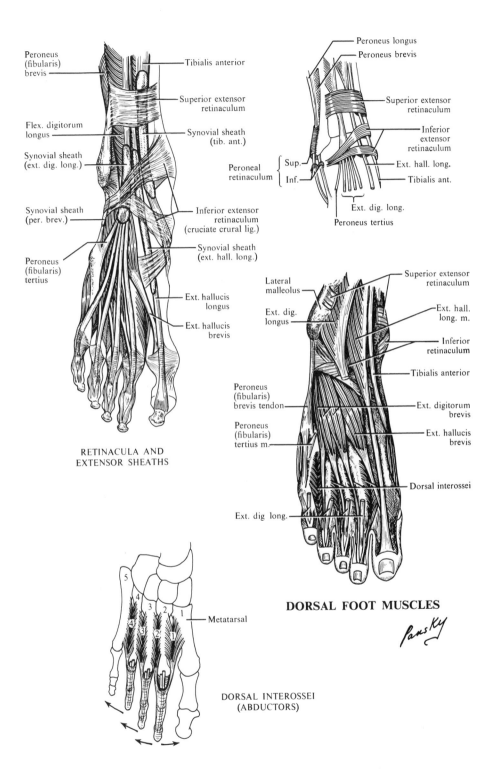

Peroneus (fibularis) brevis

Tibialis anterior

Flex. digitorum longus

Superior extensor retinaculum

Synovial sheath (tib. ant.)

Synovial sheath (ext. dig. long.)

Synovial sheath (per. brev.)

Inferior extensor retinaculum (cruciate crural lig.)

Peroneus (fibularis) tertius

Synovial sheath (ext. hall. long.)

Ext. hallucis longus

Ext. hallucis brevis

RETINACULA AND EXTENSOR SHEATHS

Peroneus longus

Peroneus brevis

Superior extensor retinaculum

Inferior extensor retinaculum

Ext. hall. long.

Peroneal retinaculum { Sup. Inf.

Tibialis ant.

Ext. dig. long.

Peroneus tertius

Lateral malleolus

Ext. dig. longus

Superior extensor retinaculum

Ext. hall. long. m.

Inferior retinaculum

Tibialis anterior

Peroneus (fibularis) brevis tendon

Ext. digitorum brevis

Peroneus (fibularis) tertius m.

Ext. hallucis brevis

Dorsal interossei

Ext. dig. long.

DORSAL FOOT MUSCLES

Pansky

Metatarsal

DORSAL INTEROSSEI (ABDUCTORS)

-459-

204. EXTENSOR AND PERONEAL RETINACULA, SHEATHS, AND ARTERIES ON DORSUM OF FOOT

I. **Extensor retinacula** hold extensor tendons in place
A. SUPERIOR (transverse crural) across leg above ankle. Consists of transverse fibers from the medial side of tibia to the anterior side of fibula
B. INFERIOR (cruciate crural) has 2 parts
   1. Superficial: Y-shaped, stem attached to lateral side of calcaneus, splits after passing over the extensor digitorum longus muscle
     a. Upper limb passes upward and medialward to attach to the medial malleolus
     b. Lower limb passes medially and downward over foot to insert on first cuneiform and sole of foot
   2. Deep: beneath the stem of the Y and passes deep to the peroneus tertius and extensor longus muscles

II. **Peroneal (fibular) retinacula** hold peroneal muscles in place
A. SUPERIOR from lateral malleolus to fascia of back of leg and lateral side of calcaneus
B. INFERIOR overlies tendons on lateral calcaneus and is attached to the bone on either side of the tendons. Joins superficial part of the inferior extensor retinaculum

III. **Synovial sheaths.** For tibialis anterior, extends from above superior extensor retinaculum to interval between the limbs of the inferior extensor retinaculum. For extensor digitorum longus and extensor hallucis longus muscles, the sheath begins just below the superior extensor retinaculum, the former extending to the level of the base of the fifth metatarsal, the latter to the base of the first metatarsal. Both peronei lie in a common sheath extending 4 cm above and below the tip of the malleolus

IV. **Arteries**
A. DORSALIS PEDIS passes down dorsum of foot to the proximal end of the first intermetatarsal space and terminates as the *first dorsal metatarsal and deep plantar arteries.* Branches:
   1. Lateral tarsal
   2. Medial tarsal
   3. Arcuate, which gives rise to *second, third, and fourth metatarsal arteries.* These subsequently split to form the *dorsal digital arteries* to the adjoining sides of the toes
   4. Terminal branches
     a. First dorsal metatarsal to both sides of big toe and adjoining sides of big and second toes
     b. Deep plantar (see p. 466)

V. **Veins** (see p. 406)

VI. **Nerves**
A. MOST OF THE DORSUM OF THE FOOT is supplied by the superficial peroneal nerve. The sural contributes to the lateral side and the saphenous to the medial side of the foot
B. ON THE TOES: the deep peroneal nerve supplies adjoining areas on the first and second toes; the superficial peroneal nerve covers the remainder of the first, second, and third toes to the second joint. Part of the fourth toe is supplied by the medial plantar nerve. Adjoining parts of the fourth and fifth plus the terminal ends of the first 3 toes receive lateral plantar branches

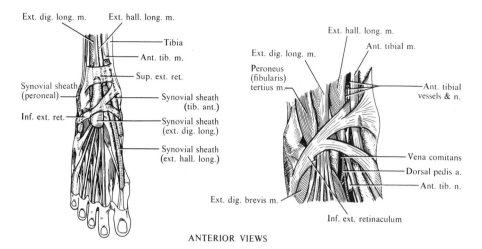

Ext. dig. long. m.
Ext. hall. long. m.
Tibia
Ant. tib. m.
Sup. ext. ret.
Synovial sheath (peroneal)
Synovial sheath (tib. ant.)
Inf. ext. ret.
Synovial sheath (ext. dig. long.)
Synovial sheath (ext. hall. long.)

Ext. hall. long. m.
Ext. dig. long. m.
Ant. tibial m.
Peroneus (fibularis) tertius m.
Ant. tibial vessels & n.
Vena comitans
Dorsal pedis a.
Ant. tib. n.
Ext. dig. brevis m.
Inf. ext. retinaculum

ANTERIOR VIEWS

DORSUM OF FOOT

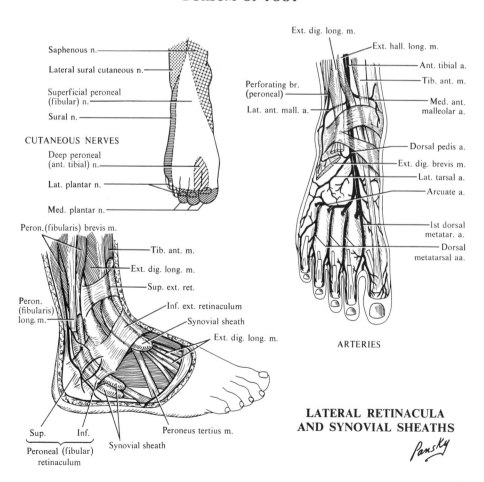

Saphenous n.
Lateral sural cutaneous n.
Superficial peroneal (fibular) n.
Sural n.

CUTANEOUS NERVES

Deep peroneal (ant. tibial) n.
Lat. plantar n.
Med. plantar n.

Peron. (fibularis) brevis m.
Tib. ant. m.
Ext. dig. long. m.
Sup. ext. ret.
Inf. ext. retinaculum
Synovial sheath
Ext. dig. long. m.
Peron. (fibularis) long. m.

Sup.   Inf.
Peroneal (fibular) retinaculum
Synovial sheath
Peroneus tertius m.

Ext. dig. long. m.
Ext. hall. long. m.
Ant. tibial a.
Tib. ant. m.
Perforating br. (peroneal)
Lat. ant. mall. a.
Med. ant. malleolar a.
Dorsal pedis a.
Ext. dig. brevis m.
Lat. tarsal a.
Arcuate a.
1st dorsal metatar. a.
Dorsal metatarsal aa.

ARTERIES

LATERAL RETINACULA AND SYNOVIAL SHEATHS

Pansky

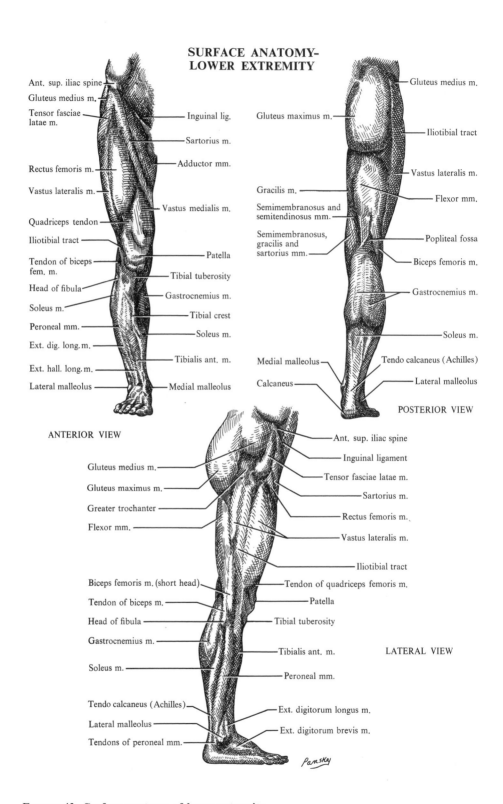

SURFACE ANATOMY-
LOWER EXTREMITY

Ant. sup. iliac spine
Gluteus medius m.
Tensor fasciae latae m.

Inguinal lig.

Sartorius m.

Adductor mm.

Rectus femoris m.

Vastus lateralis m.

Vastus medialis m.

Quadriceps tendon

Iliotibial tract

Patella

Tendon of biceps fem. m.

Tibial tuberosity

Head of fibula

Gastrocnemius m.

Soleus m.

Peroneal mm.

Tibial crest

Soleus m.

Ext. dig. long. m.

Tibialis ant. m.

Ext. hall. long. m.

Lateral malleolus

Medial malleolus

ANTERIOR VIEW

Gluteus medius m.

Gluteus maximus m.

Iliotibial tract

Vastus lateralis m.

Gracilis m.

Flexor mm.

Semimembranosus and semitendinosus mm.

Semimembranosus, gracilis and sartorius mm.

Popliteal fossa

Biceps femoris m.

Gastrocnemius m.

Soleus m.

Medial malleolus

Tendo calcaneus (Achilles)

Calcaneus

Lateral malleolus

POSTERIOR VIEW

Ant. sup. iliac spine

Inguinal ligament

Gluteus medius m.

Tensor fasciae latae m.

Gluteus maximus m.

Sartorius m.

Greater trochanter

Rectus femoris m.

Flexor mm.

Vastus lateralis m.

Iliotibial tract

Biceps femoris m. (short head)

Tendon of quadriceps femoris m.

Tendon of biceps m.

Patella

Head of fibula

Tibial tuberosity

Gastrocnemius m.

Tibialis ant. m.

LATERAL VIEW

Soleus m.

Peroneal mm.

Tendo calcaneus (Achilles)

Ext. digitorum longus m.

Lateral malleolus

Ext. digitorum brevis m.

Tendons of peroneal mm.

Pansky

FIGURE 43. **Surface anatomy of lower extremity.**

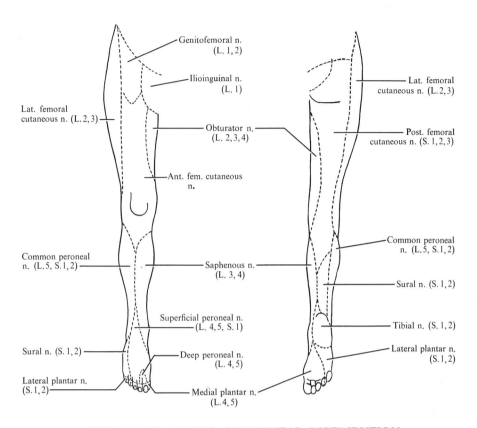

CUTANEOUS NERVES - SEGMENTAL DISTRIBUTION

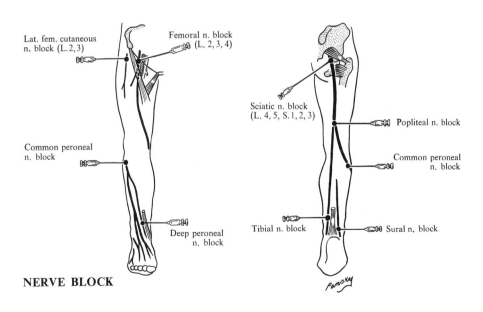

NERVE BLOCK

FIGURE 44. **Nerve blocks of lower extremity.**

205. MUSCLES OF THE PLANTAR SURFACE OF THE FOOT

I.

Name	Origin	Insertion	Action	Nerve
Flexor digitorum brevis	Medial process, tuber. calcaneus	Middle phalanx, lateral 4 toes	Flexes toes	Medial
Quadratus plantae				
Lateral head	Lateral process tuberosity calcaneus	Lateral border, long flexor tendons	Flexes toes	Lateral plantar
Medial head	Med. side calcan.			
Lumbricales				
1st	Medial side of 1st longus flexor tendon	Proximal phalanx and extensor tendons, toe 2	Flexes proximal, extends distal phalanges	Medial plantar
2nd, 3rd, and 4th	Long flexor tendons 2, 3, 4	As above for toes 3, 4, 5	See above	Lateral plantar
Abductor hallucis	Medial process, tuberosity calcaneus	Medial side, base proximal phalanx of big toe	Flexes and abducts big toe	Medial plantar
Flexor hallucis brevis	Cuboid and 3rd cuneiform	Abductor tendon Base, proximal phalanx	Flexes proximal phalanx, big toe	Medial plantar
Adductor hallucis				
Oblique head	Peroneus long. sheath Base 2, 3, 4 metatarsals	With flexor hallucis brevis	Adducts and flexes proximal phalanx, big toe	Lateral plantar
Transverse head	Capsules 3, 4, 5 metatarso-phalang. joints	Lateral side of base of proximal phalanx, big toe	See above	See above
Abductor digiti minimi	Lat. and med. processes, tuberosity calcaneus	Lateral surface, proximal phalanx, small toe	Flexes and abducts proximal phalanx, small toe	Lateral plantar
Flexor digiti minimi brevis	Peron. long. tendon Base metatarsal 5	Base of proximal phalanx, small toe	Flexes proximal phalanx, small toe	Lateral plantar
Plantar interossei (3)	Proximal shaft and base, metatarsals 3, 4, 5	Base of prox. phalanx of toes 3, 4, and 5 extensor tendons	Adducts toes Flexes proximal, extends distal part of toes 3–5	Lateral plantar

II. Special features

A. THE MUSCLES AND TENDONS of the plantar surface of foot are arranged in 4 layers
1. Layer 1: abductor hallucis, flexor digitorum brevis, and abductor digiti minimi
2. Layer 2: quadratus plantae, lumbricales, tendons of flexor digitorum longus, and flexor hallucis longus
3. Layer 3: flexor hallucis brevis, adductor hallucis, and flexor digiti minimi brevis
4. Layer 4: the interossei, tendons of tibialis posterior and peroneus longus

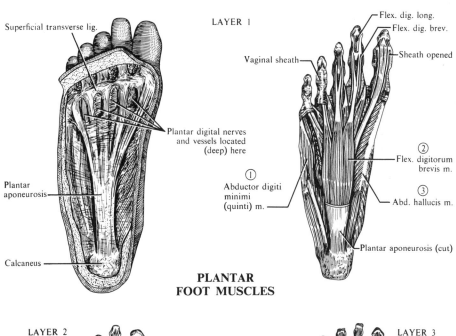

Superficial transverse lig.

Plantar digital nerves
and vessels located
(deep) here

Plantar
aponeurosis

Calcaneus

LAYER 1

Flex. dig. long.

Flex. dig. brev.

Vaginal sheath

Sheath opened

① Abductor digiti
minimi
(quinti) m.

② Flex. digitorum
brevis m.

③ Abd. hallucis m.

Plantar aponeurosis (cut)

PLANTAR
FOOT MUSCLES

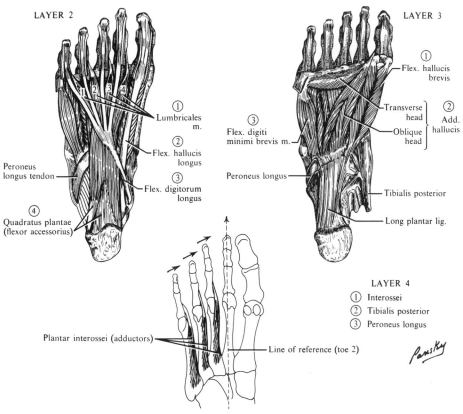

LAYER 2

① Lumbricales
m.

② Flex. hallucis
longus

③ Flex. digitorum
longus

Peroneus
longus tendon

④ Quadratus plantae
(flexor accessorius)

LAYER 3

① Flex. hallucis
brevis

Transverse
head

Oblique
head

② Add.
hallucis

③ Flex. digiti
minimi brevis m.

Peroneus longus

Tibialis posterior

Long plantar lig.

LAYER 4

① Interossei

② Tibialis posterior

③ Peroneus longus

Plantar interossei (adductors)

Line of reference (toe 2)

Pansky

206. PLANTAR APONEUROSIS, THE VESSELS AND NERVES OF THE SOLE OF THE FOOT

I. Plantar aponeurosis: thickened deep fascia of the sole. Extends from the tuberosity of the calcaneus to end in 5 slips attached to the skin over the heads of the metatarsals and into the flexor tendon sheaths (bands 2, 3, and 4 are stronger than 1 and 5)

II. Arteries

A. POSTERIOR TIBIAL runs beneath the origin of the abductor hallucis muscle and divides into 2 branches
   1. Medial plantar along the medial side of foot and medial border of big toe
   2. Lateral plantar goes obliquely lateralward to reach the base of fifth metatarsal, then arches medially to a point between the bases of the first and second metatarsals. It joins the deep branch of the dorsalis pedis artery to form the *plantar arch.* Branches:
      a. Perforating to dorsal metatarsal arteries
      b. Plantar metatarsals (4) in the interosseous spaces where it divides into *proper plantar digital arteries* to the adjacent sides of toes. Sends *anterior perforating branches* to the dorsal metatarsal arteries
      c. First plantar metatarsal supplies adjoining sides of big and second toes
      d. Proper plantar digital to the lateral side of the small toe

III. Nerves

A. CUTANEOUS
   1. The saphenous nerve (from the femoral) and the sural nerve (from both the tibial and peroneal nerves) supply small areas of skin on the medial and lateral parts, respectively, of the sole
   2. Plantar surface of the heel is supplied by the tibial nerve
   3. Most of the medial sole, including the toes to the middle of the fourth toe, are supplied by the medial plantar nerve. The medial plantar nerve accompanies the medial plantar artery and gives off branches
      a. Plantar cutaneous branches to medial side of sole
      b. Proper digital nerve to medial side of big toe
      c. Three common digital nerves, which split to give 2 proper digital nerves to adjacent sides of toes 1–4
   4. The lateral half of the fourth toe and lateral side of the foot are supplied by the lateral plantar nerve. The lateral plantar nerve is accompanied by the lateral plantar artery. The nerve gives off a superficial branch, which sends proper digital nerves to the lateral side of the little toe, and common digital nerves, which divide to supply adjacent sides of toes 4 and 5

B. MUSCULAR
   1. Medial plantar nerve to the abductor hallucis, flexor digitorum brevis, flexor hallucis brevis, and first lumbrical muscle
   2. Lateral plantar to the quadratus plantae and abductor digiti minimi muscles
      a. Superficial branch to flexor digiti minimi and the 2 interossei muscles of the fourth space
      b. Deep branch to adductor hallucis muscle, interosseous muscles of spaces 1, 2, and 3 and lumbrical muscles 2, 3, and 4

IV. Clinical: cuts of the feet may bleed profusely. Due to the numerous anastomoses between vessels, both ends of a severed artery must be ligated

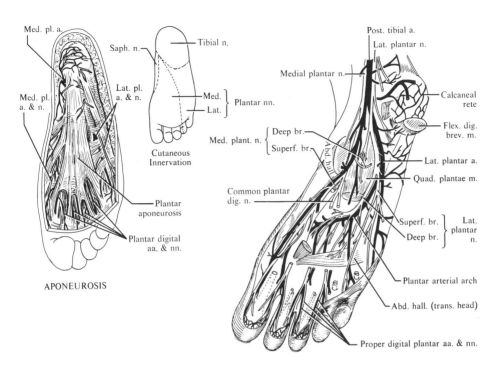

Med. pl. a.

Saph. n.

Tibial n.

Med. pl. a. & n.

Lat. pl. a. & n.

Med. } Plantar nn.
Lat. }

Cutaneous Innervation

Plantar aponeurosis

Plantar digital aa. & nn.

APONEUROSIS

Post. tibial a.

Lat. plantar n.

Medial plantar n.

Deep br.
Med. plant. n. {
Superf. br.

Calcaneal rete

Flex. dig. brev. m.

Lat. plantar a.

Quad. plantae m.

Common plantar dig. n.

Superf. br.
Deep br.
} Lat. plantar n.

Plantar arterial arch

Abd. hall. (trans. head)

Proper digital plantar aa. & nn.

VESSELS AND NERVES OF PLANTAR FOOT

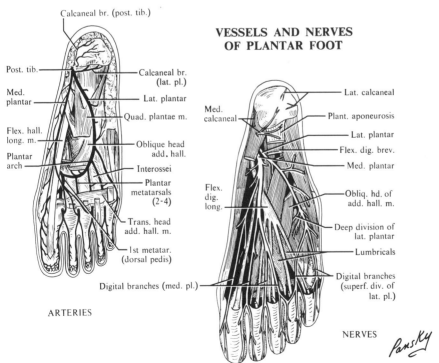

Calcaneal br. (post. tib.)

Post. tib.

Calcaneal br. (lat. pl.)

Med. plantar

Lat. plantar

Quad. plantae m.

Flex. hall. long. m.

Oblique head add. hall.

Plantar arch

Interossei

Plantar metatarsals (2-4)

Trans. head add. hall. m.

1st metatar. (dorsal pedis)

Digital branches (med. pl.)

ARTERIES

Lat. calcaneal

Med. calcaneal

Plant. aponeurosis

Lat. plantar

Flex. dig. brev.

Med. plantar

Flex. dig. long.

Obliq. hd. of add. hall. m.

Deep division of lat. plantar

Lumbricals

Digital branches (superf. div. of lat. pl.)

NERVES

Pansky

207. SURFACE ANATOMY OF LOWER EXTREMITY

I. Arteries

A. FEMORAL: with the thigh slightly flexed and laterally rotated, its course is indicated by a line beginning at a point halfway between the anterior superior iliac spine and pubic symphysis, extending downward to a point just above the adductor tubercle

B. POPLITEAL: beginning at a point on the back of the thigh just above the adductor tubercle, the line curves slightly laterally to the middle line of the thigh and then extends downward through the middle of the popliteal fossa to the apex of its inferior angle

C. POSTERIOR TIBIAL: its position can be indicated by a line drawn from the middle of the popliteal fossa to a point midway between the medial malleolus and the medial border of the tendo calcaneus

D. PERONEAL (FIBULAR): begins on posterior tibial line about 1 in. below the inferior angle of popliteal fossa, the line curving obliquely lateralward to reach the medial border of the fibula at the junction of its upper and middle thirds, thence continuing downward along the fibula, behind the lateral malleolus to end on the lateral side of the calcaneus

E. ANTERIOR TIBIAL: indicated by a line beginning at a point medial and just below the head of the fibula extending downward to a point midway between the malleoli at the front of the ankle

F. DORSALIS PEDIS: from a point midway between the malleoli at the front of the ankle to the base of the first intermetatarsal space on the dorsum of the foot

G. LATERAL PLANTAR AND PLANTAR ARCH: from a point midway between the medial malleolus and the medial border of the lower end of the tendo calcaneus, across the sole of the foot to the proximal end of the fourth intermetatarsal space; the arch continues medially across the sole of the foot to the proximal end of the first intermetatarsal space

II. Nerves

A. SCIATIC (IN THIGH): its course may be indicated by a line drawn from a point midway between the great trochanter of the femur and the ischial tuberosity to the superior angle of the popliteal fossa

B. TIBIAL: from the superior angle of the popliteal fossa down the midline of the fossa, thence following the line of the posterior tibial artery through the rest of its extent

C. COMMON PERONEAL (FIBULAR): diverges from the line of the sciatic at the superior angle of the popliteal fossa, continuing downward along the lateral boundary of the fossa (the biceps femoris) to the level of the head of the fibula where its terminal branches are given off

D. FEMORAL: enters thigh about 1.25 cm lateral to the line of the femoral artery and extends only about 2.5 cm below the inguinal ligament where it breaks up into several branches

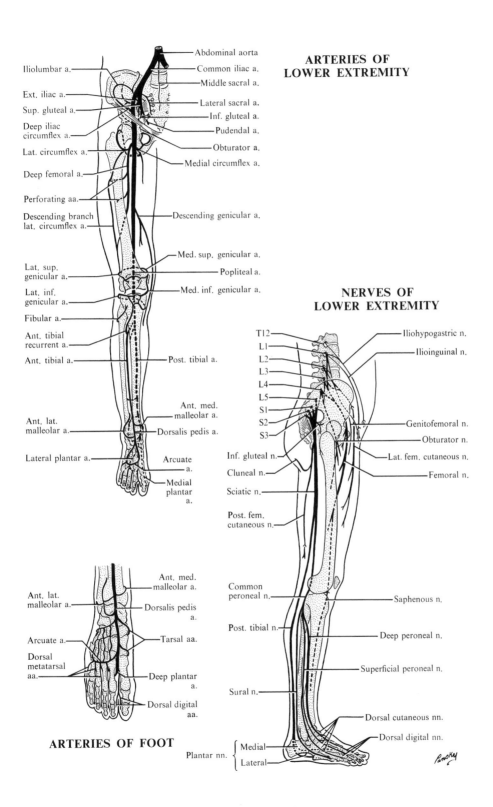

ARTERIES OF
LOWER EXTREMITY

Iliolumbar a.

Abdominal aorta
Common iliac a.
Middle sacral a.

Ext. iliac a.
Sup. gluteal a.

Lateral sacral a.
Inf. gluteal a.
Pudendal a.

Deep iliac
circumflex a.

Obturator a.

Lat. circumflex a.

Medial circumflex a.

Deep femoral a.

Perforating aa.

Descending branch
lat. circumflex a.

Descending genicular a.

Lat. sup.
genicular a.

Med. sup. genicular a.
Popliteal a.

Lat. inf.
genicular a.

Med. inf. genicular a.

Fibular a.

Ant. tibial
recurrent a.

Ant. tibial a.

Post. tibial a.

Ant. lat.
malleolar a.

Ant. med.
malleolar a.
Dorsalis pedis a.

Lateral plantar a.

Arcuate
a.
Medial
plantar
a.

NERVES OF
LOWER EXTREMITY

T12
L1
L2
L3
L4
L5
S1
S2
S3

Iliohypogastric n.
Ilioinguinal n.

Genitofemoral n.
Obturator n.
Lat. fem. cutaneous n.
Femoral n.

Inf. gluteal n.
Cluneal n.

Sciatic n.

Post. fem.
cutaneous n.

Ant. lat.
malleolar a.

Ant. med.
malleolar a.

Dorsalis pedis
a.

Arcuate a.

Tarsal aa.

Dorsal
metatarsal
aa.

Common
peroneal n.

Saphenous n.

Post. tibial n.

Deep peroneal n.

Superficial peroneal n.

Deep plantar
a.

Dorsal digital
aa.

Sural n.

Dorsal cutaneous nn.
Dorsal digital nn.

ARTERIES OF FOOT

Plantar nn. { Medial
Lateral

Panoky

-469-

208. JOINTS OF THE ANKLE AND FOOT

I. Ankle joint (talocrural)
A. TYPE: ginglymus (hinge)
B. BONES: medial malleolus of tibia, lateral malleolus of fibula, and talus
C. MOVEMENTS: dorsiflexion (extension) and plantar flexion (flexion)
D. LIGAMENTS
    1. Articular capsule
    2. Medial (deltoid) consists of 4 ligaments—the anterior and posterior tibiotalar, the tibionavicular, and the tibiocalcaneal. The anterior and posterior fibers check plantar and dorsiflexion, respectively. The anterior fibers also limit abduction
    3. Lateral ligaments: the anterior and posterior talofibular and calcaneofibular

II. Subtalar (talocalcaneal) joint (only 3 more movable intertarsal joints mentioned)
A. TYPE: arthrodial
B. BONES: talus and calcaneus
C. MOVEMENTS: gliding
D. LIGAMENTS
    1. Anterior, posterior, medial, and lateral talocalcaneal
    2. Interosseous talocalcaneal

III. Talocalcaneonavicular joint
A. TYPE: arthrodial
B. BONES: talus, navicular, and calcaneus
C. MOVEMENTS: gliding and some rotation
D. LIGAMENTS
    1. Articular capsule
    2. Dorsal talonavicular
    3. Plantar calcaneonavicular supports the head of the talus (*spring ligament*)

IV. Calcaneocuboid joint
A. TYPE: arthrodial
B. MOVEMENTS: gliding and slight rotation
C. LIGAMENTS
    1. Dorsal and plantar calcaneocuboid
    2. Bifurcated: from the calcaneus to the cuboid and navicular
    3. Long plantar

V. Tarsometatarsal joints between the bases of the metatarsals, the 3 cuneiforms, and the cuboid. Are arthrodial joints with slight gliding movements. Each has dorsal, plantar, and interosseous ligaments

VI. Metatarsophalangeal joints between the heads of the metatarsals and the bases of the proximal phalanges. Are condyloid joints, permitting flexion, extension, abduction, and adduction. Each has plantar and 2 collateral ligaments

VII. Interphalangeal joints between bases and heads of adjoining phalanges. Are ginglymus joints permitting flexion and extension. Each has plantar and collateral ligaments. The extensor aponeurosis acts as dorsal ligaments

VIII. Clinical considerations
A. SPRAINS. A sprained ankle usually is an inversion injury in which, as a result of rocking the foot toward the midline, some part of a ligament is torn

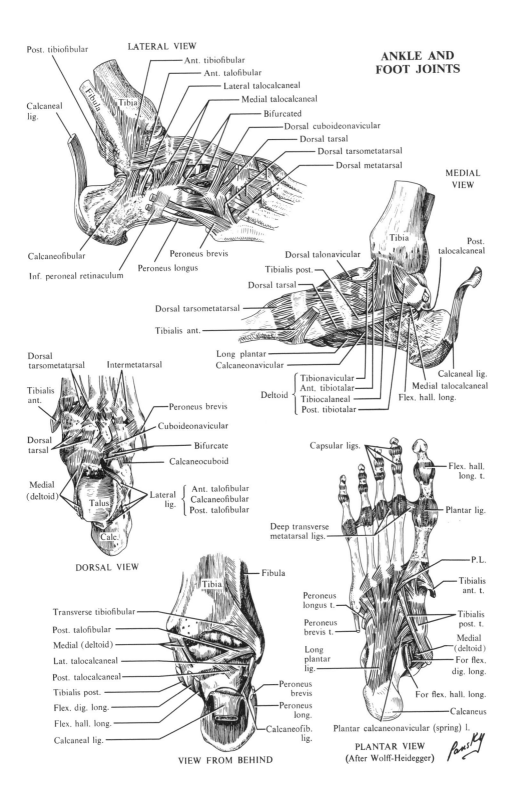

LATERAL VIEW

**ANKLE AND
FOOT JOINTS**

Post. tibiofibular

Ant. tibiofibular
Ant. talofibular
Lateral talocalcaneal
Medial talocalcaneal
Bifurcated
Dorsal cuboideonavicular
Dorsal tarsal
Dorsal tarsometatarsal
Dorsal metatarsal

Calcaneal lig.

Fibula

Tibia

MEDIAL VIEW

Calcaneofibular
Peroneus brevis
Inf. peroneal retinaculum
Peroneus longus

Tibia

Post. talocalcaneal

Dorsal talonavicular
Tibialis post.
Dorsal tarsal

Dorsal tarsometatarsal

Tibialis ant.

Long plantar
Calcaneonavicular

Deltoid {
Tibionavicular
Ant. tibiotalar
Tibiocalaneal
Post. tibiotalar

Calcaneal lig.
Medial talocalcaneal
Flex. hall. long.

Dorsal tarsometatarsal
Intermetatarsal
Tibialis ant.

Peroneus brevis
Cuboideonavicular
Bifurcate
Calcaneocuboid

Dorsal tarsal

Medial (deltoid)

Talus

Calc.

Lateral lig. {
Ant. talofibular
Calcaneofibular
Post. talofibular

DORSAL VIEW

Capsular ligs.

Flex. hall. long. t.
Plantar lig.

Deep transverse metatarsal ligs.

P.L.
Tibialis ant. t.

Fibula

Tibia

Transverse tibiofibular
Post. talofibular
Medial (deltoid)
Lat. talocalcaneal
Post. talocalcaneal
Tibialis post.
Flex. dig. long.
Flex. hall. long.
Calcaneal lig.

Peroneus longus t.
Peroneus brevis t.
Long plantar lig.

Tibialis post. t.
Medial (deltoid)
For flex. dig. long.

Peroneus brevis
Peroneus long.
Calcaneofib. lig.

For flex. hall. long.
Calcaneus

Plantar calcaneonavicular (spring) l.

VIEW FROM BEHIND

PLANTAR VIEW
(After Wolff-Heidegger)

Pansky

209. ARCHES OF THE FOOT

I. Functional considerations
A. LARGE NUMBER OF BONES allow for both great mobility and absorption of shock
B. THE TOES are important in balancing, running, and climbing

II. Structure of the arches
A. LONGITUDINAL: consists of 2 columns of bones, a medial and a lateral
   1. Medial: calcaneus, talus, navicular, 3 cuneiforms, and 3 medial metatarsals
      a. Most important: it bears most of the weight
      b. Talus and navicular are most subject to injury
   2. Lateral: calcaneus, cuboid, and 2 lateral metatarsals
      a. Balances weight. Calcaneus most subject to injury
B. TRANSVERSE: due to the fact that the second and third cuneiforms and the second through fourth metatarsals are wedge-shaped, with their base directed dorsally

III. Weight distribution and support
A. DISTRIBUTION (percentage of total body weight): 25% to calcaneus and 25% to heads of metatarsals (10% to head of first and 3.75% to the more lateral metatarsals)
B. SUPPORT: 80% stress on plantar ligaments; 20% on such muscles or tendons as tibialis posterior, peroneus (fibularis) longus, abductor and adductor hallucis for the medial longitudinal arch and adductor hallucis for the transverse arch

IV. Muscles acting on the ankle joint

Dorsiflexion	Plantar Flexion
Tibialis anterior	Gastrocnemius
Extensor digitorum longus	Soleus
Peroneus (fibularis) tertius	Plantaris
Extensor hallucis longus	Flexor hallucis longus
	Peroneus (fibularis) longus and brevis
	Tibialis posterior

V. Muscles acting on subtalar and transverse tarsal joints

Inversion and Adduction	Eversion and Abduction
Tibialis ant. and post.	Peroneus (fibularis) longus and brevis
Flexor hallucis longus	Peroneus (fibularis) tertius
Flexor digitorum	Extensor digitorum longus (lateral part)

VI. Clinical considerations
A. FLATFOOT (pes planus): a condition in which the medial arch of the foot is nonexistent due to the misalignment of the bones and a stretching of the ligaments
B. TALIPES. All the deformities of the foot can be classified here. For example, talipes varus consists of a foot twisted upon itself at the midtarsal joint so that in walking the dorsum of the foot is directed forward and the sole backward. This by itself is not common, but when combined with talipes equinus (in which the heel is abnormally elevated, and all the weight is placed on the balls of the toes), it is the most common deformity
C. HALLUX VALGUS (bunion). The large toe deviates laterally making the head of the first metatarsal prominent
D. HAMMER TOE is an abnormal contraction

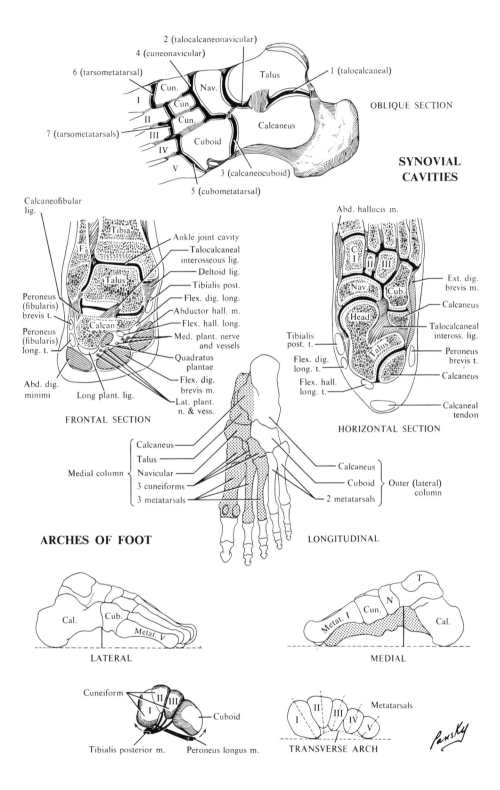

2 (talocalcaneonavicular)

4 (cuneonavicular)

6 (tarsometatarsal)

1 (talocalcaneal)

Talus

Cun.

Nav.

I

Cun.

II

Cun.

7 (tarsometatarsals)

III

Calcaneus

IV

Cuboid

V

3 (calcaneocuboid)

5 (cubometatarsal)

OBLIQUE SECTION

SYNOVIAL CAVITIES

Calcaneofibular lig.

Abd. hallucis m.

Tibia

Ankle joint cavity

Talocalcaneal interosseous lig.

Deltoid lig.

C

I

II

III

Talus

Tibialis post.

Nav.

Cub.

Ext. dig. brevis m.

Flex. dig. long.

Peroneus (fibularis) brevis t.

Abductor hall. m.

Head

Calcaneus

Flex. hall. long.

Peroneus (fibularis) long. t.

Calcan.

Med. plant. nerve and vessels

Tibialis post. t.

Talus

Talocalcaneal inteross. lig.

Quadratus plantae

Flex. dig. long. t.

Peroneus brevis t.

Abd. dig. minimi

Long plant. lig.

Flex. dig. brevis m.

Flex. hall. long. t.

Calcaneus

Lat. plant. n. & vess.

Calcaneal tendon

FRONTAL SECTION

HORIZONTAL SECTION

Calcaneus

Talus

Medial column

Navicular

Calcaneus

3 cuneiforms

Cuboid

Outer (lateral) column

3 metatarsals

2 metatarsals

ARCHES OF FOOT

LONGITUDINAL

Cal.

Cub.

Metat. V

T

N

Cun.

Metat. I

Cal.

LATERAL

MEDIAL

Cuneiform

II

III

I

Cuboid

II

III

Metatarsals

I

IV

V

Tibialis posterior m.

Peroneus longus m.

TRANSVERSE ARCH

Pansky

-473-

210. MUSCLE–NERVE RELATIONSHIP: FUNCTIONAL

I. Specific cases of nerve injury

A. NERVES to iliacus and psoas muscles: No superficial signs, but, in walking, pelvis does not swing when the extremity is fixed

B. FEMORAL: signs and symptoms are wasting of the anterior thigh; when the thigh is flexed, there is "leg drop"—the inability to extend the leg; loss of sensation on the anterior thigh and the medial leg; weakness in flexion of the thigh due to loss of action of rectus femoris, sartorius, and pectineus muscles

C. OBTURATOR: the thigh cannot be adducted, and there is some loss of sensation on its upper medial part

D. SUPERIOR GLUTEAL: there is weakness in abduction of thigh

E. INFERIOR GLUTEAL: there are wasting of the buttock and weakness in extension of thigh, as in climbing

F. SCIATIC: wasting of all hamstrings and muscles of leg and foot; except for flexion of thigh and adduction of thigh, no other activity in the lower extremity would be possible. Sensation loss on leg and foot
   1. Tibial division: if severed in the upper popliteal fossa, there would be loss of flexion of toes and foot, no inversion of foot, wasting of calf, loss of sensation on sole of foot and terminal parts of both surfaces of toes
   2. Common peroneal (fibular): if damaged near point of division from sciatic, will result in "foot drop," loss of extension of all toes, and a loss of sensation on the side of the leg

II. Anatomy of the straight-leg-raising test

A. THE INVESTING SHEATHS OF THE SCIATIC NERVE (like all nerves) are continuous with the coverings of the C.N.S. Thus, the epineurium of the sciatic nerve is continuous with the dura mater. If the epineurium is pulled upon, some degree of tension will be transmitted to the dura mater and to the nerve roots. Clinically, one method of exerting tension on the nerve roots and the dura is to exert a pull on the sciatic nerve by means of the straight-leg-raising test. The nerve is a close posterior relation of the hip joint, and the 2 branches of the nerve are close posterior relations of the knee joint. Furthermore, the sciatic nerve is 2 cm broad and its supporting sheaths are strong. Thus, when the hip is flexed while the knee is extended, as in the straight-leg-raising test, tension is transmitted back to the L4, L5, S1, S2, and S3 nerve roots and to the dura mater. The clinician may elicit pain not only from the nerve roots and dura but also from other structures. Thus, clinical experience is necessary for the interpretation of the test results

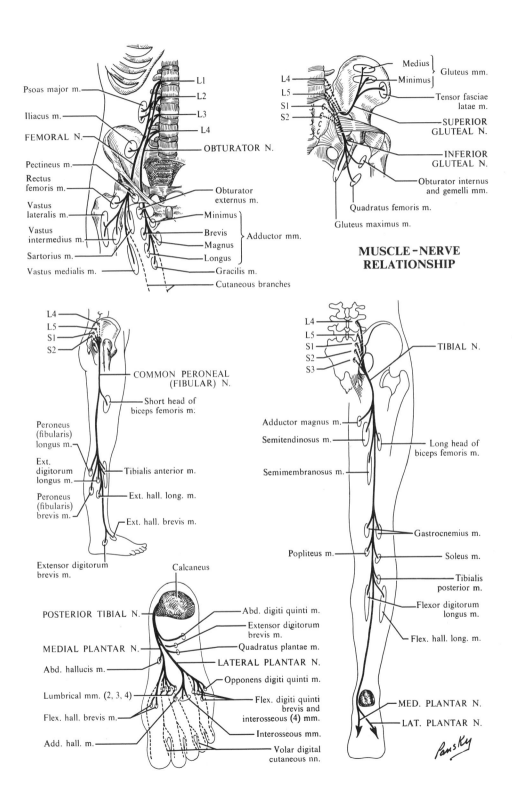

Psoas major m.
Iliacus m.
FEMORAL N.
Pectineus m.
Rectus femoris m.
Vastus lateralis m.
Vastus intermedius m.
Sartorius m.
Vastus medialis m.

L1
L2
L3
L4
OBTURATOR N.
Obturator externus m.
Minimus
Brevis
Magnus
Longus
} Adductor mm.
Gracilis m.
Cutaneous branches

Medius
Minimus
} Gluteus mm.
L4
L5
S1
S2
Tensor fasciae latae m.
SUPERIOR GLUTEAL N.
INFERIOR GLUTEAL N.
Obturator internus and gemelli mm.
Quadratus femoris m.
Gluteus maximus m.

MUSCLE-NERVE RELATIONSHIP

L4
L5
S1
S2
COMMON PERONEAL (FIBULAR) N.
Short head of biceps femoris m.
Peroneus (fibularis) longus m.
Ext. digitorum longus m.
Peroneus (fibularis) brevis m.
Tibialis anterior m.
Ext. hall. long. m.
Ext. hall. brevis m.
Extensor digitorum brevis m.

L4
L5
S1
S2
S3
TIBIAL N.
Adductor magnus m.
Semitendinosus m.
Semimembranosus m.
Long head of biceps femoris m.
Gastrocnemius m.
Soleus m.
Popliteus m.
Tibialis posterior m.
Flexor digitorum longus m.
Flex. hall. long. m.

Calcaneus
POSTERIOR TIBIAL N.
MEDIAL PLANTAR N.
Abd. hallucis m.
Lumbrical mm. (2, 3, 4)
Flex. hall. brevis m.
Add. hall. m.
Abd. digiti quinti m.
Extensor digitorum brevis m.
Quadratus plantae m.
LATERAL PLANTAR N.
Opponens digiti quinti m.
Flex. digiti quinti brevis and interosseous (4) mm.
Interosseous mm.
Volar digital cutaneous nn.

MED. PLANTAR N.
LAT. PLANTAR N.

Pansky

-475-

211. SUMMARY OF MUSCLE-NERVE RELATIONSHIPS IN LOWER EXTREMITY

Segment	Muscle	Named Nerve
L2–L3	Adductor brevis	Obturator
L2–L3	Adductor longus	Obturator
L2–L3	Gracilis	Obturator
L2–L3	Sartorius	Femoral
L2–L4	Iliacus	Femoral
L2–L4	Psoas	Nn. to psoas
L3–L4	Obturator externus	Obturator
L3–L4	Pectineus	Femoral
L3–L4	Quadriceps femoris	Femoral
L3–S1	Adductor magnus	Obturator and sciatic
L4–L5	Tibialis anterior	Deep peroneal
L4–S1	Gluteus medius	Superior gluteal
L4–S1	Gluteus minimus	Superior gluteal
L4–S1	Inferior gemellus	N. to quadratus femoris
L4–S1	Lumbricales	Medial and lateral plantar
L4–S1	Plantaris	Tibial
L4–S1	Popliteus	Tibial
L4–S1	Quadratus femoris	N. to quadratus femoris
L4–S1	Tensor fascia latae	Superior gluteal
L4–S2	Biceps femoris	Sciatic
L4–S2	Semimembranosus	Sciatic
L4–S2	Semitendinosus	Sciatic
L5–S1	Extensor digitorum longus	Deep peroneal
L5–S1	Extensor hallucis longus	Deep peroneal
L5–S1	Peroneus brevis	Superficial peroneal
L5–S1	Peroneus longus	Peroneal
L5–S1	Tibialis posterior	Tibial
L5–S2	Gluteus maximus	Inferior gluteal
L5–S2	Obturator internus	N. to obturator internus
L5–S2	Soleus	Tibial
L5–S2	Superior gemellus	N. to obturator internus
L5–S2	Biceps femoris	Sciatic
S1–S2	Abductor digiti minimi	Lateral plantar
S1–S2	Abductor hallucis	Medial plantar
S1–S2	Adductor hallucis	Lateral plantar
S1–S2	Extensor digitorum brevis	Deep peroneal
S1–S2	Extensor hallucis brevis	Deep peroneal
S1–S2	Flexor digiti minimi	Lateral plantar
S1–S2	Flexor digitorum brevis	Medial plantar
S1–S2	Flexor digitorum longus	Tibial
S1–S2	Flexor hallucis brevis	Medial plantar
S1–S2	Flexor hallucis longus	Tibial
S1–S2	Gastrocnemius	Tibial
S1–S2	Interossei	Lateral plantar
S1–S2	Piriformis	N. to piriformis
S1–S2	Quadratus plantae	Lateral plantar

APPENDIX I. DEFINITIONS OF ANATOMIC TERMS

Alveolus: bony socket of a tooth; pouch in lung air sac.

Ampulla: dilatation of a canal or duct.

Annulus: ring-shaped opening.

Antebrachium: forearm.

Antrum: any nearly closed cavity, particularly one with bony walls.

Aponeurosis: expanded tendon for the attachment of a flat muscle.

Artery (a.): vessel carrying blood from the heart through the body.

Articulation: connection between bones.

Autonomic nervous system: for the innervation of smooth muscle, heart muscle, and glands, consisting of a craniosacral (parasympathetic) and thoracolumbar (sympathetic) portion.

Axilla: arm pit.

Azygos: without a yoke; certain vessels or nerves not in pairs.

Belly: fleshy part of a muscle.

Body: broadest or longest mass of a bone.

Bone: inflexible structure composing skeleton.

Brachium: arm.

Bursa: literally, a purse; a small sac with fluid found in the fascia under skin, muscles, or tendons.

Capillary: anatomic unit connecting the arterial and venous systems; minute vessel, functional unit of the circulatory system.

Cartilage: substance from which some bone ossifies; gristle.

Cell: structural and functional body unit.

Central nervous system (C.N.S.): brain and spinal cord.

Condyle: polished articular surface, usually rounded.

Cornua: plural of cornu, a horn.

Costal: relating to a rib or costa. Costal cartilage.

Crest: ridge or border.

Cribriform: resembling a sieve.

Deltoid: shaped like the Greek letter delta.

Diaphysis: shaft of a long cylindrical bone.

Diarthrosis: movable joint.

Diverticulum: outpocketing or sac; the cecum.

Eminence: low convexity just perceptible.

Endocrine: internal secretion without the use of glandular ducts.

Epicondyle: elevation near a condyle.

Epiphyseal plate (line): growth center for elongation of bone, found between shaft and extremities of the bone.

Epiphysis: extremity or head of a long bone.

Exocrine: secretion discharged by way of a duct system.

Facet: small articular area, often a pit.

Falciform: sickle-shaped (falx: a sickle).

Fascia: fibrous envelopment of muscle structures and other tissues.

Foramen: hole, perforation.

Fornix: vaultlike space.

Fossa: shallow depression.

Fundus: base.

Ganglion: group of nerve cell bodies outside the central nervous system.

Genu: knee

Gingiva: gum.

Glenoid: having the form of a shallow cavity.

Head: enlarged round end of a long bone; knob.

Inguinal: belonging to or near the thigh, or inguen.

Insertion: relatively movable part of a muscle attachment.

Joint: connection between bones.

Jugular: belonging to the neck, or jugulum.

Lacrimal: to do with tears, or lacrymae.

Lamella: little plate or thin layer.

Lamina: plate or layer.

Ligament: fibrous tissue binding bones together or holding tendons and muscles in place.

Linea: line.

Lingual: belonging to the tongue.

Lymph vessels: like veins but walls are thinner and valves more numerous; drain tissue spaces.

Macula: spot.

Malar: belonging to the cheek.

Mastoid: shaped like a breast.

Mesentery: double layer of peritoneum (mesothelium).

Muscle: contractile organ capable of producing movement.

Neck: constriction of a bone near head; connecting head and body.

Nerve (n.): group of fibers outside the central nervous system.

Neuron: nerve cell body plus its processes.

Nucleus: group of nerve cell bodies within the central nervous system.

Omentum: fold of peritoneum connecting abdominal viscera with the stomach.

Organ: 2 or more tissues grouped together to perform a highly specialized function.

Origin: relatively fixed part of a muscle attachment.

Peripheral nervous system (P.N.S.): cerebrospinal nerves and the peripheral parts of the autonomic nervous system.

Process: projection (can be grasped with fingers).

Protuberance: swelling (can be felt under fingers).

Ramus: platelike branch of a bone; branch of a vessel or nerve.

Ramus communicans: nerve branch from the anterior root of a spinal nerve to the sympathetic chain of ganglia; white—nerve to chain; gray—chain back to spinal nerve.

Raphe: seam; the union of two parts in a line.

Shaft: body of a long bone.

Sheath: protective covering.

Spine: pointed projection of sharp ridge.

Suture: interlocking of teethlike edges.

Symphysis: union of right and left sides in the midline.

System: group of organs acting together to perform a highly complex but specialized function, e.g., nervous, skeletal, muscular, circulatory, respiratory, digestive, urinary, endocrine, and reproductive.

Tendon: fibrous tissue securing a muscle to its attachment.

Teres: round.

Thenar: relating to the palm or sole.

Tissue: differentiation and specialization of groups of cells bound together to perform a special function, e.g., epithelial, connective, muscular, and nervous.

Trochanters: 2 processes on the upper part of the femur below its neck.

Trochlea: spool-shaped articular surface.

Tubercle: small bump (can be felt under finger).

Tuberosity: large and conspicuous bump.

Uncinate: hooked; process shaped like a hook.

Vein (v.): vessel returning blood to the heart.

APPENDIX II. ESSENTIAL TERMS OF DIRECTION AND MOVEMENT

Abduction (abd.): draws away from midline.

Adduction (add.): draws toward the midline.

Anatomic position: standing erect with arms at the sides and palms of the hands turned forward.

Anterior (ant.) or ventral (vent.): situated before or in front of.

Corrugator: that which wrinkles skin, draws skin in.

Deep: farther from the surface (in a solid form).

Depressor: that which lowers.

Distal (dist.): farther from the root.

Dorsal (dors.): toward the rear, back; also back of hand and top of foot.

Erector: that which draws upward.

Evert (ever.): turn outward (as foot at ankle joint).

Extension (ext.): straightening.

External (extern.): outside (refers to wall of cavity or hollow form).

Flexion (flex.): bending or angulation.

Frontal (front.) or coronal (coron.): vertical; at right angles to sagittal; divides body into anterior and posterior parts.

Horizontal (horiz.): at right angles to vertical.

Inferior (inf.): lower, farther from crown of head.

Internal (int.): inside (refers to wall of cavity or hollow form).

Inverted (invert.): turned inward (as foot at ankle joint).

Lateral (lat.): farther from midline (or center plane).

Levator (lev.): that which raises.

Longitudinal (longit.): refers to long axis.

Medial (med.): nearer to midline (or center plane).

Median: midway, being in the middle.

Midline: divides body into a right and left side.

Midsagittal: vertical plane at midline dividing body into right and left halves.

Oblique: slanting.

Palmar (palm.) or volar (vol.): palm side of hand.

Plantar (plant.): sole side of foot.

Posterior (post.) or dorsal (dors.): rear or back.

Pronator (pronat.): that which turns palm of hand downward.

Prone: forearm and hand turned palm side down; body lying face down.

Proximal (prox.): nearer to limb root.

Rotator (rotat.): that which causes to revolve.

Sagittal (sagit.): vertical plane or section dividing body into right and left portions.

Sphincter: that which regulates closing of aperture.

Superficial (superf.): nearer to surface (refers to solid form).

Superior (sup.): upper, nearer to crown of head.

Supinator (supinat.): that which turns palm of hand upward.

Supine: forearm and hand turned palm side up; body lying face up.

Tensor (tens.): that which draws tight.

Transverse (trans.): at right angles to long axis; body divided into upper and lower parts.

Ventral (vent.) or anterior (ant.): situated before or in front of.

Vertical (vert.): refers to long axis in erect position.

Volar (vol.) or palmar (palm.): palm side of hand.

APPENDIX III. CLASSIFICATION OF JOINTS

I. Immovable joints (synarthroses)

A. SUTURE: interlocking of bones along their saw-toothed edges, e.g., joints of cranium

B. SCHINDYLESIS: a thin plate of bone is received into a cleft or fissure formed by the separation of 2 laminae in another bone, e.g., vomer in fissure between maxillae

C. GOMPHOSIS: insertion of a conical process into a socket, e.g., roots of teeth with alveolae of jaws

D. SYNCHONDROSIS: where the connecting medium is cartilage, e.g., sternocostal joint

II. Slightly movable joints (amphiarthroses)

A. SURFACES CONNECTED BY FIBROCARTILAGE (symphysis), e.g., joints of spine (intervertebral disks)

B. SURFACES UNITED BY AN INTEROSSEOUS LIGAMENT (syndesmosis), e.g., inferior tibiofibular articulation

III. Freely movable (synovial or lubricated, diarthroses)

A. GLIDING JOINT (arthrodia): bones glide face to face limited by restraining ligaments, e.g., intercarpal, intertarsal joints

B. HINGE JOINT (ginglymus): movement about a transverse axis only, e.g., elbow (humero-ulnar) joint, knee joint

C. SADDLE JOINT: increases the extent of hinge-joint movement by adding an axis of movement perpendicular to transverse axis, e.g., joint of thumb with wrist bone (carpometacarpal)

D. CONDYLOID (ellipsoidal): a modified ball-and-socket joint in which the opposed surfaces are ellipsoidal and not spherical. Increases extent of saddle joint by permitting circular movement that describes a cone, but vertical rotation is impossible, e.g., wrist joint

E. PIVOT JOINT (trochoid): a cylindrical form moving within a complete or partial ring (or a ring moving about the cylinder); only a vertical axis is present, e.g., joints of the forearm (radioulnar)

F. BALL-AND-SOCKET JOINT (enarthroses): provides freest movement by means of a spherical head in a cuplike cavity; adds to wide play of condyloid joint, the vertical axis of rotation, e.g., hip joint (at acetabulum), shoulder joint

APPENDIX IV. REFERENCES

Anderson, J. E.: *Grant's Atlas of Anatomy,* 7th ed. Williams & Wilkins Company, Baltimore, 1978.

Anson, B. J., and McVay, C. B.: *Surgical Anatomy,* 5th ed. W. B. Saunders Company, Philadelphia, 1971.

Basmajian, J. V. (ed.): *Grant's Method of Anatomy,* 9th ed. Williams & Wilkins Company, Baltimore, 1975.

Basmajian, J. V.: *Primary Anatomy,* 7th ed. Williams & Wilkins Company, Baltimore, 1976.

Gardner, E.; Gray, D. J.; and O'Rahilly, R.: *Anatomy: A Regional Study of Human Structure,* 4th ed. W. B. Saunders Company, Philadelphia, 1975.

Gardner, W. D., and Osburn, W. A.: *Structure of the Human Body,* 2nd ed. W. B. Saunders Company, Philadelphia, 1973.

Goss, C. M. (ed.): *Gray's Anatomy of the Human Body,* 29th Amer. ed. Lea & Febiger, Philadelphia, 1973.

Hamilton, W. J. (ed.): *Textbook of Human Anatomy,* 2nd ed. C. V. Mosby Company, Saint Louis, 1976.

Hollinshead, W. H.: *Textbook of Anatomy,* 3rd ed. Harper & Row, Publishers, Inc., Hagerstown, Md., 1974.

House, E. L.; Pansky, B.; and Siegel, A.: *A Systemic Approach to Neuroanatomy,* 3rd ed. McGraw-Hill Book Company, New York, 1979.

Laurenson, R. D.: *An Introduction to Clinical Anatomy by Dissection of the Human Body.* W. B. Saunders Company, Philadelphia, 1968.

Lockhart, R. D.; Hamilton, G. F.; and Fyle, F. W.: *Anatomy of the Human Body,* rev. ed. J. B. Lippincott Company, Philadelphia, 1974.

Paff, G. H.: *Anatomy of the Head and Neck.* W. B. Saunders Company, Philadelphia, 1973.

Romanes, G. J. (ed.): *Cunningham's Textbook of Anatomy,* 11th ed. Oxford University Press, London, 1972.

Schadé, J. P.: *Introduction to Functional Human Anatomy: An Atlas.* W. B. Saunders Company, Philadelphia, 1974.

Snell, R. S.: *Clinical Anatomy for Medical Students.* Little Brown & Company, Boston, 1973.

Spalteholz, W., and Spanner, R.: *Atlas of Human Anatomy,* 16th ed. F. A. Davis Company Philadelphia, 1967.

Tenth International Anatomical Nomenclature Committee in Tokyo: *Nomina Anatomica,* 4th ed. Excerpta Medica Foundation, Amsterdam, 1977.

Thorek, P.: *Anatomy in Surgery,* 2nd ed. J. B. Lippincott Company, Philadelphia, 1962.

Tobin, C. E.: *Basic Human Anatomy.* McGraw-Hill Book Company, New York, 1973.

Warwick, R., and Williams, P. L. (eds.): *Gray's Anatomy,* 35th Brit. ed. W. B. Saunders Company, Philadelphia, 1973.

Woodburne, R. T.: *Essentials of Human Anatomy,* 6th ed. Oxford University Press, New York, 1978.

Zuckerman, S.: *A New System of Anatomy,* rev. ed. Oxford University Press, London, 1974.

Atlas of
Systemic Anatomy

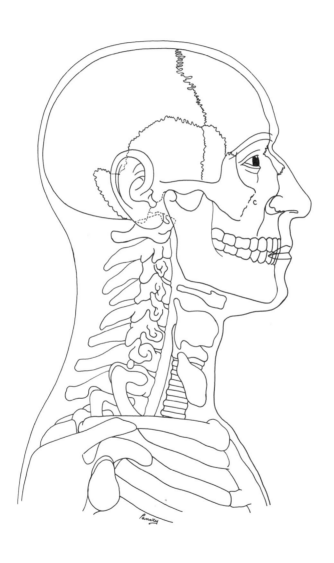

SKELETOMUSCULAR SYSTEM

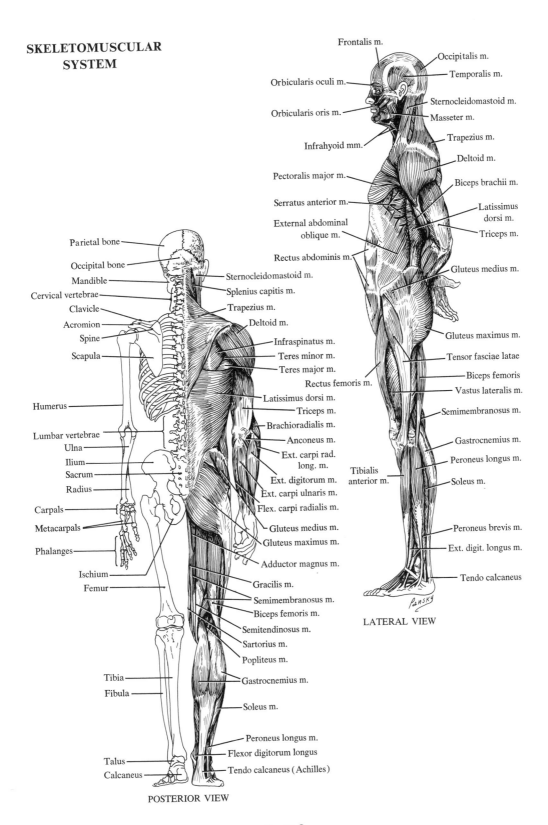

Frontalis m.
Occipitalis m.
Orbicularis oculi m.
Temporalis m.
Orbicularis oris m.
Sternocleidomastoid m.
Masseter m.
Infrahyoid mm.
Trapezius m.
Pectoralis major m.
Deltoid m.
Serratus anterior m.
Biceps brachii m.
External abdominal oblique m.
Latissimus dorsi m.
Triceps m.
Rectus abdominis m.
Gluteus medius m.
Gluteus maximus m.
Tensor fasciae latae
Biceps femoris
Rectus femoris m.
Vastus lateralis m.
Semimembranosus m.
Gastrocnemius m.
Peroneus longus m.
Tibialis anterior m.
Soleus m.
Peroneus brevis m.
Ext. digit. longus m.
Tendo calcaneus

LATERAL VIEW

Parietal bone
Occipital bone
Mandible
Cervical vertebrae
Clavicle
Acromion
Spine
Scapula
Sternocleidomastoid m.
Splenius capitis m.
Trapezius m.
Deltoid m.
Infraspinatus m.
Teres minor m.
Teres major m.
Rectus femoris m.
Latissimus dorsi m.
Triceps m.
Brachioradialis m.
Anconeus m.
Ext. carpi rad. long. m.
Ext. digitorum m.
Ext. carpi ulnaris m.
Flex. carpi radialis m.
Gluteus medius m.
Gluteus maximus m.
Adductor magnus m.
Gracilis m.
Semimembranosus m.
Biceps femoris m.
Semitendinosus m.
Sartorius m.
Popliteus m.
Gastrocnemius m.
Soleus m.
Peroneus longus m.
Flexor digitorum longus
Tendo calcaneus (Achilles)

Humerus
Lumbar vertebrae
Ulna
Ilium
Sacrum
Radius
Carpals
Metacarpals
Phalanges
Ischium
Femur
Tibia
Fibula
Talus
Calcaneus

POSTERIOR VIEW

Plate I

-485-

SKELETOMUSCULAR SYSTEM

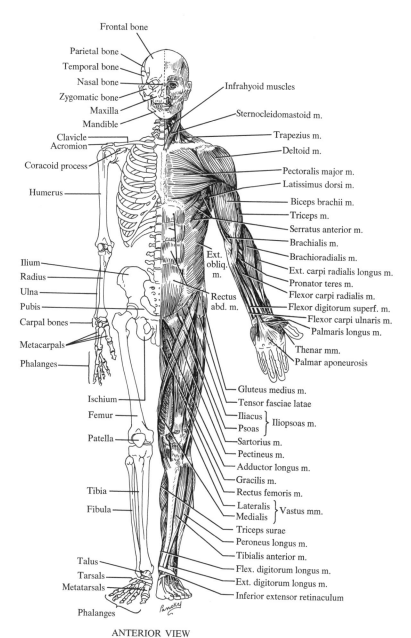

Frontal bone

Parietal bone

Temporal bone

Nasal bone

Zygomatic bone

Maxilla

Mandible

Clavicle

Acromion

Coracoid process

Humerus

Ilium

Radius

Ulna

Pubis

Carpal bones

Metacarpals

Phalanges

Ischium

Femur

Patella

Tibia

Fibula

Talus

Tarsals

Metatarsals

Phalanges

Infrahyoid muscles

Sternocleidomastoid m.

Trapezius m.

Deltoid m.

Pectoralis major m.

Latissimus dorsi m.

Biceps brachii m.

Triceps m.

Serratus anterior m.

Brachialis m.

Brachioradialis m.

Ext. carpi radialis longus m.

Pronator teres m.

Flexor carpi radialis m.

Flexor digitorum superf. m.

Flexor carpi ulnaris m.

Palmaris longus m.

Thenar mm.

Palmar aponeurosis

Ext. obliq. m.

Rectus abd. m.

Gluteus medius m.

Tensor fasciae latae

Iliacus ⎱ Iliopsoas m.
Psoas ⎰

Sartorius m.

Pectineus m.

Adductor longus m.

Gracilis m.

Rectus femoris m.

Lateralis ⎱ Vastus mm.
Medialis ⎰

Triceps surae

Peroneus longus m.

Tibialis anterior m.

Flex. digitorum longus m.

Ext. digitorum longus m.

Inferior extensor retinaculum

ANTERIOR VIEW

PLATE II

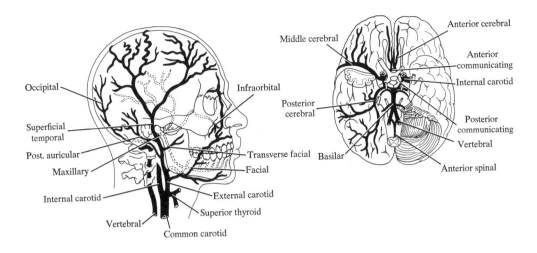

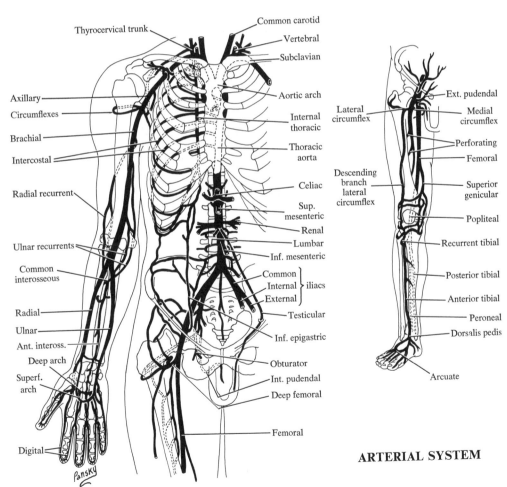

ARTERIAL SYSTEM

Plate III

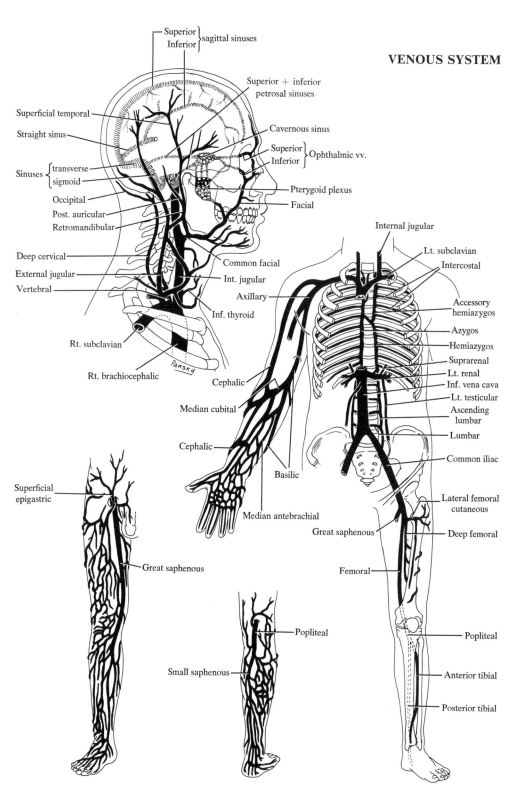

VENOUS SYSTEM

Superior | sagittal sinuses
Inferior |

Superior + inferior petrosal sinuses

Superficial temporal

Straight sinus

Cavernous sinus

Superior | Ophthalmic vv.
Inferior |

Sinuses { transverse
sigmoid

Pterygoid plexus

Occipital

Facial

Post. auricular

Retromandibular

Deep cervical

Common facial

External jugular

Int. jugular

Vertebral

Axillary

Inf. thyroid

Rt. subclavian

Pansky

Rt. brachiocephalic

Internal jugular

Lt. subclavian

Intercostal

Accessory hemiazygos

Azygos

Hemiazygos

Suprarenal

Lt. renal

Inf. vena cava

Lt. testicular

Ascending lumbar

Lumbar

Common iliac

Cephalic

Median cubital

Cephalic

Basilic

Median antebrachial

Superficial epigastric

Great saphenous

Lateral femoral cutaneous

Great saphenous

Deep femoral

Femoral

Popliteal

Popliteal

Small saphenous

Anterior tibial

Posterior tibial

PLATE IV

LYMPHATIC SYSTEM

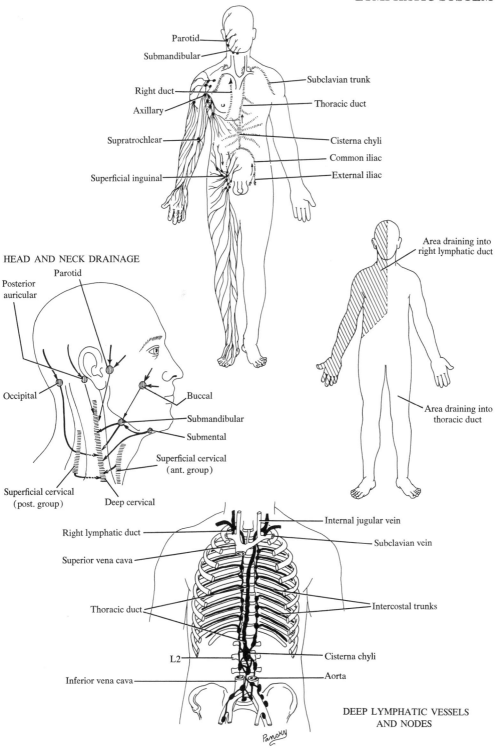

Parotid
Submandibular
Subclavian trunk
Right duct
Thoracic duct
Axillary
Supratrochlear
Cisterna chyli
Common iliac
Superficial inguinal
External iliac

Area draining into
right lymphatic duct

Area draining into
thoracic duct

HEAD AND NECK DRAINAGE

Parotid
Posterior
auricular
Occipital
Buccal
Submandibular
Submental
Superficial cervical
(ant. group)
Superficial cervical
(post. group)
Deep cervical

Internal jugular vein
Right lymphatic duct
Subclavian vein
Superior vena cava
Intercostal trunks
Thoracic duct
L2
Cisterna chyli
Aorta
Inferior vena cava

Panoky

DEEP LYMPHATIC VESSELS
AND NODES

PLATE V

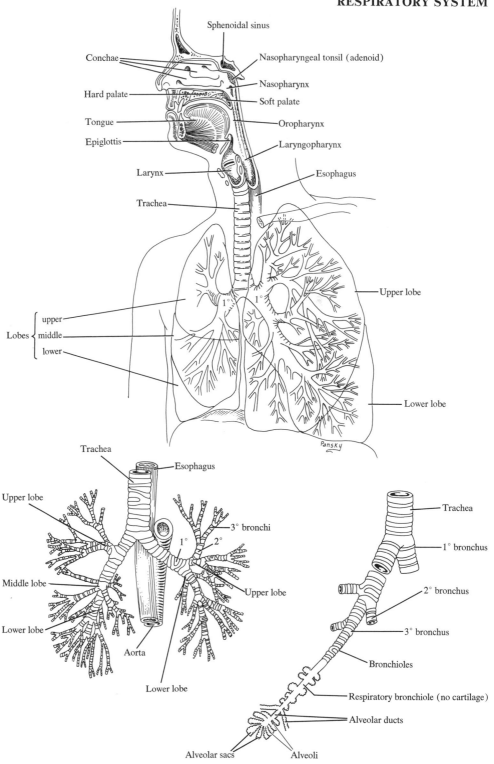

Sphenoidal sinus

Conchae

Nasopharyngeal tonsil (adenoid)

Nasopharynx

Hard palate

Soft palate

Tongue

Oropharynx

Epiglottis

Laryngopharynx

Larynx

Esophagus

Trachea

Upper lobe

1°

1°

Lobes { upper / middle / lower

Lower lobe

Pansky

Trachea

Esophagus

Upper lobe

3° bronchi

1° 2°

Trachea

1° bronchus

Middle lobe

Upper lobe

2° bronchus

Lower lobe

3° bronchus

Aorta

Bronchioles

Lower lobe

Respiratory bronchiole (no cartilage)

Alveolar ducts

Alveolar sacs

Alveoli

PLATE VI

– 490 –

DIGESTIVE SYSTEM

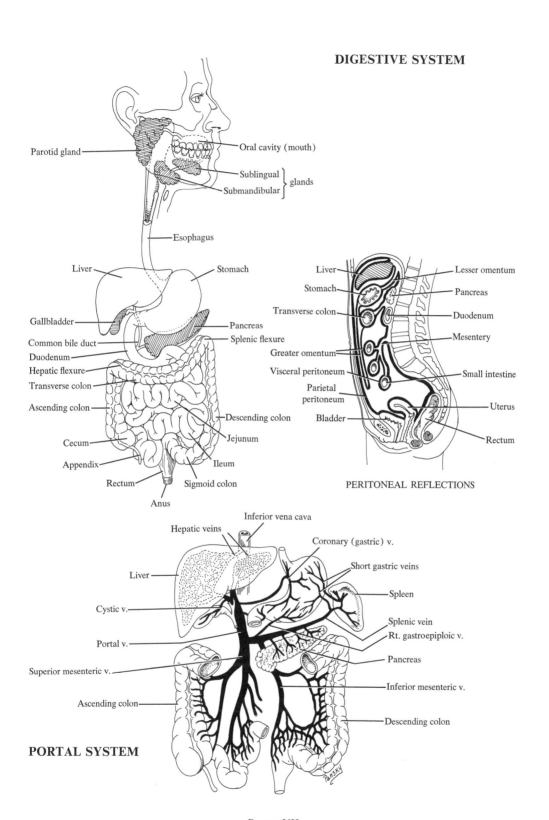

Parotid gland

Oral cavity (mouth)

Sublingual
Submandibular } glands

Esophagus

Liver

Stomach

Gallbladder

Common bile duct
Duodenum
Hepatic flexure
Transverse colon

Pancreas
Splenic flexure

Ascending colon

Descending colon

Jejunum

Cecum

Appendix

Ileum

Rectum

Sigmoid colon

Anus

PERITONEAL REFLECTIONS

Liver

Lesser omentum

Stomach

Pancreas

Transverse colon

Duodenum

Mesentery

Greater omentum

Visceral peritoneum

Small intestine

Parietal
peritoneum

Bladder

Uterus

Rectum

PORTAL SYSTEM

Inferior vena cava

Hepatic veins

Coronary (gastric) v.

Short gastric veins

Liver

Spleen

Cystic v.

Splenic vein
Rt. gastroepiploic v.

Portal v.

Pancreas

Superior mesenteric v.

Inferior mesenteric v.

Ascending colon

Descending colon

PLATE VII

URINARY SYSTEM

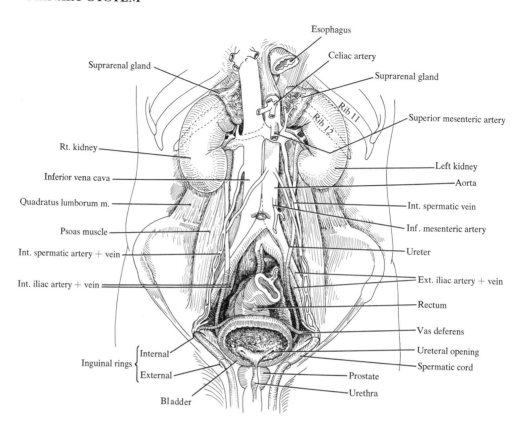

Esophagus
Celiac artery
Suprarenal gland
Suprarenal gland
Rib 11
Rib 12
Superior mesenteric artery
Rt. kidney
Inferior vena cava
Quadratus lumborum m.
Psoas muscle
Int. spermatic artery + vein
Int. iliac artery + vein
Left kidney
Aorta
Int. spermatic vein
Inf. mesenteric artery
Ureter
Ext. iliac artery + vein
Rectum
Vas deferens
Ureteral opening
Spermatic cord
Inguinal rings { Internal / External }
Prostate
Urethra
Bladder

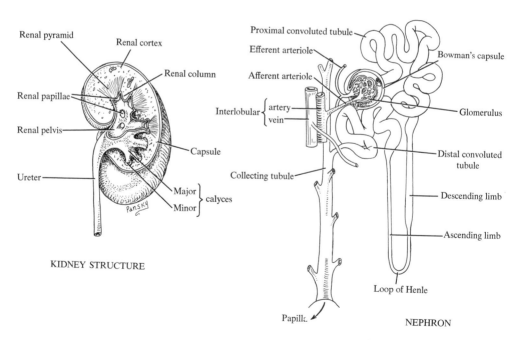

Renal pyramid
Renal cortex
Renal column
Renal papillae
Renal pelvis
Capsule
Ureter
Major } calyces
Minor

PANSKY

KIDNEY STRUCTURE

Proximal convoluted tubule
Efferent arteriole
Afferent arteriole
Bowman's capsule
Interlobular { artery / vein }
Glomerulus
Distal convoluted tubule
Collecting tubule
Descending limb
Ascending limb
Loop of Henle
Papilla
NEPHRON

PLATE VIII

-492-

REPRODUCTIVE SYSTEM

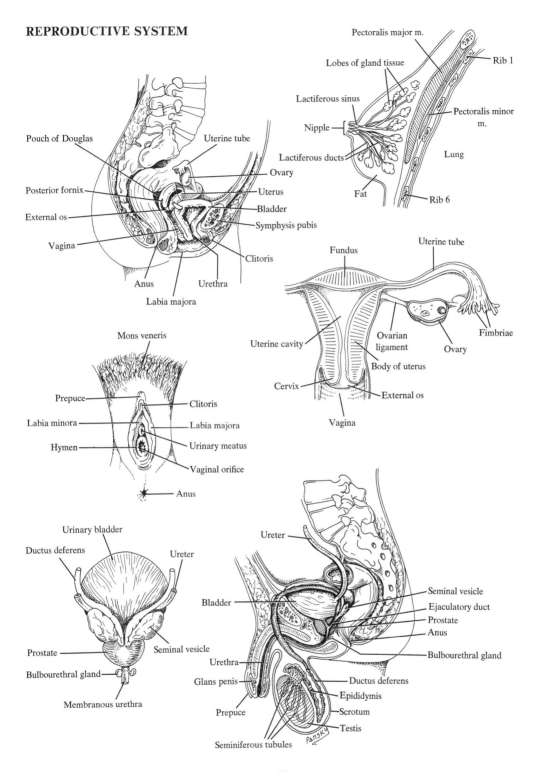

Pectoralis major m.

Lobes of gland tissue

Rib 1

Lactiferous sinus

Pectoralis minor m.

Nipple

Lactiferous ducts

Lung

Fat

Rib 6

Pouch of Douglas

Uterine tube

Ovary

Posterior fornix

Uterus

External os

Bladder

Vagina

Symphysis pubis

Clitoris

Anus

Urethra

Labia majora

Fundus

Uterine tube

Uterine cavity

Ovarian ligament

Fimbriae

Body of uterus

Ovary

Cervix

External os

Vagina

Mons veneris

Prepuce

Clitoris

Labia minora

Labia majora

Hymen

Urinary meatus

Vaginal orifice

Anus

Urinary bladder

Ductus deferens

Ureter

Ureter

Bladder

Seminal vesicle

Ejaculatory duct

Prostate

Anus

Prostate

Seminal vesicle

Bulbourethral gland

Bulbourethral gland

Urethra

Ductus deferens

Membranous urethra

Glans penis

Epididymis

Prepuce

Scrotum

Testis

Seminiferous tubules

PLATE IX

-493-

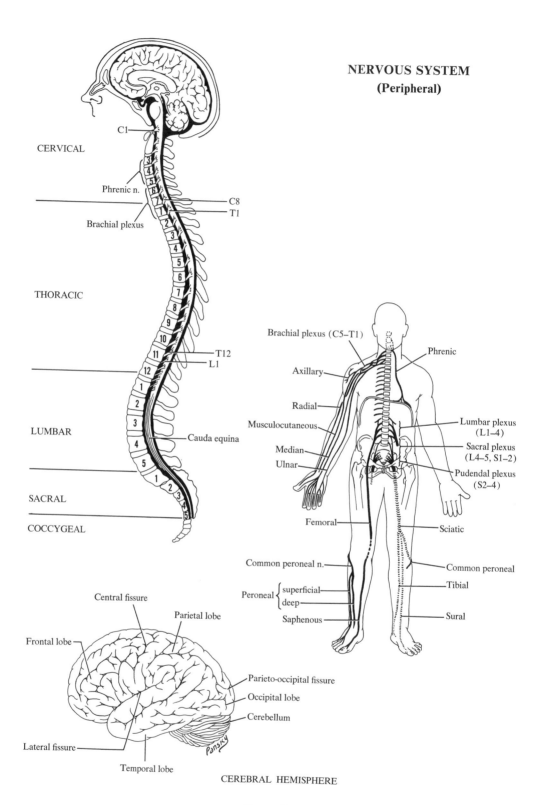

NERVOUS SYSTEM
(Peripheral)

CERVICAL

C1

Phrenic n.

C8
T1

Brachial plexus

THORACIC

T12
L1

LUMBAR

Cauda equina

SACRAL

COCCYGEAL

Brachial plexus (C5–T1)

Phrenic

Axillary

Radial

Musculocutaneous

Median

Ulnar

Lumbar plexus (L1–4)

Sacral plexus (L4–5, S1–2)

Pudendal plexus (S2–4)

Femoral

Sciatic

Common peroneal n.

Common peroneal

Peroneal { superficial
           { deep

Tibial

Saphenous

Sural

Central fissure

Parietal lobe

Frontal lobe

Parieto-occipital fissure

Occipital lobe

Cerebellum

Lateral fissure

Pansky

Temporal lobe

CEREBRAL HEMISPHERE

PLATE X

CRANIAL NERVES

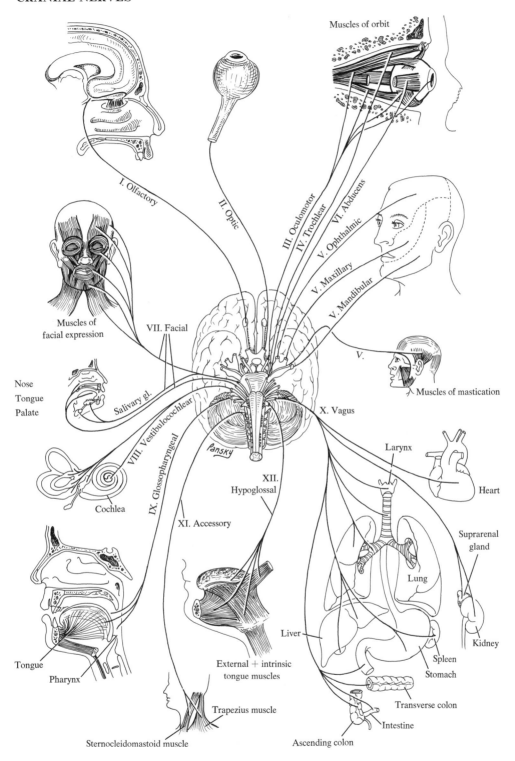

Muscles of orbit

I. Olfactory

II. Optic

III. Oculomotor

IV. Trochlear

VI. Abducens

V. Ophthalmic

V. Maxillary

V. Mandibular

V.

Muscles of mastication

Muscles of
facial expression

VII. Facial

Nose
Tongue
Palate

Salivary gl.

VIII. Vestibulocochlear

IX. Glossopharyngeal

Cochlea

XI. Accessory

XII.
Hypoglossal

X. Vagus

Larynx

Heart

Suprarenal
gland

Lung

Kidney

Liver

Spleen

Stomach

Transverse colon

Intestine

Ascending colon

Tongue

Pharynx

External + intrinsic
tongue muscles

Trapezius muscle

Sternocleidomastoid muscle

Pansky

PLATE XI

-495-

AUTONOMIC NERVOUS SYSTEM

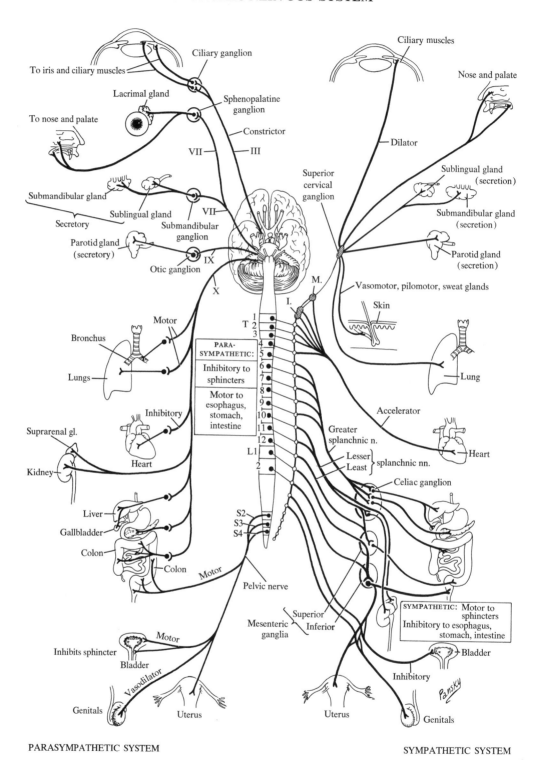

PARASYMPATHETIC SYSTEM

SYMPATHETIC SYSTEM

PLATE XII

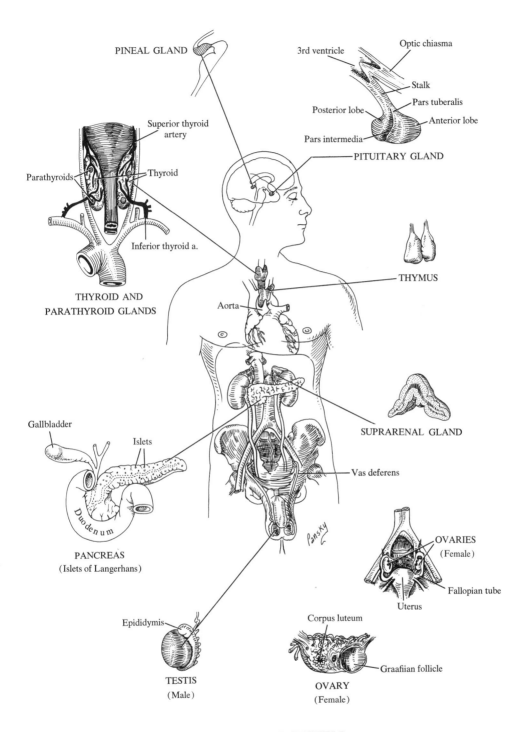

PINEAL GLAND

3rd ventricle — Optic chiasma

Stalk

Posterior lobe — Pars tuberalis

Pars intermedia — Anterior lobe

PITUITARY GLAND

Superior thyroid artery

Parathyroids — Thyroid

Inferior thyroid a.

THYROID AND PARATHYROID GLANDS

THYMUS

Aorta

SUPRARENAL GLAND

Gallbladder

Islets

Vas deferens

Duodenum

PANCREAS
(Islets of Langerhans)

OVARIES
(Female)

Fallopian tube

Uterus

Epididymis

Corpus luteum

TESTIS
(Male)

Graafiian follicle

OVARY
(Female)

Pansky

ENDOCRINE SYSTEM

PLATE XIII

INDEX

Figures in **boldface** type refer to pages on which illustrations appear.

Brain, **89**, 90, **91, 95**
Breast. *See* Glands, mammary
Breathing, 306
Bregma, **7,** 8
Broad ligament, **379,** 380, **381**
Bronchi, 276, 278, **279,** 298, **299**
  obstruction, 298
Bronchograms, **280, 281**
Bronchopulmonary segments, 278, **279**
Bulb, jugular, 38, **39**
  ocular (eyeball), 108, **109**
  olfactory, 84, **85,** 132, **133**
  penile, **369, 394, 395, 398, 399**
  vestibular, **391,** 392, **393**
Bulla, ethmoidalis, 122, **123,** 126
Bundle, atrioventricular of His, 296
Bunion (hallux valgus), 468
Bursa(e), biceps femoris muscle, 442
  gastrocnemius muscle, 442, **443**
  infrapatellar, 442, **443**
  olecranon, 230, **231**
  omental, 322, **323,** 332, **333**
  popliteus muscle, 442, **443**
  prepatellar, 442, **443**
    subcutaneous, 442, **443**
  radial, 246, **247**
  sartorius muscle, 442
  semimembranosus muscle, 442, **443**
  subacromial, **193,** 208, **209**
  subdeltoid, **193,** 208, **209**
  subscapular, **193, 209**
  suprapatellar, 442, **443**
  ulnar, **239,** 246, **247**
Bursitis, ischial tuberosity, 410
  knee, 442

Cage, thoracic, 266, **267**
Calcaneus, 452, **453, 454, 473**
Calvarium, 12
Calyx(ces), renal, 352, **353**
Canal(s), adductor, **417,** 418, **419, 423**
  anal, 372, **373,** 374
  apical, dental root, **67**
  central, spinal cord, **173**
  cervix, 378, **379, 381,** 382, **385**
  condyloid, 10, **13**
  facial, hiatus, 12, **13**
    prominence, 118, **119**
  femoral, 418
  hyaloid, **109**
  hypoglossal, 10, 12, **13**
  incisive, 10, **11**
  infraorbital (foramen), 4, **5, 7,** 104
  inguinal, 320
  obturator, **413**
  optic (foramen), 12, **13,** 104, **105**
  pterygopalatine, **11**
  pudendal (Alcock's), 376, **377,** 390
  sacral, 150, **151,** 175
  semicircular, **117,** 120, **121**
  vertebral, 146, **147**

Canaliculi, lacrimal, 100, **101, 103**
Capitate bone, **183,** 234, **235**
Capitulum, humerus, 194, **195**
Capsule, internal, 88, **89**
  kidney, 352
  tonsil, 68
Caput medusae, 314, 348
Cardia, stomach, 326, **327**
Cardiac nerves, **283,** 296, **297**
Cardiac plexus, 296, **297**
Carina, trachea, **279**
Carotid sheath, 36, **37**
Carpal bones, 234, **235, 237**
Cartilage(s), alar, 122, **123**
  articular, **439, 441**
  arytenoid, 52, **53**
  auricular, 116
  corniculate, 52, **53**
  costal, 266, **267,** 268, **269**
  cricoid, 52, **53**
  cuneiform, 52
  epiglottic, 52, **53**
  larynx, 52, **53**
  nose, 122, **123, 125**
  thyroid, **51,** 52, **53**
  tracheal, **49, 51, 53,** 298, **299**
Caruncle(s), lacrimal, 100, **101, 103**
  sublingual, **61**
Cataract, 108
Cauda equina, **173,** 174
Cavernous bodies of penis, **395, 398, 399**
Cavernous sinus, **75,** 76, **77, 113**
Cavity, nasal, 124, **125, 127**
  pelvic, **411, 413**
  pericardial, 286, **287**
  peritoneal, greater (main), 322, **323**
    lesser (omental bursa), 322, **323**
  pleural, 274, 276, **277**
    drainage, 274
  tympanic (middle ear), 118, **119**
Cecum, 338, **339, 341**
Cells, air, ethmoid, *See* Sinus(es)
  mastoid, 118, **119**
Cementum, 66, **67**
Cerebellum, 80, **81,** 84, **85,** 88, **89**
Cerebral vascular occlusion, 90
Cerebrospinal fluid, 94
Cerebrum, 80, **81,** 82, **83,** 84, **85, 95**
Cervix uteri, 378, **379, 381**
Chalazion, 100
Chambers of eye, 108, **109**
Chest, normal, x-ray, **293**
Chiasma, optic, 80, 81, 84
Choanae, **11,** 122, 124
Choked disk, 110
Chorda tympani, 26, **27,** 30, **31,** 58, **59,** 60, **61**
Chordae tendineae, 288, **289**
Choroid, 108, **109**
Chylothorax, 304
Ciliary body, 108, **109**
  ganglion, **107, 113,** 114, **115**
Circle, arterial, cerebral (Willis), **79,** 90, **91**

Nerve(s) [*Cont.*]
meningeal, cranial, 74
mental, 14, **15**
musculocutaneous, 184, **185**, 198, **199**, 202, **203**
mylohyoid, 30, **31, 33, 43**
nasal, 30, 132, **133**
nasociliary, 112, **113**, 132
nasopalatine, 30, 66, 132, **133**
obturator, 404, **405**, 422, **423**
to obturator internus and superior gemellus, 426, 430, **431**
occipital, greater, 14, **15**, 16, **17, 45**
lesser, 14, **15**, 16, **17, 45**
third, 14, **15, 17, 55**
oculomotor, 106, **107**, 112, **113**, 114, **115**, 134, **136**
olfactory, 84, 132, **133**, 134, **136**
ophthalmic, 112, **113**, 114, **115**
optic, 108, **109**, 134, **136**
sheath, 112
palatine, 10, 30, 66, **67**
palmar cutaneous, median, 184, **185**, 220, **221, 241**
ulnar, 184, **185**, 220, **221, 241**
parasympathetic, abdomen, 328, **329**, 386, **387**
head, 26, **27**, 28, **29**, 30, **31**, 58, **59, 113**, 114, **115, 133**
pelvis, **369**, 384, 386, **387**, 390
pelvic splanchnics (nervi erigentes), **387**
perforating cutaneous, 430
perineal, **384**, 390, **395**, 398
peroneal, common, 430, **431**, 436, **437, 449**
deep, 404, **405, 449**, 460
superficial, 404, **405, 449**, 460
petrosal, deep, 26, 66, **67**
greater, 26, **27, 67**
lesser, 28, **29**
pharyngeal, glossopharyngeal, 48, **137**
superior cervical sympathetic ganglion, 54, **55**
vagus, 48, **137**
phrenic, **43, 199**, 294, **295, 303**, 364
plantar, lateral, 466, **467, 469**
medial, 466, **467, 469**
plexuses. *See* Plexus(es), nerve
polyp, nasal, 124
presacral (superior hypogastric plexus), **369**, 386, **387**
proper digital of fingers, 240, **241, 249**, 250, **251**
pterygoid canal (vidian), 66, 114, **115**
pterygopalatine, 66, **67**, 114
pudendal, 384, **387**, 390, **391, 405**, 426, **427**, 430, **431**
to quadratus femoris and inferior gemellus muscles, 424, 430, **431**
radial arm, 184, **185**, 198, **199**, 212, **213**, 216, **223**, 254
deep, forearm, 216, **217, 221**, 224, **225**
superficial, 216, **217**, 220, **221, 239, 249**
wrist, **239**, 240, **249**
rami communicantes, gray, 477
white, 477

rectal, 374, 386, **387**, 390
recurrent laryngeal (inferior laryngeal), **47, 51, 295, 299, 301**
saphenous, 404, **405**, 422, **423**, 460, **461**, 466
scapular, dorsal, 198, **199, 203, 207**
sciatic, 426, **427**, 430, **431, 437**
scrotal, 384, **395, 405**
spinal, formation, 142, **143**
testing of, **178, 179**
splanchnic, lumbar, 386, **387**
pelvic, **387**
thoracic, 302, **303, 305**, 328, **329**
statoacoustic (VIII, acoustic), **117**, 120, **121**, 135
to subclavius, 196, 198, **199**
subcostal, 142, 384
suboccipital (C1), **55**, 142
subscapular, 198, **199**, 202, **203**
supraclavicular, 14, 44, **45, 143**
supraorbital, 14, **15, 17**, 112, **113**
suprascapular, 198, **199**, 204, **207**
supratrochlear, 14, **15, 17**, 112, **113**
sural, 404, **405**, 460, **461**
cutaneous, lateral, **405, 461**
sympathetic, abdominal, 328, **329, 369**, 386, **387**
cervical, 54, **55**
pelvic, 386, **387**
thoracic, 302, **303, 305**
temporal, 26, **29**
deep, 30, **31**
to tensor tympani muscle, 30, **31**, 118
to tensor veli palatini muscle, 30, **31**, 66
thoracic, long, 198, **199**, 202, **203**
thoracodorsal, 198, **199**, 202, **203**
tibial, 430, **431**, 436, **437, 449**
transverse cervical (colli), 14, **15**, 44, **45**
trigeminal, mandibular division, 14, **15**, 30, **31**, 134, **136**
maxillary division, 132, **133**, 134, **136**
ophthalmic division, 112, **113**, 134, **136**
trochlear, **89**, 106, **107**, 112, **113**, 134, **136**
ulnar, arm, 198, **199, 203**, 212, **213**, 256, **257**
deep branch, 240, **241**, 254, **255**
digital branches, 184, **185**, 240, **241**, 248, **249**
dorsal cutaneous branch, **185, 241**, 248, **249**
forearm, 216, 220, **221**, 256
hand, **185**, 240, **241, 247**, 256, **257**
palmar cutaneous branch, 184, **185**, 240, **241**
wrist, **185, 221**, 238, **239, 241**
vagus, **135, 137**
abdomen, 328, **329**
neck, **43, 47, 55, 61, 81, 89**
thorax, **295**, 296, **297, 299**, 300, **301**, 302, **303**
to vastus lateralis muscle, 416
vestibular (VIII), 80, **81, 117**, 120, **121**, 134, **137**
zygomatic, **31**
zygomaticofacial, 14, **15, 31**
zygomaticotemporal, 14, **15, 17, 31**
Nerve blocks, 174, **175**, 229, 240, **463**
Nervi erigentes (pelvic splanchnics), 368, **369**, 386, **387**
Nervous system, autonomic, 477
central, 477